Ecological and Human Dimensions of AI-Based Supply Chain

Atour Taghipour
Normandy University, France & Solbridge, South Korea

IGI Global
Scientific Publishing
Publishing Tomorrow's Research Today

Vice President of Editorial Melissa Wagner
Managing Editor of Acquisitions Mikaela Felty
Managing Editor of Book Development Jocelynn Hessler
Production Manager Mike Brehm
Cover Design Phillip Shickler

Published in the United States of America by
IGI Global Scientific Publishing
701 East Chocolate Avenue
Hershey, PA, 17033, USA
Tel: 717-533-8845
Fax: 717-533-8661
Website: https://www.igi-global.com E-mail: cust@igi-global.com

Library of Congress Cataloging-in-Publication Data

Names: Taghipour, Atour, 1974- editor.
Title: Ecological and human dimensions of AI-based supply chain / edited by Atour Taghipour.
Description: Hershey, PA : IGI Global, [2025] | Includes bibliographical references and index. | Summary: "The book likely aims to explore the intersection of artificial intelligence (AI) with supply chain management, emphasizing both ecological sustainability and human factors"-- Provided by publisher.
Identifiers: LCCN 2024048486 (print) | LCCN 2024048487 (ebook) | ISBN 9798369374788 (hardcover) | ISBN 9798369374795 (paperback) | ISBN 9798369374801 (ebook)
Subjects: LCSH: Supply chain management. | Business logistics--Data processing. | Artificial intelligence.
Classification: LCC HD38.6 .E26 2025 (print) | LCC HD38.6 (ebook) | DDC 658.500285--dc23/eng/20241210
LC record available at https://lccn.loc.gov/2024048486
LC ebook record available at https://lccn.loc.gov/2024048487

British Cataloguing in Publication Data
A Cataloguing in Publication record for this book is available from the British Library.

Table of Contents

Detailed Table of Contents

Supply chains, the intricate networks connecting raw materials, manufacturing, distribution, and consumption, have evolved into complex systems that significantly influence organizational performance. This chapter is a literature review that delves deeper into the conceptualization of supply chain innovation, its theoretical foundations, key dimensions, research findings, and contemporary advancements. It identifies research gaps and provides actionable insights for researchers and practitioners seeking to drive innovation in their supply chain operations.

In recent years, the rise of digital transformation in logistics and supply chain management (SCM) has become quintessential for organizations; accelerated all the more with how COVID-19 induced an upheaval across global supply chains. The chapter discusses how the pandemic-induced disruptions were beyond what traditional supply chain models, largely manual processes, and data silos could accommodate. The need for strong digital infrastructure is also apparent through the use of next-gen technologies including blockchain, AI/ML, and intelligent automation. These technologies provide real-time visibility, predictive analytics, and improved coordination. This chapter examines the reasons why digital transformation is necessary, what it affords, and the challenges that include legacy system trouble, management of data issues, and cybersecurity and skills gaps. It finishes by offering some strategic guidelines for successful implementation of digital in SCM.

The convergence of big data and AI has transformed supply chain management, enabling organizations to operate with efficiency and adaptability. Big data delivers vast amounts of real-time information from diverse sources, including customer orders, supplier details, and market trends, while AI processes this data to identify patterns and make predictive analyses that optimize inventory management, forecast demand, and enhance logistics. This synergy empowers supply chain managers to make proactive decisions, reducing delays and stockouts, while supporting just-in-time inventory practices that align supply levels with consumer demand. In parallel, big data has reshaped marketing by providing deep insights into consumer behaviors and preferences, allowing for the development of tailored marketing strategies that enhance customer experiences. Despite the wealth of literature, there remains a gap in comprehensive studies that integrate its implications for personalized marketing, ethical considerations, and technical challenges.

Industry 4.0 transformations are exacerbating the skills gap, but they also present technological solutions to help bridge this disparity. Successfully taking advantage of Industry 4.0 to reduce the skills gap requires preparedness and the implementation of strategic initiatives by key actors. This chapter provides an overview of research that offers potential solutions for addressing the skills gap in the context of Industry 4.0. The methodology used is a systematic literature review. Fourteen solutions have been identified for addressing the skills gap among the three main stakeholders: governments, employers, and HR techs. These initiatives were categorized into 37 sub-themes. Based on primary and secondary themes, mappings were created to align the identified solutions from the literature, providing a roadmap for future research efforts. In the end, comprehensive recommendations were made for addressing the skills gap within the Industry 4.0 landscape, tailored to each respective stakeholder.

Big data, also known as mega data, encompasses large and complex datasets that traditional data management tools cannot adequately process. Over the years, the importance of big data has expanded across various sectors, including healthcare, finance, retail, and notably, sports. A core objective of this study is to clarify why big data has become indispensable in the sports industry. To illustrate this, the authors provide real-world examples from case studies and scientific literature that reveal how big data has revolutionized sports. Through data-driven analysis and predictive capabilities, teams are now able to make more informed and strategic decisions, optimize the performance of athletes, and boost fan engagement.

This project leverages AI and web development to create a disaster risk management platform that predicts, monitors, and responds to both natural and man-made disasters. Using AI algorithms, it processes data from sources like weather patterns and real-time sensors to forecast disasters and assess risks. With interactive dashboards and real-time alerts, the platform aids emergency responders and policymakers, enhancing preparedness and recovery efforts. By empowering users, it fosters safer, more resilient communities.

Modern AI methodologies are changing the face of warfare by enabling militaries to enhance their operational capabilities and, more importantly, their decision-making in executing the operations. Advanced technology with machine learning, natural language processing, and autonomous systems enable real-time data analysis, predictive modeling, and increased situational awareness on the battlefield. With such problematic challenges facing the military forces, AI integration provides strategic advantages over military departments for rapid responses to emerging threats and resources distribution. However, this evolution raises issues on accountability, bias, and implication of autonomy in decision-making for combat scenarios. Overall,

Deepfakes, driven by advanced generative AI models like GANs and VAEs, create highly realistic synthetic media, ranging from manipulated faces and voices to entirely fabricated videos. Originally developed for creative expression and entertainment, deepfakes now raise significant concerns, including political manipulation, privacy violations, and the spread of false information. This chapter delves into the technology behind deepfakes, exploring the models, techniques, and tools used to produce them, while also examining emerging detection methods designed to identify such synthetic content. As research continues to address the risks of deepfakes, the chapter also considers broader societal implications, such as the erosion of trust in media and public discourse. It tackles deeper ethical issues around deception, consent, and the blurred lines between reality and fabrication. By analyzing evolving legal frameworks and real-world case studies, the chapter advocates for a balanced approach to harnessing deepfake technology responsibly, ensuring its benefits outweigh its potential harms.

The integration between the internet of things (IoT) and artificial intelligence (AI) for real-time supply chain monitoring is streamlining and enhancing how businesses control and optimize their supply chains. This chapter explores the collaborative capabilities of IoT and AI in maximizing supply chain visibility, operational efficiency, and decision-making processes. IoT devices such as sensors, GPS trackers, and RFID tags yield real-time data on various parameters such as location, condition, and the status of the goods throughout their entire journey in the supply chain. This aids with real-time tracking and condition monitoring. Alternatively, incorporating AI algorithms helps analyze historical data, market trends, and various external factors and provides a demand forecast report, which helps in inventory management. Also, AI algorithms can work on traffic patterns to analyze the shortest route, helping to make efficient use of resources, leading to cost savings.

The increasing complexity of global supply chains necessitates advanced analytical tools to address uncertainty, optimize decision-making, and incorporate sustainability goals. This chapter explores the application of Markov models in supply chain management, with a particular focus on their integration into human and ecological dimensions. Markov decision processes (MDPs) are examined for their ability to optimize resource allocation and dynamic decision-making under uncertainty, while hidden Markov models (HMMs) are highlighted for their utility in understanding latent state dynamics. Finally, Markov chain Monte Carlo (MCMC) methods are discussed for their robust probabilistic modeling capabilities, particularly in scenarios involving high-dimensional data. The chapter also delves into the challenges of integrating these models into real-world supply chain systems. Finally, future research directions are outlined, emphasizing the role of AI-driven Markov frameworks in advancing sustainable, adaptive, and human-centric supply chain solutions.

Collection and delivery points (CDPs) are crucial for addressing last-mile logistics challenges, particularly with the rapid growth of e-commerce. This study examines factors influencing e-consumers' preferences for CDPs in Morocco, using deep learning (DL) algorithms to develop a predictive model. A dataset of 3,000 responses, including age, income, and accessibility to CDPs, was used to train the models. Convolutional neural networks (CNN), long short-term memory (LSTM), and FT-transformer were employed to classify preferences. Hyperparameter tuning optimized these models, with the FT-transformer achieving 92% accuracy, outperforming CNN and LSTM. The findings offer actionable insights for logistics providers to better meet consumer needs, enhancing satisfaction and operational efficiency. The chapter also discusses the study's limitations and suggests future research directions in last-mile logistics.

An Expert System in AI is software designed to mimic a human expert's decision-making abilities by applying logic to a vast knowledge base. It provides recommendations or decisions similar to those of professionals in fields like medicine, law, or engineering. The system consists of a knowledge base with domain-specific information and an inference engine that applies rules, often through "if-then" logic, to draw conclusions. Its strengths include efficient, continuous analysis and reliable advice, free from human fatigue or error, especially when expertise is scarce, such as in medical diagnoses. The system also allows non-experts to input queries via a user interface, offering real-time support based on predefined knowledge. Some systems include explanation features to clarify decision-making processes.

This chapter examines the impact of artificial intelligence (AI) on workforce dynamics and employment within supply chains. As AI technologies become increasingly integrated into supply chain operations, they are reshaping job roles, skill requirements, and organizational structures. The chapter explores the dual nature of AI's influence, highlighting both the opportunities for efficiency and innovation and the challenges related to job displacement and workforce adaptation. Through case studies and analysis of current trends, the chapter provides insights into how businesses can strategically manage the transition toward AI-enhanced supply chains. Emphasis is placed on the importance of upskilling and reskilling initiatives to ensure workforce resilience and sustainability in this new era.

Preface

The integration of artificial intelligence (AI) and other cutting-edge technologies has become a significant force driving transformation across industries worldwide, especially in the realm of supply chain management. As editors of *Ecological and Human Dimensions of AI-Based Supply Chain*, we aim to offer an insightful exploration of the rapidly evolving landscape of AI in supply chain operations, alongside its profound implications on environmental sustainability and workforce dynamics. This book addresses the urgent need for innovation in supply chains amidst global disruptions, particularly catalyzed by events such as the COVID-19 pandemic. Through the contributions of experts from various fields, we strive to offer a comprehensive understanding of how AI not only optimizes logistical and operational efficiencies but also impacts human factors, such as critical thinking, career paths, and ethical considerations in the workplace.

As the chapters of this book unfold, they cover a diverse range of topics, from digital transformation and AI-powered logistics innovations to the pressing need for sustainable practices in supply chains. Each chapter tackles the ecological and human dimensions of AI, shedding light on the implications for business practices, consumer behavior, and environmental sustainability. The contributions present real-world case studies, theoretical frameworks, and practical solutions for overcoming challenges faced by supply chains in a fast-paced, technologically driven world. Through these analyses, we hope to inspire further research and facilitate the adoption of AI tools that align with sustainability goals, while balancing the challenges they may pose to the workforce.

As you navigate through the chapters, you will gain valuable insights into how AI technologies can be harnessed to revolutionize supply chains, improve decision-making processes, and foster innovation. At the same time, we address the ethical, environmental, and societal implications that accompany this digital shift. Our collective aim is to offer readers a nuanced perspective on AI's role in supply chains, providing not only academic insights but also actionable strategies for industry leaders, policymakers, and researchers aiming to navigate this dynamic terrain.

CHAPTER 1: ADVANCING SUPPLY CHAIN INNOVATION: THEORETICAL FOUNDATIONS, CONTEMPORARY DEVELOPMENTS, AND FUTURE DIRECTIONS

Overview

This chapter offers a comprehensive literature review on the evolution of supply chain innovation (SCI), exploring its theoretical foundations and current developments. It highlights how globalization, technological advances, and changing consumer preferences are driving innovation in supply chains. By investigating the key dimensions of SCI, the chapter identifies research gaps and provides practical recommendations for both researchers and supply chain practitioners to stay ahead in this rapidly changing landscape.

CHAPTER 2: DIGITAL TRANSFORMATION IN LOGISTICS AND SUPPLY CHAIN MANAGEMENT PRACTICES

Overview

The chapter examines the critical role of digital transformation in modern logistics and supply chain management, particularly in response to the disruptions caused by the COVID-19 pandemic. Focusing on the adoption of technologies like blockchain, AI/ML, and intelligent automation, it explores how these tools improve visibility, coordination, and efficiency. The chapter also addresses the challenges organizations face, such as legacy systems and cybersecurity concerns, and offers strategic guidelines for successful digital transformation in supply chains.

CHAPTER 3: NAVIGATING THE ARTIFICIAL INTELLIGENCE ERA: IMPACTS OF ARTIFICIAL INTELLIGENCE ON YOUTH CRITICAL THINKING, CAREER PATHS, AND ETHICAL CHALLENGES

Overview

This chapter explores the societal impact of AI on younger generations, focusing on critical thinking, career choices, and mental health. It discusses the concerns around AI dependency, especially in education, and its potential to stifle independent reasoning. The chapter also emphasizes the ethical considerations associated with

AI's widespread use, recommending ethical AI practices and proposing frameworks for encouraging balanced engagement with technology.

CHAPTER 4: REVOLUTIONIZING THE SUSTAINABLE SUPPLY CHAIN OF MEDICAL PRODUCTS WITH AI-DRIVEN DRONE DELIVERY

Overview

AI-powered drone technology is presented as a game-changer in sustainable healthcare supply chains. This chapter explores the integration of AI and drones in medical product delivery, enhancing route optimization, sustainability, and cost efficiency. Through a proposed mathematical framework, the chapter addresses the benefits of autonomous drones, such as reducing emissions and energy consumption, while also outlining future advancements in AI-driven drone systems for healthcare logistics.

CHAPTER 5: THE FUTURE OF MARKETING: INTEGRATING BIG DATA AND AI FOR PERSONALIZED CUSTOMER ENGAGEMENT

Overview

Focusing on the synergy between Big Data and AI, this chapter discusses their transformative effect on supply chain management and marketing. By harnessing Big Data for insights into consumer behavior, organizations can optimize inventory, forecast demand, and enhance logistics. Additionally, AI is shown to personalize marketing strategies, creating tailored customer experiences. The chapter concludes by identifying gaps in the literature and offering insights on the challenges and ethical considerations surrounding AI and Big Data in marketing.

CHAPTER 6: THE ROLE OF GOVERNMENTS, EMPLOYERS, AND PLATFORMS TO COPE WITH SKILL GAPS IN INDUSTRY 4.0

Overview

This chapter analyzes the growing skills gap caused by the adoption of Industry 4.0 technologies. It examines how governments, employers, and platforms can collaborate to address these gaps through strategic initiatives. By reviewing existing literature, the chapter presents 14 potential solutions for bridging the skills gap, categorizing them into themes that offer a roadmap for future research and practical applications aimed at reskilling the workforce in the age of Industry 4.0.

CHAPTER 7: IMPORTANCE OF AI AND BIG DATA IN THE WORLD OF SPORTS

Overview

AI and Big Data are increasingly revolutionizing the sports industry. This chapter highlights how data-driven analysis is enhancing decision-making in sports, from performance optimization to fan engagement. By providing examples from case studies and literature, the chapter demonstrates how sports teams and organizations use Big Data to gain a competitive edge and optimize athlete performance.

CHAPTER 8: AI-POWERED WEB SOLUTION FOR DISASTER RISK MANAGEMENT

Overview

The chapter focuses on the use of AI-driven web platforms to predict, monitor, and manage both natural and man-made disasters. By utilizing AI algorithms to analyze weather patterns and sensor data, the platform improves preparedness and response efforts. It also discusses the broader societal benefits of such platforms in enhancing disaster resilience and community safety through real-time alerts and interactive dashboards.

CHAPTER 9: AI METHODOLOGIES IN UPCOMING MODERN WARFARE SYSTEMS

Overview

This chapter delves into the role of AI in modern warfare, emphasizing its impact on decision-making and military operations. By examining AI technologies such as machine learning, natural language processing, and autonomous systems, the chapter highlights how they improve real-time analysis, situational awareness, and resource allocation. The chapter also raises ethical issues related to AI in combat, including autonomy in decision-making and the potential for bias.

CHAPTER 10: EVALUATING GENDER BIAS IN COMPUTATIONAL AI-BASED APPROACHES: A CASE ANALYSIS OF LINKEDIN

Overview

This chapter critically examines the gender biases inherent in AI algorithms, using LinkedIn's AI systems as a case study. By analyzing the data and decision-making processes behind AI applications, the chapter explores the ethical implications of AI in recruitment, hiring, and workplace dynamics. It offers insights into the challenges and necessary measures to ensure fairness and inclusivity in AI-based systems.

CHAPTER 11: EXPLORING INFORMATION TECHNOLOGY (IT) ACCEPTANCE AMONG SUPPLY CHAIN STAKEHOLDERS IN MOROCCO'S PHARMACEUTICAL INDUSTRY

Overview

This chapter investigates the factors influencing IT adoption in the pharmaceutical supply chain in Morocco. Using a quantitative approach, it identifies the barriers to digital transformation in supply chains and highlights the importance of collaboration and information sharing. The findings provide practical implications for improving IT acceptance and fostering digital transformation within the Moroccan pharmaceutical sector.

CHAPTER 12: EXTENDED REALITY AND SECURITY

Overview

Exploring the intersection of extended reality (XR) technologies and security concerns, this chapter discusses the vulnerabilities introduced by XR's integration of the physical and digital worlds. It examines how AI, edge computing, and 5G are transforming XR's potential while increasing its susceptibility to cyber threats. The chapter offers insights into emerging security measures necessary to protect sensitive data in XR environments.

CHAPTER 13: IMPACT OF CIRCULAR ECONOMY ON SUPPLY CHAIN MANAGEMENT: AN EXPLORATORY STUDY

Overview

This chapter explores the integration of circular economy practices in supply chain management, using case studies of two companies. It highlights the role of business ecosystems in fostering circular practices, their impact on supply chain performance, and the challenges faced by organizations adopting such practices. The findings contribute to both theoretical advancements and practical strategies for companies aiming to implement more sustainable supply chain models.

CHAPTER 14: GENERATIVE AI IN DEEPFAKES

Overview

This chapter addresses the technological, ethical, and societal implications of deepfakes created by generative AI models like GANs. It explains the techniques behind deepfakes and explores their potential for misinformation, political manipulation, and privacy violations. Additionally, the chapter discusses the detection methods emerging to counter deepfake content and the broader implications for trust in media and public discourse.

CHAPTER 15: IOT AND AI INTEGRATION FOR REAL-TIME SUPPLY CHAIN MONITORING

Overview

This chapter explores how the integration of IoT and AI is enhancing real-time monitoring in supply chains. By using sensors, GPS, and RFID tags, IoT provides real-time data on goods, while AI helps analyze trends, forecast demand, and optimize resource allocation. The chapter demonstrates how this combination improves visibility, efficiency, and decision-making in supply chain management.

CHAPTER 16: INTEGRATING MARKOV MODELS IN SUPPLY CHAIN MANAGEMENT EXPLORING HUMAN AND ECOLOGICAL DIMENSIONS

Overview

Focusing on Markov models, this chapter explores their application in supply chain management with an emphasis on human and ecological factors. It examines how these models optimize decision-making under uncertainty and discusses their role in resource allocation, sustainability, and ecological performance. The chapter also outlines the future potential of AI-driven Markov models in adapting to dynamic supply chain environments.

CHAPTER 17: PREDICTING E-CONSUMER PREFERENCES FOR COLLECTION & DELIVERY POINTS USING DEEP LEARNING

Overview

This chapter presents a deep learning model for predicting e-consumer preferences for collection and delivery points in Morocco, using data such as age, income, and accessibility. By analyzing these preferences with deep learning algorithms, the study offers valuable insights for improving logistics and customer satisfaction in last-mile delivery systems.

CHAPTER 18: SYSTEM EXPERTS IN AI

Overview

This chapter explores AI-driven expert systems, which simulate human expertise in various domains. It examines how these systems use knowledge bases and inference engines to provide recommendations and decision-making support. The chapter highlights the potential of expert systems to offer reliable advice in fields such as healthcare and law, emphasizing their advantages in efficiency and continuous analysis.

CHAPTER 19: THE AI REVOLUTION IMPLICATIONS FOR WORKFORCE DYNAMICS AND EMPLOYMENT IN SUPPLY CHAINS

Overview

This chapter analyzes how AI is reshaping workforce dynamics within supply chains. By exploring both the opportunities and challenges that AI presents—such as job displacement and the need for new skills—the chapter provides a roadmap for businesses to navigate workforce adaptation and ensure sustainable growth through reskilling and upskilling initiatives.

CHAPTER 20: THE USE OF IOV FOR VEHICLE ROUTING OPTIMIZATION: A SYSTEMATIC LITERATURE REVIEW

Overview

This chapter reviews the literature on the Internet of Vehicles (IoV) and its impact on vehicle routing optimization. By exploring the contributions of IoV to performance in areas like energy, economics, and environmental sustainability, it offers insights into how IoV enhances vehicle routing and traffic management, with a focus on improving logistics efficiency.

CONCLUSION

In closing, the contributions to *Ecological and Human Dimensions of AI-Based Supply Chain* underscore the transformative potential of AI in reshaping supply chains in the 21st century. As businesses continue to innovate and adapt to emerging technologies, the insights offered in this book provide crucial guidance on leveraging AI while maintaining ethical integrity and environmental responsibility. From the use of AI-driven drones for medical deliveries to the integration of machine learning algorithms for real-time supply chain monitoring, the chapters highlight diverse applications that promise to redefine industry standards and operational practices.

However, as much as AI presents opportunities for growth and efficiency, it also raises critical questions about the future of work, data privacy, and bias in algorithmic decision-making. These ethical challenges demand careful consideration and collaborative efforts from researchers, industry stakeholders, and policymakers to ensure that the adoption of AI aligns with the broader goals of human welfare and ecological sustainability. The book stresses the importance of upskilling and reskilling workers, safeguarding against potential job displacement, and ensuring that AI's benefits are equitably distributed.

We trust that the insights shared in this volume will spark further discussions and inspire future research in this rapidly evolving field. As the landscape of AI in supply chain management continues to grow, it is essential that we keep in mind the dual objective of advancing technological capabilities while ensuring the long-term sustainability of both the planet and its people. We hope that the knowledge gained from this book will serve as a foundation for creating innovative, sustainable, and inclusive supply chains in the years to come.

Chapter 1
Advancing Supply Chain Innovation:
Theoretical Foundations, Contemporary Developments, and Future Directions

Atour Taghipour
https://orcid.org/0000-0003-1687-8881
Normandy University, France & Soldbridge, South Korea

XiaoWen Lu
Normandy University, France

ABSTRACT

Supply chains, the intricate networks connecting raw materials, manufacturing, distribution, and consumption, have evolved into complex systems that significantly influence organizational performance. This chapter is a literature review that delves deeper into the conceptualization of supply chain innovation, its theoretical foundations, key dimensions, research findings, and contemporary advancements. It identifies research gaps and provides actionable insights for researchers and practitioners seeking to drive innovation in their supply chain operations.

INTRODUCTION

In today's interconnected global economy, supply chains represent the backbone of industries, linking raw material sourcing, manufacturing processes, distribution networks, and end-user consumption. These intricate networks have evolved into

DOI: 10.4018/979-8-3693-7478-8.ch001

multifaceted systems that significantly influence organizational performance, sustainability, and competitive advantage. As globalization accelerates, technological advancements reshape industries, and consumer preferences continuously evolve, supply chain operations face unprecedented challenges and opportunities. These factors underscore the critical need for Supply Chain Innovation (SCI), a transformative approach that incorporates novel technologies, strategies, and practices to improve efficiency, resilience, and sustainability across supply chain systems (Dadsena et al., 2021).

The concept of SCI transcends incremental improvements, advocating for ground breaking changes that enable organizations to meet the demands of dynamic market environments. From blockchain-enhanced traceability and artificial intelligence-driven forecasting to agile practices for rapid response, SCI encompasses a broad spectrum of innovations that empower businesses to adapt and thrive. The urgency for such innovation is further magnified by external pressures such as global disruptions, heightened sustainability requirements, and the rise of digital transformation (Taghipour & Merimi, 2021).

One pivotal study in the field is Wong and Ngai's (2016) comprehensive review and other contributions, which map the trajectory of SCI research. Their work provides an extensive overview of the theoretical perspectives, methodological approaches, and empirical findings that have shaped the domain. By analyzing the progression of SCI over nearly two decades, Wong and Ngai highlighted key dimensions of innovation, identified areas of growth, and paved the way for further exploration.

Building upon this foundational research, the present literature review seeks to delve deeper into the evolving landscape of SCI. It begins by exploring the conceptualization of SCI, unpacking its theoretical underpinnings and essential dimensions. Additionally, it examines key findings from prior studies, sheds light on contemporary advancements, and evaluates the practical implications of emerging technologies and strategies. Beyond this, the review identifies critical gaps in the existing body of knowledge, providing actionable insights for both researchers and practitioners eager to foster innovation within their supply chain operations.

CONCEPTUALIZING SUPPLY CHAIN INNOVATION

A clear and comprehensive understanding of Supply Chain Innovation (SCI) is crucial for its successful implementation within modern organizations. Despite varying definitions across the literature, a central theme consistently emerges: SCI involves the integration of novel ideas, technologies, and business models that lead to significant improvements in supply chain performance. These innovations are designed to address the ever-evolving challenges within the global supply chain landscape and aim to increase efficiency, resilience, and adaptability. As businesses face pressures to meet the demands of a rapidly changing market environment, SCI emerges as a strategic imperative that drives sustainable and competitive advantage (Taghipour et al., 2023; Taghipour, 2024).

In the context of SCI, four key dimensions stand out, each of which contributes uniquely to the advancement of supply chain operations. These dimensions are not mutually exclusive, but rather interrelated, working together to foster holistic transformation within supply chain systems.

1. Technological Innovation

Technological innovation is one of the cornerstones of SCI. The advent of advanced technologies such as Internet of Things (IoT), Artificial Intelligence (AI), blockchain, and robotics has revolutionized supply chain management, making it more transparent, automated, and efficient. IoT enables real-time data collection and monitoring of assets, inventory, and shipments, providing unprecedented visibility into the supply chain. AI and machine learning can predict demand fluctuations, optimize routes, and automate decision-making processes, leading to enhanced forecasting accuracy and responsiveness. Blockchain technology, with its ability to ensure traceability and security, enhances transparency and trust within the supply chain, particularly in industries where provenance and authenticity are critical. Robotics and automation streamline manufacturing processes, reduce human error, and accelerate production times, improving overall efficiency (Ju et al., 2016).

The continuous development and integration of these technologies help organizations build more resilient, agile, and responsive supply chains (Taghipour, 2023). By leveraging these innovations, businesses can achieve cost savings, reduce operational risks, and better serve customer demands.

2. Process Innovation

Process innovation within the supply chain focuses on improving operational efficiency by rethinking and refining existing workflows, systems, and procedures. It encompasses strategies that aim to streamline processes, reduce lead times, cut costs, and eliminate waste. Lean methodologies, Six Sigma, and other operational excellence techniques have become integral to process innovation. Lean focuses on eliminating waste and maximizing value by improving flow and reducing inefficiencies, while Six Sigma targets defect reduction and process standardization to ensure high-quality outcomes (Tay & Loh, 2022).

By adopting these methodologies, organizations can achieve significant cost reductions and operational improvements. For example, faster product delivery times, lower inventory holding costs, and minimized stockouts are all direct results of process innovations. The continuous focus on process improvement helps supply chains remain competitive in a rapidly evolving global market by ensuring they operate efficiently and effectively.

3. Organizational Innovation

Organizational innovation in SCI refers to the reorganization of structures, roles, and relationships within a company to enhance collaboration, agility, and responsiveness to market demands. Traditional hierarchical structures may no longer suffice in a rapidly changing business environment. To foster innovation and improve performance, companies must embrace more flexible, decentralized, and dynamic organizational models. This often involves rethinking team structures, leadership approaches, and communication flows to ensure cross-functional collaboration and faster decision-making (Adenekan et al., 2024).

For example, agile methodologies have found their place within supply chain organizations, enabling teams to quickly adapt to changing customer needs or supply chain disruptions. By empowering teams to make decisions, encourage creativity, and work collaboratively across departments, organizations can increase their responsiveness to new opportunities and threats.

4. Business Model Innovation

Business model innovation within the context of SCI involves developing new models that create value not just for the organization but also for customers, suppliers, and other stakeholders. The traditional supply chain business model, which often relied on linear, transactional relationships, is being increasingly replaced by more innovative, value-driven approaches. New business models are emerging, such as subscription-based models, platform-based models, and those grounded in the circular economy (Nayal et al., 2022).

Subscription-based models allow companies to secure recurring revenue streams while ensuring a consistent demand for products or services. Platform-based models, such as those seen with the rise of online marketplaces, enable businesses to create ecosystems that connect customers and suppliers, facilitating more efficient transactions and interactions. Circular economy models, which focus on reducing waste by reusing and recycling materials, offer a sustainable approach to supply chain operations that helps conserve resources and reduce environmental impact.

Each of these business model innovations enables supply chains to add more value to customers while driving sustainability and profitability. By adopting innovative business models, organizations can not only enhance their financial performance but also differentiate themselves in a crowded marketplace.

KEY DIMENSIONS OF SUPPLY CHAIN INNOVATION

Supply chain innovation (SCI) encompasses a broad array of activities designed to enhance the efficiency, agility, and sustainability of supply chains (Taghipour & Beneteau-Piet, 2020; Loivet et al., 2020; Zoghlami et al., 2016). Wong and Ngai's (2016) comprehensive framework categorizes SCI into three primary dimensions: logistics-oriented, marketing-oriented, and technology-oriented innovations. These dimensions each target different aspects of supply chain performance, but they are inherently interconnected, requiring a holistic approach to fully realize the benefits of SCI. Each dimension plays a vital role in improving operational processes and aligning supply chains with market demands, while also ensuring that businesses stay competitive in an increasingly complex global environment.

Logistics-Oriented Innovations focus on optimizing the physical and informational flows within the supply chain. These innovations are central to improving the efficiency of core logistics activities such as inventory management, transportation, and warehousing. Advanced inventory management systems, including technologies like RFID and barcode scanning, have revolutionized the way companies monitor and control their stock. These systems provide real-time visibility into inventory levels, which helps reduce stockouts, improve stock replenishment, and optimize inventory levels to match demand. For transportation management, logistics-oriented innovations emphasize the optimization of transport routes, modes, and carrier selection to minimize costs and improve delivery times (Mbiatem et al., 2018). Technologies such as GPS tracking and telematics enable better route planning and real-time vehicle performance monitoring, which ultimately improves fuel efficiency, reduces environmental impact, and enhances service levels. Furthermore, warehouse management systems (WMS) have automated critical warehouse operations, such as order picking, packing, and shipping, optimizing storage space, improving accuracy in order fulfillment, and enhancing overall warehouse productivity. Collaborative logistics, another key aspect of this dimension, focuses on fostering stronger collaboration among supply chain partners. By sharing information and coordinating activities such as planning, forecasting, and replenishment through tools like Collaborative Planning, Forecasting, and Replenishment (CPFR), businesses can align their operations more closely, reduce inventory levels, and improve overall supply chain responsiveness (Cauchois et al., 2017; Taghipour & Frayret, 2011).

Marketing-Oriented Innovations seek to align supply chain activities with market demands and customer expectations. This dimension focuses on understanding customer preferences and ensuring that supply chains are agile enough to meet those needs. One important area is customer-centric design, where companies involve customers in the product development process, ensuring that the products created align with their needs, desires, and preferences. This approach enhances customer satisfaction and fosters loyalty. Additionally, mass customization enables firms to offer tailored products and services at scale, leveraging flexible supply chain configurations that allow them to produce products that meet individual customer specifications without compromising efficiency. Another important facet of marketing-oriented innovation is reverse logistics, which focuses on the processes involved in recovering products from customers for reuse, recycling, or disposal (Taghipour et al., 2015). This is an increasingly important consideration as sustainability becomes a growing concern for both businesses and consumers. Efficient reverse logistics processes help reduce waste, recover valuable materials, and enhance the environmental sustainability of supply chains. Finally, supply chain visibility is crucial for marketing-oriented innovation, as providing real-time insights into supply chain operations enables companies to improve customer service, track products, and reduce lead times (Nadia et al., 2021). Enhanced visibility helps companies respond to customer demands more quickly, ensuring they maintain competitive advantage in fast-moving markets.

Technology-Oriented Innovations drive the transformation of supply chain operations through the application of cutting-edge technologies. One of the most influential innovations in this area is the Internet of Things (IoT), which enables the collection of real-time data on product movement, inventory levels, and equipment performance. IoT devices provide critical insights into the operational health of a supply chain, allowing for better forecasting, tracking, and decision-making. Another key technology-oriented innovation is artificial intelligence (AI) and machine learning (ML), which optimize supply chain decision-making by analyzing large volumes of data to predict demand, forecast inventory needs, and detect potential disruptions. These technologies can significantly improve the accuracy of supply chain predictions, reduce human error, and enable proactive problem-solving. Additionally, blockchain technology has emerged as a game-changer in enhancing supply chain transparency and traceability. By creating an immutable digital ledger, blockchain ensures that every transaction and movement of goods is securely recorded, providing an unprecedented level of security, accountability, and transparency in supply chain operations. Lastly, robotics and automation are transforming supply chain operations by automating repetitive tasks and improving efficiency. From robotic process automation in warehouses to the use of drones for inventory management, these technologies streamline operations, reduce the need for manual labor, and increase overall throughput and accuracy.

When organizations effectively integrate these three dimensions: logistics, marketing, and technology, they can create more responsive, efficient, and sustainable supply chains that not only meet customer needs but also position the business for long-term success. However, it is crucial to understand that these dimensions are not isolated; they often overlap and interconnect. For instance, logistics innovations such as transportation management are enhanced by technology-oriented innovations like GPS tracking and AI algorithms, while marketing-oriented innovations like mass customization depend on logistics systems that can handle flexible product configurations and rapid changes in demand. A successful SCI strategy requires a comprehensive and integrated approach that considers how these various dimensions interact and contribute to the overall performance of the supply chain. Through this holistic view, companies can achieve greater efficiency, sustainability, and resilience, ultimately driving innovation across their entire supply chain ecosystem (Taghipour & Frayret, 2013a; Glaa et al., 2014).

EXPLORING EMERGING TECHNOLOGIES

As the landscape of global supply chains continues to evolve, the integration of cutting-edge technologies is becoming increasingly vital in driving efficiency, improving decision-making, and optimizing operations. Among the most promising of these emerging technologies are quantum computing, digital twins, and artificial intelligence (AI) coupled with machine learning (ML). Each of these technologies holds significant potential to revolutionize how supply chains function, offering new solutions to age-old problems while creating opportunities for unprecedented innovation.

Quantum Computing for Revolutionizing Supply Chain Optimization: Quantum computing, a rapidly advancing field, has the potential to reshape the future of supply chain optimization in ways that were once thought impossible. Unlike classical computing, which relies on binary bits to process data, quantum computing uses quantum bits (qubits) that can exist in multiple states simultaneously, enabling it to solve complex problems at exponentially faster speeds. This makes it particularly suitable for tackling the optimization challenges that supply chains face, such as routing, inventory management, and demand forecasting (Taghipour, 2020).

Quantum computing could provide breakthroughs in optimization problems that involve a vast number of variables and constraints, such as supply chain network design, transportation scheduling, and resource allocation. For instance, a quantum computer could process vast amounts of data in real-time, allowing organizations to make better decisions about production schedules, transportation routes, and inventory levels. Moreover, quantum computing could also enhance predictive models, improving the accuracy of demand forecasting and minimizing stockouts or excess inventory. The power of quantum algorithms could ultimately lead to more efficient, cost-effective, and agile supply chains, transforming industries across the board, from manufacturing to logistics (Mittal et al., 2024).

Digital Twins for Enhancing Visibility and Simulation in Supply Chains: Digital twins, virtual replicas of physical assets, processes, or systems, are quickly emerging as a key technology in the optimization of supply chains. By creating a real-time digital representation of the supply chain, businesses can simulate, monitor, and optimize their operations with unprecedented accuracy. This technology allows companies to virtually model their entire supply chain, from raw material sourcing to final delivery, providing real-time insights into performance, potential risks, and opportunities for improvement (Moshood et al., 2021).

The use of digital twins enables businesses to simulate various scenarios and evaluate the potential outcomes of different strategies without disrupting real-world operations. This capability allows supply chain managers to predict the impact of changes—such as a new supplier, a different route, or an alternative production method—before implementing them. Moreover, digital twins can help in detecting inefficiencies or bottlenecks in the supply chain by continuously monitoring the flow of goods and services. By using this data to create a virtual "twin" of the entire system, companies can optimize their processes in real-time, improving resilience, reducing costs, and enhancing customer satisfaction.

Artificial Intelligence and Machine Learning for Transforming Predictive Analytics and Forecasting: Artificial intelligence (AI) and machine learning (ML) are already playing a pivotal role in reshaping supply chain operations. By harnessing vast amounts of historical and real-time data, AI and ML algorithms can make predictions and offer recommendations that improve decision-making across the entire supply chain. AI-powered systems can forecast demand with a high degree of accuracy, optimizing inventory levels and reducing the risk of overstocking or stockouts. Additionally, ML models can help identify patterns in customer behavior, enabling companies to better anticipate future needs and trends (Bharadiya,2023).

One of the most significant applications of AI and ML in supply chains is predictive analytics. By analyzing historical data, AI algorithms can predict disruptions, changes in demand, or shifts in market conditions. This predictive capability enables companies to respond proactively rather than reactively, allowing them to adjust their operations to avoid potential issues or capitalize on emerging opportunities. In logistics, AI can also optimize routes, reduce fuel consumption, and improve delivery times, while ML algorithms can be used to identify anomalies or inefficiencies in the supply chain, improving overall operational performance.

Moreover, AI and ML are also instrumental in anomaly detection, identifying outliers or unusual patterns in data that could indicate fraud, operational issues, or supply chain disruptions. For example, machine learning algorithms can analyze purchasing patterns to detect fraudulent activity or identify potential risks in the supply chain, such as supplier failure or transportation delays. These capabilities empower businesses to address problems before they escalate, ensuring smoother, more reliable operations.

THEORETICAL FOUNDATIONS OF SUPPLY CHAIN INNOVATION

The field of Supply Chain Innovation (SCI) has drawn upon a diverse range of theoretical perspectives to understand its key drivers, mechanisms, and outcomes. Each of these perspectives contributes unique insights into how SCI can shape organizational performance and competitive advantage in an increasingly complex and dynamic business environment. This section explores some of the most widely applied and influential theories in the study of SCI: the Resource-Based View (RBV), Dynamic Capabilities Theory (DCT), Transaction Cost Economics (TCE), and Social Network Theory (SNT). These theories provide a deeper understanding of the various ways SCI can be harnessed to enhance supply chain performance, drive innovation, and foster long-term sustainability (Lu & Taghipour, 2023).

1. Resource-Based View (RBV)

The Resource-Based View (RBV) posits that firms gain competitive advantage through the acquisition and utilization of valuable, rare, inimitable, and non-substitutable resources and capabilities. According to RBV, resources such as unique technologies, intellectual property, skilled personnel, and organizational knowledge serve as the foundation for a firm's competitive advantage. In the context of SCI, innovation within the supply chain can be viewed as a strategic resource that helps firms differentiate themselves in the market and outperform competitors (Huang et al., 2023).

By developing and leveraging innovative technologies, operational practices, and business models, organizations can establish capabilities that are difficult for rivals to replicate. For instance, adopting advanced digital technologies such as AI, blockchain, and IoT can create unique capabilities in supply chain visibility, automation, and real-time data analytics, which provide a significant advantage over competitors who have not yet embraced such technologies (Taghipour et al., 2022). Similarly, process innovations like lean manufacturing or just-in-time practices can lead to improved efficiency and cost-effectiveness, further enhancing a firm's resource base. Through the application of SCI, firms can continuously improve and refine their resource base, reinforcing their competitive position in the market.

2. Dynamic Capabilities Theory (DCT)

The Dynamic Capabilities Theory (DCT) focuses on a firm's ability to sense, seize, and shape opportunities in response to rapid changes in the external environment. DCT emphasizes the importance of organizational agility and adaptability—skills that are essential for firms to survive and thrive in the face of constant technological advancements, market disruptions, and evolving consumer preferences. In this theory, SCI is seen as a manifestation of a firm's dynamic capabilities, enabling organizations to respond flexibly and effectively to changes in the supply chain environment (Gupta et al., 2020, July).

SCI allows firms to continuously assess and adjust their strategies, processes, and technologies in response to environmental changes. For instance, when disruptions occur, such as supply shortages, shifts in demand, or new regulatory requirements, firms with dynamic capabilities can quickly reconfigure their supply chains to maintain resilience and ensure continuity. Additionally, dynamic capabilities allow organizations to not only adapt to changes but also to proactively shape opportunities for innovation, improving their ability to identify emerging trends and capitalize on them. Therefore, SCI plays a key role in enhancing a firm's dynamic capabilities, enabling it to remain competitive and adaptable in a rapidly changing global market.

3. Transaction Cost Economics (TCE)

Transaction Cost Economics (TCE) examines the costs associated with economic transactions, focusing on the governance mechanisms that firms use to manage these costs. These costs may include negotiation, monitoring, and enforcement costs that arise when firms interact with external partners. TCE emphasizes that firms seek to minimize these transaction costs by designing efficient governance structures for their supply chains. SCI can play a critical role in reducing transaction costs by improving coordination, information sharing, and trust among supply chain partners (Patil et al., 2024).

For example, the adoption of advanced technologies like blockchain can reduce the costs associated with verifying and tracking transactions by providing a secure, transparent, and immutable ledger for all parties involved. Similarly, adopting collaborative practices and real-time information sharing platforms reduces the need for costly monitoring and negotiation, as parties can access accurate and up-to-date information at any time. These innovations enhance trust and cooperation among supply chain partners, ultimately lowering transaction costs. By fostering better coordination and communication, SCI helps firms streamline operations, improve supply chain visibility, and create stronger, more reliable partnerships that are essential for long-term success.

4. Social Network Theory (SNT)

Social Network Theory (SNT) focuses on the structure and dynamics of social relationships within organizations and between external stakeholders. SNT explores how the strength, density, and directionality of relationships influence knowledge flow, collaboration, and innovation within a network. SCI, in this context, is facilitated by strong social networks that enable knowledge sharing, collaboration, and trust-building among supply chain partners, employees, and other stakeholders (Liu et al., 2024).

In supply chains, the ability to access and share knowledge quickly is crucial for innovation. SNT emphasizes that firms with strong social ties—whether internally among employees or externally with suppliers, customers, and other partners—are better able to innovate and respond to challenges. Social networks enable the exchange of tacit knowledge, such as best practices, industry insights, and technological advancements, which are often difficult to codify and disseminate through formal channels. Additionally, by fostering trust and collaboration among network members, social networks contribute to more efficient decision-making and problem-solving, both of which are vital for driving SCI. The cultivation of strong relationships and networks within the supply chain thus plays a central role in fostering innovation and ensuring the success of supply chain initiatives.

CONTEMPORARY ADVANCEMENTS IN SUPPLY CHAIN INNOVATION

Since Wong and Ngai's seminal review, supply chain innovation (SCI) has evolved considerably, driven by technological advancements, globalization, and shifting consumer expectations. As industries adapt to these changes, several contemporary developments are reshaping the landscape of supply chain management, pushing organizations toward smarter, more sustainable, and adaptable supply chains. One of the most transformative developments in SCI is the rise of smart supply chains, made possible by the integration of technologies like the Internet of Things (IoT), blockchain, and artificial intelligence (AI) (Taghipour et al., 2023b). These intelligent systems are enhancing the visibility, traceability, and responsiveness of supply chains, offering a range of key features that optimize operations. For instance, IoT devices enable real-time tracking of shipments, inventory, and equipment, providing organizations with valuable insights into the movement and condition of goods, while also improving supply chain performance by reducing inefficiencies. Additionally, AI-powered predictive analytics are increasingly being used to forecast demand, identify potential disruptions, and optimize inventory levels, leading to more accurate decision-making and proactive problem-solving. Blockchain technology plays a pivotal role in improving collaboration and trust among supply chain partners by facilitating secure, transparent information sharing, reducing the risk of fraud, and enhancing overall supply chain governance. Furthermore, automation technologies such as robotic process automation (RPA) and autonomous vehicles are driving significant improvements in efficiency by reducing labor costs, enhancing operational performance, and minimizing human error, thus fostering a more streamlined supply chain environment.

In addition to technological advancements, sustainability-driven innovations have emerged as a core focus for businesses aiming to meet the growing demand for environmentally and socially responsible practices. Sustainability is increasingly seen as a critical concern, not only for consumers but also for businesses that must adapt to stricter environmental regulations and societal expectations. Supply chain innovation plays a key role in driving sustainable practices across industries. One of the key areas of focus within sustainable supply chain practices is the adoption of circular economy principles, which aim to minimize waste and maximize resource efficiency. This involves designing products for reuse, recycling, and recovery, which not only reduces waste but also helps companies conserve valuable resources. Another important aspect is sustainable sourcing, where businesses prioritize responsible sourcing of materials from suppliers who adhere to ethical and social sustainability practices. This commitment to sustainability extends to green logistics, which includes strategies to reduce carbon emissions, optimize transport routes, and adopt energy-efficient vehicles. By prioritizing sustainability in every aspect of the supply chain, from sourcing to distribution, organizations can significantly reduce their environmental impact and meet the growing consumer demand for sustainable products. Furthermore, ethical sourcing ensures that supply chains operate responsibly by ensuring that workers are treated fairly and that the environmental impact of production is minimized.

Another significant trend reshaping SCI is the growing impact of globalization on supply chain dynamics. While globalization has led to more complex and interconnected supply chains, it has also introduced challenges that require tailored strategies. Regional factors such as cultural differences, regulatory variances, and infrastructure limitations can have a significant impact on how supply chains operate in different parts of the world. This has led to a shift toward more region-specific SCI practices, particularly in emerging markets, where unique challenges and opportunities demand innovative approaches. For example, organizations are increasingly turning to local sourcing, leveraging local suppliers to reduce transportation costs, shorten lead times, and contribute to the growth of local economies. Additionally, the complexity of global trade has made the need for agile supply chains even more critical. In today's fast-paced business environment, companies need supply chains that are flexible and adaptable to rapidly changing demand and supply conditions, ensuring they can respond quickly to disruptions or shifts in consumer preferences. Finally, the ongoing digital transformation is playing a crucial role in enhancing supply chain efficiency and transparency, with businesses adopting digital technologies to streamline operations, monitor performance in real-time, and create more responsive systems (Merimi & Taghipour, 2021).

As these contemporary advancements continue to evolve, organizations can create more resilient, sustainable, and customer-centric supply chains. The integration of emerging technologies, along with a focus on sustainability and the recognition of regional diversity in global supply chains, will enable businesses to respond to the challenges of today's fast-changing environment (Taghipour, 2014). Future research should explore the implications of these trends, including how organizations can effectively integrate these technologies and practices into their existing supply chains, while also identifying new opportunities for innovation that can drive the next phase of supply chain transformation. By doing so, businesses can build supply chains that not only meet current market demands but are also positioned to thrive in an increasingly complex and dynamic global landscape.

ADDRESSING ETHICAL CONSIDERATIONS

Ethical sourcing has emerged as a critical issue within modern supply chains, as businesses are increasingly expected to consider the broader social and environmental implications of their sourcing practices. Ethical sourcing involves ensuring that the materials and products are procured under conditions that promote fairness, labor rights, and environmental sustainability. This includes safeguarding the welfare of workers throughout the supply chain, ensuring fair wages, and preventing exploitative labor practices such as child labor or forced labor. Moreover, businesses are now more than ever being held accountable for the environmental impact of their sourcing decisions, from the carbon footprint associated with transportation and production to the sustainable use of natural resources. Companies that adopt ethical sourcing practices often engage in fair trade certifications or align with industry standards to guarantee that their supply chains do not contribute to exploitation or environmental degradation. This growing awareness of ethical sourcing not only reflects changing consumer expectations but also signals the increasing importance of responsible business practices in maintaining brand integrity and trust.

Data privacy and security concerns are another critical aspect of Supply Chain Innovation (SCI) that businesses must address. With the rapid advancement of digital technologies and the increasing reliance on real-time data, supply chains are generating vast amounts of sensitive information, ranging from customer preferences and financial transactions to operational and inventory data. The collection, analysis, and sharing of this data raise significant privacy and security issues. Organizations must implement robust data protection mechanisms to safeguard against cyber threats, breaches, and unauthorized access to sensitive information. The increasing use of cloud-based platforms and interconnected systems further heightens the risk of data leaks or cyberattacks that could jeopardize the confidentiality of both organizational and customer data. In order to mitigate these risks, supply chains must integrate strong cybersecurity measures, comply with data privacy regulations, and establish clear protocols for the ethical use and sharing of data across the supply chain network. Addressing these concerns is not only vital for maintaining trust with customers and partners but also for ensuring compliance with legal and regulatory frameworks governing data privacy (Taghipour et al., 2023c).

The social impact of Supply Chain Innovation (SCI) also warrants careful consideration, as the implementation of new technologies and practices can have significant effects on workers, communities, and broader societal structures. One of the key social concerns is job displacement. As automation, artificial intelligence, and robotics become more prevalent within supply chains, there is growing concern about the potential loss of jobs, particularly in industries where manual labor has historically been essential. The rise of automation may lead to the displacement of workers who lack the skills to transition into more technology-driven roles, thus exacerbating income inequality and social stratification. On the other hand, SCI also presents opportunities for reskilling and upskilling workers, creating new job categories, and improving workplace conditions. Furthermore, the adoption of sustainable supply chain practices can have a positive social impact by fostering community development, supporting local economies, and ensuring that supply chains contribute to societal well-being. SCI initiatives that prioritize inclusivity, fairness, and community development can help build stronger, more resilient supply chains that not only drive business success but also benefit society at large. Therefore, companies must consider the long-term social impact of their SCI efforts, striving to create positive outcomes for workers and communities while mitigating the risks associated with technological disruption and inequality (Hastuti, 2023).

RESEARCH FINDINGS AND EMERGING TRENDS

A growing body of research has been dedicated to examining the multifaceted nature of Supply Chain Innovation (SCI) and its transformative impact on the operational, financial, and strategic success of organizations. As the global business landscape continues to evolve, companies increasingly recognize the need to innovate and enhance their supply chains to remain competitive and sustainable. Several key findings and emerging trends have emerged from this research, each shedding light on the crucial elements that drive SCI and shape its outcomes.

1. The Impact of SCI on Performance

One of the most consistent findings across the literature is the positive impact of SCI on various performance metrics. Numerous studies have demonstrated that organizations that actively engage in supply chain innovation tend to experience significant improvements in their operational performance, financial performance, and customer satisfaction. These improvements stem from the ability of SCI to streamline processes, reduce inefficiencies, and enhance flexibility (Belhadi et al., 2024).

Operational Performance: SCI enables organizations to optimize their supply chain processes, leading to faster production times, reduced inventory costs, and better demand forecasting. For example, the adoption of real-time data analytics, predictive modeling, and advanced logistics technologies can significantly reduce delays and bottlenecks in the supply chain, improving operational efficiency.

Financial Performance: Research has shown that firms that embrace SCI tend to achieve higher profitability and a better return on investment (ROI). Innovation in areas such as automation, robotics, and advanced inventory management systems can help companies cut operational costs while improving scalability and responsiveness to market changes.

Customer Satisfaction: As customer expectations continue to rise, the importance of delivering high-quality products on time and with increased transparency becomes critical. SCI helps companies provide superior customer experiences through more efficient delivery, personalized services, and improved product quality (Mahfod et al., 2019). By improving both operational and service aspects, SCI directly contributes to increased customer loyalty and satisfaction.

2. The Role of Technology in Driving SCI

Technological advancements have been at the core of supply chain innovation. The rapid evolution of digital technologies has provided companies with new tools and capabilities to enhance their supply chain operations. However, successful adoption and integration of these technologies are not without challenges. Research highlights that while technological innovations have tremendous potential, their success depends heavily on organizational readiness and effective change management strategies (Attaran, 2020).

Technological Advancements: Technologies such as Internet of Things (IoT), Artificial Intelligence (AI), blockchain, and robotics have proven to be transformative in supply chain management. IoT enables real-time tracking of inventory and goods, AI helps with predictive analytics for demand forecasting, blockchain provides transparency and traceability, and robotics automates repetitive tasks. These innovations lead to greater efficiency, accuracy, and visibility across the supply chain, facilitating quicker decision-making and reducing operational risk.

Organizational Readiness: Research has shown that the implementation of new technologies in supply chains requires careful planning and a commitment to developing the necessary infrastructure (Taghipour & Frayret, 2013b). Organizations must invest in training, resource allocation, and fostering a culture that is open to change. Technological adoption can be a disruptive process, and without adequate preparation, organizations may face resistance from employees and stakeholders, undermining the success of the innovation.

Effective Change Management: A key finding from the literature is that successful technology adoption is contingent upon strong change management strategies. This includes involving key stakeholders in the decision-making process, ensuring proper training for employees, and continuously monitoring the implementation process. When managed correctly, the adoption of new technologies can greatly enhance an organization's ability to compete and adapt to changing market conditions (Atilgan & McCullen 2011).

3. The Importance of Collaboration in SCI

Collaboration among supply chain partners is another critical component of successful SCI. Research has consistently shown that building strong, cooperative relationships within and across organizations in the supply chain results in significant advantages in terms of innovation, efficiency, and flexibility (Rodríguez et al., 2020).

Collaborative Networks: The complexity of modern supply chains requires close cooperation between suppliers, manufacturers, distributors, and retailers. A collaborative approach helps to ensure that all parties work together toward common goals, reducing the risk of delays, misunderstandings, or conflicts. Collaborative networks foster the sharing of resources, knowledge, and best practices, leading to innovations that benefit all participants in the supply chain (Bititci, et al., 2004; Tamym et al., 2021).

Information Sharing: One of the key elements of collaboration is the exchange of information between supply chain partners. Research emphasizes that transparency and the flow of real-time data are vital to improving decision-making, forecasting, and problem-solving. For instance, sharing information on inventory levels, customer demand, and potential disruptions allows supply chain partners to adjust their plans more efficiently, reducing risk and improving overall performance (Basana et al., 2024).

Joint Initiatives: Developing joint initiatives, whether it's co-developing new products, shared technological investments, or mutual sustainability goals can yield substantial benefits for all parties involved. Such initiatives strengthen trust, align objectives, and foster a deeper sense of partnership. Collaborative efforts in sustainability, for example, can lead to cost savings, better resource utilization, and improved social responsibility.

4. The Need for Sustainability in SCI

Sustainability has emerged as a critical consideration in supply chain innovation. As companies face increasing pressure from regulators, consumers, and investors to adopt more environmentally and socially responsible practices, integrating sustainability into supply chain strategies has become essential. SCI can contribute to sustainability efforts by introducing processes and practices that minimize environmental impact, promote social responsibility, and ensure ethical sourcing (Javed et al., 2024).

Environmental Impact: Supply chains are major contributors to environmental challenges, including carbon emissions, waste, and resource depletion. Through SCI, companies are adopting more sustainable practices, such as using renewable energy sources, optimizing transportation routes, and adopting circular economy models. Innovations such as smart logistics and green technologies can help reduce the environmental footprint of supply chain activities (Oliveira-Dias et al., 2024).

Social Responsibility: SCI also addresses the social aspect of sustainability, focusing on improving labor practices, ensuring fair wages, and fostering safe working conditions. Ethical sourcing practices, such as ensuring that suppliers adhere to social standards and regulations, are becoming an essential part of supply chain innovation (Le et al., 2024).

Ethical Sourcing: Another important element of sustainability in SCI is ensuring that products are sourced ethically. Companies are increasingly scrutinizing their supply chains to ensure that raw materials are sourced from suppliers who adhere to environmental, social, and ethical standards. The integration of sustainability into the supply chain not only meets regulatory requirements but also enhances brand reputation and fosters consumer trust (Ahmed et al., 2024; Mubarik et al., 2024).

FUTURE DIRECTIONS AND PRACTICAL IMPLICATIONS

While significant progress has been made in Supply Chain Innovation (SCI) research, several areas warrant further investigation to fully understand its potential and maximize its impact. One such area is the role of human factors in the implementation of SCI. While technological innovations often take center stage, the human element plays a crucial role in ensuring the success of SCI initiatives. Leadership, organizational culture, and the skills of employees are key determinants in adopting and leveraging new technologies, processes, and business models. Effective leadership fosters a culture that supports innovation, while well-trained employees are essential in executing and optimizing new practices. Moreover, the ability of a company's workforce to adapt to technological changes can significantly influence the successful implementation of SCI, making human factors an important focus for future research.

Another area that requires deeper exploration is the impact of Industry 4.0 on supply chains. The convergence of cutting-edge technologies such as Artificial Intelligence (AI), the Internet of Things (IoT), big data, and robotics is reshaping how supply chains operate. These technologies enable greater connectivity, automation, and data-driven decision-making, leading to more agile, responsive, and efficient supply chains. However, understanding the specific ways these technologies intersect and how they can be harmoniously integrated into existing supply chain systems remains a complex challenge. Future research could further investigate how these technologies can be best leveraged, as well as the barriers organizations face in adopting them. Understanding these dynamics will be crucial for businesses seeking to stay competitive in an increasingly digital and interconnected world.

Additionally, the role of small and medium-sized enterprises (SMEs) in SCI adoption is another important area that requires attention. While much of the research on SCI has focused on large corporations, SMEs face unique challenges in implementing SCI strategies. They often lack the resources and expertise of larger firms, making it more difficult to adopt advanced technologies and processes. However, SMEs are also critical players in global supply chains, and their success in innovation can have a profound impact on the overall system. Research could explore the specific barriers SMEs face in adopting SCI, such as financial constraints, limited access to technology, and the lack of skilled personnel. Addressing these challenges will be key to enabling SMEs to contribute to and benefit from SCI.

The ethical implications of SCI are also a critical concern. As supply chains become more complex and globalized, ethical considerations related to labor practices, environmental impact, and data privacy must be carefully addressed. For example, the use of technologies like big data and AI raises concerns about data security and privacy, while the environmental impact of supply chains continues to be a growing concern. Research in this area could explore how SCI initiatives can be designed and implemented with ethical considerations in mind, ensuring that organizations do not sacrifice their social and environmental responsibilities in the pursuit of innovation. Future studies could also explore frameworks for integrating ethical standards into SCI strategies, helping businesses make responsible decisions while driving innovation.

For practitioners, there are several actionable insights that can guide the implementation of SCI in their organizations. First, it is crucial to develop a clear innovation strategy that aligns SCI initiatives with the company's overall business objectives. Identifying specific goals, such as improving supply chain efficiency, reducing costs, or enhancing sustainability, can help focus efforts and provide a clear roadmap for innovation. Second, fostering a culture of innovation within the organization is essential. Encouraging creativity, experimentation, and risk-taking helps employees feel empowered to generate and implement new ideas, driving continuous improvement. Building this culture requires leadership commitment and a willingness to embrace change.

Another important aspect is to invest in technology and talent. Organizations must acquire the necessary tools and skills to drive innovation effectively. This includes not only investing in cutting-edge technologies but also ensuring that employees possess the skills and knowledge to use them effectively. Companies should also provide ongoing training and development to help their workforce keep pace with technological advancements. Furthermore, building strong relationships with suppliers and customers is crucial for successful SCI implementation. Collaborative efforts can lead to the identification and implementation of innovative solutions that benefit all parties in the supply chain. Strong partnerships can also help organizations address challenges and unlock new opportunities for innovation.

Lastly, organizations must monitor and evaluate the performance of their SCI initiatives. Continuous assessment allows businesses to measure the impact of innovation efforts, identify areas for improvement, and adjust as needed. By tracking key performance indicators (KPIs) such as operational efficiency, cost savings, and customer satisfaction, companies can ensure that their SCI strategies are delivering tangible results and driving long-term success.

By embracing SCI, organizations can enhance their competitiveness, improve customer satisfaction, and build sustainable supply chains that thrive in the 21st century. The key to success lies in understanding the emerging trends, addressing the challenges, and seizing the opportunities that SCI presents. As research continues to evolve in this field, both academics and practitioners can contribute to shaping a future where supply chains are more innovative, resilient, and sustainable than ever before.

CONCLUSION

We conducted a critical review building upon the foundational work of Wong and Ngai's seminal review, while also incorporating insights from other recent studies, to better understand the evolution and ongoing challenges associated with Supply Chain Innovation (SCI). Their comprehensive review laid the groundwork for identifying the major dimensions and drivers of SCI, yet the field continues to evolve as new technological advancements and shifts in global market dynamics reshape how supply chains operate. Our review delves into these evolving aspects by addressing the gaps that still persist in the current body of knowledge, such as the integration of emerging technologies, the need for greater collaboration across supply chain networks, and the shift towards more sustainable practices. By exploring these gaps, we aim to provide a clearer roadmap for future research in SCI, encouraging scholars to focus on key areas where innovation can lead to more resilient, efficient, and adaptive supply chains.

As the global business environment becomes increasingly complex and interconnected, the need for innovative approaches in supply chain management is more pressing than ever. Future research in SCI can significantly contribute to the development of supply chains that not only improve operational efficiency and resilience but also have a broader societal impact. By addressing the critical gaps identified in the literature, scholars can uncover new strategies for integrating cutting-edge technologies, such as AI, blockchain, and IoT, into supply chain processes in ways that improve transparency, sustainability, and decision-making. Moreover, the need for more robust frameworks for collaboration between supply chain partners will be critical in overcoming challenges related to global disruptions, such as the COVID-19 pandemic or natural disasters, which have underscored the vulnerabilities in traditional supply chain models.

Furthermore, the role of innovation in supply chains extends far beyond operational excellence; it can also be leveraged to tackle some of the most pressing global challenges. As climate change, social inequality, and economic instability continue to affect global markets, supply chains are increasingly viewed as a powerful tool for creating positive environmental and social impact. Through the adoption of sustainable practices, such as circular economy models, eco-efficient technologies, and ethical sourcing, SCI can drive significant reductions in environmental footprints and promote social responsibility across the entire value chain. By embedding sustainability into their innovation strategies, organizations can contribute to the global effort to combat climate change, reduce inequality, and promote economic stability. Moreover, by driving innovation in supply chains, businesses can not only strengthen their competitive advantage but also create long-term value for all stakeholders, including customers, employees, communities, and the environment.

In conclusion, SCI is no longer just about improving internal processes or enhancing competitiveness; it has the potential to be a transformative force in addressing the world's most urgent challenges. By continuing to innovate and improve supply chain systems, businesses can contribute to a more sustainable and equitable global economy. As future research continues to explore and address the gaps identified in the literature, it will pave the way for more advanced, resilient, and socially responsible supply chains, ensuring that innovation serves not only business goals but also the broader objective of fostering a sustainable and just global society.

REFERENCES

Adenekan, O. A., Solomon, N. O., Simpa, P., & Obasi, S. C. (2024). Enhancing manufacturing productivity: A review of AI-Driven supply chain management optimization and ERP systems integration. *International Journal of Management & Entrepreneurship Research*, 6(5), 1607–1624. DOI: 10.51594/ijmer.v6i5.1126

Ahmed, H., Al Bashar, M., Taher, M. A., & Rahman, M. A. (2024). Innovative Approaches To Sustainable Supply Chain Management In The Manufacturing Industry: A Systematic Literature Review. Global Mainstream Journal of Innovation, Engineering & Emerging Technology, 3(2), 1-13.

Atilgan, C., & McCullen, P. (2011). Improving supply chain performance through auditing: A change management perspective. *Supply Chain Management*, 16(1), 11–19. DOI: 10.1108/13598541111103467

Attaran, M. (2020, July). Digital technology enablers and their implications for supply chain management. In Supply Chain Forum *International Journal (Toronto, Ont.)*, 21(3), 158–172.

Basana, S., Malelak, M., Suprapto, W., Siagian, H., & Tarigan, Z. (2024). The impact of SCM integration on business performance through information sharing, quality integration and innovation system. *Uncertain Supply Chain Management*, 12(1), 435–448. DOI: 10.5267/j.uscm.2023.9.008

Belhadi, A., Mani, V., Kamble, S. S., Khan, S. A. R., & Verma, S. (2024). Artificial intelligence-driven innovation for enhancing supply chain resilience and performance under the effect of supply chain dynamism: An empirical investigation. *Annals of Operations Research*, 333(2), 627–652. DOI: 10.1007/s10479-021-03956-x PMID: 33551534

Bharadiya, J. P. (2023). The role of machine learning in transforming business intelligence. *International Journal of Computing and Artificial Intelligence*, 4(1), 16–24. DOI: 10.33545/27076571.2023.v4.i1a.60

Bititci, U. S., Martinez, V., Albores, P., & Parung, J. (2004). Creating and managing value in collaborative networks. *International Journal of Physical Distribution & Logistics Management*, 34(3/4), 251–268. DOI: 10.1108/09600030410533574

Cauchois, R., Taghipour, A., Kang, D., Zoghlami, N., & Abed, M. (2017). Advanced solutions for a supply chain with stochastic information. *Journal of Advanced Management Science*, 5(1), 9–13. DOI: 10.18178/joams.5.1.9-13

Dadsena, K. K., Mathiyazhagan, K., & Taghipour, A. (2021). Analysis of barriers for the build the resilient supply chain networks post-COVID-19. In *Managing Supply Chain Risk and Disruptions: Post COVID-19* (pp. 79–89). Springer International Publishing. DOI: 10.1007/978-3-030-72575-4_7

Glaa, B., Zoghlami, N., & Taghipour, A. (2014, May). Outsourcing and off-shoring healthcare services: A way to cut the costs and improve quality. In 2014 international conference on advanced logistics and transport (ICALT) (pp. 189-194). IEEE.

Gupta, S., Modgil, S., Gunasekaran, A., & Bag, S. (2020, July). Dynamic capabilities and institutional theories for Industry 4.0 and digital supply chain. In Supply Chain Forum *International Journal (Toronto, Ont.)*, 21(3), 139–157.

Hastuti, R., & Syafruddin, . (2023). Ethical considerations in the age of artificial intelligence: Balancing innovation and social values. *West Science Social and Humanities Studies*, 1(02), 76–87. DOI: 10.58812/wsshs.v1i02.191

Huang, K., Wang, K., Lee, P. K., & Yeung, A. C. (2023). The impact of industry 4.0 on supply chain capability and supply chain resilience: A dynamic resource-based view. *International Journal of Production Economics*, 262, 108913. DOI: 10.1016/j.ijpe.2023.108913

Javed, A., Basit, A., Ejaz, F., Hameed, A., Fodor, Z. J., & Hossain, M. B. (2024). The role of advanced technologies and supply chain collaboration: During COVID-19 on sustainable supply chain performance. *Discover Sustainability*, 5(1), 46. DOI: 10.1007/s43621-024-00228-z

Ju, K. J., Park, B., & Kim, T. (2016). Causal relationship between supply chain dynamic capabilities, technological innovation, and operational performance. Management and Production Engineering Review, 7.

Le, T. T., Tran, P. Q., Lam, N. P., Tra, M. N. L., & Uyen, P. H. P. (2024). Corporate social responsibility, green innovation, environment strategy and corporate sustainable development. *Operations Management Research : Advancing Practice Through Research*, 17(1), 114–134. DOI: 10.1007/s12063-023-00411-x

Liu, W., Liu, T., Tang, O., Lee, P. T. W., & Chen, Z. (2024). The impact of digital supply chain announcements disclosing corporate social responsibility information on stock market value. *Industrial Management & Data Systems*, 124(2), 724–760. DOI: 10.1108/IMDS-03-2023-0189

Loivet, W., Taghipour, A., & Kang, D. S. (2020). The rise of green supply chain management: Between complexity and necessity. *J. Econ. Bus. Manag*, 8(1), 1–7. DOI: 10.18178/joebm.2020.8.1.604

Lu, X., & Taghipour, A. (2023). A French Guide for Businesses and Governments to Integrate the Concept of Sustainable Development Into Their Supply Chains. In *Government Impact on Sustainable and Responsible Supply Chain Management* (pp. 1–12). IGI Global. DOI: 10.4018/978-1-6684-9062-4.ch001

Mahfod, J., Canel-Depitre, B., & Taghipour, A. (2019). Quality function deployment-ELECTRE in supplier evaluation. Journal of Advanced Management Science, 7(4).

Mbiatem, J., Taghipour, A., & Canel-Depitre, B. (2018, August). Supplier selection approaches for decision makers. In *Proceedings of the 8th International Conference on Information Communication and Management* (pp. 108-112).

Merimi, M., & Taghipour, A. (2021). Accelerating the digitalization of the supply chain: An empirical research about COVID-19 crisis. In *Digitalization of decentralized supply chains during global crises* (pp. 1–24). IGI Global. DOI: 10.4018/978-1-7998-6874-3.ch001

Mittal, S., Koushik, P., Batra, I., & Whig, P. (2024). AI-Driven Inventory Management for Optimizing Operations With Quantum Computing. In Quantum Computing and Supply Chain Management: A New Era of Optimization (pp. 125-140). IGI Global.

Moshood, T. D., Nawanir, G., Sorooshian, S., & Okfalisa, O. (2021). Digital twins driven supply chain visibility within logistics: A new paradigm for future logistics. *Applied System Innovation*, 4(2), 29. DOI: 10.3390/asi4020029

Mubarik, M. S., & Khan, S. A. (2024). Digital Supply Chain and Sustainability Challenges. In The Theory, Methods and Application of Managing Digital Supply Chains (pp. 133-143). Emerald Publishing Limited. DOI: 10.1108/978-1-80455-968-020241009

Nadia, E. N., Beatrice, C. D., & Atour, T. (2021). The determinants of hotel room rates in Beirut: A hedonic pricing model. International Journal of Trade. *Economics and Finance*, 12(2), 33–42.

Nayal, K., Kumar, S., Raut, R. D., Queiroz, M. M., Priyadarshinee, P., & Narkhede, B. E. (2022). Supply chain firm performance in circular economy and digital era to achieve sustainable development goals. *Business Strategy and the Environment*, 31(3), 1058–1073. DOI: 10.1002/bse.2935

Oliveira-Dias, D., Maqueira-Marín, J. M., Moyano-Fuentes, J., & Tortorella, G. (2024). How does the technology environment impact supply chain strategy and operational performance? *Journal of Manufacturing Technology Management*.

Patil, K., Garg, V., Gabaldon, J., Patil, H., Niranjan, S., & Hawkins, T. (2024). Firm performance in digitally integrated supply chains: A combined perspective of transaction cost economics and relational exchange theory. *Journal of Enterprise Information Management*, 37(2), 381–413. DOI: 10.1108/JEIM-09-2022-0335

Rodríguez, M. Á., Alemany, M. M. E., Boza, A., Cuenca, L., & Ortiz, Á. (2020). Artificial intelligence in supply chain operations planning: Collaboration and digital perspectives. In Boosting Collaborative Networks 4.0: 21st IFIP WG 5.5 Working Conference on Virtual Enterprises, PRO-VE 2020, Valencia, Spain, November 23–25, 2020 *Proceedings*, 21, 365–378.

Taghipour, A. (2014). Improving the plan of a manufacturing network with non-integrated business units. *International Journal of Applied Logistics*, 5(2), 1–11. DOI: 10.4018/ijal.2014040101

Taghipour, A. (Ed.). (2020). *Demand forecasting and order planning in supply chains and humanitarian logistics*. IGI Global.

Taghipour, A. (Ed.). (2023). *Government Impact on Sustainable and Responsible Supply Chain Management*. IGI Global. DOI: 10.4018/978-1-6684-9062-4

Taghipour, A., Abed, M., & Zoghlami, N. (2015, May). Design for remanufacturing respecting reverse logistics processes: A review. In 2015 4th International Conference on Advanced Logistics and Transport (ICALT) (pp. 299-304). IEEE. DOI: 10.1109/ICAdLT.2015.7136602

Taghipour, A., & Beneteau-Piet, C. (2020). Sustainable supply chain management performance. *International Journal of Innovation, Management and Technology*, 11(6), 165–169. DOI: 10.18178/ijimt.2020.11.6.897

Taghipour, A., Foukolaei, P. Z., Ghaedi, M., & Khazaei, M. (2023a). Sustainable Multi-Objective Models for Waste-to-Energy and Waste Separation Site Selection. *Sustainability (Basel)*, 15(22), 15764. DOI: 10.3390/su152215764

Taghipour, A., & Frayret, J. M. (2011). Supply chain coordination planning: A review. Work paper, École polytechnique de Montréal.

Taghipour, A., & Frayret, J. M. (2013a, May). An algorithm to improve operations planning in decentralized supply chains. In *2013 International Conference on Advanced Logistics and Transport* (pp. 100-103). IEEE. DOI: 10.1109/ICAdLT.2013.6568442

Taghipour, A., & Frayret, J. M. (2013b). Coordination of operations planning in supply chains: A review. *International Journal of Business Performance and Supply Chain Modelling*, 5(3), 272–307. DOI: 10.1504/IJBPSCM.2013.055729

Taghipour, A., Lu, X., & Smahi, M. (2022). Opportunities and challenges for society 5.0 in procurement. *Blockchain Technology in Supply Chain Management for Society*, 5(0), 91–100. DOI: 10.1201/9781003177432-9

Taghipour, A., & Merimi, M. (2021, March). Digital transformation of supply chains during crisis: covid-19. In 11th Annual International Conference on Industrial Engineering and Operations Management, IEOM (pp. 7493-7501). DOI: 10.46254/AN11.20211298

Taghipour, A., Padash, A., Etemadi, V., Khazaei, M., & Ebrahimi, S. (2024). Sustainable and Circular Hotels and the Water–Food–Energy Nexus: Integration of Agrivoltaics, Hydropower, Solar Cells, Water Reservoirs, and Green Roofs. *Sustainability (Basel)*, 16(5), 1985. DOI: 10.3390/su16051985

Taghipour, A., Ramezani, M., Khazaei, M., Roohparvar, V., & Hassannayebi, E. (2023b). Smart transportation behavior through the COVID-19 pandemic: A ride-hailing system in Iran. *Sustainability (Basel)*, 15(5), 4178. DOI: 10.3390/su15054178

Taghipour, A., Sohrabi, A., Ghaedi, M., & Khazaei, M. (2023c). A robust vaccine supply chain model in pandemics: Case of Covid-19 in Iran. *Computers & Industrial Engineering*, 183, 109465. DOI: 10.1016/j.cie.2023.109465

Tamym, L., Benyoucef, L., Moh, A. N. S., & El Ouadghiri, M. D. (2021). A big data based architecture for collaborative networks: Supply chains mixed-network. *Computer Communications*, 175, 102–111. DOI: 10.1016/j.comcom.2021.05.008

Tay, H. L., & Loh, H. S. (2022). Digital transformations and supply chain management: A Lean Six Sigma perspective. *Journal of Asia Business Studies*, 16(2), 340–353. DOI: 10.1108/JABS-10-2020-0415

Wong, D. T., & Ngai, E. W. (2019). Critical review of supply chain innovation research (1999–2016). *Industrial Marketing Management*, 82, 158–187. DOI: 10.1016/j.indmarman.2019.01.017

Zoghlami, N., Taghipour, A., Merlo, C., & Abed, M. (2016). Management of divergent production network using decentralised multi-level capacitated lot-sizing models. *International Journal of Shipping and Transport Logistics*, 8(5), 590–604. DOI: 10.1504/IJSTL.2016.078686

Chapter 2
Digital Transformation in Logistics and Supply Chain Management Practices

Hafiz Wasim Akram
https://orcid.org/0000-0002-9116-5913
Dhofar University, Oman

ABSTRACT

In recent years, the rise of digital transformation in logistics and supply chain management (SCM) has become quintessential for organizations; accelerated all the more with how COVID-19 induced an upheaval across global supply chains. The chapter discusses how the pandemic-induced disruptions were beyond what traditional supply chain models, largely manual processes, and data silos could accommodate. The need for strong digital infrastructure is also apparent through the use of next-gen technologies including blockchain, AI/ML, and intelligent automation. These technologies provide real-time visibility, predictive analytics, and improved coordination. This chapter examines the reasons why digital transformation is necessary, what it affords, and the challenges that include legacy system trouble, management of data issues, and cybersecurity and skills gaps. It finishes by offering some strategic guidelines for successful implementation of digital in SCM.

DOI: 10.4018/979-8-3693-7478-8.ch002

1. INTRODUCTION

The pandemic has been a wakeup call for industries globally - clearly underscoring the need to re-think current supply chain models (Zhao, 2021). The original outbreak stalled production and trade, causing significant disruptions in the global supply chains that are closely linked to Chinese manufacturing processes and raw materials. Businesses aimed at narrow margins and low inventory reserves were also caught off guard by the Crona pandemic. This emphasised the need for a move away from efficient-centred approaches to resilience-based supply chain methods. The havoc that COVID-19 wreaked re-emphasized the necessity of powerful and flexible digital setups which can shift their ways at verging on lightening speed in response to unforeseen circumstances. Reimaging their operations to be more profitable than before, organisations in every industry has realised the importance of investing into next-gen technologies including blockchain, AI (Artificial Intelligence), ML (Machine Learning) and smart automation (Kataria et al., 2023). Now the foundation of next-gen supply chains taking real-time visibility, predictive analytics and seamless co-ordination across different nodes in the supply chain are these technologies (Morgan et al., 2021).

Digital transformation in the logistics and supply chain management sector is not just about implementing new technology, but involves a re-imagining of business processes to increase elasticity and more responsiveness (Awwad et al., 2022). Digital solutions enable routine tasks to be automated and human capabilities supplemented with insights driven by AI, while creating ecosystems that are inter-connected for real-time decision-making. One of the key reasons for automation is to reduce human intervention in low-value-added tasks. Some repetitive processes can be executed by technologies like chatbots and robotic process automation (RPA), trying to allow a less numerous workforce for more creative tasks. For example, automating operations like invoice processing and order creation does more than reduce errors - it increases efficiency and customer satisfaction. A growing number of support technologies like drones and autonomous guided vehicles (AGVs) carry out high-precision duties or hazardous processes. These technologies improve the safety of operations as well as provide consistency and accuracy in execution. Robotics in warehouses are the perfect example for automation transforming logistics operations that improve fulfilment times. The application of AI and advanced analytics to this data in turn facilitates cognitive decision-making (Cauchois et al., 2017; Jarrahi, 2018). Powered by big data, cognitive decision centers (CDCs) turn this raw material into insights the can serve as triggers for early warning provisions or even to anticipate disruptions well in advance of their occurrence (Talari et al., 2022). Preventive insight also helps supply chain planners optimize inventory management and production planning accomadating additional 'what-if' scenarios using a digital model cited above These technologies empower companies to move from a reactive state of supply chain management towards more proactive measures where they can quickly adapt to the changing market environment.

2. THE NEED FOR DIGITAL TRANSFORMATION

In the face of a fast-paced, globalized market, the traditional supply chain model - typified by manual processes and closed data integration - is not only outdated but approaches obsolescence. Key factors which emphasis on the necessity for digital transformation include:

Figure 1. Supply Chain Cycle

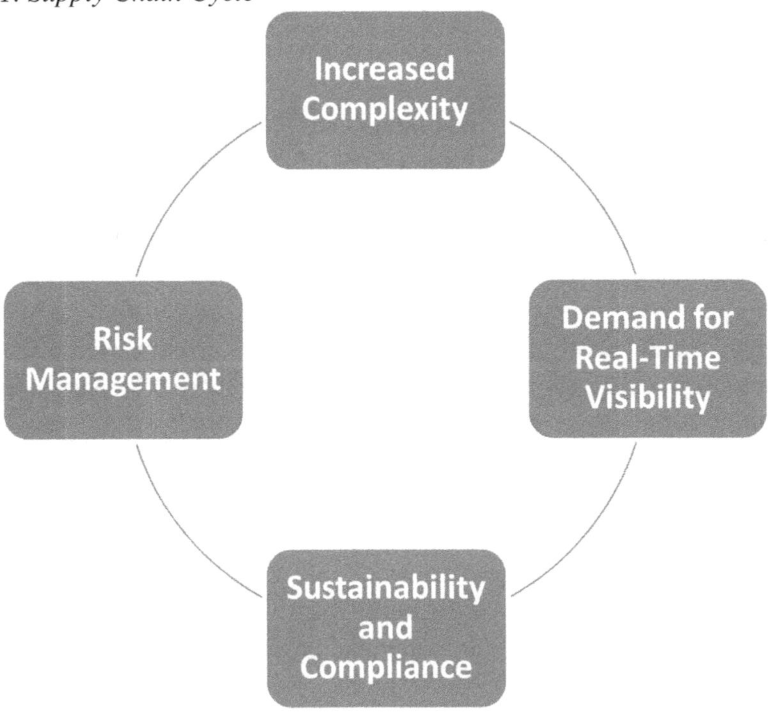

Source: Based on literature review

- **Increased Complexity**: Modern supply chains are complex networks including multiple stakeholders across different geographies and regulatory environments. Doing this manually is inefficient and error-prone.

- **Demand for Real-Time Visibility**: Both consumers and businesses now demand information about the status or location in real-time, so they can check product availability, shipment status & estimated delivery time. That means digital systems that span the entire supply chain.
- **Risk Management**: The COVID-19 pandemic shed light on exposed vulnerabilities and made a case for appropriate risk management. Predictive analytics through digital tools provide insights to challenges companies might experience and prepare them in advance for any potential disruptions.
- **Sustainability and Compliance**: Regardless of your segment, growing regulatory pressures and consumer expectations around sustainable practices are creating a need for more supply chain transparency and accountability.

Many of the biggest names in industry today have applied digital transformation to improve how they manage their supply chains (Taghipour & Beneteau-Piet, 2020; Taghipour, 2023). When it comes to digital transformation, Amazon's supply chain is a prime example. The company is riding on the success of their highly technical approach in every operation they have integrated (Aversa et al., 2021). And 200,000 robots working alongside human employees in Amazon warehouses efficiently pick, pack and ship orders all over the world (Volodymyr and Oksana, 2020). This eliminates the common errors and speed processing times. Amazon uses AI to stock and manage inventory using demand predictions workflows, where the main goal is to have an item in its warehouse just before a customer orders so it can be shipped directly from there. This eliminates stockouts and reduces excess inventory. The data analytics tools at Amazon's disposal can track supply chain performance in real time, deliver insights directly on the spot. This feedback loop allows for proactive and constant improvement efforts to take place throughout their warehouse operations.

Walmart has adopted the blockchain technology to increase transparency and traceability in their supply chain, mainly focused on food safety (Ellahi et al., 2023). Walmart can practically trace the steps of its food products from farm to store in near real-time via IBM's Food Trust blockchain. This way, if there is a foodborne illness outbreak, Walmart can identify the contamination source and quickly remove those harmful products from store shelves to protect shoppers. Walmart's commitment to supply-chain transparency helps ensure the safety and quality of its products in front of consumers.

Given that Maersk is the largest container shipping company worldwide, it represents a unique logistical challenge. Now the company is doing digital transformation to reduce its paperwork. Maersk uses IoT devices to track the status of goods in transit (Durlik et al., 2023). Temperature, relative humidity and location are all monitored by sensors to guarantee you sturdy transportation. TradeLens, a blockchain-powered platform that would digitize the shipping process has been co-developed by Maersk and IBM (Nguyen et al., 2023). TradeLens provides transparency, cuts paperwork and increases collaboration between supply-chain stakeholders.

Another example of how predictive analytics has helped a global consumer goods giant to increase its supply chain agility is P&G. P&G has successfully integrated advanced analytics technologies into its supply chain operations to forecast demand, optimize production schedules, and address potential disruptions. In particular, P&G uses predictive models that evaluate historical sales data, seasonal variations, and emerging market trends to forecast demand with high precision (Baisya, 2024). In this way, P&G can maintain optimal inventory levels and reduce the risks of stockouts and overstocking. Moreover, data insights help P&G improve production schedules and develop optimal manufacturing resource allocations which make the company's production more efficient (Akram et al., 2024a-b). In addition, using predictive analytics, P&G became better prepared for potential disruptions such as supply shortages and logistical problems, enabling the company to take preventative measures (Shen and Sun, 2023).

Figure 2. Benefits of Digital Transformation in Supply Chain

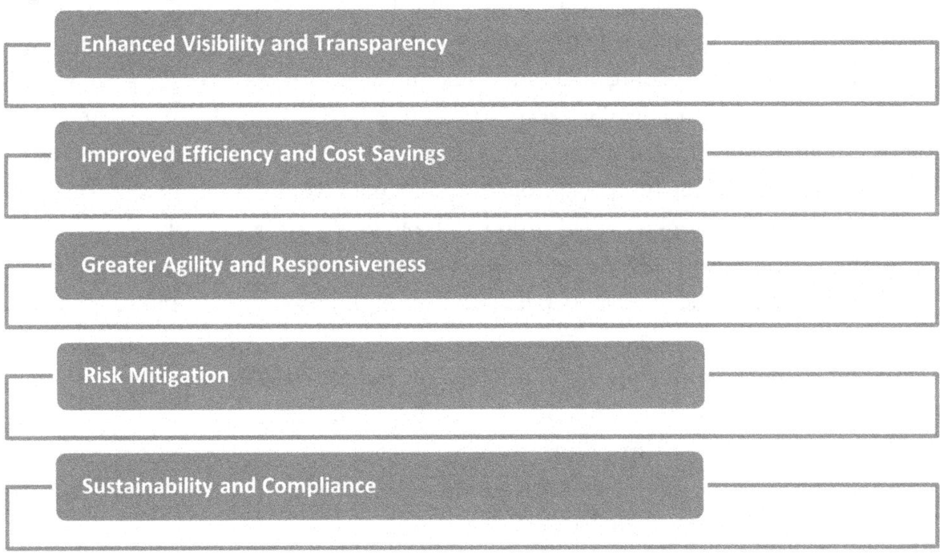

Source: Based on literature review

Digital Transformation is providing numerous benefits to supply chain management of the companies mentioned above. One of the biggest benefits that digital tools offer are real-time insights into all facets of supply chain from raw material procurement to final delivery. This visibility results in better decision-making and accountability. Automation as well as data analytics simplifies operations, greatly minimising human effort and scope of errors from manual handling to improved resource utilisation resulting in substantial cash savings and productivity advancements. Also, the organisations quickly respond as and when market dynamics, customer expectations or prospective disruptions in their industry have an impact. This agility is essential to maintaining a competitive edge. Using predictive analytics and real-time monitoring, organisations can identify potential threats in advance which would help them should there be any disruption of service due to some unforeseen circumstance (Sudan et al., 2023). Digital tools also make it easier to measure and report on sustainability metrics, which in turn helps companies comply with regulations and satisfy consumer demand for socially conscious practices.

3. EIGHT KEY DIGITAL TECHNOLOGIES IN LOGISTICS AND SCM

3.1. Robotics and Automation

Supply chain operations are one area that have benefited substantially from robotics and automation in order to improve efficiency and reduce the labor costs. They do things like picking, packing, sorting and moving inventory inside of warehouses. As an example, Amazon uses robotic systems in fulfillment centers to manage operations quicker (Bentalha, 2024). And mobile robotslike Kiva's units, which work autonomously in warehouses to navigate warehouse floors and collect items for human workers to fulfill orders (Chuah et al., 2023). Companies are also increasingly utilizing collaborative robots (cobots), which operate alongside human workers (Liu et al., 2024). Robots help perform high precision automated tasks, as they can carry out monotonous and manual jobs which over time reduce efficiency of human workers.-lowering errors-increasing output. BMW is putting in place completely digitalized factories, and robotic operations are leading the way to build a product through it till inspection.

3.2. Blockchain Technology

Blockchains provide a public ledger system, for which records of transactions are shared across a network rather than gathered in one place (Khalid et al., 2023). It benefits SCM to improve its traceability, lower chances of fraud and enable greater transparency. By providing a tamper-proof ledger of transactions, blockchain enables real-time tracking of products from origin to consumer. This is especially useful in industries like food and pharma where product integrity (and authenticity) are critical. IBM and Maersk joined hands to develop a blockchain platform for global trade (Wong et al., 2023) is one of the examples. It helps in that it standardizes and digitizes the shipping process, ensuring all parties operate on a single record of transactions for each shipment (creating digital versions of bills-of-lading), so paper work is basically reduced to null and information flows faster than before.

Figure 3. Digital Technologies in Logistics and SCM Chart

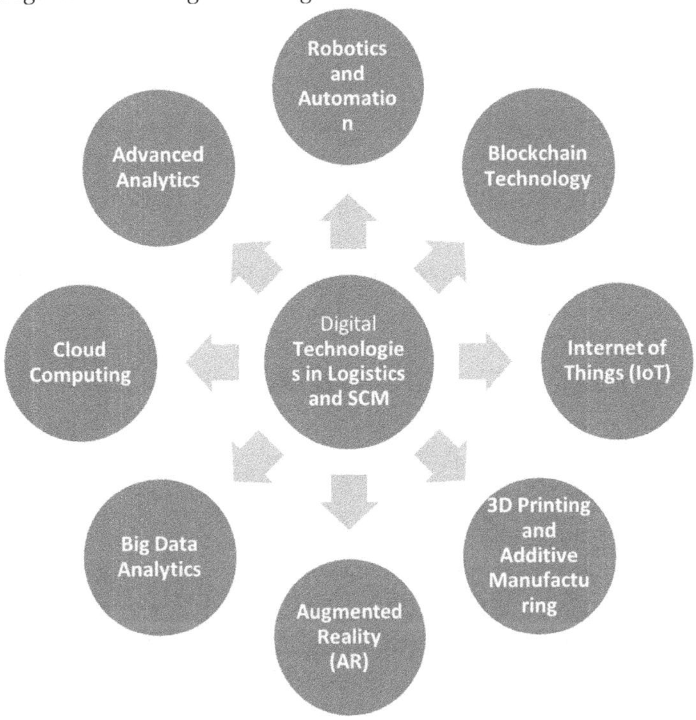

Source: Based on literature review

3.3. 3D Printing and Additive Manufacturing

Additive manufacturing or 3D printing means the direct production of items from digital models by building them up, layer-by-layer (Prashar et al., 2023). Local production option reduces warehousing and supply chains thereby transforming the SCM (Akbari et al., 2024). 3D printing is enabling companies like GE and Siemens to make complex on-demand components that cut lead times and inventory costs (Rama Murthy et al., 2023). This further improves customer satisfaction and response to market changes while enabling rapid and efficient customization of products unlike larger manufacturers that typically only know how mass-produce identical items.

Nike, for instance, uses 3D printing to rapidly prototype footwear which enables a faster turnaround in design iteration and innovation (Jiang et al., 2023)

3.4. Internet of Things (IoT)

The IoT is a broad concept of interconnected communication and information-sharing between multiple devices via the internet grid (Zhang et al., 2023). In supply chain management, there are numerous IoT devices, including sensors, RFID tags, and GPS trackers, installed on goods carriers (Helo andThai, 2024; Taj et al., 2023). They provide real-time tracking of the location and quality of produced and shipped goods. It is vital in optimizing transportation routes, minimizing delays, and securing the safety and quality of goods, particularly foods and other perishable products. For example, DHL use IoT technology to oversee packages that require certain temperature settings, guaranteeing that the transportation method maintains ideal circumstances (Pajic et al., 2024). Implementing Internet of Things (IoT) technology for predictive maintenance of machinery and vehicles can effectively minimize downtime and prolong the lifespan of essential assets.

3.5. Augmented Reality (AR)

This technology superimposes data over the real world to create a rich visual context for information that is relevant in physical and digital interactions (Mendoza-Ramírez et al., 2023). In SCM area, AR is used especially for inventory management and operational circumstances such as training or maintenance. In fact, AR-enabled devices like smart glasses offer warehouse workers with the real-time whereabouts of products to address this problem removing mistake in picking while increasing efficiency simultaneously (Porter and Heppelmann, 2017). Retail has been an early beneficiary, with IKEA among the retailers to use AR so shoppers can see how products will look in their homes before buying them, improving shopping experience (and therefore decreasing returns). It also helps in training new employees where step by step instructions, and visual guidance makes complex tasks easily achievable.

3.6. Big Data Analytics

Big data analytics refers to analyzing massive amounts of data to identify patterns, correlations, and insights that can be used to make key decisions (Tiwari et al., 2018). In the SCM space, big data analytics is used widely in forecasting, inventory management, and risk management. Analyzing structured and unstructured big data enables companies such as Walmart to develop predictive analytics models that help in predicting buying trends for customers (Yafooz et al., 2020). As a result, when predicting likely purchases for its items, the firm can adjust the stock levels. Predictive analysis also enables the company to predict possible disruptions and take measures to avoid them.

3.7. Cloud Computing

Another enabling technology that supports seamless information access and exchange among the stakeholders in the supply chain is cloud computing. According to Singh et al. (2021), cloud computing is a model that allows all-time available computing resources and data storage over the internet. By extension, Singh confirms that exact nature of cloud computing supports real-time interaction among the stakeholders in the entire supply chain management process. To this end, the SCM platforms operating on the cloud system can report real-time statistics of every inventory and order status and even track the shipments foot by foot (Handfield and Linton, 2017). The implication is better alignment and optimal coordination in the case of a point of disruption or interference. Unilever, for example, uses cloud-based systems to align the global operations of the supply chain to enhance efficiency and optimize operations on the go (Rasheed, 2024).

3.8. Advanced Analytics

SCM is also being revolutionized with advanced analytics and the advent of machine learning, AI capable enough to churn out insights from data that would not only be deeper but also provide options for making decisions autonomously. Through processing lots of historical data then providing actionable insights in almost real-time, AI could be used to optimize logistics routes. AI is fantastic for UPS to work out the most efficient routes, which means they use less fuel and offer better delivery times (Sharma, 2023). It also helps to target inefficiencies and areas for improvement within the supply chain, which results in cost effectiveness & improved performance.

4. TEN KEY CHALLENGES AND BARRIERS TO DIGITAL TRANSFORMATION

Digital transformation in logistics and supply chain (SC) management is a major step towards improving productivity, vehicle operating cost reduction & customer service. While the above has resulted in some changes, not to mention improvements of course, it is expected that integrating digital technologies like IoT, artificial intelligence (AI) blockchain and Big Data analytics can do wonders for logistics operations. Yet this sort of transformation is easier said than done, and many obstacles stand in the way before an organization can fully achieve its potential benefits.

4.1. Legacy Systems and Infrastructure

One of the most obvious and striking barriers to digital transformation in logistics and supply chain management is the existing legacy systems and obsolete infrastructure (Sobb et al., 2022). Considering the massive logistics sector, several companies and industries still use old technologies that were not designed to work hand in hand with contemporary digital products. Many of the implementations work in silo, meaning they are unable to ensure the seamless exchange of data and visualize various operational aspects under one roof (Armengaud et al., 2020). There is a high likelihood of a technical difficulty or incompatibility. Several organizations just cannot afford the high-entry level of ROI because digitization requires the funds to install and connect new and old data and operational structures. In addition, there are so many consolidation points, such as data gathering, visualization and analysis, and training. Also, legacy systems create data silo; information exists in dozens of hard-to-reach databases, making access and comprehensive insights nearly impossible.

4.2. Data Management and Quality

According Epiphaniou et al. (2020) and Kern (2021), efficient data management is vital for the success of digital transformation in logistics and supply chain management. Nevertheless, there are numerous obstacles when it comes to overseeing and preserving data of exceptional quality.

Data Overload: The Logistics produces a large amount of data. Processing these data and gaining resultant insights is a formidable challenge for organizations.

Data Accuracy: It is hard to maintain the same data across different tables within a dataset. Poor and incorrect information is another leading culprit for flawed decisions, as well as ineffective logistics operations.

Real-Time Data: Ensuring real-time visibility of the data is important in Supply Chain Management to work effectively. Yet, most companies do not have systems in place to ensure current information can reach across the supply chain.

4.3. Cybersecurity Concerns

The role digital technologies play in logistics and supply chain management raised more focus on cybersecurity. Securing sensitive data and systems from cyber threats is crucial.

Data Security: Supply chain and logistics operations require information from different business entities, including customer data and proprietary business data. The security of this data is a top priority.

Compliance: There are several data protection regulations like the GDPR (General Data Protection Regulation) and these rules lays down stringent mandates pertaining to how data is used, managed and stored in a company. As you can imagine, maintaining compliance with these regulations is not straightforward (in fact, quite the opposite) and requires significant resources to do so.

System Vulnerabilities: Digital structures may also be open to hacker attack, Malware infections and all types of cyber-assault It is important to have strong cyber security in place and in this case there are real risks that need protection.

4.4. Skills and Talent Gap

Digital logistics and supply chain management cannot be successful without human resources having the ability to use these technologies.

Technical Expertise: The professionals with the knowledge of emerging technologies, including AI, IoT and blockchain are hard to look for. This skills gap can impede the successful utilisation and proliferation of digital solutions.

Continuous Learning: The world is changing so rapidly regarding technology that we have to learn continuously and adapt. Organizations need to spend on constant training and development of its workforce for continuous upgrades in technology.

4.5. Change Management

In addition, this transformation requires significant changes in the processes and systems which a company has built over decades or centuries. Meeting this transition well is new big problem.

Resistance to Change: Employees and management may resist common or systems processes changes that are not part of familiar workflow. Getting past this resistance will help systems thinkers realize their potential and equip them with the necessary communication strategies to engage more individuals when discussing change.

Cultural Shift: The process of digital transformation requires a change in culture that emphasizes innovation and ongoing development. Establishing a culture that readily accepts and encourages change and promotes innovation might present difficulties.

Training: As digital transformation is nothing without technology so employees training of these new technologies and processes are one the key things for the successful implementation. But standard training can take time and be expensive.

4.6. Financial Constraints

Financial investment could/869+.lbe difficult for many organizations when it comes to digital transformation.

Investment Costs: The investments are going to be into the digital technologies, infrastructure and training upfront. Organizations with a smaller team, in particular, can struggle to allocate resource.

ROI Uncertainty: The outcomes of digital transformation initiatives are not set at the start, leading stakeholders to be hesitant towards major investments. One has to show what the long-term value could be, to get those key stakeholders on board.

4.7. Interoperability Issues

For a smooth sailing of digital transformation in logistics and supply chain management, it is crucial to make sure there is interoperability between various systems & technologies.

Standardization: While much of the other issues focus on specific companies and products, a lack of industry-wide standards is a significant challenge. It refers not only to compatibility between different applications and systems but also the ability to exchange data seamlessly.

Vendor Lock-In: Many companies depend on their primary digital solutions vendors. This can create a lack of flexibility and create so high costs. Thus, organizations must ensure that have open systems to avoid skilled vendor lock-in.

4.8. Supply Chain Complexity

While supply chains are already complex systems, digital transformation makes this layer even thicker because of how complicated global supply chains are.

Global Supply Chains: The transformation needs to establish a reliable connection across entire global supply chain networks, which are extremely elementary due to different regulations, technologies, and even work processes unique to the region.

Coordination: Effective supply chain management requires the coordination of multiple actors, including suppliers and manufacturers as well as logistics providers. They have to work in tandem and with great precision, making it difficult to achieve the level of communication required for digital transformation.

4.9. Customer Expectations

Digital technologies are evolving exponentially and so as the customer expectations from logistics supply chain.

Rising Expectations: Customers expect real-time information, quick turnarounds and customized products. Meeting these demands creates need for advanced digital tools and capabilities.

Transparency: Enabling transparency across the supply chain is key to meeting customer demands (Brun et al., 2020). This level of transparency need not be the property solely maintained by governments through heavy technological investment and inter-agency coordination.

4.10. Regulatory and Compliance Issues

Logistics and supply chain management are set to continue evolving, serving up a complex regulatory environment that complicates digital transformation.

Evolving Regulations: Keeping pace with changing regulations across geos is a nightmare. So, it is important for the organisations to make sure that their digital solutions are complying with these regulations (Annoni et al., 2023).

Compliance Costs: Organizations that need to comply with multiple regulatory requirements may end up finding compliance costs increasing and adding complexity to their digital transformation initiatives. Resource must invested in this activity to monitor and govern compliance properly, as with any other governance responsibility.

5. FUTURE OUTLOOK AND RECOMMENDATIONS

Digital technologies have considerable potential to transform the future of logistics and supply chain management. The introduction of the Internet of Things, artificial intelligence, blockchain, and big data analytics to the sector has sparked interest in the ability to become increasingly efficient, transparent and agile. However, the realization of the benefits brings challenges and implies a strategic operational approach.

For system transformation to be successful, it is necessary to have a clear and well-defined digital strategy. This will help organizations understand the major objectives and highlight which exact challenge they want to address through the implementation of digital technologies. In this regard, it is crucial to define the planned outcome, state the technologies that an organization wants to incorporate, the extent and specifics of utilization, and stages that the implementation should go through. A proper digital strategy will help organizations have a sense of a road-map and remain within the organization's objectives. By investing in scalable and interoperable technologies, organizations can have a flexible solution - but not too flexible to avoid the pitfalls of vendor lock-in. Scalable solutions can scale with the business, allowing for larger data volumes and more operational needs. On the other hand, interoperable technology ensures integration with current systems or digital tools paving a way for a cohesive and allied digital ecosystem.

Digital transformation is a journey, not a project. With every changing technology in the world, organisations need to have an innovative and a continuous improvement culture within themselves so as to stay abreast with cutting edge tech; evovling consumer demands. This means encouraging experimentation, promoting cross-functional collaboration and rewarding out-of-the-box thoughts. Positive leadership helps guide the culture change, it leads by example from top down; ensures every member of staff is engaged and motivated to want things done a different way. As the digital world continues to become more prevalent so too does cybersecurity, and organizations will be placed unto different scales of achievement based upon what they deliver. The need of the hour is to cover such instances through a robust cybersecurity strategy to ensure sensitive data, and systems are protected from cyber threats. This may involve putting in place proper securities such as encryption, multi-factor authentication and security audits. Moreover, companies need to keep the data of their customers confidential as well as fulfil requirements related with a law like GDPR (General Data Protection Regulation) in order for them not losing faiths from prospective buyers.

Access to high quality data is the foundation of any successful digital transformation effort. With that, organisations need to implement data management practices by identifying how they set policies and rules over who can have permission for ensuring accuracy, consistency completeness of the quality. This includes the development of data governance frameworks, as well as cleansing and enriching of data. In addition, using advanced analytics and AI-driven insights harness the power of data to act on it while making decisions and enhancing operational performance. A digital workforce with the required skills and expertise is a must to implement these technologies successfully. To combat this, companies must encourage and take part in training and development programs designed to upskill their workers. Including technical training in the use of digital tools, and more general services for data literacy or a change in thinking with respect to digital. Further still, partnerships with schools and industry can develop digitally talented talent through a digital skills pipeline.

Working with technology providers and consultants can speed up the journey to digital transformation. When you leverage technology partners, they bring targeted expertise and market knowledge with access to new solutions. These can be technology-implementing collaborations, integrated-tech initiatives in future-innovations and many more. Partnerships need to be driven by carefulness, and organizations should only pair up with partners that carry the right skill sets their strategic direction as well a successful track record of digital transformation projects. Success in digital transformation initiatives leads to creating key performance indicators (KPIs) which organizations should use for tracking progress over time. These KPIs should match the strategic goals of your organization and encompass all necessary elements of change including operational efficiency, customer satisfaction or financial performance among others. They can see where they need to improve, what data is driving decision making and make those changes as we go along.

6. CONCLUSION

For the modern business world, digital transformation in logistics and supply chain management is today a matter of inevitability. The events of recent years, such as the COVID-19 pandemic and its devastating impact on global supply chains, have highlighted the urgency of this shift. The pandemic revealed significant weaknesses and drawbacks of traditional supply chain management' reliance on manual processes and data blocks and the siloed nature of the supply chain operations. Since most of the world's manufacturing was located in China, which contained the worst of the outbreak in its Wuhan province, the initial spread of the virus had disastrous consequences for production and trade. Most companies were running very low stocks and could not recover from the first months of the crisis in time to adapt to the new reality with its delayed shipments and longer delivery times. As a result, the world's corporations realized that business as usual was no longer viable, and they need to shift away from efficiency models to more resilient supply chains (Dadsena et al., 2021). As a result, they concluded that they would have to develop a digital infrastructure capable of quick pivoting, and technologies like blockchains, artificial intelligence, machine learning, and intelligent automation would be integral to implementing this vision.

Digital transformation is not just about a collection of new technologies entering the enterprise; it requires rethinking business processes to better facilitate agility and responsiveness. It automates repetitive tasks, augment the human capabilities with AI driven insights, create interconnected ecosystems for real-time decisioning. Automation is key in minimizing human intervention on menial aspects, and re-sourcing humans to a more strategic role. Chatbots and Robotic Process Automation

(RPA) can also be used in enhancing efficiency of redundant processes. Assistance technologies like drones and Automated Guided Vehicles (AGVs), makes sure secure working & accurate performance in execution. By leveraging the power of AI and advanced analytics to augment businesses, it drives cognitive decision-making processes, thereby helping in anticipating disruptions, inventory management optimization and improved production planning.

The complexity of globalised supply chains, real-time visibility requirements, risk management strategies and sustainability & compliance are key factors underpinning the need for digital transformation. Amazon, Walmart Maersk and P&G have been considered digital transformation pioneers, using top tier technology to revolutionise their supply chains. Amazon, for example employs more than 200k robots in its warehouse to help humans pick pack and ship the orders at an unimaginable speed. In the case of Walmart, blockchain technology is used to share data easily and accurately in food safety and Maersk stores IOT devices for monitoring cargo condition when transporting. P&G uses predictive analytics for demand forecasting and production optimization.

However, there are many challenges and obstacles to clear before these benefits can be fully realized. There are many hurdles to overcome. Addressing these challenges necessitates a structured digital strategy, investment in scalable and inter-operable technologies, fostering a culture of innovation and continuing improvement, prioritising cyber security and data privacy; focusing on quality & analytics or improving the recruitment process with experienced employees from partnering technology partners.

REFERENCES

Akbari, M., Kok, S. K., Hopkins, J., Frederico, G. F., Nguyen, H., & Alonso, A. D. (2024). The changing landscape of digital transformation in supply chains: Impacts of industry 4.0 in Vietnam. *International Journal of Logistics Management*, 35(4), 1040–1072. DOI: 10.1108/IJLM-11-2022-0442

Akram, H. W., Abbas, H., & Ciddikie, D. (2024a). Green Manufacturing: A Comprehensive Exploration of Sustainable Practices and Industry 4.0 Integration. In Green Manufacturing for Industry 4.0 (pp. 159-178). Productivity Press.

Akram, H. W., Abbas, H., & Sulaiman, M. A. B. A. (2024b). Logistics & supply chain resilience in context of Oman. In *Reference module in social sciences*. Elsevier. DOI: 10.1016/B978-0-443-13701-3.00055-4

Annoni, A., Nativi, S., Çöltekin, A., Desha, C., Eremchenko, E., Gevaert, C. M., Giuliani, G., Chen, M., Perez-Mora, L., Strobl, J., & Tumampos, S. (2023). Digital earth: Yesterday, today, and tomorrow. *International Journal of Digital Earth*, 16(1), 1022–1072. DOI: 10.1080/17538947.2023.2187467

Armengaud, E., Fruhwirth, M., Rothbart, M., Weinzerl, M., & Zembacher, G. (2020). Digitalization as an opportunity to remove silo-thinking and enable holistic value creation. Systems Engineering for Automotive Powertrain Development, 1-28.

Aversa, P., Haefliger, S., Hueller, F., & Reza, D. G. (2021). Customer complementarity in the digital space: Exploring Amazon's business model diversification. *Long Range Planning*, 54(5), 101985. DOI: 10.1016/j.lrp.2020.101985

Awwad, A. S., Ababneh, O. M. A., & Karasneh, M. (2022). The mediating impact of IT capabilities on the association between dynamic capabilities and organizational agility: The case of the Jordanian IT sector. *Global Journal of Flexible Systems Managment*, 23(3), 315–330. DOI: 10.1007/s40171-022-00303-2 PMID: 37521253

Baisya, R. (2024). Supply Chain and Logistics Management: An Integrated Approach. Taylor & Francis.

Bentalha, B. (2024). Employees and Robots in Amazon's Robotic Mobile Fulfillment Systems: A Netnographic Analysis of a Supply Chain Transformation. In Blockchain Applications for Smart Contract Technologies (pp. 188-207). IGI Global. DOI: 10.4018/979-8-3693-1511-8.ch009

Brun, A., Karaosman, H., & Barresi, T. (2020). Supply chain collaboration for transparency. *Sustainability (Basel)*, 12(11), 4429. DOI: 10.3390/su12114429

Cauchois, R., Taghipour, A., Kang, D., Zoghlami, N., & Abed, M. (2017). Advanced solutions for a supply chain with stochastic information. *Journal of Advanced Management Science*, 5(1), 9–13. DOI: 10.18178/joams.5.1.9-13

Chuah, Y. R., Goh, A. V., & Ooi, H. T. (2023). The adoption of automation in Schenker Penang Logistics Centre 5 (plc5) warehouse: A case study (Doctoral dissertation, UTAR).

Dadsena, K. K., Mathiyazhagan, K., & Taghipour, A. (2021). Analysis of barriers for the build the resilient supply chain networks post-COVID-19. In *Managing Supply Chain Risk and Disruptions: Post COVID-19* (pp. 79–89). Springer International Publishing. DOI: 10.1007/978-3-030-72575-4_7

Durlik, I., Miller, T., Cembrowska-Lech, D., Krzemińska, A., Złoczowska, E., & Nowak, A. (2023). Navigating the sea of data: A comprehensive review on data analysis in maritime IoT applications. *Applied Sciences (Basel, Switzerland)*, 13(17), 9742. DOI: 10.3390/app13179742

Ellahi, R. M., Wood, L. C., & Bekhit, A. E. D. A. (2023). Blockchain-based frameworks for food traceability: A systematic review. *Foods*, 12(16), 3026. DOI: 10.3390/foods12163026 PMID: 37628025

Epiphaniou, G., Bottarelli, M., Al-Khateeb, H., Ersotelos, N. T., Kanyaru, J., & Nahar, V. (2020). Smart distributed ledger technologies in Industry 4.0: Challenges and opportunities in supply chain management. Cyber Defence in the Age of AI, Smart Societies and Augmented Humanity, 319-345.

Handfield, R., & Linton, T. (2017). The LIVING supply chain: The evolving imperative of operating in real time. John Wiley & Sons.

Helo, P., & Thai, V. V. (2024). Logistics 4.0–digital transformation with smart connected tracking and tracing devices. *International Journal of Production Economics*, 275, 109336. DOI: 10.1016/j.ijpe.2024.109336

Jarrahi, M. H. (2018). Artificial intelligence and the future of work: Human-AI symbiosis in organizational decision making. *Business Horizons*, 61(4), 577–586. DOI: 10.1016/j.bushor.2018.03.007

Jiang, Y., Tran, T. H., & Williams, L. (2023). Machine learning and mixed reality for smart aviation: Applications and challenges. *Journal of Air Transport Management*, 111, 102437. DOI: 10.1016/j.jairtraman.2023.102437

Kataria, D., Walid, A., Daneshmand, M., Dutta, A., Enright, M. A., Gu, R., . . . Polk, C. (2023, November). Artificial Intelligence And Machine Learning. In 2023 IEEE Future Networks World Forum (FNWF) (pp. 1-69). IEEE.

Kern, J. (2021). The digital transformation of logistics: A review about technologies and their implementation status. The digital transformation of logistics: Demystifying impacts of the fourth industrial revolution, 361-403.

Khalid, M. I., Ehsan, I., Al-Ani, A. K., Iqbal, J., Hussain, S., Ullah, S. S., & Nayab, . (2023). A comprehensive survey on blockchain-based decentralized storage networks. *IEEE Access : Practical Innovations, Open Solutions*, 11, 10995–11015. DOI: 10.1109/ACCESS.2023.3240237

Liu, L., Guo, F., Zou, Z., & Duffy, V. G. (2024). Application, development and future opportunities of collaborative robots (cobots) in manufacturing: A literature review. *International Journal of Human-Computer Interaction*, 40(4), 915–932. DOI: 10.1080/10447318.2022.2041907

Mendoza-Ramírez, C. E., Tudon-Martinez, J. C., Félix-Herrán, L. C., Lozoya-Santos, J. D. J., & Vargas-Martínez, A. (2023). Augmented reality: Survey. *Applied Sciences (Basel, Switzerland)*, 13(18), 10491. DOI: 10.3390/app131810491

Morgan, J., Halton, M., Qiao, Y., & Breslin, J. G. (2021). Industry 4.0 smart reconfigurable manufacturing machines. *Journal of Manufacturing Systems*, 59, 481–506. DOI: 10.1016/j.jmsy.2021.03.001

Nguyen, S., Leman, A., Xiao, Z., Fu, X., Zhang, X., Wei, X., Zhang, W., Li, N., Zhang, W., & Qin, Z. (2023). Blockchain-Powered Incentive System for JIT Arrival Operations and Decarbonization in Maritime Shipping. *Sustainability (Basel)*, 15(22), 15686. DOI: 10.3390/su152215686

Pajic, V., Andrejic, M., & Chatterjee, P. (2024). Enhancing cold chain logistics: A framework for advanced temperature monitoring in transportation and storage. *Mechatron. Intell Transp. Syst*, 3(1), 16–30.

Porter, M. E., & Heppelmann, J. E. (2017). Why every organization needs an augmented reality strategy. HBR'S 10 MUST, 85, 2-18.

Prashar, G., Vasudev, H., & Bhuddhi, D. (2023). Additive manufacturing: Expanding 3D printing horizon in industry 4.0. *International Journal on Interactive Design and Manufacturing*, 17(5), 2221–2235. DOI: 10.1007/s12008-022-00956-4

Rama Murthy, S., Huang, J., & Velu, C. (2023). Business Model Innovation in Additive Manufacturing Equipment Sector. In *Springer Handbook of Additive Manufacturing* (pp. 43–54). Springer International Publishing. DOI: 10.1007/978-3-031-20752-5_3

Rasheed, H. (2024). Consideration of Cloud-Web-Concepts for Standardization and Interoperability: A Comprehensive Review for Sustainable Enterprise Systems, AI, and IoT Integration. *Journal of Information Technology and Informatics*, 3(2).

Sharma, R. (2023). The transformative power of AI as future GPTs in propelling society into a new era of advancement. *IEEE Engineering Management Review*, 51(4), 215–224. DOI: 10.1109/EMR.2023.3315191

Shen, Z. M., & Sun, Y. (2023). Strengthening supply chain resilience during COVID-19: A case study of JD. com. *Journal of Operations Management*, 69(3), 359–383. DOI: 10.1002/joom.1161

Singh, N., Raza, M., Paranthaman, V. V., Awais, M., Khalid, M., & Javed, E. (2021). Internet of Things and cloud computing. In *Digital Health* (pp. 151–162). Academic Press. DOI: 10.1016/B978-0-12-818914-6.00013-2

Sobb, T., Turnbull, B., & Moustafa, N. (2020). Supply chain 4.0: A survey of cyber security challenges, solutions and future directions. *Electronics (Basel)*, 9(11), 1864. DOI: 10.3390/electronics9111864

Sudan, T., Taggar, R., Jena, P. K., & Sharma, D. (2023). Supply chain disruption mitigation strategies to advance future research agenda: A systematic literature review. *Journal of Cleaner Production*, 425, 138643. DOI: 10.1016/j.jclepro.2023.138643

Taghipour, A. (Ed.). (2023). *Government Impact on Sustainable and Responsible Supply Chain Management*. IGI Global. DOI: 10.4018/978-1-6684-9062-4

Taghipour, A., & Beneteau-Piet, C. (2020). Sustainable supply chain management performance. *International Journal of Innovation, Management and Technology*, 11(6), 165–169. DOI: 10.18178/ijimt.2020.11.6.897

Taj, S., Imran, A. S., Kastrati, Z., Daudpota, S. M., Memon, R. A., & Ahmed, J. (2023). IoT-based supply chain management: A systematic literature review. *Internet of Things : Engineering Cyber Physical Human Systems*, 24, 100982. DOI: 10.1016/j.iot.2023.100982

Talari, G., Cummins, E., McNamara, C., & O'Brien, J. (2022). State of the art review of Big Data and web-based Decision Support Systems (DSS) for food safety risk assessment with respect to climate change. *Trends in Food Science & Technology*, 126, 192–204. DOI: 10.1016/j.tifs.2021.08.032

Tiwari, S., Wee, H. M., & Daryanto, Y. (2018). Big data analytics in supply chain management between 2010 and 2016: Insights to industries. *Computers & Industrial Engineering*, 115, 319–330. DOI: 10.1016/j.cie.2017.11.017

Volodymyr, M., & Oksana, O. (2020). World Trends in Warehousing Logistics. Intellectualization of logistics and Supply Chain Management, (2), 32-50.

Wong, S., Yeung, J. K. W., Lau, Y. Y., & Kawasaki, T. (2023). A case study of how Maersk adopts cloud-based blockchain integrated with machine learning for sustainable practices. *Sustainability (Basel)*, 15(9), 7305. DOI: 10.3390/su15097305

Yafooz, W. M., Bakar, Z. B. A., Fahad, S. A., & Mithun, M. (2019). A. (2020). Business intelligence through big data analytics, data mining and machine learning. In Data Management, Analytics and Innovation *Proceedings of ICDMAI*, 2, 217–230.

Zhang, G., Yang, Y., & Yang, G. (2023). Smart supply chain management in Industry 4.0: The review, research agenda and strategies in North America. *Annals of Operations Research*, 322(2), 1075–1117. DOI: 10.1007/s10479-022-04689-1 PMID: 35531562

Zhao, S. (2021). Rhetoric and reality of China's global leadership in the context of COVID-19: implications for the US-led world order and liberal globalization. In *China's Big Power Ambition under Xi Jinping* (pp. 3–18). Routledge. DOI: 10.4324/9781003198871-1

Chapter 3
Navigating the Artificial Intelligence Era:
Impacts of Artificial Intelligence on Youth Critical Thinking, Career Paths, and Ethical Challenges

Atour Taghipour
https://orcid.org/0000-0003-1687-8881
Normandy University, France & Soldbridge, South Korea

XiaoWen Lu
Normandy University, France

ABSTRACT

Artificial intelligence (AI) has rapidly evolved from a theoretical concept to an integral part of daily life. This chapter examines AI's impact on the development of critical thinking, consumption habits, career prospects, mental health, and ethical considerations among younger generations. The easy accessibility of AI-generated responses raises concerns that youth may increasingly depend on AI, potentially diminishing independent reasoning skills. This reliance has led educational systems to implement AI-detection tools to limit overuse in assignments, aiming to promote students' critical thinking and academic integrity. The chapter concludes with recommendations for ethical AI use and suggests perspectives for future developments.

DOI: 10.4018/979-8-3693-7478-8.ch003

INTRODUCTION

History and Definition of Artificial Intelligence

Today, we are witnessing the impressive evolution of artificial intelligence, even though it originated much earlier than commonly believed (the history of artificial intelligence began in 1943 with the publication of the article "A Logical Calculus of Ideas Immanent in Nervous Activity" by Warren McCullough and Walter Pitts). Today, AI plays a significant role in daily life. John McCarthy defines artificial intelligence as "the science and engineering of making intelligent machines."

AI is now deeply integrated into society, from the innovation sector to the automotive industry and even space exploration. The rapid growth of artificial intelligence continues to accelerate. According to Thierry Manant, the number of AI users increased by over 60% in 2023, indicating a strong, ongoing evolution.

Increasing Use of AI by Young Generations

In everyday life, AI is widely used by French people, particularly among younger generations. According to a study by EdTech company Diplomeo, 79% of 16-25-year-olds use artificial intelligence in their studies, making it an almost essential daily tool. Of these users, 46% admit to using ChatGPT, the most well-known AI tool, for purposes such as answering questions, assisting with cover letters, and exploring educational options.

One could even say that, for them, this "robot" has become more legitimate than a real person's input. Generally, younger generations are known for their strong attachment to new technologies; it's often said they were born with these tools, accessing information instantly and unfiltered with just a click, without much need to think critically.

Issues Related to the Use of AI in Education

Could the recent surge in artificial intelligence risk leading young people to find all the answers they seek without engaging in reflection or developing critical thinking skills? This concern is already visible in the national education system, which has implemented AI detection tools to address the increasing use of AI by students in their writing and even on exams.

The purpose of this article is to examine and address the following question: How is AI influencing the consumption habits and career perspectives of Generations Z and Alpha?

To achieve this, after exploring the applications of AI in supply chains, we will explore AI's impact on consumption habits and education (while defining who these generations are), examine career perspectives, and assess the physiological impact of AI and the ethical challenges it poses. Finally, we will address the central question in the conclusion.

AI AND SUPPLY CHAINS

AI has become essential in modernizing and optimizing supply chains. Here are several key applications of AI in supply chains that are transforming how companies manage logistics, inventory, and production:

One of the first application of AI is demand forecasting and planning. AI algorithms analyze historical data, market trends, and external factors like weather, seasonal variations, and economic indicators to predict product demand accurately. This helps companies minimize overstock and understock situations, reducing waste and increasing profitability (Taghipour & Frayret, 2013a, and 2013b; Zoghlami et al., 2016). Advanced predictive analytics allow for more agile and responsive planning, making it easier to adjust to sudden shifts in demand (Cauchois et al., 2017; Dadsena et al., 2021; Loivet et al., 2020; Taghipour, 2020; Taghipour, 2014)

Inventory management is also a domain of contribution for AI. AI can automate inventory tracking in real time, helping companies maintain optimal stock levels. Machine learning models predict when to reorder items and identify patterns in demand, minimizing storage costs and reducing the risk of product obsolescence (Merimi, & Taghipour, 2021; Taghipour& Frayret, 2011). By using AI-powered tools to monitor inventory, companies can increase accuracy, streamline operations, and reduce manual work.

Other application of AI is supplier relationship management. AI helps evaluate and select suppliers based on performance metrics such as delivery time, quality, and cost. By analyzing data on supplier performance and market conditions, companies can optimize procurement processes and build stronger, data-driven partnerships (Mbiatem et al., 2018; Nadia et al., 2021). AI can also predict supplier risks by monitoring geopolitical, economic, and environmental data, allowing companies to switch to backup suppliers if needed.

AI can also greatly contribute to warehouse management by automation. Robotics and AI-driven systems can automate repetitive tasks within warehouses, such as sorting, picking, packing, and shelving. Autonomous robots and drones are used for inventory audits, reducing human error and increasing efficiency (Taghipour et al., 2022; Taghipour & Merimi, 2021). AI can also optimize warehouse layouts and guide automated guided vehicles (AGVs) to maximize space and minimize travel time within the facility.

Route optimization is other contribution of AI. AI-driven route optimization tools consider traffic conditions, weather, delivery windows, and other variables to find the most efficient routes for shipments. This minimizes delivery time and fuel costs, reducing both expenses and environmental impact. Predictive analytics and machine learning allow real-time route adjustments to avoid delays, improve on-time deliveries, and enhance customer satisfaction (Taghipour et al., 2023b; Taghipour et al., 2015).

Risk management and mitigation of risk can be easier using AI. AI helps in identifying potential supply chain risks by analyzing various data sources for signs of disruptions, such as economic changes, political instability, or supplier issues. Early detection enables companies to proactively address issues and maintain continuity (Taghipour et al., 2023c). AI-powered simulation models can predict the impact of potential disruptions, helping companies plan alternative scenarios and improve resilience.

AI can help companies to control and manage quality. Machine learning algorithms and computer vision systems enable real-time quality control by inspecting goods as they are produced or packaged. These systems identify defects early, ensuring quality standards are met and reducing the costs associated with faulty products. AI helps in tracking quality trends, which can provide insights for continuous improvement and reduce production-related issues over time.

Sustainability and waste reduction can be adapted by companies using AI. AI models analyze data on energy consumption, emissions, and waste to suggest ways to reduce the supply chain's environmental footprint. Predictive maintenance helps extend equipment life, reducing waste and maintenance costs. By optimizing routes, reducing waste, and planning efficient processes, AI helps companies meet sustainability goals, lower costs, and enhance brand reputation (Lu & Taghipour, 2023; Taghipour et al., 2023a; Taghipour et al., 2024; Taghipour, 2023; Taghipour & Beneteau-Piet, 2020).

Predictive maintenance of equipment is other domain of application of AI in supply chains. Machine learning algorithms monitor equipment performance data to predict failures before they occur, allowing companies to perform maintenance only when necessary and avoiding costly downtime. Predictive maintenance improves equipment lifespan and reduces unexpected disruptions in production, enhancing overall supply chain reliability.

Customer Service management can be easier using AI. AI-driven chatbots and virtual assistants can handle customer inquiries about order status, expected delivery times, and returns, providing 24/7 support. AI-powered personalization in the supply chain helps tailor offerings to customer preferences, improving satisfaction and retention by ensuring timely and accurate order fulfillment.

Finally, AI can facilitate order fulfillment and tracking. Real-time tracking powered by AI and IoT devices provides customers and supply chain managers with live updates on the status and location of products. Automated systems improve order accuracy, reduce processing time, and allow for smoother, more efficient deliveries, which enhances the end-user experience.

By incorporating AI across various supply chain stages, companies achieve greater efficiency, reduced costs, improved resilience, and enhanced customer satisfaction. The ability to analyze vast amounts of data in real-time makes AI an invaluable tool for optimizing and modernizing global supply chains.

IMPACT OF AI ON DAILY LIFE AND DEFINITION OF ITS MOST REGULAR USERS

Consumption Habits

Content Personalization With AI

Content personalization through Artificial Intelligence (AI) is a technique used by platforms like Netflix and YouTube to deliver content recommendations tailored to individual user preferences. This personalization relies on sophisticated algorithms that analyze a vast range of data, including viewing behaviors, explicit preferences (such as ratings and likes), and previous user interactions with the platform.

Netflix, for example, employs collaborative filtering algorithms, machine learning, and interleaving techniques to refine its recommendations. These algorithms can even analyze more subtle aspects, like scrolling speed, pauses, and viewing drop-offs, to better understand user preferences (growthsetting) (Netflix Tech Blog). Additionally, Netflix incorporates external data and content tags to address challenges such as the “cold start problem” and ensure diversity in recommendations (growthsetting).

Influence on the Media Habits of Young People

Content personalization through AI has a significant impact on media consumption habits, particularly among young people. These highly personalized recommendation systems enhance user engagement and satisfaction by providing relevant content, which reduces churn rates and increases loyalty to the platform (JUMP).

For young users, who make up a large portion of the audience for these services, personalization creates a tailored viewing experience that aligns with their specific tastes and digital lifestyles. This approach helps shape their consumption habits by encouraging them to spend more time on these platforms due to a continuous stream of relevant and appealing content (Netflix Tech Blog).

Moreover, it influences the diversity of the content they consume by introducing suggestions that may not align with their explicit preferences but are predicted to be interesting for them based on similar behaviors (growthsetting).

Thus, content personalization through AI plays a crucial role in the modern media ecosystem, transforming the way young people interact with digital content and contributing to a more engaged and varied consumption experience.

Targeted advertising algorithms are advanced systems used by social media and online platforms to deliver advertisements tailored to users' preferences and behaviors. These algorithms rely on machine learning techniques and data analysis to identify user trends and habits.

For instance, Facebook and Google collect user information such as age, gender, location, and interests. They use cookies and tracking pixels to monitor users' interactions with websites and advertisements. With this data, the algorithms can predict which types of ads are most likely to interest each user. Facebook employs the FBLearner Flow system to personalize ads based on users' past behaviors, such as liked pages and clicked ads (Zhu et al., 2017). Google AdWords, on the other hand, relies on contextual and behavioral data to target ads based on users' search queries and the websites they visit.

Effects of AI on Online Purchasing Behavior

Targeted advertising has a significant impact on purchasing decisions and consumer behavior. By presenting personalized ads, these algorithms increase the likelihood that users will click on the ads and make purchases. Here are des effects on purchasing decisions:

- Increase in Conversion Rates: Targeted ads are more effective because they directly address the interests and needs of users, resulting in higher conversion rates compared to generic ads (Lambrecht & Tucker, 2013).

- Impulse Purchases: Personalized ads can encourage users to make impulse purchases by presenting products or services they find appealing due to their perceived relevance (Bleier & Eisenbeiss, 2015).

Effects on Consumer Behavior

- Creation of Needs: Targeted advertisements can create a sense of need for products that consumers had not initially planned to purchase, thereby influencing their buying priorities.
- Reinforcement of Brand Preferences: Repeated exposure to the same brands through targeted advertisements can strengthen recognition and trust in those brands (Goldfarb & Tucker, 2011).
- Privacy Concerns: The intensive data collection used to personalize advertisements can raise privacy concerns, which may affect consumer perception.

Education and Learning

In this section we first define the generations Z and Alpha. In the next section we explain the educational tools based on AI.

Generation Z: Individuals of Generation Z were born between the mid-1990s and the early 2010s. They have grown up in a world where the Internet and digital technologies are ubiquitous. This generation is often described as "digital natives" due to their familiarity with smartphones, social media, and streaming platforms from a young age.

Generation Alpha: Generation Alpha includes children born from the 2010s to the mid-2020s. Being the first generation to grow up entirely in a post-digital environment, they are exposed to smart technologies and the Internet from birth. This generation commonly uses connected devices and interacts with technology in an intuitive manner.

Presentation of AI-Based Educational Tools

Educational technologies using artificial intelligence (AI) are transforming learning by providing personalized and interactive experiences (Ivanović et al., 2022). Here are some notable examples:

- Educational Apps: Applications like Duolingo utilize AI to tailor language lessons based on individual users' progress, making learning more effective and engaging.

- Online Learning Platforms: Websites such as Coursera and Khan Academy offer online courses with personalized recommendation systems based on users' performance and interests. These platforms use AI to deliver content tailored to the specific needs of each learner.
- Virtual Tutors and Chatbots: Tools like IBM Watson Tutor provide personalized support by answering students' questions and explaining complex concepts interactively and in a way that meets each student's needs.

Advantages and Disadvantages for Young Learning Methods

Advantages:

- Personalization: AI tools enable the creation of tailored learning paths that meet the needs and pace of each student, which can enhance engagement and understanding.
- Accessibility: AI-based educational technologies make education more accessible, especially for students in remote areas or those with special educational needs.
- Instant Feedback: AI tools provide immediate feedback, allowing students to quickly correct their mistakes and progress more effectively.

Disadvantages:

- Technological Dependence: An excessive reliance on technology may hinder interpersonal skills and the ability to learn without technological assistance.
- Privacy Concerns: The collection of personal data by AI tools raises concerns regarding privacy and security.
- Inequality of Access: Unequal access to advanced technologies can exacerbate educational disparities among students based on their socio-economic environment.

CAREER PROSPECTS AND SKILLS DEVELOPMENT

Importance of Learning AI-Related Skills

With AI rapidly transforming the job market, it is essential for Generation Z and Alpha to acquire skills in coding, data analysis, and understanding AI systems (Squicciarini et al., 2021). These skills are crucial to remain competitive in a technology-driven professional environment.

Key Skills:

- Coding: Understanding the basics of programming has become essential, as many careers now require coding knowledge.
- Data Analysis: The ability to interpret and analyze data is crucial in many sectors, enabling decision-making based on accurate information.
- AI Understanding: Knowing how AI systems work and being able to work with them is increasingly important across various professional fields.

Educational Initiatives to Prepare Generations Z and Alpha

Many educational initiatives aim to prepare young people for the demands of an AI-driven job market:

- School Programs: Some educational systems are already integrating programming and computational thinking into their curricula. Initiatives like the Hour of Code project encourage young people to learn coding basics.
- Bootcamps and Online Courses: Organizations like Codecademy and Coursera offer intensive courses and bootcamps to teach coding and data analysis, making these skills accessible to everyone.
- Public-Private Partnerships: Tech companies collaborate with schools and universities to provide resources and training in AI and digital technologies. For example, Google has launched initiatives to teach coding and AI in schools.

Job Sectors Likely to be Automated by AI, and impact on Career Prospects

Artificial intelligence (AI) is rapidly transforming various sectors by automating tasks previously performed by humans. Here are some sectors where automation is particularly prominent:

- Manufacturing: Industrial robots, capable of performing repetitive tasks with precision, are increasingly used in factories. They help boost productivity while reducing labor costs.
- Transportation and Logistics: The rise of autonomous vehicles could reduce the need for truck and taxi drivers. Additionally, warehouses are using robots for inventory management and order fulfillment.
- Financial Services: AI algorithms are used for portfolio management, fraud detection, and even providing personalized financial advice.
- Customer Service: Chatbots and interactive voice response (IVR) systems automate customer interactions, handling simple requests and redirecting more complex ones to human agents.
- Healthcare: AI assists doctors by analyzing medical images and aiding in diagnosis. Robots can also perform routine tasks in hospitals, such as medication delivery.

The increasing automation in these sectors has significant implications for young people entering the job market:

- Reduction in Certain Jobs: Manual and repetitive jobs are the most at risk of automation. Young people will need to turn toward professions requiring skills that machines cannot easily replicate, such as creativity and empathy.
- Need for New Skills: Demand for skills in technology, programming, and automated systems management is growing. Young people must develop these skills to remain competitive.
- Career Evolution: Traditional career paths are changing. Young people may need to switch jobs multiple times throughout their careers and continue learning throughout their lives.

Exploration of New Professions Created by AI

Artificial intelligence not only eliminates jobs; it also creates new ones, often in technically advanced fields:

- AI Specialists: These experts develop and maintain artificial intelligence systems, creating algorithms and machine learning models to solve complex problems.
- Data Analysts: With the increase in data volumes, analysts are needed to interpret this information and help businesses make informed decisions.
- Robotics Engineers: They design, build, and maintain robots for various applications, ranging from manufacturing to healthcare.

- Chatbot Developers: These professionals create virtual assistants capable of answering user questions and providing automated customer service.

To prepare for the opportunities created by AI, young people need to acquire specific skills and pursue appropriate training:

- University Programs: Degrees in computer science, data science, and engineering are essential for building a solid foundation. Many universities now offer specialized courses in AI.
- Online Courses and Certifications: Platforms like Coursera, edX, and Udacity provide courses and certifications in AI and data science, making them accessible to everyone.
- Bootcamps and Workshops: Intensive short-term programs allow individuals to quickly gain practical skills in programming and data analysis.
- Hands-On Experience: Working on real projects, participating in hackathons, and interning at tech companies is crucial for preparing for careers in AI.

PSYCHOSOCIAL IMPACT

Analysis of the Impact of Social Media and AI-Based Applications on Young People's Mental Health

Social media and AI-powered applications significantly impact young people's mental health (Shahzad et al., 2024). Heavy use of these platforms can influence users both positively and negatively.

Negative Impact:

- Anxiety and Depression: Studies show that excessive social media use can exacerbate symptoms of anxiety and depression. Constant comparison with others and exposure to idealized content can lead to feelings of inadequacy and personal dissatisfaction.
- Cyberbullying: Online anonymity and easy access to a large audience increase the risk of cyberbullying, which can have devastating effects on young people's mental health.
- Addiction: AI algorithms designed to maximize user engagement can make young people dependent on social media, reducing real-life social interactions and overall well-being.

Two examples are Facebook and Instagram: While these platforms offer benefits for social connection, they have been criticized for their negative impact on young people's mental health due to social pressure and comparisons with other users.

Positive Impact:

- Mental Health Support: Some applications use AI to provide mental health support. For example, chatbots like Woebot offer advice based on cognitive behavioral therapy principles, helping users manage anxiety and depression. This application uses AI to offer techniques for managing stress and depression. Studies have shown that using Woebot can reduce depression symptoms in as little as two weeks.
- Supportive Communities: Social media can also provide supportive spaces where young people can share their experiences and receive encouragement and advice from peers. For example, Headspace and Calm, are meditation and mindfulness apps use AI to personalize meditation sessions based on users' specific needs, helping to reduce stress and improve overall well-being.

Effects of AI on Social Interactions and Interpersonal Relationships

Artificial intelligence is profoundly altering social interactions and interpersonal relationships in the following ways (Hohenstein et al., 2023):

Chatbots, used by customer service and messaging platforms, offer instant, constant interaction. However, they can also diminish the quality of human interactions by replacing interpersonal conversations with automated exchanges. In addition to this, Virtual Reality (VR) enables immersive interactions that can simulate in-person meetings. For example, applications like VRChat allow users to meet and interact in virtual environments, transforming traditional modes of communication. For example, platforms like WhatsApp, Messenger, and Snapchat use algorithms to suggest contacts, groups, and content, thus influencing social interactions. They promote quick and frequent communication, though it is often superficial. Other examples are the applications like Pokémon Go use AR to encourage in-person social interactions by bringing users together in real-world spaces to achieve shared goals.

Considering these examples, we can see that AI-based technologies have varied impacts on young people's mental health and social interactions. While some applications and tools provide valuable support and enhance well-being, others can exacerbate mental health issues and reduce the quality of human interactions.

ETHICAL CHALLENGES AND RESPONSIBILITY

Issues Related to the Collection of Personal Data by AI-Based Applications

AI-powered applications collect significant amounts of personal data to enhance their services (Meurisch & Mühlhäuser, 2021). However, this data collection poses several challenges regarding privacy protection.

- Privacy Intrusion: AI applications collect sensitive personal information, such as browsing habits, shopping preferences, social interactions, and even biometric data. This often-intrusive collection sometimes occurs without fully informed user consent.
- Security Risk: Collected data is vulnerable to cyberattacks. Data breaches can expose sensitive personal information, endangering user security and confidentiality.
- Third-Party Use: Personal data collected by applications can be shared or sold to third parties, like advertisers or business partners, often without users' full awareness. This can lead to intrusive advertising and the commercial exploitation of personal data.

The Role of Parents and Educators in Protecting Young People's Privacy

Parents and educators play a crucial role in safeguarding young people's privacy in the digital age.

- Privacy Education: Parents and educators should raise awareness among young people about the importance of online privacy. This includes understanding the risks of sharing personal information and learning good practices for digital security.
- Use of Parental Controls: Parents can use parental control tools to monitor and limit young people's access to certain applications. These tools can help prevent data collection risks and ensure safer use of AI-based technologies.
- Encouraging Critical Thinking: Educators should encourage young people to develop critical thinking toward the technologies they use, including understanding the privacy policies and terms of use of applications.

Risks of Bias in AI Algorithms and Impact on Young People

AI algorithms can perpetuate and amplify existing biases, leading to unintentional or systematic discrimination (Ferrer et al., 2021).

- Unequal Representation: Algorithms trained on biased data can replicate these biases, resulting in discriminatory outcomes. For example, facial recognition systems may have higher error rates for people of color due to a lack of diversity in the training data.
- Systematic Discrimination: Algorithmic biases can also influence decisions in critical areas like education, employment, and criminal justice, negatively impacting young people from marginalized communities.

Measures to Ensure Fair and Inclusive AI

- Data Diversification: It is crucial to diversify the datasets used to train algorithms to reduce biases. This includes collecting data representative of various populations and contexts.
- Transparency and Accountability: Companies and developers must be transparent about how their algorithms work and make decisions. They should also be accountable for the impacts of their technologies and establish mechanisms to correct identified biases.
- Stakeholder Inclusion: Involving diverse stakeholders, including marginalized groups, in the development and assessment of AI technologies can help identify and address potential biases before system deployment.

In summary, protecting privacy and addressing biases in AI algorithms are essential for the mental health and future prospects of young people. Parents, educators, developers, and regulators must work together to create a safer and fairer digital environment (Hernández, 2024).

CONCLUSION

In this article, we explored several key aspects related to the impact of artificial intelligence (AI) on younger generations, particularly Generation Z and Generation Alpha. We examined privacy protection issues, risks of bias and discrimination in algorithms, and the crucial role of parents and educators in safeguarding young people. The collection of personal data by AI-based applications raises significant privacy and security concerns. Additionally, biases within algorithms can lead to systemic discrimination, negatively affecting young people from marginalized communities.

RECOMMENDATIONS

To better protect young people and ensure the ethical and responsible use of AI, several measures can be implemented:

Educators and Parents

- Raise awareness among young people about online privacy issues and teach them good digital security practices.
- Use parental control tools to monitor and limit young people's access to certain applications.
- Encourage young people to develop critical thinking about the technologies they use.
- Policy Makers:
- Establish strict regulations regarding the collection and use of personal data by AI-based applications.
- Promote policies that encourage diversity and inclusion in AI technology development.

Companies Developing AI Technologies

- Diversify the datasets used to train algorithms to reduce biases.
- Be transparent about how algorithms function and take responsibility for the impacts of their technologies.
- Involve diverse stakeholders, including marginalized groups, in the development and evaluation of AI technologies.

FUTURE PERSPECTIVES

AI will continue to evolve rapidly, bringing significant changes in various aspects of society. For Generation Z and Generation Alpha, these developments will present both opportunities and challenges.

Potential Developments in AI

- Advances in AI could transform the education, health, employment, and communication sectors, creating new opportunities but also new challenges regarding privacy and ethics.
- Young people will need to acquire AI-related skills, such as coding and data analysis, to remain competitive in the future job market.

Ethical and Responsible Approach

- It is crucial to develop and use AI in an ethical and responsible manner to avoid risks of bias and discrimination.
- Companies, governments, and educational institutions must collaborate to ensure that AI benefits everyone in an equitable and inclusive way.

Finally, it is important to note that AI is now an integral part of our daily lives. The fact that this article was created with the help of artificial intelligence illustrates the impact and omnipresence of this technology in our everyday existence.

REFERENCES

Acemoglu, D., & Restrepo, P. (2018). Artificial Intelligence, Automation, and Work. NBER Working Paper No. 24196.

Bleier, A., & Eisenbeiss, M. (2015). The Impact of Relevance on the Effectiveness of Online Advertising. *Journal of Advertising*, 44(3), 320–328.

Brynjolfsson, E., & McAfee, A. (2014). *The Second Machine Age: Work, Progress, and Prosperity in a Time of Brilliant Technologies*. W. W. Norton & Company.

Cauchois, R., Taghipour, A., Kang, D., Zoghlami, N., & Abed, M. (2017). Advanced solutions for a supply chain with stochastic information. *Journal of Advanced Management Science*, 5(1), 9–13. DOI: 10.18178/joams.5.1.9-13

Dadsena, K. K., Mathiyazhagan, K., & Taghipour, A. (2021). Analysis of barriers for the build the resilient supply chain networks post-COVID-19. In *Managing Supply Chain Risk and Disruptions: Post COVID-19* (pp. 79–89). Springer International Publishing. DOI: 10.1007/978-3-030-72575-4_7

Ferrer, X., Van Nuenen, T., Such, J. M., Coté, M., & Criado, N. (2021). Bias and discrimination in AI: A cross-disciplinary perspective. *IEEE Technology and Society Magazine*, 40(2), 72–80. DOI: 10.1109/MTS.2021.3056293

Glaa, B., Zoghlami, N., & Taghipour, A. (2014, May). Outsourcing and off-shoring healthcare services: A way to cut the costs and improve quality. In 2014 international conference on advanced logistics and transport (ICALT) (pp. 189-194). IEEE.

Goldfarb, A., & Tucker, C. (2011). Online Display Advertising: Targeting and Obtrusiveness. *Marketing Science*, 30(3), 389–404. DOI: 10.1287/mksc.1100.0583

Growthsetting. (2023). How Netflix Enhances User Experience with AI Recommendations.

Hernández, E. G. (2024). Towards an ethical and inclusive implementation of artificial intelligence in organizations: a multidimensional framework. arXiv preprint arXiv:2405.01697.

Hohenstein, J., Kizilcec, R. F., DiFranzo, D., Aghajari, Z., Mieczkowski, H., Levy, K., Naaman, M., Hancock, J., & Jung, M. F. (2023). Artificial intelligence in communication impacts language and social relationships. *Scientific Reports*, 13(1), 5487. DOI: 10.1038/s41598-023-30938-9 PMID: 37015964

JUMP. (2024). *The Netflix Effect: Shaping the future of personalization in streaming*Recent Advances and Case Studies.

Lambrecht, A., & Tucker, C. (2013). When Does Retargeting Work? Information Specificity in Online Advertising. *JMR, Journal of Marketing Research*, 50(5), 561–576. DOI: 10.1509/jmr.11.0503

Loivet, W., Taghipour, A., & Kang, D. S. (2020). The rise of green supply chain management: Between complexity and necessity. *J. Econ. Bus. Manag*, 8(1), 1–7. DOI: 10.18178/joebm.2020.8.1.604

Lu, X., & Taghipour, A. (2023). A French Guide for Businesses and Governments to Integrate the Concept of Sustainable Development Into Their Supply Chains. In *Government Impact on Sustainable and Responsible Supply Chain Management* (pp. 1–12). IGI Global. DOI: 10.4018/978-1-6684-9062-4.ch001

Mahfod, J., Canel-Depitre, B., & Taghipour, A. (2019). Quality function deployment-ELECTRE in supplier evaluation. Journal of Advanced Management Science, 7(4).

Manyika, J., Chui, M., Miremadi, M., Bughin, J., George, K., Willmott, P., & Dewhurst, M. (2017). *A Future That Works: Automation, Employment, and Productivity*. McKinsey Global Institute.

Mbiatem, J., Taghipour, A., & Canel-Depitre, B. (2018, August). Supplier selection approaches for decision makers. In *Proceedings of the 8th International Conference on Information Communication and Management* (pp. 108-112).

Merimi, M., & Taghipour, A. (2021). Accelerating the digitalization of the supply chain: An empirical research about COVID-19 crisis. In *Digitalization of decentralized supply chains during global crises* (pp. 1–24). IGI Global. DOI: 10.4018/978-1-7998-6874-3.ch001

Meurisch, C., & Mühlhäuser, M. (2021). Data protection in AI services: A survey. *ACM Computing Surveys*, 54(2), 1–38. DOI: 10.1145/3440754

Nadia, E. N., Beatrice, C. D., & Atour, T. (2021). The determinants of hotel room rates in Beirut: A hedonic pricing model. International Journal of Trade. *Economics and Finance*, 12(2), 33–42.

Netflix Research. (2024). Discovering Creative Insights in Promotional Artwork.

Netflix TechBlog. (2024). Innovating Faster on Personalization Algorithms at Netflix Using Interleaving.

Shahzad, M. F., Xu, S., Lim, W. M., Yang, X., & Khan, Q. R. (2024). Artificial intelligence and social media on academic performance and mental well-being: Student perceptions of positive impact in the age of smart learning. *Heliyon*, 10(8), e29523. DOI: 10.1016/j.heliyon.2024.e29523 PMID: 38665566

Squicciarini, M., & Nachtigall, H. (2021). Demand for AI skills in jobs: Evidence from online job postings.

Taghipour, A. (2014). Improving the plan of a manufacturing network with non-integrated business units. *International Journal of Applied Logistics*, 5(2), 1–11. DOI: 10.4018/ijal.2014040101

Taghipour, A. (Ed.). (2020). *Demand forecasting and order planning in supply chains and humanitarian logistics*. IGI Global.

Taghipour, A. (Ed.). (2023). *Government Impact on Sustainable and Responsible Supply Chain Management*. IGI Global. DOI: 10.4018/978-1-6684-9062-4

Taghipour, A., Abed, M., & Zoghlami, N. (2015, May). Design for remanufacturing respecting reverse logistics processes: A review. In 2015 4th International Conference on Advanced Logistics and Transport (ICALT) (pp. 299-304). IEEE. DOI: 10.1109/ICAdLT.2015.7136602

Taghipour, A., & Beneteau-Piet, C. (2020). Sustainable supply chain management performance. *International Journal of Innovation, Management and Technology*, 11(6), 165–169. DOI: 10.18178/ijimt.2020.11.6.897

Taghipour, A., Foukolaei, P. Z., Ghaedi, M., & Khazaei, M. (2023a). Sustainable Multi-Objective Models for Waste-to-Energy and Waste Separation Site Selection. *Sustainability (Basel)*, 15(22), 15764. DOI: 10.3390/su152215764

Taghipour, A., & Frayret, J. M. (2011). Supply chain coordination planning: A review. Work paper, École polytechnique de Montréal.

Taghipour, A., & Frayret, J. M. (2013a, May). An algorithm to improve operations planning in decentralized supply chains. In *2013 International Conference on Advanced Logistics and Transport* (pp. 100-103). IEEE. DOI: 10.1109/ICAdLT.2013.6568442

Taghipour, A., & Frayret, J. M. (2013b). Coordination of operations planning in supply chains: A review. *International Journal of Business Performance and Supply Chain Modelling*, 5(3), 272–307. DOI: 10.1504/IJBPSCM.2013.055729

Taghipour, A., Lu, X., & Smahi, M. (2022). Opportunities and challenges for society 5.0 in procurement. *Blockchain Technology in Supply Chain Management for Society*, 5(0), 91–100. DOI: 10.1201/9781003177432-9

Taghipour, A., & Merimi, M. (2021, March). Digital transformation of supply chains during crisis: covid-19. In 11th Annual International Conference on Industrial Engineering and Operations Management, IEOM (pp. 7493-7501). DOI: 10.46254/AN11.20211298

Taghipour, A., Padash, A., Etemadi, V., Khazaei, M., & Ebrahimi, S. (2024). Sustainable and Circular Hotels and the Water–Food–Energy Nexus: Integration of Agrivoltaics, Hydropower, Solar Cells, Water Reservoirs, and Green Roofs. *Sustainability (Basel)*, 16(5), 1985. DOI: 10.3390/su16051985

Taghipour, A., Ramezani, M., Khazaei, M., Roohparvar, V., & Hassannayebi, E. (2023b). Smart transportation behavior through the COVID-19 pandemic: A ride-hailing system in Iran. *Sustainability (Basel)*, 15(5), 4178. DOI: 10.3390/su15054178

Taghipour, A., Sohrabi, A., Ghaedi, M., & Khazaei, M. (2023c). A robust vaccine supply chain model in pandemics: Case of Covid-19 in Iran. *Computers & Industrial Engineering*, 183, 109465. DOI: 10.1016/j.cie.2023.109465

West, D. M. (2018). *The Future of Work: Robots, AI, and Automation*. Brookings Institution Press.

Zhu, H., Huberman, B. A., & Lu, Q. (2017). To Personalize or Not to Personalize: A Risk Management Perspective. *Journal of Management Information Systems*, 34(2), 578–607.

Zoghlami, N., Taghipour, A., Merlo, C., & Abed, M. (2016). Management of divergent production network using decentralised multi-level capacitated lot-sizing models. *International Journal of Shipping and Transport Logistics*, 8(5), 590–604. DOI: 10.1504/IJSTL.2016.078686

Chapter 4
Revolutionizing the Sustainable Supply Chain of Medical Product With AI–Driven Drone Delivery:
Innovations in Supply Chain Management Through Autonomous UAV Technology in Healthcare

M. Gowsalya
https://orcid.org/0000-0001-8756-1246
Vellore Institute of Technology, India

Mayuri K. N.
Vellore Institute of Technology, India

ABSTRACT

The integration of artificial intelligence with drone technology is transforming sustainable supply chain management in the field of healthcare, addressing significant difficulties in medical product delivery. This chapter explores the connections between AI and drones, emphasizing developments such as autonomous navigation, real-time monitoring, and enhanced path planning to optimize delivery routes and increase sustainability. AI-integrated drone technology helps to promote sustainability by reducing emissions and energy consumption while providing cost-effective alternatives to existing transportation modes. A mathematical framework is proposed to optimize cost, time, emissions, and energy consumption, resulting in efficient and

DOI: 10.4018/979-8-3693-7478-8.ch004

sustainable healthcare supply chains. Future improvements, such as autonomous drone swarms, renewable energy integration, and improved AI-driven decision making, promise to significantly transform this field. By utilizing the transformative potential of AI-powered drone delivery systems improves sustainability and scalability in the healthcare sector.

1. INTRODUCTION

The dynamic environment of healthcare logistics requires novel solutions to ensure the timely, cost-effective, and long-term delivery of medical supplies. Sustainable supply chain management (SSCM) is critical in solving these difficulties because it optimizes resource consumption while minimizing environmental effects. SSCM integrates logistics, procurement, and operations to reduce waste and emissions while increasing system resilience. Drone delivery systems have further changed healthcare logistics, allowing for faster and more precise delivery of crucial supplies including blood products, vaccines, and diagnostics. These unmanned aerial vehicles (UAVs) extend beyond traditional obstacles such as traffic congestion and geographic distance, making them extremely useful in emergency response and pandemic management scenarios.

AI is another game-changing technology that is transforming healthcare through predictive analytics, operational optimization, and personalized care. In logistics, artificial intelligence improves decision-making, improves operations, and assures optimal resource allocation. When paired with drone delivery, AI-driven solutions provide new levels of accuracy and efficiency, allowing for autonomous navigation, real-time monitoring, and demand forecasting. This research investigates the convergence of supply chain management, drones, and artificial intelligence in healthcare logistics. It explores the revolutionary potential of AI-powered drone delivery systems, examines worldwide case studies, and indicates significant problems and opportunities. A mathematical model is proposed to optimize essential characteristics such as cost, time, energy usage, and emissions, thereby establishing the groundwork for sustainable and equitable healthcare supply chains. This study emphasizes the importance of technology in addressing systemic inefficiencies and promoting innovation in healthcare delivery.

1.1. The Role of SSCM in Healthcare

Sustainable Supply Chain Management (SSCM) plays a critical role in healthcare by ensuring the efficient, cost-effective, and environmentally sustainable delivery of medical products and services. It encompasses the planning, implementation, and monitoring of supply chain activities to optimize resource utilization while minimizing waste and emissions. SSCM in healthcare is essential for addressing challenges like time-sensitive deliveries, cold chain management for perishable items, and ensuring equitable distribution of medical resources, particularly in underserved areas. By integrating sustainability into the supply chain, healthcare organizations can improve patient outcomes, reduce environmental impact, and enhance system resilience, particularly during emergencies or pandemics.

1.2. The Role of Drone Delivery in Healthcare

Drone delivery systems are transforming healthcare logistics by enabling rapid, reliable, and cost-effective transportation of medical supplies, especially in remote or inaccessible regions. Drones are capable of delivering critical items such as blood products, vaccines, and diagnostic samples with precision and speed, bypassing traditional barriers like traffic congestion and geographic challenges. They are especially useful in emergency response scenarios, disaster relief operations, and pandemic management, where timely delivery is crucial. Drones also support cold chain management by maintaining temperature-sensitive medical products during transit. By enhancing accessibility and efficiency, drone delivery systems are bridging gaps in healthcare delivery and saving lives.

1.3. The Role of Artificial Intelligence (AI) in Healthcare

Artificial Intelligence (AI) is revolutionizing healthcare by enhancing decision-making, improving operational efficiency, and enabling personalized care. AI applications in healthcare range from predictive analytics for disease outbreaks to diagnostic tools powered by machine learning algorithms. In logistics, AI optimizes supply chain operations, predicts demand variations, and ensures efficient resource allocation. In medical imaging, AI aids in accurate and timely diagnosis, it supports virtual consultations and remote monitoring. By leveraging large datasets and advanced algorithms, AI is driving innovation in healthcare, improving patient outcomes, and addressing systemic inefficiencies.

1.4. The Role of AI-Driven Drone Delivery in Healthcare

AI-driven drone delivery represents a cutting-edge innovation at the intersection of technology and healthcare. By combining AI with drone capabilities, healthcare systems can achieve uncommon levels of efficiency, accuracy, and accessibility in medical supply logistics. AI enables 2 autonomous navigation, obstacle detection, and route optimization, ensuring timely and cost-effective deliveries. This technology is particularly valuable in emergencies, pandemic responses, and rural healthcare, where drones can quickly transport critical supplies like vaccines, blood, and medicines. Additionally, AI-driven drones support sustainability goals by minimizing energy consumption and emissions. As a transformative solution, AI-powered drone delivery is enhancing healthcare resilience and equity while driving advancements in Sustainable Supply Chain Management.

2. RELATED WORKS

(Addy, 2023) explores role of AI in enhancing vaccine supply chain management, with a focus on Ghana and the broader West African healthcare sector. It highlights successful implementations, such as Rwanda's AI-powered Zipline drones for efficient vaccine delivery and India's eVIN system for real-time tracking, which significantly reducing stockouts. The WHO's Smart Vaccination Management System (SVMS) is also noted for promoting equitable vaccine distribution using predictive analytics. The abstract effectively demonstrates how AI can optimize efficiency, reduce wastage, and improve decision-making, offering a compelling case for its adoption in Ghana to enhance healthcare outcomes. (Aliahmadi et al., 2022) discuss an AIoT in creating sustainable supply chains within today's complex global market. It emphasizes the need for integrating IoT and AI to optimize processes and achieve sustainability goals. The study explores key dimensions and components of AIoT-based supply chains and presents a conceptual framework for an intelligent supply chain model. Findings underscore the necessity of investing in AIoT for sustainable benefits while addressing challenges such as IT security, workforce skill gaps, and integrated information sharing.

(Khadem et al., 2023) Reducing healthcare costs and enhancing care quality are universal priorities for healthcare systems. The impact of globalization on healthcare outsourcing and offshoring, exploring their potential to achieve these goals. It also analyses the cost-effectiveness of healthcare interventions, comparing outcomes between domestic and outsourced services. (Damoah et al., 2021) AI-enhanced medical drones in Ghana's healthcare supply chain (HSC) and their contributions to sustainable development. It addresses a research gap by applying a corporate social responsibility (CSR) lens to explore the impact of these drones. Findings highlight their significant contributions to Sustainable Development Goals (SDGs) by reducing carbon emissions, lowering mortality rates, and improving socio-economic conditions. Additionally, the study underscores the long-term corporate sustainability benefits for implementing firms. Overall, the study effectively demonstrates how AI-driven drones enhance HSC efficiency while supporting broader social and environmental objectives. (Rejeb et al., 2021) The impact of AI on supply chain management, focused on its ability to enhance efficiency, reduce costs, and optimize decision-making. It provides an overview of key AI technologies like machine learning, robotics, and natural language processing and examines their application across various supply chain stages. The study also addresses challenges, ethical considerations associated with AI adoption and highlights AI's potential to revolutionize supply chain operations and advance intelligent automation.

(Glaa et al., 2014) explores AI and ML in the digital transformation of SCM and highlights how they optimize demand forecasting, logistics, and inventory management, driving efficiency, cost reduction, and supply chain resilience. It also discusses the challenges, ethical issues related to AI and ML adoption, including data privacy and workforce impacts. It offers valuable insights for academics and industry professionals, emphasizing the growing importance of these technologies in shaping the future of SCM. (Singh, 2023) reviews a systematic literature exploring the potential and challenges of drones (UAVs) in supply chain management (SCM) and logistics. The key advantages such as improved delivery efficiency, cost reduction, enhanced flexibility, and sustainability, as well as their utility in humanitarian logistics. It also categorizes the challenges into technical, organizational, safety, and regulatory domains. The research examines real-life drone applications and proposes directions for future studies. It is concise and effectively summarizes the scope, contributions, and implications for advancing drone integration in SCM and logistics.

(Lu & Taghipour, 2023) discuss sustainable development is a new type of development that meets current needs while ensuring that future generations can meet their own. In this framework, sustainable growth must be at the intersection of three pillars: economic, social, and environmental. The implementation of sustainable development in the production process, as well as cost changes, are followed by new internal corporate social responsibility mechanisms. Governments, businesses, and scholars can use the findings of this study to include sustainable development in their decisions. The study (Loivet et al., 2020) explores the origins of the green supply chain, which has significantly altered the organizational structure. It also addresses various theories used to evaluate the adoption of green supply chain management (GSCM). It also focuses on the internal and external factors of GSCM deployment and the significance of performance. The research (Taghipour & Beneteau-Piet, 2020) on SSCM focuses on solutions aimed at sustainable development using the "triple bottom line" approach, which incorporates economic, environmental, and social issues. The focus is on monitoring organizational performance using SSCM to match business practices with sustainability goals.

(Taghipour, 2020) Decentralized supply chains often suffer from limited information sharing, leading to myopic, locally focused decisions. Demand Forecasting and Order Planning in Supply Chains and Humanitarian Logistics addresses these issues with insights on AI, forecasting, and operations, offering solutions for improved coordination and competitiveness. (Taghipour et al., 2023a) addressing challenges in vaccine supply chains by proposing a three-objective optimization model to allocate mobile and fixed vaccine injection centres. The objectives include minimizing costs, maximizing route reliability, and reducing unmet demand. The model effectively optimizes vaccine distribution in a large city like Tehran, covering various supply chain levels. Results highlight the robust probability model's superiority in stability and resilience to uncertainty compared to the fuzzy probability model.

Supply Chain Management and Operations Planning (Cauchois et al., 2017; Taghipour and Frayret, 2011; Taghipour and Frayret, 2013a; Taghipour and Frayret, 2013b; Taghipour, 2014) explores advanced methodologies and frameworks for enhancing supply chain coordination, decision-making, and operational planning, particularly in decentralized and stochastic environments. It emphasizes improving manufacturing networks and addressing the complexities of non-integrated business units to optimize supply chain performance.

The studies (Dadsena et al., 2021; Taghipour et al., 2024; Taghipour et al., 2023b; Taghipour, 2023) focus on building resilient and sustainable supply chains post-COVID-19 by addressing barriers and integrating innovative solutions like agrivoltaics, hydropower, and green roofs. They also highlight multi-objective models for waste-to-energy strategies and the importance of government policies in promoting responsible supply chain management. The impact of digital transformation. (Merimi and Taghipour, 2021; Taghipour and Merimi, 2021; Taghipour et al., 2022) during crises, particularly COVID-19, and the opportunities brought by Society 5.0 for procurement processes. It showcases the acceleration of supply chain digitalization and the adoption of blockchain technology to enhance resilience and adaptability.

(Mahfod et al., 2019; Mbiatem et al., 2018) advanced methodologies for supplier evaluation and selection explore frameworks for assisting decision-makers in choosing optimal suppliers through systematic and structured methods. The study (Taghipour et al., 2023c) analyzes changes in transportation behavior in Tehran's during COVID-19 using social network analysis. It ranks districts based on centrality measures and offers recommendations for developing sustainable, behavior-based smart transportation solutions in urban settings.

The study (Zoghlami et al., 2016) explores how incentive systems can enable coordination in divergent supply networks without requiring strategic information sharing. It provides a general approach to improving operations planning and network performance while maintaining organizational autonomy. The study (Nadia et al., 2021) uses hedonic pricing and OLS regression to analyze factors influencing hotel room rates in Beirut, revealing key impacts of star ratings, location, and amenities. The findings guide hotel managers in optimizing pricing strategies and market positioning. The literature review on design for remanufacturing, (Taghipour et al., 2015) emphasizes its role in achieving cost-effective and environmentally sustainable manufacturing. It highlights design strategies that facilitate disassembly, testing, and reassembly, offering a foundation for future research in green manufacturing.

3. SUSTAINABLE SUPPLY CHAIN MANAGEMENT WITH AI-DRIVEN DRONE DELIVERY IN HEALTHCARE

Sustainable Supply Chain Management (SSCM) in healthcare focuses on ensuring the seam less delivery of medical products while minimizing environmental and operational inefficiencies. Integrating AI-driven drones into SSCM addresses critical challenges such as geographic in accessibility, time-sensitive deliveries, and demand fluctuations. By leveraging AI for route optimization, obstacle detection, and autonomous navigation, drones enhance the efficiency and sustainability of healthcare logistics. These systems not only ensure the timely delivery of essential medical supplies but also reduce carbon emissions, operational costs, and dependency on traditional transportation methods. AI-driven drone delivery exemplifies an innovative approach to achieving healthcare equity, resilience, and sustainability in supply chain operations.

3.1. The Need for AI-Driven Drone Delivery in Healthcare

Healthcare delivery systems often face significant challenges in ensuring timely access to critical supplies, particularly in remote or underserved regions. Factors like traffic congestion, geographic barriers, and infrastructure limitations can delay the delivery of life-saving medical products. Integrating AI-driven drones addresses these challenges by offering a faster, more reliable, and sustainable alternative to conventional logistics methods. AI-driven drones are particularly vital in:

3.1.1. **Emergency Response:** Drones enable rapid delivery of critical medical supplies such as blood products, vaccines, and medications to accident sites and disaster-stricken areas, significantly reducing response times and saving lives in emergencies.

3.1.2. **Pandemic Management**: During pandemics, drones efficiently transport diagnostic samples, vaccines, and personal protective equipment, ensuring swift distribution to high-demand and remote areas while minimizing human contact.

3.1.3. **Equity in Healthcare**: Drones overcome geographic barriers by reaching remote and underserved regions with limited healthcare infrastructure, ensuring equitable access to essential medical resources and improving healthcare outcomes. By combining speed, precision, and environmental sustainability, AI-driven drones are becoming an indispensable tool in modern healthcare delivery systems.

3.2. Integrating Artificial Intelligence With Drones for Optimal Healthcare Delivery

Integrating AI with drones enhances their operational capabilities, making them more effective for healthcare delivery. AI algorithms enable drones to navigate complex environments, optimize routes, and ensure delivery precision. Key aspects of this integration include:

3.2.1. **Autonomous Navigation**: Autonomous navigation is an important aspect of AI-powered drone delivery systems, allowing drones to function autonomously without the need for on-going human interaction. AI algorithms process data from on board sensors, cameras, and GPS systems to guide drones through complex environments. This capability ensures that drones can adapt to dynamic conditions, such as changing weather or airspace restrictions, and maintain efficient flight paths. In healthcare, autonomous navigation significantly reduces the time required for critical deliveries, enhancing response times during emergencies and ensuring consistent operations in remote or resource limited areas.

3.2.2. **Route Optimization**: AI-powered route optimization enhances the efficiency of drone deliveries by calculating the fastest and most energy-efficient paths. Using real-time data on weather, airspace regulations, and traffic conditions, AI algorithms dynamically adjust routes to minimize travel time and resource consumption. For healthcare logistics, this translates into cost-effective operations and timely delivery of time-sensitive products such as blood, vaccines, or medications. Optimized routes also contribute to sustainability goals by reducing energy usage and emissions.

3.2.3. **Real-Time Monitoring:** Real-time monitoring enabled by AI ensures transparency and reliability in drone operations. AI systems continuously track drone locations, performance metrics, and payload status, providing live updates to healthcare providers and logistics managers. This capability enhances accountability and allows for immediate interventions in case of technical issues or route deviations. In the healthcare context, real-time monitoring ensures the safe and timely arrival of critical supplies, improving trust and operational efficiency.

3.2.4. **Demand Forecasting**: AI-driven demand forecasting leverages historical data, seasonal trends, and real-time inputs to predict variations in demand for medical supplies. By identifying patterns and potential surges in need, healthcare supply chains can proactively allocate resources and schedule drone deliveries accordingly. This capability is particularly valuable during pandemics, natural disasters, or vaccination campaigns, where demand spikes are common. Accurate forecasting ensures that healthcare facilities are adequately stocked, reducing wastage and improving patient care.

3.3. Core Components of an AI-Driven Drone Delivery System

An AI-driven drone delivery system for healthcare comprises several interdependent components that work together to achieve efficient and reliable operations. These include:

3.3.1. **Drones (UAVs)**: Drones, or Unmanned Aerial Vehicles (UAVs), are the core physical component of an AI-driven delivery system. Equipped with advanced sensors, cameras, and GPS systems, drones are capable of stable flight and precise navigation. For healthcare applications, drones must handle a variety of medical payloads, including temperature-sensitive products like vaccines and blood. Their design ensures reliability, even under challenging weather conditions or in remote areas, making them ideal for critical healthcare deliveries.

3.3.2. **AI Algorithms**: AI algorithms are the intelligence behind the system, enabling drones to operate autonomously and efficiently. Machine learning models process data for navigation, obstacle detection, and route optimization, ensuring precise and reliable deliveries. These algorithms also facilitate real-time decision-making, such as rerouting to avoid unexpected obstacles or adjusting for weather conditions, thus enhancing overall system efficiency and safety.

3.3.3. **Communication Systems**: Reliable communication networks, such as 5G or satellite connectivity, are essential for seamless data exchange between drones, control canters, and ground operators. These systems ensure real-time monitoring, provide live updates on drone status and payload, and support remote troubleshooting. In healthcare, strong communication infrastructure is critical for ensuring uninterrupted operations, particularly during emergencies or in remote regions with limited connectivity.

3.3.4. **Control Centres**: Control centers serve as the operational hub for managing drone activities. These can be ground-based or cloud-based systems that oversee flight paths, monitor drone performance, and respond to contingencies. In healthcare logistics, control centers play a vital role in coordinating multiple drones, ensuring timely deliveries, and addressing any operational challenges that arise in real-time.

3.3.5. **Battery and Energy Management**: Efficient energy systems are crucial for the sustainable operation of drones. Advanced battery technologies provide the necessary power for flight while optimizing energy consumption to align with sustainability goals. AI systems monitor battery health and charge levels, ensuring drones complete their deliveries without interruptions. This is particularly important in healthcare, where uninterrupted operations can mean the difference between life and death.

3.3.6. **Payload Management:** Payload management systems ensure the safe transportation of medical supplies. Drones are equipped with specialized compartments designed to securely handle a variety of payloads, including fragile or temperature-sensitive products. Cold chain management features, such as insulated containers or cooling systems, maintain the integrity of items like vaccines or biological samples during transit, ensuring they reach their destination in optimal condition.

3.3.7. **Regulatory Compliance Framework:** Compliance with aviation regulations, data privacy laws, and cyber security standards is critical for the legal and safe operation of drones. This includes obtaining necessary permissions for airspace use, ensuring data encryption for secure communication, and adhering to healthcare-specific regulations like those governing the transport of medical products. Robust compliance frameworks build trust and enable the scalable deployment of drone delivery systems.

3.3.8. **Data Analytics Platform**: A data analytics platform powered by AI provides valuable insights into system performance, including delivery times, energy efficiency, and drone reliability. These analytics support continuous improvement by identifying bottlenecks, optimizing operations, and predicting maintenance needs. In healthcare logistics, this ensures a responsive and adaptable supply chain capable of meeting dynamic demands.

3.3.9. **User Interface:** An intuitive user interface allows healthcare providers to interact seamlessly with the system. Through this platform, users can request deliveries, track shipments in real time, and receive updates on drone status. A user-friendly interface ensures efficient coordination between medical facilities and logistics teams, enabling timely and accurate delivery of critical supplies

4. CASE STUDIES

4.1. Rwanda

Rwanda has been a pioneer in drone delivery, partnering with Zipline to address healthcare challenges. Fixed-wing drones are used to deliver critical medical supplies, such as blood and vaccines, to remote health facilities. The system operates from distribution hubs and has significantly reduced delivery times, making healthcare more accessible. Despite its success, high operational costs remain a challenge.

4.2. Ghana

Ghana adopted Zipline's drone delivery system to improve healthcare logistics, including COVID-19 vaccine distribution. With multiple distribution centers, drones have expanded healthcare access to underserved regions, reducing delivery times from hours to minutes. The project demonstrates scalability in improving medical supply chains but shares challenges like cost sustainability.

4.3. USA

The USA has explored diverse applications for drones, led by companies like Amazon Prime Air, UPS Flight Forward, and Wing. These programs focus on delivering medical supplies, retail goods, and food in urban and suburban areas. Regulatory barriers, airspace management, and public acceptance are all challenges. However, trial operations in North Carolina and Virginia demonstrate the potential for speedier, more effective logistics.

4.4. India

India has initiated drone delivery trials through programs like Medicine from the Sky, focusing on delivering vaccines, medicines, and diagnostics to remote and inaccessible regions. States like Telangana and Arunachal Pradesh have led pilot projects using GPS-enabled drones. While promising, the system faces regulatory, infrastructure, and weather-related challenges, underscoring the need for localized solutions and government support.

5. PROBLEM STATEMENT: UAV-INTEGRATED SUSTAINABLE SUPPLY CHAIN MANAGEMENT (SSCM)

In recent years, the integration of UAVs into supply chain operations has emerged as a transformative solution to address the challenges of transportation and delivery in SSCM. This approach leverages UAVs to achieve faster, more efficient, and environmentally friendly delivery of medical supplies, especially in geographically distributed areas. However, this innovation presents unique challenges that necessitate the development of robust optimization models.

The focus of this research is on transportation and delivery problems associated with UAV-integrated SSCM, particularly in addressing: Route Optimization determines the most efficient routes for UAVs to minimize total travel time, energy consumption, and operational costs while adhering to constraints such as payload capacity, range limitations, and battery life. Ensuring timely delivery of medical products, including pharmaceuticals, vaccines, and emergency supplies, to a network of demand locations such as hospitals, clinics, pharmacies, and home locations. Sustainability objectives minimizing the environmental impact of UAV operations by reducing CO_2 emissions and optimizing energy consumption.

Figure 1. UAV integrated Supply Chain Network

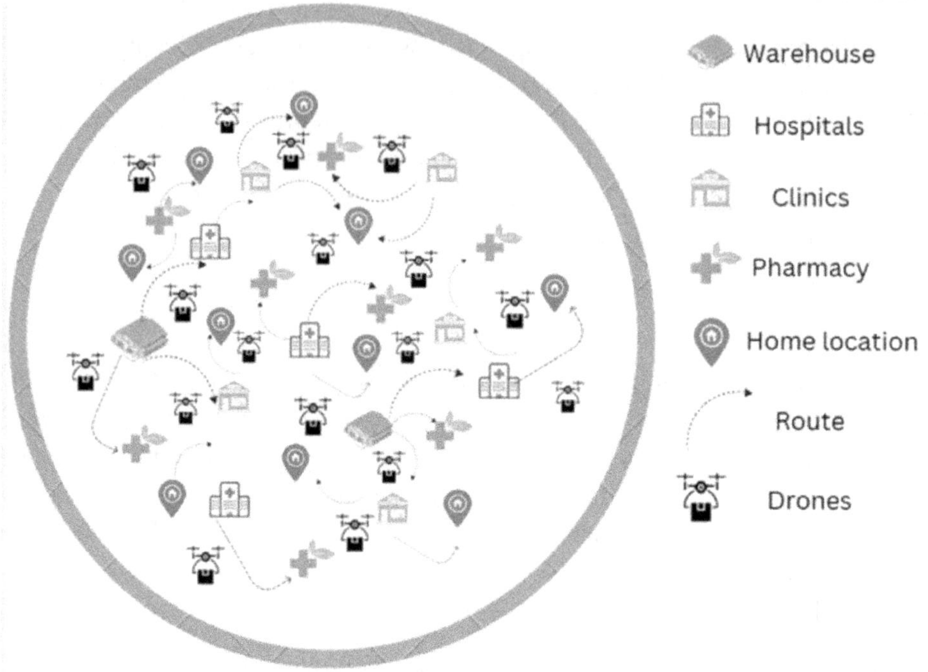

The scenario presents a comprehensive framework for integrating UAVs into a **SSCM** system, specifically targeting the efficient transportation and delivery of medical products. This system is designed to address critical challenges in healthcare logistics by leveraging drone technology to ensure timely, cost-effective, and environmentally sustainable delivery services across a 100 km radius is considered, consisting of multiple supply points and demand locations. The supply points include medical warehouses, hospitals, and pharmacies, which serve as the sources for distributing essential medical supplies like medications, vaccines, and emergency items. The demand locations comprise geographically distributed points, including clinics, hospitals, and home locations, each with varying delivery demands and priorities. High-priority locations, such as hospitals, require urgent deliveries within few hours, while clinics and home deliveries have moderate and low-priority demands, respectively. A heterogeneous fleet of drones is employed for deliveries, classified into small, medium, and large types, based on their payload capacity, range, speed, battery life, energy consumption, and CO_2 emissions. Each drone type operates under specific constraints, such as limited battery range, maximum payload capacity, and distance per charge.

The delivery system is further constrained by factors such as delivery deadlines, battery limitations requiring drones to return to supply points for charging, and the need to operate within their specified range and payload limits. To simulate the scenario, precomputed distance matrices are used, along with randomized demand quantities, depending on the type of demand location. The UAV fleet consists of small, medium, and large drones. The primary objective is to optimize the delivery system to minimize transportation time, operational costs, and environmental impact while ensuring all deliveries are completed on time. This involves designing efficient routes, maximizing fleet utilization, and reducing energy consumption and CO_2 emissions. The scenario provides a structured framework for testing optimization models and algorithms, ensuring they are applicable to real-world medical supply chain logistics while promoting sustainability and operational efficiency.

6. MATHEMATICAL FORMULATION

6.1. Objective Function

The objective is to minimize factors such as time, cost carbon emission and energy consumption.

Minimize $\boldsymbol{Z} = w_1 \boldsymbol{C} + w_2 \boldsymbol{T} + w_3 \boldsymbol{CE} + w_4 \boldsymbol{EC}$ ... (1)

where

$T = travel\ time(TT) + hovering\ time(HT) + material\ handling\ time(MHT)$

$C = initial\ cost(IC) + carrying\ cost(CC) + operational\ cost(OC)$

$CE = distance * emission/km$

$EC = distance * energy\ usage/km$

Subject to the Constraints:

Supply capacity $\geq$ Demand ...(2)

Payload carried by the drone $\leq$ Maximum drone capacity ...(3)

Travel speed of the drone $\leq$ Maximum allowable speed ...(4)

Distance travelled by the drone $\leq$ Maximum range of the drone ...(5)

Total energy consumption $\leq$ Drone battery capacity ...(6)

TT + HT + MHT $\leq$ Maximum allowable time ...(7)

Distance to each demand location $\leq$ Coverage radius of the drone ...(8)

Emissions generated per km $\leq$ Maximum permissible emission level ...(9)

Energy used per km $\leq$ Maximum permissible energy consumption ...(10)

6.2. Advanced Techniques and Algorithms

Optimization Techniques:

- Particle Swarm Optimization (PSO): A population-based metaheuristic inspired by social behaviour like bird flocking. Particles (candidate solutions) move through the search space, guided by their best-found position and the swarm's best-found position. Effective for continuous optimization problems like finding optimal flight paths.
- Particle Swarm Optimization (MOPSO): An extension of PSO designed to handle problems with multiple conflicting objectives (e.g., minimizing time *and* emissions). It maintains a set of non-dominated solutions (the Pareto front) representing trade-offs between objectives. Crucial for UAV delivery where balancing time, cost, and environmental impact is essential.
- Ant Colony Optimization (ACO): A metaheuristic inspired by the foraging behaviour of ants. Artificial ants traverse the search space, leaving pheromone trails that guide subsequent ants toward better solutions. Suitable for discrete optimization problems like finding the best sequence of deliveries.
- Simulated Annealing (SA): A probabilistic metaheuristic inspired by the annealing process in metallurgy. It explores the search space by accepting both better and worse solutions with a decreasing probability, allowing it to escape local optima. Useful for complex optimization problems with rugged search spaces.
- Nearest Neighbour (NN): A simple greedy heuristic that starts at a depot and iteratively adds the nearest unvisited delivery point to the route. Fast but often produces suboptimal solutions. Useful as a baseline for comparison or for very quick initial route planning.
- Clustering: While not strictly an optimization algorithm itself, clustering techniques (e.g., k-means) group delivery points into clusters. This simplifies the routing problem by dividing the delivery area into smaller, more manageable regions.
- Genetic Algorithms (GA): Evolutionary algorithms inspired by natural selection. They use genetic operators like selection, crossover, and mutation to evolve a population of solutions towards better ones. Effective for handling complex constraints and multiple objectives.
- Tabu Search (TS): A metaheuristic that explores the solution space by iteratively moving from one solution to another. It uses a "tabu list" to prevent revisiting recently explored solutions, helping to avoid cycling and explore new areas of the search space.
- Evolutionary Algorithms (EA): A broader class of algorithms that includes GAs. They are inspired by biological evolution and use mechanisms like selection, mutation, and recombination to evolve solutions.

- Non-dominated Sorting Genetic Algorithm II (NSGA-II): A popular multi-objective GA that uses non-dominated sorting to rank solutions and crowding distance to maintain diversity in the Pareto front.

AI Techniques:

- Reinforcement Learning (RL): An agent learns to interact with an environment by taking actions and receiving rewards or penalties. Suitable for dynamic environments where conditions change (e.g., weather, traffic). Enables UAVs to learn optimal flight paths and adapt to real-time information.
- Unsupervised Learning (USL): Learns from unlabelled data to discover hidden patterns and structures. Clustering is a key unsupervised learning method. Useful for customer segmentation, demand forecasting, and airspace management.
- Machine Learning (ML): A broad field encompassing algorithms that allow computers to learn from data without explicit programming. Includes supervised, unsupervised, and reinforcement learning.
- Supervised Learning (SL): Learns from labeled data
    - Regression: Predicts continuous values (e.g., delivery time, fuel consumption).
    - Classification: Predicts categorical values (e.g., weather conditions, delivery success/failure).
- Deep Learning (DL): Uses artificial neural networks with multiple layers (deep neural networks) to learn complex patterns from data.
    - Convolutional Neural Networks (CNN): Particularly effective for image and video processing, useful for obstacle detection and navigation using onboard cameras.
    - Recurrent Neural Networks (RNN): Designed to handle sequential data, such as time series. Can be used for predicting demand patterns or analyzing sensor data from UAVs.
    - Long Short-Term Memory Networks (LSTM): A type of RNN that can handle long-term dependencies in sequential data, making them suitable for complex time series forecasting and analysis.

Figure 2. Advanced Solution Techniques

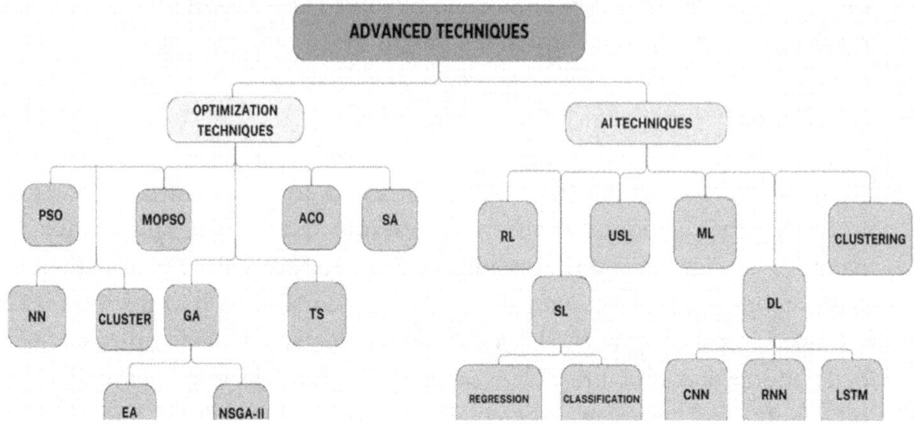

7. CHALLENGES

7.1. Regulatory and Legal Challenges

Drone operations face challenges from evolving regulations that vary across regions, complicating compliance for cross-border supply chains. Safe integration into shared airspace alongside manned aircraft demands sophisticated coordination to avoid collisions and congestion. AI driven drones also collect sensitive data, raising concerns about compliance with data protection laws like GDPR and HIPAA. Ensuring secure data handling, encryption, and adherence to privacy standards is essential for maintaining trust. Together, these complexities highlight the need for robust frameworks to ensure legal, safe, and efficient drone operations globally.

7.2. Technological Challenges

Technological challenges for drones include limitations in battery life and payload capacity, hindering long-range deliveries and the transport of heavy or temperature-sensitive medical supplies until battery and drone efficiency improve. AI system reliability is another concern, as algorithms require vast training data and may falter in dynamic environments, leading to errors or reduced performance. Communication failures, especially in remote areas with weak network infrastructure, can disrupt operations, delay deliveries, or cause crashes. Continuous refinement of AI systems and resilient communication networks, such as 5G, are crucial for reliable drone operations. Addressing these challenges is vital for the effective use of drones in healthcare logistics.

7.3. Environmental Constraints

Environmental constraints for drones include sensitivity to adverse weather conditions like strong winds, rain, and extreme temperatures, which can compromise flight stability, battery efficiency, and payload safety, leading to delays or failures in medical deliveries. Additionally, drones face challenges navigating difficult terrains such as mountainous regions, dense urban environments, or areas with poor GPS coverage. Even advanced AI systems may struggle in harsh or infrastructure-limited settings, affecting operational reliability. These environmental factors necessitate robust designs and adaptive systems to ensure drones can perform effectively in diverse conditions. Addressing these issues is crucial for reliable and timely healthcare logistics.

7.4. Cost and Investment Concerns

Cost and investment concerns in AI-driven drone delivery include high upfront expenses for procurement, infrastructure, and system integration, which can be a barrier for smaller healthcare providers or those in developing regions. On-going operational costs, such as maintenance, battery replacements, and training, further add to financial challenges, especially in high-demand or logistically complex areas. Expanding networks to cover larger regions requires significant investment in drones, infrastructure, and technology upgrades, making scalability difficult. These challenges highlight the need for cost-effective solutions and financial support to ensure broader adoption.

7.5. Cold Chain Integrity for Medical Supplies

Maintaining cold chain integrity for medical supplies in drones involves ensuring temperature sensitive items, like vaccines and blood, stay within required ranges. This demands specialized cooling or insulation systems, adding complexity and cost to drone design. Continuous monitoring is essential, and failures in sensors or alert systems can harm the safety and quality of the medical supplies. The reliability of these systems is crucial to preserving the integrity of sensitive products during delivery. Effective cold chain management remains a top concern in AI-driven drone logistics.

7.6. AI Model Limitations

AI models in drones face limitations in data quality and availability, especially in remote areas where inconsistent data can affect decision-making and cause delays. Bias in AI systems, arising from incomplete or skewed datasets, can hinder their adaptability and accuracy in diverse environments. Additionally, optimizing drone operations for energy efficiency, speed, and sustainability presents challenges as AI must balance competing factors like energy consumption, emissions, and delivery timeliness. Achieving this balance requires sophisticated algorithms that can adapt to dynamic conditions. These limitations impact the overall performance and reliability of AI-driven drone systems.

7.7. Public Perception and Acceptance

Public perception of AI-driven drones is shaped by safety concerns, including the risks of mid-air collisions, mechanical failures, and privacy violations, which can slow adoption in healthcare. Noise pollution, especially in urban areas, and privacy issues related to drones with cameras or sensors also contribute to distrust. Additionally, many healthcare providers are hesitant to trust AI systems due to concerns about technological failures or errors in critical operations. Overcoming these challenges requires transparent communication, rigorous testing, and education to build trust in drone safety and AI reliability. Addressing these issues is key to wider acceptance and adoption.

7.8. Sustainability challenges

Sustainability challenges with drones include their reliance on non-renewable energy sources like fossil fuels for backup power or unsustainable batteries, which can undermine environmental goals. Additionally, drones contribute to the growing issue of electronic waste, particularly due to the short lifespan of components like batteries. If not properly recycled or disposed of, these drones could negatively affect the environment, offsetting their potential benefits. Ensuring clean energy sources and responsible disposal is essential for aligning drone operations with sustainability objectives. These inconsistencies must be addressed for drones to truly contribute to sustainable logistics.

8. APPLICATION

8.1. Rapid Emergency Response

AI-powered drones enable fast and reliable delivery of critical medical supplies during emergencies, ensuring timely aid. They are especially effective in delivering blood, organs, vaccines, and antivenoms in life-threatening situations. For instance, drones can bypass traffic and inaccessible roads to deliver life-saving supplies to rural or disaster-hit areas, significantly reducing response times and saving lives.

8.2. Improved Accessibility to Remote Areas

AI-integrated drones bridge logistical gaps by reaching remote and infrastructure-deficient regions. These drones use AI-based route optimization to navigate rough terrains and inaccessible locations. Programs in countries like Rwanda and Ghana demonstrate this by successfully delivering blood and vaccines to remote hospitals and clinics, ensuring equitable access to healthcare.

8.3. Cold Chain Management for Perishables

AI-driven drones equipped with advanced cold chain technology ensure the safe delivery of temperature-sensitive medical products, such as vaccines, blood, and insulin. Real-time monitoring and adaptive AI control maintain optimal temperatures during transit, preventing spoilage and preserving the viability of critical supplies like COVID-19 vaccines.

8.4. Disaster Relief and Humanitarian Aid

During natural disasters or crises, AI-powered drones deliver medical supplies to affected areas where traditional logistics are disrupted. Predictive algorithms help identify safe routes between destruction or challenging conditions. For example, drones have supported post-earthquake relief efforts in Nepal and Haiti, delivering life-saving medical aid efficiently.

8.5. Pandemic Management

Drones equipped with AI play a crucial role in addressing health crises by scaling up supply chains for medical essentials like test kits, personal protective equipment (PPE), and vaccines. AI algorithms prioritize deliveries based on urgency and geographical needs, ensuring resources reach areas with the highest demand during pandemics like COVID-19.

8.6. Real-Time Surveillance and Inventory Management

AI-integrated drones assist in inventory audits and stock management at healthcare facilities. They monitor blood banks and rural health centres to ensure timely replenishment of medical supplies. By analyzing consumption patterns, AI systems create optimized delivery schedules, reducing instances of shortages or overstocking.

8.7. Reduced Environmental Impact

AI-optimized drones help minimize the environmental footprint of healthcare logistics by re placing traditional fuel-based vehicles for short to medium-distance deliveries. Energy-efficient flight paths and load balancing, powered by AI, enhance sustainability and significantly reduce carbon emissions associated with medical supply chains.

8.8. Delivery of Diagnostic Samples

Drones expedite the transport of diagnostic samples, such as blood or biopsy specimens, to centralized laboratories, enabling faster testing and results. AI-driven real-time tracking ensures samples are delivered within their viability window, reducing delays in diagnosis and improving patient outcomes.

8.9. Integration With Telemedicine

Drones complement telemedicine services by delivering medications, diagnostic kits, or test samples directly to patients. AI systems coordinate drone schedules with telemedicine appointments, ensuring timely deliveries that enhance remote healthcare services, especially in areas with limited access to medical facilities.

8.10. Cost-Effective Healthcare Logistics

AI-powered drones reduce operational costs by minimizing the need for human resources and fuel consumption. Cost optimization algorithms identify economical routes and efficient schedules for bulk deliveries to multiple locations, making healthcare supply chains more affordable and sustainable over time.

9. ADVANTAGES AND DISADVANTAGES

9.1. Advantages

9.1.1. **Speed and Efficiency**: AI-driven drones significantly enhance the speed of medical supply deliveries, particularly in emergencies or remote locations. By bypassing traditional transportation constraints, drones ensure rapid response during critical situations, reducing life-threatening delays. For example, Zip line's drone delivery service in Rwanda has reduced transportation times for blood products from hours to minutes, saving countless lives in emergency scenarios.

9.1.2. **Enhanced Accessibility:** AI-driven drones are adept at reaching areas that are inaccessible due to disasters, poor infrastructure, or geographic isolation. This capability ensures that underserved and remote regions receive essential medical supplies, bridging the healthcare access gap. For instance, drones have been used in India to deliver vaccines to remote mountainous areas, providing equitable healthcare support.

9.1.3. **Environmental Sustainability**: Drones produce significantly fewer carbon emissions compared to traditional fuel-based logistics vehicles, making them a greener alternative. AI optimizes energy consumption during operations, further reducing their environmental impact. These features align with global sustainability goals, helping to lower the carbon footprint of healthcare supply chains.

9.1.4. **Cost Efficiency** While the initial investment in drone technology is high; the long-term operational costs are considerably lower. Drones reduce expenses associated with fuel, labour-intensive processes, and vehicle maintenance, offering financial savings over time. These cost benefits are particularly valuable in optimizing healthcare supply chains for resource-limited settings.

9.1.5. **Real-Time Monitoring and Decision-Making:** AI algorithms provide real-time tracking and monitoring of drone deliveries, ensuring transparency and minimizing risks of loss or damage. These systems can adapt delivery routes dynamically based on live weather, traffic, or other environmental conditions, improving accuracy and reliability in healthcare logistics.

9.1.6. **Improved Cold Chain Management**: AI-integrated drones equipped with temperature-controlled compartments preserve the integrity of sensitive medical supplies such as blood, plasma, and vaccines. This capability minimizes spoilage and ensures that critical supplies remain viable during transit, addressing one of the most challenging aspects of healthcare logistics.

9.1.7. **Improved Supply Chain Resilience**: AI-driven drones provide robust logistics solutions during disasters or pandemics when traditional transport methods may fail. For example, during earthquakes, floods, or health crises like COVID-19, drones have been deployed to deliver essential medical supplies, strengthening the healthcare system's ability to respond to emergencies.

9.1.8. **Enhanced Healthcare Delivery in Urban Areas:** In dense urban environments, drones bypass traffic congestion to ensure timely delivery of medical supplies. For instance, drones have been used to transport blood or organs between hospitals in metropolitan areas, reducing delays and improving outcomes for time-critical procedures.

9.1.9. **Precision and Accuracy in Deliveries**: AI algorithms enable drones to optimize flight paths and deliver supplies with accuracy. This ensures that healthcare facilities, especially in rural or isolated areas, receive exactly what they need without misdeliveries, improving the overall efficiency and reliability of healthcare supply chains.

9.1.10. **Scalability and Adaptability:** AI-driven drone systems are flexible, allowing for the expansion of logistics networks to meet varying demands, such as seasonal vaccination drives or emergencies. Their adaptability ensures that supply chains remain responsive to evolving healthcare needs, enhancing operational efficiency and reach.

9.1.11. **Contribution to Public Health Goals:** The timely and reliable delivery of medical supplies through AI-driven drones supports broader public health initiatives, such as vaccination campaigns and emergency medical supply chains. These contributions improve healthcare accessibility, operational efficiency, and overall population health outcomes.

9.2. Disadvantages

9.2.1. **Regulatory Challenges**: Extremely strict and insufficient regulations governing drone operations across different regions pose significant barriers to their deployment. Compliance with varying airspace usage rules and privacy laws complicates implementation, delaying the scalability and effectiveness of drone based sustainable supply chain management (SSCM) systems.

9.2.2. **High Initial Costs**: AI-driven drone systems require substantial investment in advanced infrastructure, AI software, and high-performance drone technology. These financial demands can act as a barrier, particularly for low-resource healthcare organizations or systems in developing regions.

9.2.3. **Limited Payload and Range**: Current drone technology is restricted by payload capacity and battery life, limiting its ability to handle bulk or long-distance deliveries. These constraints reduce the feasibility of using drones for large-scale or heavy medical supply logistics, making them suitable only for smaller, localized operations.

9.2.4. **Weather Dependency**: Drones are highly vulnerable to adverse weather conditions such as high winds, heavy rain, or snow. These environmental factors can disrupt operations, making the technology less reliable in regions with unpredictable or extreme weather patterns, potentially delaying critical medical deliveries.

9.2.5. **Data Privacy and Cyber security Risks**: AI-driven drones process and transmit critical healthcare logistics data, which is vulnerable to unauthorized access. This raises concerns about the security of patient data and the system's stability, requiring strong cyber security measures to prevent hazards.

9.2.6. **Energy and E-Waste Concerns:** The short lifespan of drone batteries, as well as component decay, all contribute to electronic waste. This creates a sustainability problem, since drones' environmental benefits may be overshadowed by their contribution to e-waste and reliance on non-renewable energy sources.

9.2.7. **Social and Ethical Concerns**: Concerns about safety, noise pollution, and AI decision-making are some of the factors contributing to the public's mistrust of drone technology. The community's acceptance and uptake of drones for healthcare logistics may be slowed by these moral and social issues, especially in urban or residential regions.

9.2.8. **Technical Failures:** AI algorithms and drone systems are liable to technical issues such as software bugs, hardware malfunctions, or sensor inaccuracies. These failures can result in delivery delays, lost cargo, or compromised operations, undermining trust and efficiency in SSCM solutions.

10. FUTURE SCOPE

The future of AI-driven drone delivery in sustainable supply chain management will feature transformative advancements such as autonomous swarm systems for large-scale synchronized deliveries, managed by AI algorithms to optimize routes and reduce conflicts. Enhanced ma chine learning models will handle dynamic variables like real-time weather, demand, and traffic, while integration with IoT and smart cities will enable seamless communication with health care facilities and urban infrastructure, including drone corridors and charging stations. Drones will expand cold chain capabilities for transporting temperature-sensitive medical products with precise AI-driven monitoring and adaptive cooling. Sustainability efforts will drive the adoption of renewable energy-powered drones, supported by AI-optimized energy consumption. Hybrid air-ground delivery systems will combine drones with autonomous ground vehicles for last-mile logistics, enhancing reach. Unified regulatory frameworks will facilitate broader adoption, complemented by blockchain technology for secure and transparent tracking. AI-driven predictive maintenance will improve reliability, while modular drones will revolutionize disaster relief by prioritizing critical deliveries. Personalized medical deliveries will align with tailored healthcare needs, supported by enhanced AI decision-making for adaptive operations. 5G and satellite connectivity will ensure real-time data exchange, even in remote regions. Cross-industry collaborations between healthcare, logistics, and technology sectors will spur innovation, and a focus on equity will prioritize underserved regions, promoting inclusive and resilient supply chains worldwide.

11. CONCLUSION

AI-driven drone delivery systems represent a ground-breaking innovation in SSCM, offering unparalleled benefits for healthcare logistics by addressing speed, efficiency, accessibility, and environmental sustainability. These systems have proven their potential in emergency response, pandemic management, and disaster relief, ensuring timely and precise delivery of critical medical supplies, even in remote and underserved regions. Despite significant challenges, including regulatory complexities, high costs, and technological constraints, advancements in AI and drone technology, coupled with supportive regulatory frameworks and cross-industry collaborations, are paving the way for wider adoption. By incorporating features like renewable energy, cold chain management, and IoT integration, the future of AI-driven drone delivery systems holds immense promise in creating resilient and sustainable healthcare supply chains. This transformative approach not only supports global health initiatives but also aligns with broader goals of equity and environmental responsibility, ensuring a more inclusive and efficient healthcare logistics framework.

Declaration of Competing Interests: The authors declare no conflicts of interest.

Funding Source: Not applicable

REFERENCES

Addy, A. (2023). Artificial intelligence in the supply chain management for vaccine distribution in the West African healthcare sector with a focus on Ghana. *International Journal of Legal Science Innovation, 10.*

Aliahmadi, A., Nozari, H., & Ghahremani-Nahr, J. (2022). Aiot-based sustainable smart supply chain framework. *International Journal of Innovation in Management. Economics and Social Sciences*, 2(2), 28–38.

Cauchois, R., Taghipour, A., Kang, D., Zoghlami, N., & Abed, M. (2017). Advanced solutions for a supply chain with stochastic information. *Journal of Advanced Management Science*, 5(1), 9–13. DOI: 10.18178/joams.5.1.9-13

Dadsena, K. K., Mathiyazhagan, K., & Taghipour, A. (2021). Analysis of barriers for the build the resilient supply chain networks post-COVID-19. In *Managing Supply Chain Risk and Disruptions: Post COVID-19* (pp. 79–89). Springer International Publishing. DOI: 10.1007/978-3-030-72575-4_7

Damoah, I. S., Ayakwah, A., & Tingbani, I. (2021). Artificial intelligence (ai)-enhanced medical drones in the healthcare supply chain (HSC) for sustainability development: A case study. *Journal of Cleaner Production*, 328, 129598. DOI: 10.1016/j.jclepro.2021.129598

Glaa, B., Zoghlami, N., & Taghipour, A. (2014, May). Outsourcing and off-shoring healthcare services: A way to cut the costs and improve quality. In *2014 international conference on advanced logistics and transport (ICALT)* (pp. 189-194). IEEE. DOI: 10.1109/ICAdLT.2014.6864116

Khadem, M., Khadem, A., & Khadem, S. (2023). Application of artificial intelligence in supply chain revolutionizing efficiency and optimization. *International Journal of Industrial Engineering and Operational Research*, 5(1), 29–38.

Loivet, W., Taghipour, A., & Kang, D. S. (2020). The rise of green supply chain management: Between complexity and necessity. *J. Econ. Bus. Manag*, 8(1), 1–7. DOI: 10.18178/joebm.2020.8.1.604

Lu, X., & Taghipour, A. (2023). A French Guide for Businesses and Governments to Integrate the Concept of Sustainable Development Into Their Supply Chains. In *Government Impact on Sustainable and Responsible Supply Chain Management* (pp. 1–12). IGI Global. DOI: 10.4018/978-1-6684-9062-4.ch001

Mahfod, J., Canel-Depitre, B., & Taghipour, A. (2019). Quality function deployment-ELECTRE in supplier evaluation. *Journal of Advanced Management Science, 7*(4).

Mbiatem, J., Taghipour, A., & Canel-Depitre, B. (2018, August). Supplier selection approaches for decision makers. In *Proceedings of the 8th International Conference on Information Communication and Management* (pp. 108-112).

Merimi, M., & Taghipour, A. (2021). Accelerating the digitalization of the supply chain: An empirical research about COVID-19 crisis. In *Digitalization of decentralized supply chains during global crises* (pp. 1–24). IGI Global. DOI: 10.4018/978-1-7998-6874-3.ch001

Nadia, E. N., Beatrice, C. D., & Atour, T. (2021). The determinants of hotel room rates in Beirut: A hedonic pricing model. *International Journal of Trade. Economics and Finance*, 12(2), 33–42.

Rejeb, A., Rejeb, K., Simske, S. J., & Treiblmaier, H. (2021). Drones for supply chain management and logistics: A review and research agenda. *International Journal of Logistics*, 26(6), 708–731. DOI: 10.1080/13675567.2021.1981273

Singh, P. K. (2023). Digital transformation in supply chain management: Artificial intelligence (AI) and machine learning (ML) as catalysts for value creation. *International Journal of Supply Chain Management*, 12(6), 57–63. DOI: 10.59160/ijscm.v12i6.6216

Taghipour, A. (2014). Improving the plan of a manufacturing network with non-integrated business units. *International Journal of Applied Logistics*, 5(2), 1–11. DOI: 10.4018/ijal.2014040101

Taghipour, A. (Ed.). (2020). *Demand forecasting and order planning in supply chains and humanitarian logistics*. IGI Global.

Taghipour, A. (Ed.). (2023). *Government Impact on Sustainable and Responsible Supply Chain Management*. IGI Global. DOI: 10.4018/978-1-6684-9062-4

Taghipour, A., Abed, M., & Zoghlami, N. (2015, May). Design for remanufacturing respecting reverse logistics processes: A review. In *2015 4th International Conference on Advanced Logistics and Transport (ICALT)* (pp. 299-304). IEEE. DOI: 10.1109/ICAdLT.2015.7136602

Taghipour, A., & Beneteau-Piet, C. (2020). Sustainable supply chain management performance. *International Journal of Innovation, Management and Technology*, 11(6), 165–169. DOI: 10.18178/ijimt.2020.11.6.897

Taghipour, A., Foukolaei, P. Z., Ghaedi, M., & Khazaei, M. (2023b). Sustainable Multi-Objective Models for Waste-to-Energy and Waste Separation Site Selection. *Sustainability (Basel)*, 15(22), 15764. DOI: 10.3390/su152215764

Taghipour, A., & Frayret, J. M. (2011). Supply chain coordination planning: A review. *Work paper, École polytechnique de Montréal.*

Taghipour, A., & Frayret, J. M. (2013a, May). An algorithm to improve operations planning in decentralized supply chains. In *2013 International Conference on Advanced Logistics and Transport* (pp. 100-103). IEEE. DOI: 10.1109/ICAdLT.2013.6568442

Taghipour, A., & Frayret, J. M. (2013b). Coordination of operations planning in supply chains: A review. *International Journal of Business Performance and Supply Chain Modelling*, 5(3), 272–307. DOI: 10.1504/IJBPSCM.2013.055729

Taghipour, A., Lu, X., & Smahi, M. (2022). Opportunities and challenges for society 5.0 in procurement. *Blockchain Technology in Supply Chain Management for Society*, 5(0), 91–100. DOI: 10.1201/9781003177432-9

Taghipour, A., & Merimi, M. (2021, March). Digital transformation of supply chains during crisis: covid-19. In *11th Annual International Conference on Industrial Engineering and Operations Management, IEOM* (pp. 7493-7501). DOI: 10.46254/AN11.20211298

Taghipour, A., Padash, A., Etemadi, V., Khazaei, M., & Ebrahimi, S. (2024). Sustainable and Circular Hotels and the Water–Food–Energy Nexus: Integration of Agrivoltaics, Hydropower, Solar Cells, Water Reservoirs, and Green Roofs. *Sustainability (Basel)*, 16(5), 1985. DOI: 10.3390/su16051985

Taghipour, A., Ramezani, M., Khazaei, M., Roohparvar, V., & Hassannayebi, E. (2023c). Smart transportation behavior through the COVID-19 pandemic: A ride-hailing system in Iran. *Sustainability (Basel)*, 15(5), 4178. DOI: 10.3390/su15054178

Taghipour, A., Sohrabi, A., Ghaedi, M., & Khazaei, M. (2023a). A robust vaccine supply chain model in pandemics: Case of Covid-19 in Iran. *Computers & Industrial Engineering*, 183, 109465. DOI: 10.1016/j.cie.2023.109465

Zoghlami, N., Taghipour, A., Merlo, C., & Abed, M. (2016). Management of divergent production network using decentralised multi-level capacitated lot-sizing models. *International Journal of Shipping and Transport Logistics*, 8(5), 590–604. DOI: 10.1504/IJSTL.2016.078686

Chapter 5
The Future of Marketing:
Integrating Big Data and AI for Personalized Customer Engagement

XiaoWen Lu
Normandy University, France

Atour Taghipour
https://orcid.org/0000-0003-1687-8881
Normandy University, France & Soldbridge, South Korea

ABSTRACT

The convergence of big data and AI has transformed supply chain management, enabling organizations to operate with efficiency and adaptability. Big data delivers vast amounts of real-time information from diverse sources, including customer orders, supplier details, and market trends, while AI processes this data to identify patterns and make predictive analyses that optimize inventory management, forecast demand, and enhance logistics. This synergy empowers supply chain managers to make proactive decisions, reducing delays and stockouts, while supporting just-in-time inventory practices that align supply levels with consumer demand. In parallel, big data has reshaped marketing by providing deep insights into consumer behaviors and preferences, allowing for the development of tailored marketing strategies that enhance customer experiences. Despite the wealth of literature, there remains a gap in comprehensive studies that integrate its implications for personalized marketing, ethical considerations, and technical challenges.

DOI: 10.4018/979-8-3693-7478-8.ch005

INTRODUCTION

The intersection of Big Data, artificial intelligence (AI), and supply chain management has revolutionized how organizations operate, providing insights that drive efficiency, reduce costs, and enhance customer satisfaction (Cauchois et al., 2017; Dadsena et al., 2021; Loivet et al., 2020). Big Data provides vast amounts of real-time information from various sources (Taghipour, 2020; Taghipour, 2014; Merimi, & Taghipour, 2021; Taghipour& Frayret, 2011) like customer orders (Nadia et al., 2021; Taghipour et al., 2022), supplier details (Mahfod et al., 2019; Mbiatem et al., 2018), shipping logistics (Zoghlami et al., 2016), and market trends. AI processes and analyzes this data quickly, identifying patterns and making predictions that help companies optimize inventory management (Taghipour & Frayret, 2013a, and 2013b), forecast demand (Taghipour et al., 2023a; Taghipour et al., 2024), and streamline logistics (Taghipour & Merimi, 2021; Taghipour et al., 2023b). For instance, AI algorithms can predict when and where stock shortages may occur, allowing supply chain managers to make proactive adjustments (Taghipour et al., 2015; Taghipour et al., 2023c), reducing delays and stockouts (Lu & Taghipour, 2023). Additionally, AI-powered systems support just-in-time (JIT) inventory practices by aligning supply levels closely with customer demand (Taghipour, 2023), improving responsiveness while minimizing waste (Taghipour & Beneteau-Piet, 2020). The synergy between Big Data and AI empowers supply chain management to become more adaptive, resilient, and efficient, allowing companies to navigate uncertainties and remain competitive in a rapidly changing marketplace.

In the modern world, Big Data has become a fundamental asset, reshaping various sectors, including marketing. To fully appreciate Big Data's significance and its impact on personalized marketing strategies, it's crucial to define key terms and their context. "Data" refers to factual information—such as measurements or statistics—used as a basis for reasoning, discussion, or calculation (Merriam-Webster, 2024). Data can be discrete or continuous values that convey quantities, qualities, facts, or simply symbols to be interpreted (Wikipedia, 2024). Today, Big Data is distinguished by three main characteristics, known as the "3Vs": Volume (the massive amount of data generated every second), Variety (diversity of data types—structured, semi-structured, and unstructured), and Velocity (the rapid generation and processing of data).

Marketing strategy is defined as an overall plan for reaching marketing objectives, including target market analysis, product positioning, and the implementation of suitable tactics. These strategies are essential for companies aiming to differentiate from competitors and attract customers effectively.

Big Data has revolutionized marketing by allowing for a detailed analysis of consumer behaviors and preferences. Today, Big Data is regarded as a crucial asset for companies striving to enhance performance and offer personalized customer experiences. With its capacity to process massive quantities of data in real-time, it provides marketers with invaluable insights to better understand their audience and tailor campaigns (Gartner, 2023).

Our research explores how Big Data reshapes personalized marketing strategies and the ethical and technical implications involved. Despite numerous studies on Big Data, few examine these three facets together comprehensively. For example, Bouillot (2017) discusses Big Data's role in optimizing customer relations but not its broad impact on marketing strategies. Abakouy and Berrada (2018) demonstrate Big Data's efficacy in personalizing email marketing but focus on a specific application rather than on overall strategic transformations.

While some studies address Big Data's ethical concerns, they often do so in isolation. Chiapello and Godefroy (2017) highlight transparency issues regarding personal data use by companies but don't examine how these concerns affect strategic transformation in marketing. Similarly, Mikal and Hurst (2022) emphasize online research ethics but focus primarily on healthcare.

Technical and organizational challenges are also discussed in the literature. Dussart (2017) notes that Big Data's effectiveness hinges on companies' ability to handle vast amounts of data, which requires strong infrastructure and specialized skills. However, these studies lack a comprehensive analysis of Big Data's strategic impact on marketing.

In combining these perspectives, it's clear that researching Big Data's transformative effect on personalized marketing is relevant and needed. This research will offer an integrated analysis of marketing strategy transformation through Big Data, considering ethical and technical implications. Structured to examine Big Data's effect on personalized marketing, ethical/legal implications, and the technical/organizational hurdles for successful integration, this approach will provide a holistic understanding and practical solutions for companies seeking to optimize their marketing strategies in the Big Data era.

Big Data and Personalized Marketing

The era of Big Data has transformed marketing strategies by enabling companies to analyze vast amounts of data to better understand and anticipate consumer needs. Personalized marketing, which relies on leveraging these data insights, allows companies to create tailored customer experiences, thereby optimizing engagement and loyalty. To understand this transformation, it is essential to define the concepts of Big Data and personalized marketing and to explore how these two elements interact to create more effective marketing strategies.

Big Data refers to the collection, analysis, and utilization of large volumes of complex and diverse data generated at high speed. These data can come from a variety of sources, such as social media, online transactions, IoT sensors, and more. As defined by Merriam-Webster (2024), data are factual information used as a basis for reasoning, discussion, or calculation, and they may consist of discrete or continuous values that convey information about quantities, qualities, facts, statistics, or interpretable sequences (Wikipedia, 2024). Big Data's primary characteristics, known as the "3 Vs," include Volume, Variety, and Velocity: Volume refers to the massive amounts of data generated every second; Variety, to the diversity of data types (structured, semi-structured, unstructured); and Velocity, to the speed at which data are generated and must be processed.

Personalized marketing aims to create tailored marketing experiences for each customer based on data collected about their past behaviors, preferences, and interactions. Kotler and Keller (2016) describe a marketing strategy as an overarching plan to achieve a company's marketing goals, involving market analysis, product and service positioning, and implementation of targeted tactics. Leveraging Big Data allows companies to segment their markets more precisely, tailoring messages and offers to each segment. This not only enhances campaign effectiveness but also improves the customer experience by making each interaction more relevant and personalized.

Big Data and personalized marketing are closely interconnected. Utilizing Big Data enables companies to gain deeper insights into customer behavior and preferences, making it possible to develop highly customized marketing strategies. For example, Coca-Cola uses Big Data to refine its operations and marketing through its Bottling Investments Group (BIG), leading to market share gains and significant revenue growth (The Coca-Cola Company, 2022). Similarly, predictive analysis in email marketing leverages Big Data to segment mailing lists, personalize messages, and improve open and conversion rates (Abakouy & Berrada, 2018). By analyzing behavioral and transactional data, companies can anticipate customer needs, delivering products or services precisely when they are most needed.

Case Study: Amazon

Amazon is a prime example of using Big Data to transform its marketing strategy and deliver a highly personalized user experience. The e-commerce platform employs sophisticated algorithms to analyze customers' purchasing and browsing data, considering various factors like items added to the cart, past purchases, and even search queries. This approach enables Amazon to create tailored product recommendations, boosting cross-selling and upselling opportunities.

According to the study "Big Data at Amazon: Business Benefits and Threats," published in the Journal of Business Research, Amazon collects data from multiple sources, including user interactions with Alexa, its intelligent voice assistant. Data collected through these voice interactions are leveraged to improve the accuracy of recommendations and personalize the user experience further. For instance, if a user frequently requests recipes via Alexa, Amazon might recommend specific kitchen utensils or ingredients.

Personalized product recommendations are one of Amazon's most familiar Big Data applications. By analyzing purchasing behaviors and search histories, Amazon anticipates customer needs and suggests relevant products, accounting for approximately 35% of the company's annual sales.

Additionally, the study "Leveraging Big Data for Business Value: Insights from Amazon," published in the International Journal of Information Management, shows that Amazon uses Big Data to optimize its logistics operations. By analyzing customer location data and purchase histories, the company can determine optimal locations for its distribution centers and streamline delivery routes, thereby reducing costs and delivery times.

In conclusion, Amazon exemplifies how Big Data can be used to personalize the user experience and enhance customer engagement. This capacity to analyze and utilize large-scale data allows Amazon to stay competitive and foster customer loyalty, underscoring Big Data's substantial impact on personalized marketing strategies.

These case studies clearly show how Big Data can transform marketing strategies to provide more relevant and personalized customer experiences. However, while many aspects of personalized marketing through Big Data have been explored, an integrated perspective is still lacking in the current literature. A comprehensive analysis will deepen our understanding of the dynamics at play and offer practical solutions for companies looking to optimize their marketing strategies in the Big Data era.

CAREER PROSPECTS AND SKILLS DEVELOPMENT

Netflix is a prime example of using Big Data to transform its marketing strategy and provide a highly personalized user experience. The streaming platform leverages sophisticated algorithms to analyze subscriber viewing data, considering various factors such as the type of content watched, time of day, device used, and even scenes viewed multiple times. This approach enables Netflix to create personalized recommendations, which account for over 75% of viewing activity on the platform.

According to the study "Netflix and Big Data: Transforming Streaming Services" published in the Journal of Data Science, Netflix collects both explicit data, such as user ratings and reviews, and implicit data, like viewing behaviors and search histories. These insights are then used to refine recommendations and enhance user engagement. For example, if a user frequently watches science fiction films, Netflix's algorithm will suggest more movies from that genre.

Furthermore, the article "Personalization Algorithms in Online Streaming Services: A Case Study of Netflix" in the International Journal of Computer Science and Information Technology reveals that Netflix also uses Big Data for original content production. Before greenlighting a new series or film, Netflix analyzes data to assess its potential success. For instance, the decision to produce House of Cards was based on analyses showing that viewers enjoyed films featuring Kevin Spacey and works directed by David Fincher.

Netflix also uses data to personalize trailers and thumbnails displayed to users. Based on viewing preferences, a user might see a thumbnail highlighting a particular actor they like, increasing the likelihood of clicking on the content.

In conclusion, Netflix perfectly illustrates how Big Data can be used to personalize the user experience and enhance customer engagement. This ability to analyze and leverage large-scale data allows Netflix to remain competitive and retain subscribers, demonstrating the significant impact of Big Data on personalized marketing strategies.

ETHICAL AND LEGAL IMPLICATIONS

The use of Big Data in personalized marketing strategies provides considerable advantages in understanding and meeting customer needs. However, it also raises significant ethical and legal concerns, primarily related to data privacy, transparency in the use of consumer information, and corporate responsibility regarding data collection and exploitation practices. Analyzing these implications is essential to ensure the ethical and legal use of Big Data in marketing.

Data Privacy

Data privacy is a major concern in the use of Big Data. Companies collect massive amounts of personal data on their customers, including sensitive information such as purchasing habits, personal preferences, and even location data. According to Chiapello and Godefroy (2017), consumers are increasingly worried about how their data is collected, stored, and used. These concerns are heightened by data breaches, which can have serious consequences for the affected individuals.

Therefore, companies must adopt stringent practices to protect data privacy. This includes using encryption technologies to secure data, implementing strict data management policies, and ensuring that only authorized personnel have access to sensitive information. Additionally, companies should be transparent about how they collect and use consumer data, clearly informing consumers of their practices and obtaining explicit consent.

Transparency and Consent

Transparency is another crucial ethical dimension in the use of Big Data. Consumers must be clearly and understandably informed about the nature of the data collected, the reasons for its collection, and how it will be used. According to the ethical guidelines of Mikal and Hurst (2022), transparency is essential for establishing a trusting relationship between businesses and consumers.

User consent is also a fundamental aspect. Companies must obtain informed consent from consumers before collecting and using their personal data. This involves providing detailed information about data collection practices and offering consumers the option to refuse the collection or use of their data for specific purposes.

Corporate Responsibility

Companies have the responsibility to ensure that their data collection and usage practices comply with ethical and legal standards. This includes adherence to regulations such as the General Data Protection Regulation (GDPR) in Europe, which imposes strict obligations on companies regarding the protection of personal data. According to Chiapello and Godefroy (2017), companies must also be prepared to be accountable for their practices and to address consumer concerns regarding privacy and data usage.

Violations of data privacy can have significant legal and financial consequences for companies. Penalties may be imposed for non-compliance with regulations, and companies may also face lawsuits from affected consumers. Therefore, it is essential for companies to establish robust internal control mechanisms to monitor and assess their data collection and usage practices.

Legal Challenges

The legal challenges related to the use of Big Data in marketing also include the need to navigate a complex and constantly evolving regulatory landscape. Data protection laws and regulations vary from country to country, which can complicate compliance for companies operating globally. For example, the GDPR in Europe imposes strict requirements regarding consent, transparency, and individuals' rights to access and delete their data.

Furthermore, the rapid evolution of Big Data technology and analytical methods poses additional challenges. Regulators must constantly adapt legal frameworks to account for new technologies and practices. This requires close collaboration between companies, regulators, and researchers to ensure that regulations remain relevant and effective in protecting consumer rights while allowing for innovation.

In conclusion, the ethical and legal implications of using Big Data in personalized marketing are vast and complex. While the benefits of Big Data are undeniable, it is crucial to ensure that data collection and usage practices comply with ethical and legal standards. Companies must adopt transparent and responsible practices to gain consumer trust and avoid negative consequences associated with data privacy violations. This analysis highlights the importance of further study to understand and address these challenges, thus justifying our research problem.

TECHNICAL AND ORGANIZATIONAL CHALLENGES

The integration of Big Data into personalized marketing strategies is not without challenges. Companies must overcome significant technical and organizational obstacles to fully leverage the benefits offered by Big Data. These challenges include managing and analyzing large quantities of data, establishing adequate infrastructures, and acquiring the necessary skills. By exploring these aspects, we will highlight the inherent complexities of integrating Big Data into marketing practices.

Data Management and Analysis

Big Data is characterized by the massive volume, variety, and velocity of the collected data. Effectively managing and analyzing this data requires advanced technological infrastructures. According to Dussart (2017), the effectiveness of Big Data largely depends on companies' ability to manage and analyze large amounts of data. Companies must adopt technologies such as Hadoop and Spark for distributed data processing, allowing them to handle significant volumes efficiently.

However, adopting these technologies poses considerable technical challenges. Companies must ensure that their systems can process data in real time and integrate various data sources. Additionally, data quality is crucial for obtaining reliable insights. Companies must therefore implement rigorous data cleaning and validation processes to eliminate errors and inconsistencies. Hernández and García (2021) emphasize the importance of integrating Big Data technologies with data warehouses to overcome some of the limitations of traditional systems.

Infrastructures and Technologies

Integrating Big Data requires robust and scalable infrastructures. Traditional data warehouses are often insufficient to meet Big Data needs. As Hernández and García (2021) point out, integrating Big Data technologies with data warehouses helps overcome some of these limitations. For example, Hadoop allows for the storage and processing of unstructured data at scale, while traditional data warehouses can be used for structured data.

Moreover, machine learning and artificial intelligence technologies play a crucial role in data analysis. These technologies enable the identification of patterns and insights that would not be detectable by traditional analytical methods. Implementing these technologies requires significant investments in hardware and software, as well as technical expertise for deployment and maintenance.

Skills Acquisition

One of the biggest organizational challenges is acquiring the necessary skills to leverage Big Data. Skills in data science, data analysis, and database management are essential for the effective use of Big Data. According to Dussart (2017), there is a talent shortage in these fields, making it difficult for companies to recruit and retain qualified professionals.

Companies must invest in training and skill development for their personnel to bridge this gap. This can include internal training programs, partnerships with academic institutions, and encouraging continuous learning. Additionally, companies may outsource certain analytical functions to specialized providers to address the lack of internal skills.

Internal Coordination and Corporate Culture

Integrating Big Data also requires effective coordination between different organizational units. Marketing, information technology (IT), and data management departments must collaborate closely to ensure a consistent and effective use of data. This collaboration can be hindered by organizational silos and cultural barriers.

To overcome these obstacles, companies must promote a data-driven corporate culture. This involves raising awareness among all employees about the importance of Big Data and fostering a collaborative approach. Business leaders must play a key role in demonstrating their commitment to data use and facilitating communication between departments.

In conclusion, the technical and organizational challenges associated with integrating Big Data into personalized marketing strategies are numerous and complex. Companies must invest in advanced technological infrastructures, develop the necessary skills, and promote a data-driven corporate culture to overcome these obstacles. By proactively addressing these challenges, companies can realize the full potential of Big Data to transform their marketing strategies and enhance the customer experience. These considerations justify the need for a thorough quantitative study to explore these aspects and validate our research problem.

CONCLUSION

The use of Big Data in personalized marketing represents a major revolution in how companies interact with their customers. Through our exploration of the various technical, ethical, and organizational aspects, it becomes clear that the potential of Big Data to transform marketing strategies is immense, but it is not without challenges.

The first part of our analysis demonstrated how Big Data enables advanced personalization of marketing strategies, providing accurate, real-time insights into consumer behaviors. Examples from Coca-Cola, Netflix, and Amazon illustrate how companies can use Big Data to enhance customer engagement and optimize performance. For instance, Coca-Cola utilizes Big Data to improve its operations and marketing strategies, leading to market share gains and significant revenue growth (The Coca-Cola Company, 2022). Similarly, Netflix analyzes viewing habits to offer personalized recommendations, increasing user engagement and subscriber loyalty. Amazon, on its part, leverages Big Data to customize product recommendations and optimize its logistics operations, contributing to an exceptional customer experience and increased sales.

The second part highlighted the ethical and legal concerns associated with the use of Big Data. Studies show that data privacy and transparency are crucial aspects that companies must manage carefully. Consumers are increasingly aware of their rights and demand greater transparency and rigorous protection of their personal data from companies. Chiapello and Godefroy (2017) emphasize the ethical challenges related to the use of algorithms in marketing, highlighting the need for greater corporate responsibility in data management. Mikal and Hurst (2022) underline the importance of ethical guidelines for online research, which is relevant for digital marketing that heavily relies on analyzing consumers' online behaviors.

Finally, the third part explored the technical and organizational challenges related to the integration of Big Data. Companies must overcome significant obstacles in data management and analysis, technological infrastructure, and skill acquisition. Dussart (2017) notes that adopting Big Data requires robust and scalable infrastructures as well as effective data management. Hernández and García (2021) highlight that integrating Big Data technologies with data warehouses helps overcome some limitations of traditional systems, but this requires significant investments in hardware, software, and skills. Companies must also promote a data-driven corporate culture to succeed in this transformation, which involves effective coordination among different organizational units and fostering a collaborative approach.

In addition to this analysis, a promising research avenue would be to explore the impact of artificial intelligence (AI) on personalized marketing. While Big Data provides the necessary data, AI can offer even more advanced analytical and decision-making capabilities. Studying how AI can be integrated with Big Data to create even more dynamic and responsive marketing strategies could open exciting new perspectives for companies. This synergy could not only enhance the effectiveness of marketing campaigns but also address ethical and technical challenges by automating data management processes in a transparent and secure manner. Furthermore, AI can help predict market trends and personalize customer interactions more accurately, thus increasing the overall effectiveness of marketing strategies.

REFERENCES

Abakouy, K., & Berrada, I. (2018). L'analyse prédictive des campagnes email marketing avec le Big Data Mining. *Actes du Colloque International VSST*, Toulouse, France.

Bouillot, P. (2017). *Gestion de la relation client: Total relationship management, Big data et Marketing mobile*. HAL Archives Ouvertes.

Cauchois, R., Taghipour, A., Kang, D., Zoghlami, N., & Abed, M. (2017). Advanced solutions for a supply chain with stochastic information. *Journal of Advanced Management Science*, 5(1), 9–13. DOI: 10.18178/joams.5.1.9-13

Chiapello, E., & Godefroy, C. (2017). Big Data, algorithmes et marketing: Rendre des comptes. *Statistique et Société*, 5(3), 17–28.

Dadsena, K. K., Mathiyazhagan, K., & Taghipour, A. (2021). Analysis of barriers for the build the resilient supply chain networks post-COVID-19. In *Managing Supply Chain Risk and Disruptions: Post COVID-19* (pp. 79–89). Springer International Publishing. DOI: 10.1007/978-3-030-72575-4_7

Dussart, C. (2017). *Le développement du Big Data signe-t-il la fin des études marketing? ProQuest Dissertations Publishing The Coca-Cola Company. (2022). Bottling Investments Group (BIG)*. Coca-Cola Investor Relations.

Glaa, B., Zoghlami, N., & Taghipour, A. (2014, May). Outsourcing and off-shoring healthcare services: A way to cut the costs and improve quality. In 2014 international conference on advanced logistics and transport (ICALT) (pp. 189-194). IEEE.

Hernández, J. E., & García, J. R. (2021). Data Warehouse with Big Data Technology for Higher Education. *Procedia Computer Science*, 181, 123–130.

Loivet, W., Taghipour, A., & Kang, D. S. (2020). The rise of green supply chain management: Between complexity and necessity. *J. Econ. Bus. Manag*, 8(1), 1–7. DOI: 10.18178/joebm.2020.8.1.604

Lu, X., & Taghipour, A. (2023). A French Guide for Businesses and Governments to Integrate the Concept of Sustainable Development into Their Supply Chains. In *Government Impact on Sustainable and Responsible Supply Chain Management* (pp. 1–12). IGI Global. DOI: 10.4018/978-1-6684-9062-4.ch001

Mahfod, J., Canel-Depitre, B., & Taghipour, A. (2019). Quality function deployment-ELECTRE in supplier evaluation. Journal of Advanced Management Science, 7(4).

Manyika, J., Chui, M., Miremadi, M., Bughin, J., George, K., Willmott, P., & Dewhurst, M. (2017). *A Future That Works: Automation, Employment, and Productivity*. McKinsey Global Institute.

Mbiatem, J., Taghipour, A., & Canel-Depitre, B. (2018, August). Supplier selection approaches for decision makers. In *Proceedings of the 8th International Conference on Information Communication and Management* (pp. 108-112).

Merimi, M., & Taghipour, A. (2021). Accelerating the digitalization of the supply chain: An empirical research about COVID-19 crisis. In *Digitalization of decentralized supply chains during global crises* (pp. 1–24). IGI Global. DOI: 10.4018/978-1-7998-6874-3.ch001

Mikal, J. P., & Hurst, S. (2022). Ethical guidance for conducting health research with online communities: A scoping review of existing guidance. *PLoS One*, 17(1), e0302924. PMID: 38758778

Nadia, E. N., Beatrice, C. D., & Atour, T. (2021). The determinants of hotel room rates in Beirut: A hedonic pricing model. International Journal of Trade. *Economics and Finance*, 12(2), 33–42.

Taghipour, A. (2014). Improving the plan of a manufacturing network with non-integrated business units. *International Journal of Applied Logistics*, 5(2), 1–11. DOI: 10.4018/ijal.2014040101

Taghipour, A. (Ed.). (2020). *Demand forecasting and order planning in supply chains and humanitarian logistics*. IGI Global.

Taghipour, A. (Ed.). (2023). *Government Impact on Sustainable and Responsible Supply Chain Management*. IGI Global. DOI: 10.4018/978-1-6684-9062-4

Taghipour, A., Abed, M., & Zoghlami, N. (2015, May). Design for remanufacturing respecting reverse logistics processes: A review. In 2015 4th International Conference on Advanced Logistics and Transport (ICALT) (pp. 299-304). IEEE. DOI: 10.1109/ICAdLT.2015.7136602

Taghipour, A., & Beneteau-Piet, C. (2020). Sustainable supply chain management performance. *International Journal of Innovation, Management and Technology*, 11(6), 165–169. DOI: 10.18178/ijimt.2020.11.6.897

Taghipour, A., Foukolaei, P. Z., Ghaedi, M., & Khazaei, M. (2023a). Sustainable Multi-Objective Models for Waste-to-Energy and Waste Separation Site Selection. *Sustainability (Basel)*, 15(22), 15764. DOI: 10.3390/su152215764

Taghipour, A., & Frayret, J. M. (2011). Supply chain coordination planning: A review. Work paper, École polytechnique de Montréal.

Taghipour, A., & Frayret, J. M. (2013a, May). An algorithm to improve operations planning in decentralized supply chains. In *2013 International Conference on Advanced Logistics and Transport* (pp. 100-103). IEEE. DOI: 10.1109/ICAdLT.2013.6568442

Taghipour, A., & Frayret, J. M. (2013b). Coordination of operations planning in supply chains: A review. *International Journal of Business Performance and Supply Chain Modelling*, 5(3), 272–307. DOI: 10.1504/IJBPSCM.2013.055729

Taghipour, A., Lu, X., & Smahi, M. (2022). Opportunities and challenges for society 5.0 in procurement. *Blockchain Technology in Supply Chain Management for Society*, 5(0), 91–100. DOI: 10.1201/9781003177432-9

Taghipour, A., & Merimi, M. (2021, March). Digital transformation of supply chains during crisis: covid-19. In 11th Annual International Conference on Industrial Engineering and Operations Management, IEOM (pp. 7493-7501). DOI: 10.46254/AN11.20211298

Taghipour, A., Padash, A., Etemadi, V., Khazaei, M., & Ebrahimi, S. (2024). Sustainable and Circular Hotels and the Water–Food–Energy Nexus: Integration of Agrivoltaics, Hydropower, Solar Cells, Water Reservoirs, and Green Roofs. *Sustainability (Basel)*, 16(5), 1985. DOI: 10.3390/su16051985

Taghipour, A., Ramezani, M., Khazaei, M., Roohparvar, V., & Hassannayebi, E. (2023b). Smart transportation behavior through the COVID-19 pandemic: A ride-hailing system in Iran. *Sustainability (Basel)*, 15(5), 4178. DOI: 10.3390/su15054178

Taghipour, A., Sohrabi, A., Ghaedi, M., & Khazaei, M. (2023c). A robust vaccine supply chain model in pandemics: Case of Covid-19 in Iran. *Computers & Industrial Engineering*, 183, 109465. DOI: 10.1016/j.cie.2023.109465

Zoghlami, N., Taghipour, A., Merlo, C., & Abed, M. (2016). Management of divergent production network using decentralised multi-level capacitated lot-sizing models. *International Journal of Shipping and Transport Logistics*, 8(5), 590–604. DOI: 10.1504/IJSTL.2016.078686

Chapter 6
The Role of Governments, Employers, and Platforms to Cope With the Skill Gaps in Industry 4.0

Mohammad Mirkazemi Mood
https://orcid.org/0000-0002-7886-9983
University of Tehran, Iran

Mohammadhosein Ghamary
https://orcid.org/0000-0002-2679-8271
ESC Clermont Business School, Iran

Hossein Rahmany Youshanlouei
https://orcid.org/0000-0002-2041-2534
EM Normandie Business School, France

ABSTRACT

Industry 4.0 transformations are exacerbating the skills gap, but they also present technological solutions to help bridge this disparity. Successfully taking advantage of Industry 4.0 to reduce the skills gap requires preparedness and the implementation of strategic initiatives by key actors. This chapter provides an overview of research that offers potential solutions for addressing the skills gap in the context of Industry 4.0. The methodology used is a systematic literature review. Fourteen solutions have been identified for addressing the skills gap among the three main stakeholders: governments, employers, and HR techs. These initiatives were categorized into 37

DOI: 10.4018/979-8-3693-7478-8.ch006

sub-themes. Based on primary and secondary themes, mappings were created to align the identified solutions from the literature, providing a roadmap for future research efforts. In the end, comprehensive recommendations were made for addressing the skills gap within the Industry 4.0 landscape, tailored to each respective stakeholder.

INTRODUCTION

The rapid development of technologies is changing the world of work. If we harness digital solutions and artificial intelligence wisely, their positive impact on our lives can significantly expand (Taghipour et al., 2022). However, the mismatch between the supply and demand of highly skilled labor and the persistent skills gap remains a pervasive global concern, drawing significant attention from policymakers, employers, and labor market researchers (Draissi et al., 2021). This issue consistently presents challenges to businesses and policymakers operating within the realms of science, technology, and innovation. Nonetheless, what amplifies this disparity is the emergence of the Fourth Industrial Revolution, commonly denoted as Industry 4.0 (Rana & Sharma, 2019). Indeed, the digitization of professions, coupled with artificial intelligence (AI) and associated disruptive technologies such as Blockchain, Robotics, and 3D printing, constitute the cornerstones of a more comprehensive and pervasive revolution compared to the preceding industrial revolutions. These elements are poised to initiate the next wave of disruptive innovations (Cukier, 2019; Pant & Taghipour, 2023). The adoption of Industry 4.0 heralds a new era for knowledge-intensive economies, where the integration of AI, the Internet of Things, robotics, cloud computing, and other technologies will revolutionize both business models and professions. These technologies offer avenues for the acquisition of new skills, resulting in increased specialization, innovation, heightened competition, and various other outcomes. The synergy between these emerging technologies and a proficient workforce leads to heightened productivity (Puhovichova & Jankelova, 2020; Agarwal et al., 2022, Youshanlouei et al., 2024).

The embrace of Industry 4.0, while ushering in substantial prospects for employment generation and the advancement of industries and businesses, simultaneously introduces a host of challenges. The utilization of its benefits depends on how effectively these challenges are addressed (Agarwal et al., 2022). According to a report from the World Economic Forum regarding the prospective landscape of employment, it is projected that automation driven by the digital economy will result in the obsolescence of 85 million jobs by the year 2025. However, while the report anticipates the creation of 97 million new job opportunities, these emerging roles may not necessarily align with the skillsets possessed by a significant portion of the workforce. Another study reveals that, for the year 2025, seven of the top 10 skills were not featured in the list of in-demand skills for the years 2020 and 2015. While a substantial convergence in required skills is observed between 2015 and 2020, with eight out of the top 10 skills remaining consistent across both periods (Li, 2022). The persistent skills gap, a long-standing concern within the labor market, is expected to be exacerbated by the advent of advanced technologies (Cukier, 2019). A skills gap occurs when a significant disparity exists between the skill levels required by employers and the skills possessed by current employees or job seekers. Skills, in this context, refer to a list of competencies necessary for the effective performance of a specific task (Chiarello et al., 2021; Islam, 2022).

Recent technological advancements have substantially altered the spectrum of necessary competencies and skills. This transformation anticipates a demand for 3.5 million individuals over the next decade to address specific production capacity shortages. However, due to a shortage of adequately trained professionals in the required competencies, fewer job opportunities will be occupied (Hernandez-de-Menendez et al., 2020). A crucial point in this context is the lack of synchronization between educational systems and the progress of technology and labor market evolutions. Emerging technologies, including AI, are anticipated to exert a pronounced influence on the future labor landscape, thereby amplifying the extant skills gap, a phenomenon exacerbated by the inherent inflexibility of educational systems to accommodate these advancements (Cukier, 2019). All these skill challenges in the labor market, which are accentuated by the advent of Industry 4.0, are associated with a reduction in the average worker productivity and a subsequent escalation in the average labor costs (McGuinness & Ortiz, 2016).

To tackle this global challenge, it is essential for the diverse array of stakeholders within the ecosystem to fulfill their respective roles effectively. The employment ecosystem in Industry 4.0 stands as a compelling example of a complex socio-technical system. These systems are dynamic networks where social and technical elements work in tandem. Defined by a shared objective, each component plays a crucial role in driving progress and innovation, ultimately contributing to a more efficient and harmonious work environment (Mood et al., 2024). Governments, employers, and employment-centric businesses emerge as the primary components in this sphere, wielding the capacity to ameliorate this gap through concerted efforts and initiatives. Technological changes and emerging trends have led governments to take action to address the impacts of these developments. The Supply Chain Act is an example of such intervention, regulating corporate responsibility for human rights within the supply chain (Taghipour, 2023). Distributed studies have been conducted to explore potential solutions through which governments can effectively confront the adverse repercussions of labor market transformations. In addressing this issue, a range of strategies has been proposed, including investment in and the establishment of educational opportunities for critical skills (Puckett et al., 2020), fostering social innovation and promoting collaboration between academic and industrial institutions (Draissi et al., 2021), structural reform of the bureaucratic framework within the educational system (Shi & Wang, 2022), introducing financial incentives for employee training within businesses (World Economic Forum [WEF], 2020), and the dissemination of information and statistics of emerging skills and jobs (Kriechel & Vetter, 2019). Additionally, certain studies concentrate on the prominent role played by the demand side of the labor market, specifically employers and corporations. Proposed strategies encompass the active participation of pioneering businesses in offering educational opportunities, apprenticeships, and upskilling for job seekers and students (WWF, 2020; Cukier, 2019), a long-term commitment to employee training (WFF, 2020), and the development of smart human resource management (Rana & Sharma, 2019; Pillai & Sivathanu, 2020; Puhovichova & Jankelova, 2020) are among the proposed strategies.

Furthermore, new organizational structures are emerging that enhance big data practices, improve data storage, and support advanced IT architecture. This transformation is driven by substantial changes in the enterprise ecosystem, emphasizing the necessity for innovative solutions (Merimi & Taghipour, 2021). Some studies have explored the role of intermediary businesses in the employment sector, such as platforms and other emerging technology-based business models, in mitigating the impacts of the skills gap (e.g., Prikshat et al., 2021; Johnson et al., 2020). The term "platforms" refers to digital intermediaries that, through preconfigured interfaces and structured protocols, acquire data from one party, process this data using proprietary algorithms, and subsequently transmit the organized data to the other party (Ajunwa & Greene, 2019). The significance of platforms and technology-based disruptive intermediary enterprises within the employment domain is rooted in their role as innovative business models emerging from the transformative capabilities of digital technologies. They play a pivotal role in reshaping the traditional supply-driven economy (Simsek et al., 2022). The rising preference for freelancing within the digital economy and platforms' ability to facilitate gig work arrangements (Kuhn et al., 2021), the pivotal role played by platforms in empowering user learning and enhancing employee skills (Van Alstyne & Schrage, 2016; Cukier, 2019), and their contribution to the creation of a skills development and job ecosystem (Ajunwa & Greene, 2019) are the most significant reasons underscoring the importance of the role of platforms and intermediary businesses in mitigating the skills gap in the labor market.

Research endeavors aimed at proposing solutions to narrow the skills gap have been disseminated across diverse scientific domains, encompassing areas such as policy and governance, labor relations, human resource management, education, and information technology. These studies, tailored to the particular characteristics of each field, have delineated specific actors and addressed limited facets of the issue. Nonetheless, considering the multifaceted nature of the skills gap challenge and the diverse roles played by various actors in its amelioration or exacerbation, there exists a demand for interdisciplinary research to identify solutions and foster coherence among them. Hence, it is essential to delve into the research background regarding the skills gap from various perspectives and provide a cohesive framework that encompasses the roles of all stakeholders and proposed solutions across different domains to effectively cope with the issue of the skills gap. Hence, the principal objective of this current research is the identification of key actors and initiatives for addressing the challenge of the skills gap in Industry 4.0. This will be achieved through a thorough review of the existing research literature and studies, while also striving to consolidate the findings from diverse domains relevant to this issue. In this study and the subsequent sections, the research methodology is introduced and elaborated upon as a means to address the research inquiries effectively. Following this, the study delves into the findings resulting from the literature review, followed by a comprehensive discussion. In conclusion, a summary of key insights and practical recommendations, grounded in the research's findings, is presented.

RESEARCH METHODOLOGY

The methodology employed in this study is a systematic literature review. For this purpose, the recommended five-stage process introduced by Denyer & Tranfield (2009) is employed. By this systematic literature review approach, the review process encompasses five distinct stages: Question formulation, Locating studies, Study selection and evaluation, Analysis and synthesis, and Reporting and utilizing the results. Following this, each of these phases will be expounded upon in the ensuing sections:

Question Formulation

Following the methodology proposed by Denyer & Tranfield (2009), this stage involved the formation of a panel comprising representatives from the primary stakeholders of the research issue. The purpose of this panel was to design review questions that align with the specific challenges identified by these stakeholders. Due to the extensive scope of the discussed domain, it was imperative to carefully curate a heterogeneous composition of stakeholders for the panel. However, simultaneous attention was given to prevent an undue dispersion of inquiries and straying from the research's core objectives, thus preserving the study's primary focus and its capacity to address the research questions effectively. The panel was comprised of individuals with diverse backgrounds and expertise, including academic authors, a policymaker in labor and employment, a manager overseeing a job recruitment platform, and a human resources manager from an industrial company utilizing AI technologies. The review questions were formulated within this panel, with careful consideration of the research issue, to ensure a focused search and the retrieval of relevant studies:

- Which actors play a role in addressing the skills gap issue in the Industry 4.0?
- For each of these stakeholders, what strategies or solutions have been recommended to mitigate the skills gap issue in the context of Industry 4.0?

Locating Studies

Following the formulation of the review questions and panel deliberations, keywords were identified based on these questions and decisions. These identified keywords were employed in search queries utilizing Boolean operators across various search engines, including Google, Google Scholar, and Semantic Scholar. Furthermore, comprehensive searches were conducted within the title and abstract sections of renowned databases, including Emerald, ScienceDirect, Springer, Taylor & Francis, and SAGE. The retrieved documents comprise scholarly articles, conference proceedings, book chapters, and reports from reputable international institutions. In light of the contemporary understanding of Industry 4.0, as per its current definition, and taking into account the specific context presented in the 2011 paper (Xu et al., 2021), the search for articles was constrained to the period from 2011 to 2024. Table 1 presents the keywords used for the search.

Table 1. displays the keywords employed in the search process

Term 1		Term 2		Term 3
"Bridging" or "Closing" or "filling" or "coping with" the skill Gap		Industry 4.0		Employment "policies" or "initiatives" or "strategies" or "plans" or "programs"
Skill mismatching		Fourth industrial revolution		Government "policies" or "initiatives" or "strategies" or "plans" or "programs"
Skill Shortage	AND	Disruptive technologies	AND	"Job" or "work" platforms
				HR Techs
"Under skilled" or "low skilled" or "less skilled" worker		Key Enabling Technologies (KETs)		"Recruitment" or "E-recruitment"
				Human resource "policies" or "initiatives" or "strategies" or "plans" or "programs"
		Digital transformation		Employer "policies" or "initiatives" or "strategies" or "plans" or "programs"

Study Selection and Evaluation

After the identification of keywords, an extensive search for articles was carried out by utilizing the specified search engines and databases. The search queries were executed by employing diverse combinations of phrases drawn from two columns, namely Term 1, Term 2, and Term 3, or by simultaneously utilizing all three columns. While this search strategy may result in the retrieval of multiple duplicate articles, Figureit assures the identification of the majority of pertinent publications. Following the search process, 340 publications were initially identified, and after eliminating duplications, the final count was reduced to 264. These searches were carried out with consideration of two inclusion criteria: the articles had to be in English, and their publication date had to be from 2011 onwards. Following the elimination of duplicate records, the remaining articles were subjected to an initial evaluation based on the inclusion and exclusion criteria established by the academic panel members. Consequently, 89 publications were chosen for further examination. To ensure the validity of the evaluation, each author randomly chose and re-assessed 20 articles. While reviewing the inclusion and exclusion criteria for publications, 17 articles were discovered through citation chaining. Subsequently, in the following step, a comprehensive evaluation of 106 selected publications was conducted. These publications were assessed for their quality by the academic panel members (the authors) using quality assessment criteria (QE). Any publication that achieved a score greater than 4.5 (Seth et al., 2020) was chosen for in-depth analysis and examination. In the end, a total of 67 publications were chosen for the final review and analysis. Figure 1 provides a visual representation of the research execution stages.

Figure 1. The Executed Process for Systematic Literature Review

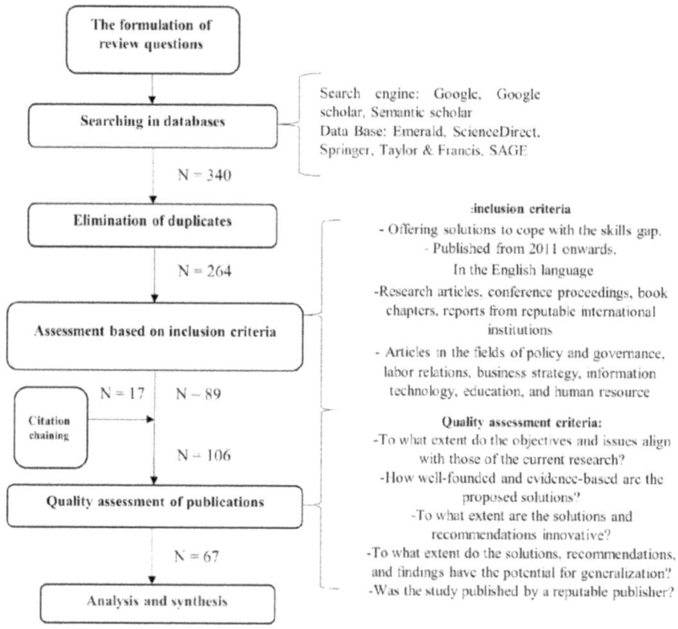

Reporting and Using the Results

After the execution of the descriptive and thematic analysis, the results are presented in this study, organized to align with the research questions and their corresponding responses. In accordance with this approach, following the exposition of findings derived from the descriptive and thematic analysis, a synthesis is presented to cohesively integrate these findings. In accordance with this approach, following the exposition of findings derived from the descriptive and thematic analysis, a map is presented to cohesively integrate these findings.

DESCRIPTIVE ANALYSIS

In this section of the paper, a descriptive analysis of the chosen publications is presented. In the sample, 83% of the selected publications comprise articles published in peer-reviewed scientific journals, 7.5% are categorized as white papers and reports, 4.5% are book chapters, and the remaining 4.5% are conference papers. During the search phase, a range of reports, white papers, and working papers were identified. Nevertheless, given the reservations about the dependability of findings from such reports (Suppatvech et al., 2019), a significant portion of these publications were omitted during the screening process. Publications are further classified based on their subject areas, most falling into interdisciplinary categories. This observation aligns with the anticipated multi-disciplinary nature of the research issue. Moreover, a noteworthy 25% of the articles are positioned within the field of Information Technology, surpassing other subject areas in terms of representation. This phenomenon could be attributed to the imperative need to offer technological solutions within the realm of Information Technology to effectively address the skills gap challenge in Industry 4.0. Within the selected publications, human resources constitute 16%, business strategy comprises 12%, labor relations comprise 8%, and education and policy account for 4%. Due to the multidimensional and comprehensive nature of articles published in the field of policy and education, a significant portion of these publications has been categorized under the multi-disciplinary classification.

In terms of methodology, 27% of the publications were conceptual. Within the methodological framework, 18% of the publications adopted survey research and an additional 18% utilized case studies. Furthermore, 16% of the articles were characterized as literature reviews, while 15% followed a design science approach and 6% employed data mining as their chosen methodology. Publications employing a design science methodology encompass studies in which a designed artifact is introduced as a solution to address the skills gap challenge.

In an alternative classification, publications have been categorized according to the actors given emphasis and their respective roles in addressing the skills gap challenge. Based on this categorization, it is notable that 30% of the publications concurrently directed their attention towards the roles of multiple stakeholders. Notably, 25% of the publications accentuated the responsibilities of employers, 24% delved into the functions of platforms and HR Techs, and 21% explored the involvement of governments. These statistics indicate that these key actors shared nearly equal prominence within the context of addressing the skills gap challenge. Considering the substantial and multifaceted role of government entities, numerous studies that have placed a heightened focus on the government's role have been categorized within the "multi-actors" group. Figures 2 and 3 in addition to table 2 succinctly illustrate a descriptive analysis of the selected articles under review.

Figure 2. The number of reviewed articles based on the year of publication

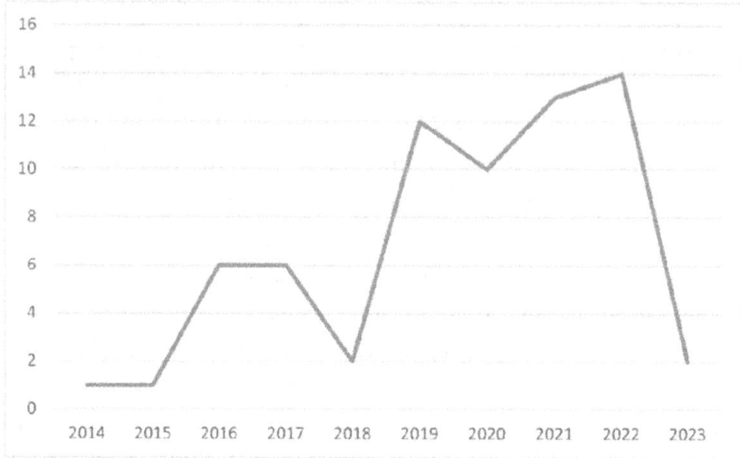

Figure 3. The number of reviewed articles based on their methodological approach

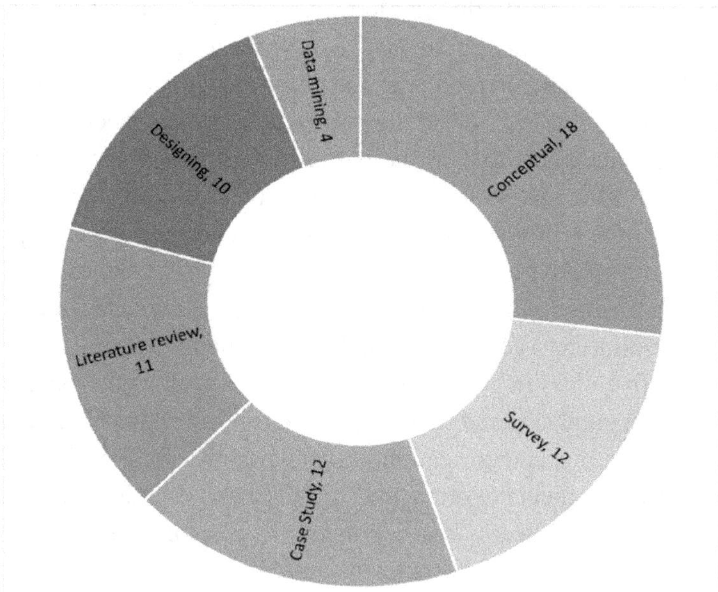

Table 2. Classification of publications based on subject areas and actors

	Governments	Employers	Platforms & HR Techs	Multi-Actors	Sum
Business Strategy	-	1	5	2	8
Human Resource	-	8	1	2	11
Education	2	-	2	-	4
Information Technology	3	2	6	6	17
Labor Relations	2	1	1	1	5
Policymaking	4	-	-	-	4
Multi-Disciplinary	3	5	1	9	18
Sum	14	17	16	20	67

THEMATIC ANALYSIS

Following data analysis and synthesis, four primary themes were identified, each encompassing multiple sub-themes. The four primary themes encompass: the governmental role, the employers' contribution, the influence of platforms and HR technologies, and collaboration among key stakeholders to cope with the skills gap. Each of these principal themes encompasses secondary-level (Level 2) and tertiary-level (Level 3) sub-themes that outline the relevant strategies with each primary actor. The findings of the thematic analysis indicate the proposal of 14 solutions, which essentially correspond to the identified sub-themes in the study, aimed at addressing the skills gap within Industry 4.0. Among the 14 proposed solutions, 4 solutions necessitate direct interaction and collaboration among the involved actors, while in the other 10 solutions, although one actor plays a primary role, their effectiveness also relies on collaboration with other stakeholders. These 14 solutions are further subdivided into more detailed sub-solutions. Figure 4 represents a mapping of the identified themes. These themes are presented across three levels. Subsequently, each of these themes is deliberated and examined.

Figure 4. Mapping of the identified primary and secondary themes from the review of selected articles

1. The role of government in coping with the skills gap

A significant segment of publications has dedicated attention to the role of the government in addressing the skills gap, particularly emphasizing the skills gap originating from Industry 4.0. These studies have identified and proposed various solutions, particularly within the realm of education, forming the sub-themes. Subsequently, attention will be directed towards these aspects:

1-1 Transformation of the Educational System

Many studies emphasize the role of government intervention in reducing the skills gap through transforming and strengthening the education system. The most significant findings in this area suggest that effecting change in the education system requires updating curricula, transforming educational structures, and emphasizing the development of soft skills within both schools and universities.

Long-term adaptation to Industry 4.0 unquestionably requires a revision of existing educational policies. This shift advocates a greater emphasis on skill mastery over the mere accumulation of qualifications. This encompasses a restructuring of the curriculum and teaching approaches to address the evolving demands of the labor market. Addressing the need for new skills in Industry 4.0 necessitates not only the enhancement of existing employees' and prospective workers' job-specific competencies but also entails updating the content within vocational education and training. The responsibility for preparing and educating the future workforce lies within educational institutions. Therefore, it is crucial to regularly review and update the curricula to integrate new developments in both science and society (Ozdemir et al., 2023; Stephany, 2022; Karakolis et al., 2022).

The educational programs within both higher education and vocational training should be crafted to equip students and learners (the future workforce) with the requisite skills, knowledge, and competencies needed for Industry 4.0. Apart from the shortcomings evident in these educational programs, research demonstrates that even in developed economies, the content and certifications of IT-focused training courses diverge significantly from the practical skills necessary in Industry 4.0. The limitations observed in governmental and official educational institutions stem from the considerable time and resources necessary to update titles and curricular content within these establishments. Furthermore, constraints in modernizing the method of providing official certifications and curricula are rooted in issues related to national regulations and educational entities. Evidence suggests that higher levels of education, particularly technical and vocational training, contribute to an individual's increased proficiency in acquiring skills (Shi & Wang, 2022; Romero-Gázquez et al., 2021; Draissi et al., 2021).

Research conducted on universities and educational institutions suggests a notable level of bureaucratization within these establishments. The culture and organizational structure of educational institutions do not evolve swiftly to accommodate the rapidly changing demands of the labor market. Decision-making in the realm of education is frequently influenced by entrenched traditions that may not readily embrace innovation. Furthermore, our present capacity to forecast long-term workforce demands is constrained, adding greater complexity to this circumstance (Cukier, 2019). As a result, policymakers must establish an appropriate legal framework aimed at increasing agility and reforming the bureaucratic structure of the educational system to address the skills gap (WEF, 2020).

However, the automation of work processes in Industry 4.0 has not only brought about the emergence of new skills but has also highlighted the importance of soft skills and other human attributes that AI is incapable of imitating or reproducing. Soft skills refer not only to communication-related abilities but also to personal qualities that enhance job performance and future employability. Today's workforce must possess not only specialized technical skills but also essential soft skills required for effective communication with their stakeholders (Cacciolatti et al., 2017).

Individuals possessing strong communication skills and high emotional intelligence play a significant role in the future economy as they serve as effective complements to technology. AI can potentially augment the relative advantage and job security of individuals exhibiting high levels of creativity, emotional intelligence, and empathy. Therefore, the workforce of the future can be classified into two distinct categories: highly skilled individuals proficient in technology-related jobs, and charismatic, innovative individuals whose appeal motivates others to collaborate with them (Cukier, 2019). Hence, in the Industry 4.0 era, the emergence of a knowledge-driven economy necessitates the cultivation of skills that promote technological innovation. This involves strengthening an educational framework that advances not just scientific and technological knowledge, but also critical thinking, decision-making, and problem-solving capabilities. In this era, employers are actively seeking individuals with critical thinking, analytical skills, planning capabilities, and strong social competencies (Abdussamad et al., 2022). Policymakers should structure university programs around the acquisition of soft skills and problem-based learning methodologies, incorporating these approaches into national educational curricula. Hence, universities must concentrate on enhancing soft skills like decision-making and problem-solving, while also emphasizing the development of analytical information skills. Furthermore, a focus on ethical conduct and responsibility towards employers and society is crucial (Islam, 2022; Cacciolatti et al., 2017).

1-2 Establishing opportunities for training in emerging skills for both current employees and job seekers

In addition to restructuring the educational system, certain researchers have suggested short-term and more instantly effective solutions within the education domain. Essentially, due to the elimination of certain jobs and the emergence of new professions and skills within Industry 4.0, some researchers argue that it is the responsibility of governments to provide equal educational opportunities at a minimal cost for everyone. Therefore, it is essential not only to reform the content and structure of the educational system but also for governments to prioritize technology training courses and essential skills relevant to Industry 4.0 using the existing potential of the education system. Consequently, offering short-term, early effective, and specialized courses while prioritizing the education of new skills, particularly Key Enabling Technologies in Industry 4.0, represents one of the most significant actions undertaken by governments in this realm.

The goal of Industry 4.0 is to establish a framework where emerging technologies interact both with each other and with employees to create an efficient and intelligent environment. However, the widespread acceptance and integration of these technologies have progressed more slowly than anticipated. The insufficient emphasis on training in technologies associated with Industry 4.0 stands as one of the main reasons contributing to this issue. The group of technologies known as Key Enabling Technologies (KETs) includes micro and nanoelectronics, nanotechnology, industrial biotechnology, advanced materials, photonics, and advanced manufacturing technologies (Romero-Gázquez et al., 2021). Hence, a preventive measure to mitigate skill mismatches within the workforce, particularly in Industry 4.0, is the education and training of individuals in such skills. The education system should ensure that immediate instruction is provided to the workforce to acquire transferable skills, facilitating their redistribution across different sectors of the economy. In this context, the collaboration between industry and academia holds significant importance in narrowing the skills gap and addressing the evolving needs of the business world (Ozdemir et al., 2023; Karakolis et al., 2022; Shi & Wang, 2022). The discussion regarding such collaborations will be addressed in the subsequent sections.

As education serves as a primary conduit linking new technologies and employment, the availability of educational opportunities directly influences how effectively present and future workers adapt to technological transformations in the workplace. Therefore, public investments aimed at establishing a substantial framework for enhancing workforce skills on a significant scale and swiftly in technologies, fostering economic empowerment, hold substantial importance. Immediate educational solutions can include government and locally subsidized learning opportunities, online educational platforms, as well as apprenticeship programs (WEF, 2020; Hughes et al., 2019). The mentioned training might also involve job searching and securing employment. Notably, the training and improvement of digital skills for both citizens and job seekers hold a crucial role in this context (Smythe et al., 2021). Due to the growing trend of online recruitment processes, Karaoglu et al. (2022) discuss how job seekers are increasingly relying on digital resources. They note that approximately 70% of individuals engage in online job searches, and leveraging the internet for job hunting significantly improves the pace of securing employment and filling vacant job positions. Enhanced digital skills can significantly impact successful job searches and contribute to more effective matching between individuals possessing specific skills and available job positions.

The idea of specialized educational programs by the government can encounter criticism. Several studies provide potential solutions to address this paradox regarding the government's role in education. The government confronts a dilemma in specialized education due to the industry's need for investment in high-level technological skills while simultaneously being required to support citizens in acquiring mid-level to lower-level skills through educational and training policies. Certainly, to address this challenge, public educational entities and universities may devise a range of programs. These should primarily concentrate on nurturing technical skills among a limited group of chosen students who might potentially transform into highly skilled innovators. Concurrently, these institutions should uphold their regular undergraduate and master's programs, which typically cater to a larger number of students (Cacciolatti et al., 2017). In recent years, universities such as the Massachusetts Institute of Technology (MIT) and Stanford have implemented a variety of programs aimed at developing graduates with the necessary skills to confront the challenges presented by Industry 4.0. Some studies (Li, 2022; Hernandez-de-Menendez et al., 2020) have focused on several of these programs. The Makerlodge program aims to educate specialists who innovatively address major societal challenges. In a similar vein, the IT Leaders for Global Operations (MITLGO) program strives to develop professionals capable of proposing strategic innovations within manufacturing enterprises. Moreover, ETH Zurich has instituted the Institute of Virtual Production to advance research in production-related domains. RWTH Aachen University also offers various academic programs in automation technology, data analytics, and decision sciences. These programs are designed to educate technical competencies in the management and analysis of vast amounts of data.

1-3 Raising Awareness and Dissemination of Accurate Information on In-Demand Jobs and Skill Outlook

The dynamic and fluid nature of jobs and their related skills in Industry 4.0 emphasizes the need for access to informative data and insights concerning demanded jobs, market information, and the necessary skill sets. Research findings suggest that governments can pursue various strategies in addressing labor market challenges. Among these strategies are the establishment or revision of job information databases, the development of job classification systems, and the implementation of national labor surveys. The concept of "labor market information" encompasses any data pertinent to understanding the scale, structure, functionality, and obstacles within the employment sphere (Kriechel & Vetter, 2019).

In the Industry 4.0 era, the ability to monitor trends and changes in the labor market is highly valuable for job seekers, employers, policymakers, and investors. Labor market participants have increasingly utilized data science and related techniques to extract fundamental patterns from large datasets for labor market analysis. As part of their critical responsibilities, governments are tasked with establishing transparent and current databases that encompass information on jobs and the labor market. For instance, a comprehensive database like O*NET in the United States provides accessible information about occupational requirements, including skills, abilities, and knowledge for the general public. The standardized data about job skills and characteristics, such as those found in databases like O*NET, serves as a crucial asset for companies, managers, and employers in obtaining essential data on the labor market's demands and necessities (Karakatsanis et al., 2017). ESCO, established by the European Commission, is a multilingual categorization of skills, competencies, qualifications, and jobs. It aims to improve the provision of information regarding skill demand in the labor market. This program aims to support individuals, employers, universities, and education providers by offering up-to-date and standardized information about skills.

The rapid technological changes, specifically in Industry 4.0, signify the need for prompt updates to databases and classification systems such as O*NET and ESCO. Recently, the European Commission has been actively exploring the use of AI to acquire insights into the development of the labor market, especially with a specialized emphasis on jobs classified within the ESCO system. JobsIreland is a public employment service available in Ireland. It offers free job advertising services to both employers and job seekers by utilizing the ESCO classification as a common language. The platforms like JobsIreland can also serve as a foundation for developing software products and online tools for skill assessments and professional career path discovery (Chiarello et al., 2021).

The establishment of these platforms to clarify labor market requirements in all nations is essential to tackle the skills gap in Industry 4.0. Furthermore, existing databases need updates and alignment with the advancements in Industry 4.0. Performing national labor surveys, which commonly adhere to the definitions and guidelines established by the International Labor Organization (ILO), can provide comprehensive insights into skill gaps and their potential remedies (Kriechel & Vetter, 2019). Regulating and merging these datasets with extensive job platform data can offer crucial insights into the state of the labor market for policymakers, employers, and researchers (Nomura et al., 2017).

1-4 Supporting Career Development Tools

Workers in the present and those in the future necessitate governmental support to adjust their professions and skills to the alterations caused by Industry 4.0. These supports can be facilitated through the development of career counseling and guidance tools. Governments have the capacity to guide both present and prospective workers by promoting and supporting the development of career and competency models. This can be achieved through the advocacy and utilization of job analysis tools and profiling in various platforms such as media, educational institutions, and career counseling centers. These strategies aim to direct individuals towards aligning their skill sets with the genuine requirements of the labor market.

Regarded as a strategic approach for recruitment, the use of career pathways informs students, parents, teachers, and educators about industries and job opportunities, providing awareness that might not have been previously recognized or could have been misunderstood. Hughes et al. (2019) emphasize the importance of two tools, career pathways, and competency models. They argue that career pathways provide students with opportunities to explore new industries and professions, allowing them to engage in real-world work experiences through internships and apprenticeships. Career pathways have the potential to effectively bridge skill gaps by identifying crucial milestones for a successful entry into the workforce. They accomplish this by linking local job demands with opportunities for advanced learning. Competency models also serve as beneficial tools in addressing skill gaps. The Career Competency Model is a descriptive tool commonly presented in visual hierarchical formats, which identifies job-related personal characteristics linked to job performance. These models enable current and prospective workers to identify and acquire the knowledge, skills, and abilities directly related to a particular job in a related industry, thus improving their proficiency (Hughes et al., 2019).

Tools such as Profiling are employed to assess an individual's skill sets, competencies, and gaps in practical experience. These tools contribute to the formulation of a comprehensive skills development plan. Additionally, gap analysis tools assist in identifying necessary measures to bridge skill gaps, monitor advancements, and create a structured training plan to understand the educational requirements. These resources exemplify essential tools for career development that could be supported by governmental initiatives to aid job seekers (Rutkauskiene & Volodzkaite, 2020).

Some specialized businesses in the field of human resources are currently developing such models and tools, but these initiatives are often undertaken in a specific and individual manner, lacking a comprehensive and overarching approach. in general, the establishment of current, authoritative resources and guidelines related to career pathways and competency frameworks in line with Industry 4.0 can profoundly assist in empowering and guiding individuals seeking employment in appropriate directions. Furthermore, governmental support for the application of career development instruments can not only guide individuals but also channel efforts and resources within educational structures and among employers. This backing holds substantial potential in bridging skill gaps and elevating the skill repertoire needed in today's professional environment.

2. The role of employers in coping with the skills gap

The second prominent theme highlighted in the chosen articles is the involvement of employers and businesses in lessening the impacts and addressing the skills gap within Industry 4.0. This main theme comprises subthemes that address potential solutions enabling employers to reduce the effects of this skills gap. Employers can adopt various approaches to address skill gaps. This study focuses on corporate strategies that have the potential for widespread application across companies, offering substantial implications in addressing skill gaps on a significant scale. These solutions, identified as subthemes, are subsequently discussed and examined in the continuation of this section.

2-1 Employee training and skill development

Despite concerted efforts to update educational programs and align them with the ever-changing demands of the labor market, technological and social advancements often are progressing so rapidly that they outpace educational systems. Therefore, numerous businesses globally are making efforts to proactively maintain the skills of their current workforce and prepare the next generation of employees with the necessary skills for the future. The findings derived from the chosen publications demonstrate that ongoing employee training and assessing the effectiveness of training are two crucial actions that need to be conducted in this domain.

Employees' reluctance to invest time and resources in upgrading or retraining themselves for the future poses a barrier for companies in closing the skills gap. However, the global prevalence of COVID-19 in recent years has imposed financial constraints on corporate budgets and mandated remote work for employees. Consequently, in certain companies, training has shifted to become a personal responsibility delegated to the employees themselves. Therefore, it is imperative for companies to establish learning opportunities for such employees, granting them access to free internet and information while also offering financial provisions such as supporting course fees for skill development (Stephany, 2022; Li, 2022).

In the Industry 4.0 landscape, companies must create an organizational milieu where employees and technology can seamlessly integrate. It's essential for these companies to prioritize the development of employee skills, especially in technical competencies, through continuous learning and shared knowledge. This strategic focus aims to enable employees to acquire the required standards and abilities to effectively engage with emerging technologies, including AI. Therefore, organizations should encourage employees to actively participate in training programs and initiatives for sharing knowledge and experiences to acquire the necessary expertise regarding new technologies (Vrontis et al., 2023; Agarwal et al., 2022). In fact, lifelong learning should be a part of an organization's strategic goals. Certainly, businesses can significantly contribute alongside technology and educational institutions to reduce the skills gap and enhance effective employment. Considerable efforts should be made to ensure these learning opportunities, such as retraining and skill upgrading, are accessible and cost-effective for employees, enabling companies to be prepared for Industry 4.0 (Li, 2022; Islam, 2022).

However, many companies view internal training as an expense rather than an investment. This viewpoint is partly due to weaknesses in training planning and needs assessment within organizations, leading to the development of courses that often fail to focus on the immediate business needs and do not synchronize with the dynamic changes within these organizations. While experiences shared by the World Economic Forum on skill gap closure projects in different nations demonstrate that the successful enhancement of skills within an organization requires a complete dedication to long-term learning. Additionally, it necessitates the organization taking responsibility for aligning its workforce development with the business strategy and corporate growth objectives (WEF, 2020).

With the financial support provided, companies reinforce the bond between an organization and its employees. Companies need to assure their employees of the advantages of acquiring new skills and embracing the learning of new technologies by offering evident and persuasive values. Collaborating with third-party training providers to deliver workplace education is an option that companies consistently pursue as a leading approach to learning new skills (Li, 2022). For more effective training outcomes, companies should prioritize the customization of employees' learning experiences. IBM's research provides a roadmap to assist managers in addressing the skills gap. It indicates that employees desire to learn skills that are uniquely aligned with their experiences, goals, and interests. Companies also seek customization. Work experiences and employee learning, designed based on both customer and market needs as well as employees' goals and interests, can assist in retaining top talents and developing the workforce of the future. Progressive companies openly display to their employees the roles and skills that are in increasing demand, enabling them to exhibit their abilities in the most critical areas. This practice assists in steering their career decisions and learning strategies (LaPrade et al., 2019).

Efficiency in training programs for addressing the skills gap cannot be assessed effectively without employing a proficient evaluation. The quality of training holds more significance than its quantity, and the emphasis on improving access to high-quality, job-relevant education is highly important. It is crucial for skill development to be an ongoing and long-term process through quality training (Nomura et al., 2017).

Evaluation, documentation, and demonstration of the positive advantages of skill training courses boost the motivation of businesses and employees to participate in Industry 4.0 skill education. Common indicators such as the number of participants, the rate of successful course completions, and learner satisfaction are commonly used for evaluating such programs. However, these indicators only measure the effectiveness of the program and the participation of individuals in these programs; they do not evaluate whether the acquired skills are applied in practice or if they have any specific business impact. To establish a stronger culture of education and learning, it is essential to first identify key performance indicators (KPIs). Assessing any educational program's effectiveness involves determining if individuals implement their acquired knowledge in practice and whether these skills positively influence their performance in their roles and within the organization (WEF, 2020). Therefore, a strong and cohesive evaluation model is essential to assess the effectiveness of training programs aimed at reducing the necessary skill gap.

2-2 Utilizing new technologies in human resources

The emergence of disruptive technologies has led to a transformation in Human Resource Management (HRM) practices within companies. The transformation, concurrent with the development of new technologies, has brought changes to the recruitment processes as the foremost element of HRM in tackling skill shortages. Reviewing chosen publications highlights that the evolution of Smart Human Resources 4.0 (SHR 4.0) and the utilization of modern methods and technologies in the recruitment process can aid employers in addressing the skill gap within Industry 4.0.

Numerous studies in the literature have stressed the necessity for companies (employers) to ready themselves for the changes inherent in Industry 4.0. The preparedness for Industry 4.0 refers to a stage at which a company fully leverages the advantages offered by Industry 4.0 technologies and effectively manages the resultant impacts across various economic and business dimensions, including labor markets and human resources. This concept encompasses changing current processes, possessing sufficient knowledge about technologies, having competent employees with appropriate skills, and their motivation to work with these technologies, all contributing to Industry 4.0 readiness (Hernandez-de-Menendez et al., 2020; Agarwal et al., 2022). Research findings indicate that a quarter of business managers lack a clear comprehension of the implications of future skill requirements brought by automation and digitalization. Furthermore, around 33% of managers believed their existing human resource infrastructure was inadequate to implement a new strategy to address the emerging skill gaps (Li, 2022). Human resource experts consider three significant global trends as disruptors in human resource management: a flexible workforce, the digitalization of business models, and AI.

Consequently, human resource processes are organizational functions requiring updates to confront the significant alterations emerging from Industry 4.0 and ensure sufficient preparedness (Minbaeva, 2020; Abdussamad et al., 2022). Management of human resources within companies is leading the effort to address the skill gap originating from Industry 4.0, and the integration of human resource automation and utilization of AI can assist in empowering companies in talent management and reducing the skill gap. The introduction of electronic human resource information systems and the analysis of big data, along with cloud computing, provide numerous opportunities to enhance processes and reduce HRM costs, including in assessing job applicants and upskilling employees (Johnson et al., 2020; Vrontis et al., 2023; Agarwal et al., 2022).

"Smart Human Resource 4.0 (SHR 4.0)" is a new concept evolving as part of the Fourth Industrial Revolution, integrating innovation in digital technologies such as the Internet of Things, big data analytics, and AI to effectively manage the next generation of employees. Smart Human Resource 4.0, based on 'people science,' presents a clear framework of the digital transformation impacting human resource functions such as recruitment, onboarding, learning, and development. Like any other change, the adoption of Smart Human Resource 4.0 by organizations will bring about a range of implementation challenges and advantages (Puhovichoval & Jankelova, 2020). The implementation of Smart Human Resources requires changes in organizational structure, leadership style, and utilized technologies. Moreover, it demands up-to-date knowledge in areas such as AI and big data, among others (Rana & Sharma, 2019).

As previously mentioned, among human resource processes, recruitment has garnered more attention in coping with the skills gap compared to other processes, according to selected studies. In terms of the utilization of emerging technologies, recruitment processes are pioneers among human resource procedures. E-recruitment is one of those initiatives that became common long before Industry 4.0. Electronic recruitment includes various digital HRM approaches like job portals, job advertisements on databases, corporate websites, the internet, as well as social network sites, and other online resources. These platforms assist in candidate screening, application management, tracking, and overseeing the recruitment system through search engines. The use of social networks can also be part of an e-recruitment model. The use of social networks such as Facebook and LinkedIn provides an opportunity to identify skilled and talented individuals to address skill gaps. Social networks utilize their networking advantages to present job opportunities and necessary skills to a wide range of job seekers. This facilitates improved access to skilled individuals and provides a very diverse space for businesses seeking to attract potentially qualified applicants. These tools have been improved by utilizing AI, enabling the identification and screening of a diverse range of skilled individuals. This is especially beneficial for large companies (Rahman et al., 2022; Johnson et al., 2020).

Thanks to technological advancements, not only have recruitment and hiring tools improved, but the methods of recruitment and hiring have also changed. The methods suggested by some studies, such as freelancing and crowdwork, can even contribute to reducing skill gaps at a larger scale. LaPrade et al. (2019) suggest that contemporary companies are facing skill challenges that cannot be adequately resolved without engaging a broader internal and external ecosystem. To stay competitive, these companies must adopt an open technological framework and collaborate with various partners to leverage the latest advancements.

Crowdworking, a specific type of crowdsourcing, represents a novel approach to structuring work. It allows companies to engage with a large number of workers, identified by varying levels of knowledge, skills, experiences, and specific backgrounds to deliver services. Crowdsourcing can be either internal or external. A crowdsourcing platform can be launched and managed by the company itself. However, there's also the option of utilizing crowdsourcing intermediaries, which essentially comprise a group that offers companies the opportunity to outsource their projects through a crowdsourcing platform (Leimeister et al., 2015). The role of these platforms in narrowing the skill gap through freelancing and crowdsourcing will be addressed in the following sections.

3. The role of platforms and HR Techs in coping with the skill gap

Another main theme identified in the review of selected studies was the role of platforms and tech-savvy businesses in the field of human resources in coping with the skill gap. Another main theme identified in the review of selected studies was the role of platforms and active tech businesses in the field of human resources in coping with the skill gap. Platforms and HR Techs, due to their facilitation in training, job matching between positions and job seekers through new technologies, and easing the development of novel employment forms such as freelancing, can offer viable solutions to cope with the skill gap in the context of Industry 4.0. In fact, Industry 4.0 technologies inherently drive the broad adoption of innovative hiring strategies by platforms and HR Techs, which can assist in balancing labor market supply and demand and reducing the skill gap (Li, 2022; Veile et al., 2022; Martins et al., 2022).

The main advantage is the large number of participants within the network or their target community, enabling each party to easily find their relevant counterpart (Ajunwa & Greene, 2019). Moreover, the study conducted by Abdul et al. (2020) demonstrates that HR Techs are widely accepted by employers due to their efficiency. Additionally, these technologies, especially in the recruitment sector, have experienced transformation through AI, blockchain, and the Internet of Things, which have the potential to improve the hiring process (Abdul et al., 2020; Sceulovs et al., 2017). In terms of education, simulations, which are referred to as AI environments, can provide a high level of interaction for job seekers and employees, enhancing learning opportunities (Vrontis et al., 2023). Currently, there is rapid growth and increased venture capital investment in areas such as text mining, advertisements and profiles, AI-based assessments and rankings, game-based assessments, robotics, linguistic analysis of candidates' written samples, and video-based interviews (Dattner et al., 2019; Upadhyay & Khandelwal, 2018). In the following, the most relevant solutions related to HR Techs and platforms for coping with the skills gap, which have received attention in selected studies and have been identified as sub-themes, will be discussed.

3-1 Skill assessment platforms and new skills training

Platforms, as a business model based on establishing interactions, are robust in creating value between producers and consumers. This platform provides an open and collaborative infrastructure for these interactions, while also establishing the governing conditions. The overall objective of platforms is to facilitate the exchange of goods, services, or social value, and even a particular type of business model, resulting in creating value for all users (Martins et al., 2022; Veile et al., 2022). Platforms have the potential to help decrease skill gaps by effectively linking diverse participants, simplifying their communication and coordination, facilitating interactions and transactions, and smoothing the path for comprehensive network effects (Veile et al., 2022; Clauss et al., 2019). Educational platforms can rapidly respond to the educational requirements for new skills that are in demand in the market, as they serve as a link between technological advancements and market demand (Simsek et al., 2022). Thus, utilizing digital platforms in education can alleviate the bureaucratic educational system's slow response to changes in market skill demands. Thus, utilizing digital platforms in education can address the bureaucratic educational system's slow response to changes in market skill demands. Such results-based and speed-centric models have motivated numerous institutions and organizations to make substantial investments in digital learning platforms. This investment aims to create cohesive and flexible ecosystems that offer high-quality content (WEF, 2020).

Some studies have referred to Massive Online Open Courses (MOOCs) as a technological educational solution and a tool to cope with the skills gap. Especially considering that MOOCs can operate on a large scale and serve as an innovative educational method for teaching and learning new skills. Although MOOC providers function as platforms and businesses, they rely on collaboration with governments and universities, establishing a mutually beneficial and win-win relationship between these platforms and academic institutions. in this relationship, MOOC providers will have access to a large database of potential purchasers of certificates and enrolled students and universities benefit from the expertise and hardware/software capabilities of platforms that assist them in delivering education at a large scale and engaging with students (Calonge & Shah, 2016).

Hence, Massive Online Open Courses (MOOCs) can be considered as online universities, providing accessible education to the general public. The objective of these interactive websites is to gather individuals who are interested in a particular subject and willing to exchange information and learn about a specific field. They are a novel and shared method for educating, developing, structuring, and sharing knowledge (Allal-Chérif et al., 2021). Studies reveal that MOOCs have a positive influence on the skill development of both graduates and employees. These courses play a pivotal role in assisting traditional graduates to rapidly enhance their skills before employment or swiftly adapt to their new roles. MOOCs have created flexible, demand-based, participatory, and timely learning opportunities, through which relevant and practical skills can be acquired (Calonge & Shah, 2016). Recommendation systems have been developed to enhance the performance of MOOCs by suggesting popular job skill training, thereby enhancing users' chances of attaining their desired careers. Some of these recommendation systems can also propose personalized career and educational needs based on a skills graph (Karakolis et al., 2022).

For better training and matching skills with job opportunities, there is a need for skill assessment. Platforms can assess and rank job seekers and their resumes based on their skills, experience, and additional credentials, guiding them toward acquiring the skills demanded by the market. (Upadhyay & Khandelwal, 2018). Platforms play a substantial role in data collection of candidates from diverse sources such as gamification and video interviews, using cutting-edge technologies like AI. They conduct an efficient and unbiased evaluation of the necessary candidates based on job descriptions (Buil et al., 2020; Oswal et al., 2021; Kim & Heo, 2022).

Cukier (2019) has introduced examples of assessment platforms. The Berke platform provides companies with the ability to assess job applicants' skills and conduct targeted evaluations to match them with current job positions. Another platform called Riipen has been embraced as a tool to cope with skill gaps. Within this platform, employers present short-term projects online, enabling students and recent graduates to showcase the skills they possess to complete these projects. Monash Talent in Australia also utilizes a series of online games to assess 50 social, emotional, and cognitive traits. Additionally, Europe's innovative educational initiative known as IN4WOOD has been introduced as a successful educational tool. It offers an easy approach for students, employees, and managers to learn, utilize, and assess skills associated with Key Enabling Technologies (KETs) in Industry 4.0. This educational tool is comprised of more than 300 instructional videos, practical applications, gamification, and assessment tests designed to evaluate the user's acquired knowledge level (Romero-Gázquez et al., 2021). Such platforms serve as a bridge between educational systems and the labor market, allowing job seekers to rapidly enhance skills and abilities that may not necessarily be fostered during formal education. Hence, these programs have the potential to help reduce the skills gap between the capabilities possessed by graduates and the skills currently required by employers (Cukier, 2019).

3-2 Job posting and matching platforms

Although the skills gap and shortage resulting from Industry 4.0 technologies are apparent, a portion of the challenges and issues faced by industries in finding skilled labor originates from the inability to identify a workforce possessing the specific skills desired by employers. Tools and technologies like talent-matching platforms can be used to alleviate the effects of transformative technologies on the labor market. Based on the identified themes, the platforms can help reduce the skill gap through two AI-augmented solutions, i.e., aggregating and posting job opportunities and matching between required skills and skill owners.

AI can be defined as the capability of machines to predict or solve problems by utilizing extensive data in complex, structured, and unstructured environments. Three components of AI—high-speed computations, utilization of big data, and advanced algorithms—enable their use in decision-making and problem-solving processes. Due to the ability of AI programs to rapidly process large volumes of data, they can effectively contribute to skill gap reduction, including the matching of skills (Oswal et al., 2021; Cukier, 2019; Upadhyay & Khandelwal, 2018; Prikshat et al., 2021). Innovations in AI suggest a wide range of new tools for identifying and assessing talents. Many of these technologies help organizations improve their ability to find the right person for the right job and to assess unsuitable candidates for improper positions faster and more cost-effectively than ever before. (Dattner et al., 2019; Hunkenschroer & Kriebitz, 2023).

The growing use of e-recruitment in companies, platforms, and HR Tech companies provides strong solutions for job posting by using AI. In addition to the traditional job postings on platforms and job portals, more job platforms are now capable of identifying relevant job seekers as soon as an employer publishes a job advertisement, informing them about suitable job postings aligned with their skills and resumes. Furthermore, by providing resume bank services, these platforms leverage AI to recommend job seekers possessing skills relevant to the employers' requirements.

Moreover, some job platforms are now capable of utilizing AI to identify, collect, and present all job posts published on the internet, as well as other websites and platforms (Hunkenschroer & Kriebitz; Rahman et al., 2022; Pillai & Sivathanu, 2020; Sceulovs et al., 2017; Enăchescu, 2016). The current advancements in AI have facilitated the extraction of job seeker and job post information without requiring active user input. These techniques intelligently extract a job seeker's qualifications and experiences directly from their resume and match this information with the qualifications and experiences required in job postings. (Hunkenschroer & Kriebitz, 2023; Oswal et al., 2021; Guo et al., 2016).

The use of these techniques allows for an appropriate understanding of the available and in-demand skills. AI, by eliminating human judgments, standardizes the screening process and increases both the efficiency and fairness of the process. This ensures that the most effective candidates are identified irrespective of gender or race (Vrontis et al., 2023; Johnson et al., 2020; Rutkauskiene & Volodzkaite, 2020). Thus, AI has evolved from a descriptive and diagnostic methodology to a prescriptive and predictive approach (Prikshat et al., 2021), a capacity that can effectively contribute to coping with the skills gap. In fact, the process of identifying the most fitting candidates in the recruitment process has been significantly impacted by advancements in AI, which have become increasingly prevalent and critical (Fritts & Cabrera, 2021). Many of the data used by matching systems are not voluntarily provided by candidates but are derived from their online presence. Semantic analysis identifies the interests, concerns, activities, and desires of individuals. Multicriteria analysis employs numerous criteria for ranking candidates. Big data analytics helps employers identify talent flows and pools within organizations and geographic regions. This tool evaluates the attractiveness of businesses for job seekers and skill holders while also assessing the appeal of job seekers for businesses (Enăchescu, 2016; Allal-Chérif et al., 2021).

Although AI-based technologies show benefits in skill identification and matching, there is uncertainty surrounding both operational and ethical concerns regarding their application within the recruitment sector. Due to the close relationship between AI-based decision-making and norms and ethical values, the use of AI in recruitment has become a topic of debate in public and academic discourse (Hunkenschroer & Kriebitz, 2023; Vrontis et al., 2023). For instance, Dattner et al. (2019) argue that there is very limited information available concerning the new generation of talent acquisition tools. Many of these tools have emerged as technological innovations rather than scientific methods or research programs. Consequently, it is not always clear what they are evaluating and whether their foundational assumptions are valid. Moreover, AI is influenced by the data it receives and learns from. Therefore, there are concerns regarding the ethical and fair nature of AI-based procedures (Karaoglu et al., 2022; Smythe et al., 2021; Kim & Heo, 2022; Nikolaou, 2021).

3-3 Gig economy-based platforms

Industry 4.0 has coincided with the pandemic and the entry of Generation Z into the labor market. Considering the significant economic and technological changes, linear careers have diminished, leading individuals not only to change jobs more frequently but also to engage more in independent or temporary work. This trend has resulted in a heightened preference for freelance work among both job seekers and employers (Minbaeva, 2020; Bellesia et al., 2019).

Today, a considerable number of the top 500 companies globally use established digital job platforms to discover freelancers, considering them as a solution for finding experts with 'the right skills for the right job at the right time' (Kuhn et al., 2021). These digital platforms, based on the gig economy, are recognized as marketplaces for exchange. These gig economy platforms accelerate competition in wages and compensation, encouraging skill owners to offer their skills at competitive prices (Clauss et al., 2019).

Using digital technologies shifts the labor market towards payment for each project, moving away from traditional permanent employment. Online freelance platforms digitally connect the workforce with individuals or businesses requiring human resources. As independent contractors or freelancers, platform workers are not committed to an employer in a long-term capacity. Rather, they are committed to short-term gigs facilitated through an application that acts neither as an agency nor an employer (Williams et al., 2021; Minbaeva, 2020). These platforms serve as intermediaries connecting skilled individuals who are not seeking full-time employment with potential employers, addressing some of the skill-related needs of these employers. Freelance platforms support two emerging forms of work, which include on-demand work and crowdwork.

On-demand work refers to a service conducted online through digital platforms but is location-based and physically delivered offline, requiring the worker to be present physically at the delivery location. Examples of On-demand work platforms include TaskRabbit, Deliveroo, and Uber. In contrast, crowdwork platforms offer tasks conducted entirely through digital platforms and are provided through transactions. Upwork and Fiverr are Examples of crowdwork platforms (Idowu & Elbanna, 2021). The concept of Crowdwork, as mentioned previously, is based on Crowdsourcing. Within this model, tasks are distributed among a large number of individuals, known as "the crowd," and members of a population function as digital labor or a collective workforce, carrying out collective tasks (Leimeister et al., 2015). Upwork, a well-known platform in this domain, primarily hosts job postings that have remained unanswered. This platform not only assists in meeting employers' skill requirements by finding expert freelancers but also enhances the skill level and availability of freelancers capable of performing these tasks or projects through investments in online educational tools (Van Alstyne & Schrage, 2016). Additionally, freelancing platforms expand the geographic distribution of job opportunities and not only contribute to improving job fairness among job seekers but also provide employers with access to the skills of job seekers worldwide, somewhat coping with regional skill gaps (Bellesia et al., 2019).

The field of crowdwork involves three main stakeholders: workers, organizations or individual employers, and intermediary platforms (online marketplaces). Due to the innovative working methods in crowdworking, both employees and employers encounter various challenges, and addressing these challenges relies on the performance of the intermediary platform (Gol et al., 2019). The intermediary platforms play a crucial role in utilizing the potential of crowdworking to cope with skill gaps. However, due to the potential for labor law violations, unfair practices, and low wages, some selected studies emphasize the necessity for regulation and increased governance of these platforms (Williams et al., 2021; Idowu & Elbanna, 2021; Gol et al., 2019; Todolí-Signes, 2017).

3-4 New technologies and innovations in recruiting

Additionally, apart from AI as a disruptive technology in the realm of recruitment, as previously discussed regarding its applications and role, innovations in the field of recruitment have been introduced by selected publications. Some of these innovations have also evolved based on AI and can contribute to coping with the skill gap. Gamification is one of these innovations.

Gamification involves digital technologies that combine serious use with an entertaining environment, designed for education, improvement, awareness, or recruitment purposes. Gamification refers to the integration of game elements into non-game contexts, such as recruitment and employee selection (Nikolaou, 2021). Serious games, another term prevalent in this domain, are simulations that replicate real environments or construct synthetic spaces. These games introduce players to AI challenges and offer complex and engaging scenarios (Buil et al., 2020; Allal-Chérif et al., 2021; Hunkenschroer & Kriebitz, 2023).

Prominent organizations, including L'Oréal, PwC, Google, the United States Army, and the UK Intelligence Service, employ gamified tools for their recruitment processes. The gamified recruitment relies on serious games to assess applicants' abilities in decision-making, improvisation, situational analysis, and proposing solutions. Gamification is a reliable and valid selection method, providing organizations with a better predictive insight into the behavior of applicants and their future job performance. Additionally, gamified recruitment makes deception and fraud more challenging. Additionally, it provides candidates with a clearer idea of the job and reduces the number of inappropriate applications (Buil et al., 2020; Nikolaou, 2021). Gamification and serious game platforms contribute to coping with skill gaps by providing better education and enhancing job seekers' skills, assisting employers in identifying and selecting the most suitable talents for specific job positions.

Smart bots are among the new technologies that can assist in reducing skill gaps by facilitating recruitment and hiring processes. A chatbot is an independent and virtual intelligent agent capable of conversing or negotiating with humans or other chatbots (Allal-Chérif et al., 2021). Chatbots are widely employed in the pre-screening of candidate interviews. Chatbots have the capability to engage in human-like conversational interactions that can record various formats such as videos, images, and documents. Chatbots can converse through text, email, and chat boxes, as they are designed to be visually appealing and empathetic (Oswal et al., 2021; Fritts & Cabrera, 2021).

These robots are designed to analyze the needs and inquiries of their real or virtual audience and offer them information, guidance, and services in response. in terms of recruitment, chatbots can reach out to potential candidates by sending messages and assessing them based on simple criteria. They can answer candidates' questions about the proposed job and the hiring company, schedule real-time interviews, and guide candidates on what the next steps in the hiring process are. Chatbots assist the employer brand of companies by enabling interaction with all candidates in the hiring process. They can engage with all applicants and effectively communicate the essential skills required for collaboration with the company. Chatbot recruitment tools offer the capability to conduct interviews with 100% of candidates 24 hours a day, 7 days a week. Consequently, they allow for the assessment of a larger quantity of candidates compared to manual methods. These chatbots can also contact recruiters and interact with them to identify the required skills of companies (Allal-Chérif et al., 2021).

Selected studies have emphasized the utilization of blockchain technology to assist in narrowing the skills gap, specifically focusing on its application within platforms. Blockchain technology has revolutionized the decentralization of business models. Blockchain technology has significant potential to cope with skill gap challenges within Industry 4.0. Previously, the use of blockchain technology in multiple industries has been discussed due to its capability to provide an intelligent, self-organizing, transparent, decentralized, and immutable system (Onik et al., 2018). Kişi (2022) argues that blockchain technology has emerged as a powerful tool for addressing these challenges and changing the recruitment process. Blockchain-based recruitment systems enhance the hiring processes by offering benefits such as increased transaction speed, improved transaction reliability, and cost reduction. Blockchain technology possesses advanced features for matching candidates' skills and performance with job requirements.

Blockchain-based HRM programs provide innovative solutions for recruitment, wage management, rewards management, occupational health and safety. In light of this technology, the implementation of programs such as automatic resume verification, the creation of smart business contracts, and the use of cryptocurrencies for international wages and salaries have made it possible to access skilled professionals and employers from all around the world. One of the most significant applications of blockchain in the recruitment process is to provide solutions for combating false or counterfeit information submitted by job seekers (Kişi, 2022; Jeong & Choi, 2019). They propose a blockchain-based system and claim that using blockchain helps reduce primary recruiting challenges such as false information, inefficient wage management, incorrect referrals, and ultimately, erroneous hiring. Assessment and validation of skill certifications is another application of blockchain in the field of recruitment. In this proposed solution, all skill certificates are archived within an employment management platform. This platform serves as a central environment for the storage and presentation of all issued certifications. All certification information is hashed and stored in the blockchain, ensuring that once stored, its contents remain unchangeable. Therefore, anyone can verify the authenticity of certificates using this blockchain (Jeong & Choi, 2019). Cloud computing technology also has features that can effectively combat the impacts of the skills gap, hence drawing attention to some studies. Cloud computing refers to the advancement, integration, and commercialization of advanced information technologies and large-scale data repositories (Wang et al., 2016).

Traditional recruitment methods mainly rely on a large volume of resumes that need to be reviewed and assessed by human resources professionals. According to Wang et al. (2016), the utilization of cloud computing enables companies to efficiently discover desired talents within a short period, consequently reducing recruitment expenses. With the use of cloud-based technologies, the modern HRM information system can rapidly update all human resources data. This is highly advantageous for companies in identifying and reaching out to individuals with the required skills. Moreover, due to the vast amount of information stored in cloud systems, employers can utilize data mining and statistical analysis to categorize candidates and develop their competency models (Wang et al., 2016). Cloud computing, in combination with the Internet of Things (IoT), has the potential to bring about numerous benefits for organizational recruitment and hiring management. The IoT can assist in sharing vast amounts of data related to job applicants and job positions. One of the most significant advantages of computing technologies within the IoT is the capacity it provides to organizations to utilize shared software resources in the cloud environment, allowing them to pay only for usage expenses. In this way, the expense of utilizing cloud-based recruitment and hiring software significantly decreases (Abdussamad et al., 2022).

Augmented Reality (AR) is suggested by certain researchers, like Ras et al. (2017), as another technology to cope with the skills gap. From their viewpoint, the utilization of augmented reality technologies for industrial training and professional development assists in coping with the skill gaps resulting from the advancements in Industry 4.0 automation. Educational technologies must be compatible with the complexities and domain parameters of Industry 4.0, and AR, through its capability of providing live guidance, training in specific skills during work, and enhancing flexibility, possesses these characteristics.

Another suggested innovation to cope with the skills gap referenced in selected studies is text mining and the use of semantic analysis. Job postings across various sectors serve as a crucial information source regarding changes in knowledge and the required skills within a specific field. This can provide valuable insights into the required skills in the labor market, aiding employers, job seekers, and policymakers in their decision-making process. Text mining utilizes algorithms such as semantic analysis to extract meaningful information from extensive volumes of textual data. Compared to manual content analysis, text mining is both less time-consuming and cost-effective (Pejic-Bach et al., 2020). These analyses are not only performed on the demand side but text mining large databases of resumes and job profiles can also provide significant information from the supply side of the skills market, assisting in measuring and decision-making regarding the existing skills gap (Shahbaz et al., 2019). Faliagka et al. (2014) introduced a novel approach using machine learning algorithms for evaluating job applicants in online recruitment systems to address issues related to candidate ranking and the implementation of semantic matching techniques. Their proposed system extracts a set of objective criteria from applicants' LinkedIn profiles and compares them semantically with job prerequisites. It also deduces their personality traits by using linguistic analysis on their blog posts. In comparison to human recruiters, these algorithms demonstrate more stability and, through the automated ranking of applicants, have the potential to accelerate and improve the efficiency of this process (Faliagka et al., 2014). They have suggested utilizing text mining techniques to analyze ESCO data regarding the emerging skill requirements resulting from Industry 4.0. This method allows for investigating and analyzing the alignment between trends in job-related databases and skill standards, like ESCO and Industry 4.0.

The utilization of academic articles as a technology data source, indicating emerging skill requirements and catching the interest of some pioneering companies, could also be employed through text mining techniques. Therefore, they can be used as resources to identify emerging skills using text-mining techniques (Chiarello et al., 2021). Karakolis et al. (2022) have also suggested a skills recommender system for selecting training courses that are valuable to the labor market. They also introduced a curriculum planning service that recommends curriculum updates based on labor market requirements. Both services utilize a text mining system that retrieves job postings from multiple online sources and employs text analysis techniques to extract skills from these postings.

4. Collaboration among key actors to cope with the skills gap

Another main theme identified from the analysis of selected articles is the collaboration and interaction among major actors to cope with the skills gap, which has been emphasized in multiple studies. Even though the main solutions or minor themes discussed in previous sections largely depend on collaboration and interaction among all actors, the solutions proposed in this section are actions that cannot be implemented without direct interaction among the players. In the following, subthemes related to this main theme are discussed.

4-1 Collaboration to address informational asymmetry and identify labor market trends

The first solution identified within this category involves narrowing the skills gap by reducing informational asymmetry between job seekers, employees, and companies. It entails collaboration among the actors to identify trends and changes within the labor market. Reducing information asymmetry to some extent is achievable through proper communication and collaboration among labor market players. Many employers have reported a noticeable gap between the competencies expected from employees and the skills possessed by graduates immediately after the completion of their studies. To close this gap, a two-way communication and information flow between employers and higher education institutions is needed. This facilitates educational systems to be informed about the competencies required by the labor market, and it enables employers to be aware of the state of the skill supply (Hughes et al., 2019). Before making changes and setting new educational curricula, further research is needed to ascertain the actual skills demanded by the labor market.

This information is crucial in guiding higher education institutions to nurture future workforces possessing such requisite skills. Hence, there is a requirement for organized collaboration between universities and industries to readjust skill goals. Such collaborations reveal the skill sets required by employers, allowing policymakers to design skill development strategies. Universities can then determine their curricula and content, while graduates can plan their careers (Cacciolatti et al., 2017). In addition to direct collaboration among actors, online resources offer valuable data to minimize information asymmetry in the labor market. Analyses of online job search engines and other recruitment services can be utilized to address information asymmetry and mismatches in the labor market (Stephany, 2022). Social media platforms including LinkedIn, Facebook, and Twitter, along with employment information websites like Glassdoor, and trend analysis sources such as Google Trends, serve as new data resources for labor market analysis and are accessible to policymakers and employers.

The rapid establishment and expansion of online job platforms significantly reduce the costs associated with accessing information for both job seekers and employers. Online global job portals such as LinkedIn, Indeed, Monster, and CareerBuilder bridge the gap between employers and job seekers in many countries. The online job portals also act as a valuable data resource for policymakers and researchers to analyze and identify the skills of the active workforce and policymaking. Furthermore, this information acts as a strong guide and motivation for job seekers, especially online freelancers, encouraging them to upskill to succeed in the labor market and acquire new abilities in line with technological changes. Local online job portals can also play a similar role in regional and local labor market policymaking. Policymakers can identify skill mismatches and gaps by analyzing this data and formulate appropriate policies based on labor supply and employer demands. The knowledge derived from analyzing this data can also be utilized to inform educational institutions and assist in curriculum adjustments to develop skills. Additionally, it can enhance job service centers to improve the connection between education and support (Stephany, 2022; Nomura et al., 2017).

4-2 Engaging Leading Companies to Enhance the Skills of Job Seekers and Students

Leading businesses, in addition to their distinctive role in identifying job trends and skill requirements in labor markets, can play a significant role in reducing the skills gap by organizing apprenticeships, workplace learning, and mentorship programs (WEF, 2020). Collaboration between higher education institutions and employers, particularly through providing integrated learning opportunities with work experience, appears to be highly effective in reducing the skills gap. Evidence suggests that individuals engaging in programs such as apprenticeships and internships have higher chances of success in the labor market (Cukier, 2019). Experiential learning should be an integral part of higher education programs. Universities should focus on enhancing students' abilities to apply theories to practical issues, and internships, especially in leading companies in the market, are a valuable step towards preparing for Industry 4.0. This approach allows students to gain a more precise understanding of their career goals by working in their intended specialization and interacting with employers and customers before graduation, consequently becoming stronger contenders for future job opportunities (Li, 2022).

The MIT LGO program is an internship-based initiative implemented by MIT in collaboration with partners such as Amazon, Caterpillar, and Boeing. This program aims to develop professionals who can suggest strategic innovations within manufacturing companies (Hernandez-de-Menendez et al., 2020). Programs like these can also be highly effective for the interest and preparedness of the future generation for Industry 4.0 careers. Children informally become familiar with various professions at early ages, but more formal exposure and hands-on experience could assist students in shaping their opinions about different career options based on more accurate information. Businesses can expose students to relevant and unbiased experiences, providing valuable information to both students and their parents about the benefits and opportunities associated with in-demand job positions. Apart from job-based internships, the schemes such as programs focusing on problem-solving through deep learning and involvement in developing innovative solutions, can play a vital role in preparing students, academics, and job seekers for Industry 4.0 careers in collaboration with the government and major corporations (Hughes et al., 2019). However, the significant role of the government lies in encouraging the involvement of leading technology businesses to embrace such collaborations and initiatives.

The lack of collaboration between governments and businesses negatively impacts the implementation of advanced skill policies in accordance with industry needs. Meanwhile, coping with the skills gap necessitates a more robust and coordinated partnership between the private and public sectors, aimed at developing more advanced skills in future employees. However, given that many companies prefer to protect their knowledge assets and business processes, providing training faces challenges and limiting it to basic or lower-level skills (Cacciolatti et al., 2017). Nevertheless, as the skills gap widens, it is crucial to elevate cooperation among businesses, educational institutions, and government. Employing a comprehensive strategy is key for individuals, governments, and businesses to collectively discover innovative, sustainable, and scalable solutions for this challenge, shifting away from individual attempts (Vrontis et al., 2023; WEF, 2020). Hence, the government must strive to obtain support from businesses through solutions such as encouraging them to fulfill their social responsibilities.

4-3 Supporting labor unions and professional communities to participate in needs assessment and continuous training

Engaging labor unions and professional communities can significantly aid in promoting the value of continuous learning and skill development programs among the workforce. This involvement can help increase awareness, acceptance, and utilization of training initiatives among employees. These entities can also use their position to encourage employers and governmental institutions to invest in educational programs aimed at addressing the challenges of technological changes and fluctuations in demand for various skills (WEF, 2020).

The collaborations and coordination at an industry level among the government, businesses, and unions ensure a more precise identification of skill needs, particularly the specific skill requirements. Mutual funds enable employers, unions, and educational institutions to generally coordinate industry-level budgets and educational requirements. Through trilateral cooperation involving the government, business representatives, and unions, supported by government funding, skill networks can assist in systematically identifying needs and organizing training courses. Overall, union advocacy for identifying companies' skill needs and promoting suitable training to meet these requirements or workplace demands at an industry level can be highly effective (McGuinness & Ortiz, 2016). Professional communities also provide professional certifications for many jobs that require skill assessment through examinations. These communities are able to recognize the necessity of enhancing the skills of their members and workforce, especially within Industry 4.0, and to keep their professional knowledge up to date. Additionally, certain professional careers require regular certification renewal to ensure those engaged in a specific profession are up-to-date with advanced technologies and current best practices (Li, 2022). Consequently, these roles play a significant part in recognizing and refreshing the knowledge of their members to reduce skill gaps in Industry 4.0.

4-4 Labor Market Regulation

Continuous regulatory updates are essential due to rapid technological progress, fostering increased awareness among both society and employees (Vrontis et al., 2023). Hence, coping with the skills gap influenced by Industry 4.0 also requires a review of employment laws and the establishment of new regulations. The legal concept of employee rights originated in the pre-internet era. The emergence of the digital age has significantly changed employment relationships, leading to substantial legal uncertainties regarding the enforcement of laws in the online space. The change in business models and their shift towards digitalization has generally revolutionized the creation and delivery of value, human resource management processes, and employment relationships across numerous organizations (Minbaeva, 2020). With the advantage of these transformations, gig economy businesses conduct their core tasks and activities entirely through classified self-employed workers via crowdwork. The new methods of organizing work made possible by technology can serve as a means to evade employment laws. In theory, these new technology companies only match the customer with the worker performing the task, a level of outsourcing that was not possible before the emergence of these new technologies.

These platform-owner companies claim that they primarily operate as a database, facilitating the connection between customers and workers and allowing the classification of workers as self-employed individuals. However, these platforms determine the working conditions for freelancers. In the absence of support from labor laws, especially regarding minimum wages and collective agreements, customers can access services at much lower costs compared to those provided by traditional organizations (Todolí-Signes, 2017; Bellesia et al., 2019). This issue can potentially challenge the balance of the labor market and pose a threat to the potential of freelancing in coping with the skills gap.

Furthermore, as previously discussed, it is often challenging for educational institutions to predict labor market demands in the coming years. Existing limitations on what should be taught could lead to inflexibility. As a result, skill mismatches that aren't resolved by market mechanisms could be targeted by regulatory policies. Public policies, including subsidy programs, tax and insurance deductions, educational taxes, grants, and crowdfunding initiatives (WEF, 2020), are examples of measures encouraging companies to invest in the training of their employees, job seekers, and apprentices (Stephany, 2022).

DISCUSSION

Findings from the descriptive and thematic analysis of selected studies confirm the assumption that the skills gap, due to its multifaceted nature, is an interdisciplinary subject. coping with this gap, especially in Industry 4.0, requires attention to the roles of various stakeholders and parties. The most significant players in this field are the government, employers, and platforms. They can assist in reducing and coping with the skills gap in Industry 4.0 through their actions, solutions, and strategies. Figure 5 shows the identified solutions, representing the same sub-themes at level 3, acting as a defensive barrier against the impacts of the skills gap resulting from Industry 4.0.

The positioning of each solution is determined by the level of involvement of the stakeholders in the specific solution. The placement of a solution within a specific domain for each stakeholder does not imply the absence of roles for other actors in that area. These positions are entirely judgmental and qualitative, established by the authors based on the knowledge acquired from selected research, and can be further explored in future studies. The central part of the figure also indicates the solutions shared by the main actors.

In light of the findings from the examination of selected studies, suggestions are offered to governments, employers, platforms, and HR tech companies to assist in alleviating the skills gap in Industry 4.0. Governments, due to their policymaking role, can have the most significant impact on reducing the skills gap in Industry 4.0. Governments, apart from the potential implementation of their own solutions and policies, also influence the feasibility and effective implementation of strategies proposed by other actors. In this way, governments can focus on the following actions to reduce the skills gap as are presented in figure 5:

Figure 5. Mapping the relationship of each actor with the identified initiatives

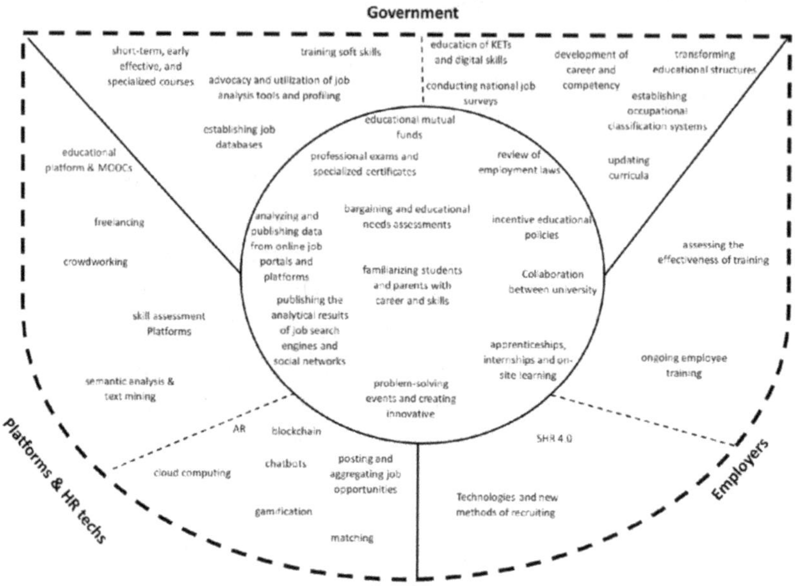

- Government and educational institution investment and collaboration in creating infrastructure, designing, developing, and validating the content of new skill-based digital education platforms for new and specialized skills;
- Supporting the development of AI-augmented digital platforms to maximize the matching of existing workforce skills with available job positions and market needs.
- Establishing regulations related to freelancing while also supporting the development of platforms to expand freelancing in response to the new generation's demands in the labor market and addressing unmet employer needs by utilizing available capacities;
- Establishing updated official resources and references on career, job competency models, and Industry 4.0-related skills by government institutions. These resources aim to guide educational system resources and efforts, along with assisting employers and job seekers in coping with the skills gap;
- Establishing comprehensive and up-to-date databases that cover job titles, job descriptions, job requirements, careers, as well as wage and salary information;
- Establishing national, local, and specialized mechanisms for skill assessment and determining skill levels in general and specifically across various sectors;

- Developing soft skills training alongside technical skills within the education system;
- Enhancing collaboration between universities, governmental institutions, and the private sector to identify labor market trends and educational needs;
- Reforming the bureaucratic structure of the education system to enhance its agility and adaptability to changes in the labor market while shifting focus towards training skills relevant to KET;
- Encouraging the participation of unions, guilds, and professional communities in needs assessment, training, and evaluating the effectiveness of skill development programs and initiatives;
- Attracting leading businesses to collaborate with universities in offering joint training courses and using the practical knowledge of industry professionals in the university;
- Attracting leading businesses to support long-term apprenticeship programs and improve the skills of job seekers through financial and legal mechanisms;
- Supporting university and leading companies' collaboration in hosting joint educational courses that utilize practical industry knowledge within the university;
- Supporting the development of intelligent human resources management in companies through training and providing technical assistance for the implementation of innovative HR technologies;
- Determine financial and tax incentives for businesses to train employees in new skills and facilitate workplace learning;

Employers and companies, as skill demanders in the labor market, also play a crucial role in reducing the skills gap. Various solutions have been suggested in the literature to help company managers minimize the negative impacts of the skills gap resulting from Industry 4.0. The majority of these solutions can be generalized and implemented across a wide range of companies, and the implementation of such strategies indirectly aids in coping with the Industry 4.0 skills gap on a broader scale and within business ecosystems. Employers and company managers can take the following actions to cope with the skills gap, leading to positive outcomes at both business and macro levels in reducing the skills gap:

- Continuous assessment of employees' skills to recognize educational needs and update their knowledge and abilities in line with the developments of Industry 4.0;
- Long-term commitment and a futurist approach to employee training;

- Providing incentives and financial aid to employees to encourage and increase their motivation to participate in training courses on key skills of Industry 4.0 and soft skills;
- Collaborating with educational platforms for personalization and utilizing the benefits and capacities of online learning;
- Assessing the practical usability of training and its impact on the performance of employees and the organization;
- Implementing Smart HR 4.0 within the organization to adapt to the advancements of Industry 4.0;
- Utilizing new recruitment methods to identify and assess talents and individuals with emerging skills;
- Collaborating with HR Tech companies in order to use the potential of new technologies in providing the required skills;
- Utilizing new work methods such as freelancing and crowdworking to use the full potential of talents and skilled individuals in the market;
- Collaborating with unions, educational institutions, and policymakers in the employment sector to provide feedback on needs and skill gaps;

Platforms and HR Tech companies can mainly assist in coping with the skills gap through facilitation. Platforms help in balancing the supply and demand of skills in the labor market and facilitate the matching process. Before Industry 4.0, intermediaries played such roles, but the emergence of platforms and HR Tech companies utilizing new technologies, particularly AI, has significantly enhanced the role of these actors in coping with the skills gap. Platforms and HR Tech companies can play an effective role in coping with the skills gap by taking the following actions:

- Developing and improving the usability of new technologies to enhance the quality of matching and identification of talents and skilled individuals;
- Assisting in reducing training costs and recruitment processes for small and medium-sized companies by providing Information Technology-based solutions;
- Increasing the customization capabilities of the educational content on learning platforms based on the needs of employers, employees, and job seekers;
- Development of tools based on new technologies for skills assessment and verification;
- Paying attention to existing ethical concerns regarding the use of AI in recruitment and hiring processes;
- Considering labor laws, fair treatment of skilled individuals, and protecting the rights of employers in gig economy-based platforms;

- Gig economy platforms collaborating with policymaking institutions to share data and help in reducing information asymmetry by publicly releasing labor market data and information;

CONCLUSION

The study aimed to identify key players and solutions to address the skills gap in Industry 4.0. It conducted a systematic literature review and identified 10 initiatives for governments, employers, and HR Tech platforms, as well as four other initiatives that require collaboration among various actors. These 14 initiatives can be further divided into 37 sub-solutions, detailing the roles of each main actor in coping with the skills gap. The study also proposed mappings to define the role of each actor, intended for researchers, policymakers, employers, and businesses in the employment, recruitment, and hiring domains. These mappings provide a clear understanding for each participant and can guide further research, particularly interdisciplinary studies. The study emphasizes the need for experimental or case studies to explore the effectiveness of these strategies in reducing the skills gap. It also highlights the multidimensional nature of the skills gap issue and the necessity for empirical studies examining the role of all actors simultaneously. Additionally, the study underlines the importance of collaboration among various actors in reducing the skills gap and suggests customization of solutions based on different countries' governance systems and economic influences. Furthermore, it emphasizes the need for more research on the role of unions and professional communities in coping with the skills gap, as well as exploring less studied solutions and contextual variables' influence on the effectiveness of the identified solutions.

The majority of these initiatives have been briefly mentioned in limited studies conceptually, and their effectiveness has been the subject of discussion and examination. However, there's a need for experimental or case studies to explore the effectiveness of these solutions in reducing the skills gap. For example, research is required to explore how the development of gig economy-based platforms and regulation in emerging labor markets can help reduce the skills gap in Industry 4.0. The findings also show that the skills gap is a multidimensional issue and requires empirical studies aimed at examining the role of all actors simultaneously in dealing with the skills gap. The study's findings also demonstrate that implementing most of the proposed solutions requires the involvement of at least two main actors. Therefore, it is essential to conduct studies on the coordination and participation mechanisms of different actors to determine each party's contribution and to implement effective initiatives for dealing with the skills gap. It appears that one of the most significant gaps in the existing literature is the insufficient attention given to collaboration among various actors in reducing the skills gap. Another issue that can be considered in future research is the customization of these solutions in different countries. These solutions will differ between developed countries and those that are developing or have emerging economies due to variations in governance systems and the private sector's influence on the economy and labor market. While the effects of Industry 4.0 are already seen in many countries, some nations are yet to face the implications of Industry 4.0 due to lagging in creating or adopting technology.

Hence, it is imperative to gain a deeper understanding of each country's or economy's stage in embracing Industry 4.0 and their readiness to address its impact. This area requires more attention in future research. There is also a lack of focus in current research on the role of unions and professional communities in addressing the skills gap. It is crucial to recognize the significance of these entities in the labor market and to devote more attention to them in future research. Additionally, there is a need to explore strategies such as studying successful national and corporate experiences in addressing the skills gap, fostering international collaborations for workforce mobility and transfer, and adapting labor laws in accordance with labor market dynamics. These solutions have received limited attention in studies and should be considered as focal points for future research, especially in developing countries. Furthermore, it would be beneficial to investigate the impact of contextual variables, such as the business ecosystem, on the effectiveness of the proposed solutions in this study for future research.

REFERENCES

Abdul, C., Wang, W., & Li, Y. (2020). The impact of technology on recruitment process. *Issues in Information Systems*, 21(4).

Abdussamad, Z., Agyei, I. T., Döngül, E. S., Abdussamad, J., Raj, R., & Effendy, F. (2022). Impact of internet of things (IOT) on human resource management: A review. *Materials Today: Proceedings*, 56, 3534–3543. DOI: 10.1016/j.matpr.2021.11.247

Agarwal, V., Mathiyazhagan, K., Malhotra, S., & Saikouk, T. (2022). Analysis of challenges in sustainable human resource management due to disruptions by Industry 4.0: An emerging economy perspective. *International Journal of Manpower*, 43(2), 513–541. DOI: 10.1108/IJM-03-2021-0192

Ajunwa, I., & Greene, D. (2019). Platforms at work: Automated hiring platforms and other new intermediaries in the organization of work. In *Work and labor in the digital age*. Emerald Publishing Limited. DOI: 10.1108/S0277-283320190000033005

Allal-Chérif, O., Aránega, A. Y., & Sánchez, R. C. (2021). Intelligent recruitment: How to identify, select, and retain talents from around the world using artificial intelligence. *Technological Forecasting and Social Change*, 169, 120822. DOI: 10.1016/j.techfore.2021.120822

Bellesia, F., Mattarelli, E., Bertolotti, F., & Sobrero, M. (2019). Platforms as entrepreneurial incubators? How online labor markets shape work identity. *Journal of Managerial Psychology*, 34(4), 246–268. DOI: 10.1108/JMP-06-2018-0269

Buil, I., Catalán, S., & Martínez, E. (2020). Understanding applicants' reactions to gamified recruitment. *Journal of Business Research*, 110, 41–50. DOI: 10.1016/j.jbusres.2019.12.041

Cacciolatti, L., Lee, S. H., & Molinero, C. M. (2017). Clashing institutional interests in skills between government and industry: An analysis of demand for technical and soft skills of graduates in the UK. *Technological Forecasting and Social Change*, 119, 139–153. DOI: 10.1016/j.techfore.2017.03.024

Chiarello, F., Fantoni, G., Hogarth, T., Giordano, V., Baltina, L., & Spada, I. (2021). Towards ESCO 4.0–Is the European classification of skills in line with Industry 4.0? A text mining approach. *Technological Forecasting and Social Change*, 173, 121177. DOI: 10.1016/j.techfore.2021.121177

Clauss, T., Harengel, P., & Hock, M. (2019). The perception of value of platform-based business models in the sharing economy: Determining the drivers of user loyalty. *Review of Managerial Science*, 13(3), 605–634. DOI: 10.1007/s11846-018-0313-0

Cukier, W. (2019). Disruptive processes and skills mismatches in the new economy: Theorizing social inclusion and innovation as solutions. *Journal of Global Responsibility*, 10(3), 211–225. DOI: 10.1108/JGR-11-2018-0079

Dattner, B., Chamorro-Premuzic, T., Buchband, R., & Schettler, L. (2019). The legal and ethical implications of using AI in hiring. *Harvard Business Review*, 25, 1–7.

Denyer, D., & Tranfield, D. (2009). Producing a systematic review. In Buchanan, D. A., & Bryman, A. (Eds.), *The Sage handbook of organizational research methods* (pp. 671–689). Sage Publications Ltd.

Draissi, Z., Zhanyong, Q., & Raguindin, P. Z. J. (2021). Knowledge mapping of skills mismatch phenomenon: A scientometric analysis. *Higher Education. Skills and Work-Based Learning*, 12(2), 271–293. DOI: 10.1108/HESWBL-02-2021-0019

Enăchescu, M. I. (2016). A prototype for an e-recruitment platform using semantic web technologies. *Journal: Informatica Economică*, 20(4), 62–75. DOI: 10.12948/issn14531305/20.4.2016.06

Faliagka, E., Iliadis, L., Karydis, I., Rigou, M., Sioutas, S., Tsakalidis, A., & Tzimas, G. (2014). On-line consistent ranking on e-recruitment: Seeking the truth behind a well-formed CV. *Artificial Intelligence Review*, 42(3), 515–528. DOI: 10.1007/s10462-013-9414-y

Fritts, M., & Cabrera, F. (2021). AI recruitment algorithms and the dehumanization problem. *Ethics and Information Technology*, 23(4), 791–801. DOI: 10.1007/s10676-021-09615-w

Gol, E. S., Stein, M. K., & Avital, M. (2019). Crowdwork platform governance toward organizational value creation. *The Journal of Strategic Information Systems*, 28(2), 175–195. DOI: 10.1016/j.jsis.2019.01.001

Guo, S., Alamudun, F., & Hammond, T. (2016). RésuMatcher: A personalized résumé-job matching system. *Expert Systems with Applications*, 60, 169–182. DOI: 10.1016/j.eswa.2016.04.013

Hernandez-de-Menendez, M., Morales-Menendez, R., Escobar, C. A., & McGovern, M. (2020). Competencies for industry 4.0. *International Journal on Interactive Design and Manufacturing*, 14(4), 1511–1524. DOI: 10.1007/s12008-020-00716-2

Hughes, C., Robert, L., Frady, K., & Arroyos, A. (2019). Middle-skill-level Employees and Technological Environments. In *Managing Technology and Middle-and Low-skilled Employees*. Emerald Publishing Limited. DOI: 10.1108/978-1-78973-077-720191002

Hunkenschroer, A. L., & Kriebitz, A. (2023). Is AI recruiting (un) ethical? A human rights perspective on the use of AI for hiring. *AI and Ethics*, 3(1), 199–213. DOI: 10.1007/s43681-022-00166-4 PMID: 35909984

Idowu, A., & Elbanna, A. (2021). Crowdworkers, social affirmation and work identity: Rethinking dominant assumptions of crowdwork. *Information and Organization*, 31(4), 100335. DOI: 10.1016/j.infoandorg.2021.100335

Islam, M. A. (2022). Industry 4.0: Skill set for employability. *Social Sciences & Humanities Open*, 6(1), 100280. DOI: 10.1016/j.ssaho.2022.100280

Jeong, W. Y., & Choi, M. (2019). Design of recruitment management platform using digital certificate on blockchain. *Journal of Information Processing Systems*, 15(3), 707–716.

Johnson, R. D., Stone, D. L., & Lukaszewski, K. M. (2020). The benefits of eHRM and AI for talent acquisition. *Journal of Tourism Futures*, 7(1), 40–52. DOI: 10.1108/JTF-02-2020-0013

Karakatsanis, I., AlKhader, W., MacCrory, F., Alibasic, A., Omar, M. A., Aung, Z., & Woon, W. L. (2017). Data mining approach to monitoring the requirements of the job market: A case study. *Information Systems*, 65, 1–6. DOI: 10.1016/j.is.2016.10.009

Karakolis, E., Kapsalis, P., Skalidakis, S., Kontzinos, C., Kokkinakos, P., Markaki, O., & Askounis, D. (2022). Bridging the gap between technological education and job market requirements through data analytics and decision support services. *Applied Sciences (Basel, Switzerland)*, 12(14), 7139. DOI: 10.3390/app12147139

Karaoglu, G., Hargittai, E., & Nguyen, M. H. (2022). Inequality in online job searching in the age of social media. *Information Communication and Society*, 25(12), 1826–1844. DOI: 10.1080/1369118X.2021.1897150

Kim, J. Y., & Heo, W. (2022). Artificial intelligence video interviewing for employment: Perspectives from applicants, companies, developer and academicians. *Information Technology & People*, 35(3), 861–878. DOI: 10.1108/ITP-04-2019-0173

Kişi, N. (2022). Exploratory research on the use of blockchain technology in recruitment. *Sustainability (Basel)*, 14(16), 10098. DOI: 10.3390/su141610098

Kriechel, B., & Vetter, T. (2019). Skills Mismatch Measurement in ETF Partner Countries. *European Training Foundation*. Retrieved in: https://www.etf.europa.eu/sites/default/files/2019-05/Skills%20mismatch%20measurement_ETF%20partner%20countries.pdf

Kuhn, K. M., Meijerink, J., & Keegan, A. (2021). Human Resource Management and the Gig Economy: Challenges and Opportunities at the Intersection between Organizational HR Decision-Makers and Digital Labor Platforms. *Research in Personnel and Human Resources Management*, 39, 1-46.

LaPrade, A., Mertens, J., Moore, T., & Wright, A. (2019). The enterprise guide to closing the skills gap: Strategies for building and maintaining a skilled workforce. *IBM Institute for Business Value*. Retrieved in: https://www.ibm.com/downloads/cas/EPYMNBJA

Leimeister, J. M., Zogaj, S., & Durward, D. (2015). New Forms of Employment and IT—Crowdsourcing. In *4th conference of the regulating for decent work network*.

Li, L. (2022). Reskilling and upskilling the future-ready workforce for industry 4.0 and beyond. *Information Systems Frontiers*, 1–16. PMID: 35855776

Martins, N., Dominique-Ferreira, S., & Lopes, C. (2022). Design and development of a digital platform for seasonal jobs: Improving the hiring process. *Journal of Global Scholars of Marketing Science*, 32(3), 452–469. DOI: 10.1080/21639159.2020.1808851

McGuinness, S., & Ortiz, L. (2016). Skill gaps in the workplace: Measurement, determinants and impacts. *Industrial Relations Journal*, 47(3), 253–278. DOI: 10.1111/irj.12136

Merimi, M., & Taghipour, A. (2021). Accelerating the digitalization of the supply chain: An empirical research about COVID-19 crisis. In *Digitalization of decentralized supply chains during global crises* (pp. 1–24). IGI Global. DOI: 10.4018/978-1-7998-6874-3.ch001

Minbaeva, D. (2021). Disrupted HR? *Human Resource Management Review*, 31(4), 100820. DOI: 10.1016/j.hrmr.2020.100820

Mood, M. M., Mohaghar, A., & Nesari, Y. (2024). Designing a method to model the socio-technical systems. *International Journal of Information and Decision Sciences*, 16(3), 284–318. DOI: 10.1504/IJIDS.2024.140185

Nikolaou, I. (2021). What is the Role of Technology in Recruitment and Selection? *The Spanish Journal of Psychology*, 24(e2), 1–6. DOI: 10.1017/SJP.2021.6 PMID: 33536110

Nomura, S., Imaizumi, S., Areias, A. C., & Yamauchi, F. (2017). Toward labor market policy 2.0: the potential for using online job-portal big data to inform labor market policies in India. *World Bank Policy Research Working Paper*, (7966). Retrieved in: https://papers.ssrn.com/sol3/papers.cfm?abstract_id=2923517

Onik, M. M. H., Miraz, M. H., & Kim, C. S. (2018, April). A recruitment and human resource management technique using blockchain technology for industry 4.0. In *Smart Cities Symposium 2018* (pp. 1-6). IET. DOI: 10.1049/cp.2018.1371

Oswal, N., Ateeq, K., & Mathew, S. (2021, April). Trends in Recruitment Information and Communication System using Artificial Intelligence in Industry 4.0. In *FEMIB* (pp. 111-118).

Ozdemir, P., Sevim, A., & Albayrak, T. (2023). Closing the gap between present and future through education: MINE-EMI project. *Case Studies on Transport Policy*, 11, 100936. DOI: 10.1016/j.cstp.2022.100936

Pant, P., & Taghipour, A. (2023). Machine learning and blockchain for 5G-enabled IIoT. In *Blockchain Applications in Cryptocurrency for Technological Evolution* (pp. 196–212). IGI Global.

Pejic-Bach, M., Bertoncel, T., Meško, M., & Krstić, Ž. (2020). Text mining of industry 4.0 job advertisements. *International Journal of Information Management*, 50, 416–431. DOI: 10.1016/j.ijinfomgt.2019.07.014

Pillai, R., & Sivathanu, B. (2020). Adoption of artificial intelligence (AI) for talent acquisition in IT/ITeS organizations. *Benchmarking*, 27(9), 2599–2629. DOI: 10.1108/BIJ-04-2020-0186

Prikshat, V., Malik, A., & Budhwar, P. (2021). AI-augmented HRM: Antecedents, assimilation and multilevel consequences. *Human Resource Management Review*, 1–18.

Puckett, J., Boutenko, V., Hoteit, L., Polunin, K., Perapechka, S., Stepanenko, A., & Bikkulova, G. (2020). Fixing the global skills mismatch. *BCG*. Retrieved in: https://www.bcg.com/publications/2020/fixing-global-skills-mismatch

Puhovichova, D., & Jankelova, N. (2020). Changes of human resource management in the context of impact of the fourth industrial revolution. *Industry 4.0, 5*(3), 138-141.

Rahman, M., Aydin, E., Haffar, M., & Nwagbara, U. (2022). The role of social media in e-recruitment process: Empirical evidence from developing countries in social network theory. *Journal of Enterprise Information Management*, 35(6), 1697–1718. DOI: 10.1108/JEIM-12-2019-0382

Rana, G., & Sharma, R. (2019). Emerging human resource management practices in Industry 4.0. *Strategic HR Review*, 18(4), 176–181. DOI: 10.1108/SHR-01-2019-0003

Ras, E., Wild, F., Stahl, C., & Baudet, A. (2017, June). Bridging the skills gap of workers in Industry 4.0 by human performance augmentation tools: Challenges and roadmap. In *Proceedings of the 10th international conference on PErvasive technologies related to assistive environments* (pp. 428-432). DOI: 10.1145/3056540.3076192

Romero-Gázquez, J. L., Cañavate-Cruzado, G., & Bueno-Delgado, M. V. (2021). IN4WOOD: A Successful European Training Action of Industry 4.0 for Academia and Business. *IEEE Transactions on Education*, 65(2), 200–209. DOI: 10.1109/TE.2021.3111696

Rutkauskiene, D., & Volodzkaite, G. (2020). Design of the Platform Solutions to Increase the Employability and E-Learning Opportunities for Low Skilled Women. In *Information and Software Technologies: 26th International Conference, ICIST 2020, Kaunas, Lithuania, October 15–17, 2020Proceedings*, 26, 334–342.

Santandreu Calonge, D., & Aman Shah, M. (2016). MOOCs, graduate skills gaps, and employability: A qualitative systematic review of the literature. *International Review of Research in Open and Distance Learning*, 17(5), 67–90.

Sceulovs, D., Shatrevich, V., & Gaile-Sarkane, E. (2017). *Evaluation of E-recruitment as a Business Model through Internet of Things Approach. Journal of Business and Economics Review* (Vol. 2). JBER.

Seth, H., Talwar, S., Bhatia, A., Saxena, A., & Dhir, A. (2020). Consumer resistance and inertia of retail investors: Development of the resistance adoption inertia continuance (RAIC) framework. *Journal of Retailing and Consumer Services*, 55, 102071. DOI: 10.1016/j.jretconser.2020.102071

Shahbaz, U., Beheshti, A., Nobari, S., Qu, Q., Paik, H. Y., & Mahdavi, M. (2019). irecruit: Towards automating the recruitment process. In *Service Research and Innovation: 7th Australian Symposium, ASSRI 2018, Sydney, NSW, Australia, September 6, 2018, and Wollongong, NSW, Australia, December 14, 2018, Revised Selected Papers 7* (pp. 139-152). Springer International Publishing.

Shi, L. P., & Wang, S. (2022). Demand-side consequences of unemployment and horizontal skill mismatches across national contexts: An employer-based factorial survey experiment. *Social Science Research*, 104, 102668. DOI: 10.1016/j.ssresearch.2021.102668 PMID: 35400384

Şimşek, T., Öner, M. A., Kunday, Ö., & Olcay, G. A. (2022). A journey towards a digital platform business model: A case study in a global tech-company. *Technological Forecasting and Social Change*, 175, 1–17. DOI: 10.1016/j.techfore.2021.121372

Smythe, S., Grotlüschen, A., & Buddeberg, K. (2021). The automated literacies of e-recruitment and online services. *Studies in the Education of Adults*, 53(1), 4–22. DOI: 10.1080/02660830.2020.1855870

Stephany, F. (2022). Closing the Digital Skill Gap: The Potential of Online Platform Data For Active Labour Markets Policies. *Zenodo*. Retrieved in: https://static.ie.edu/CGC/Stephany-%20Closing%20the%20Digital%20Skill%20Gap.pdf

Suppatvech, C., Godsell, J., & Day, S. (2019). The roles of internet of things technology in enabling servitized business models: A systematic literature review. *Industrial Marketing Management*, 82, 70–86. DOI: 10.1016/j.indmarman.2019.02.016

Taghipour, A. (Ed.). (2023). *Government Impact on Sustainable and Responsible Supply Chain Management*. IGI Global. DOI: 10.4018/978-1-6684-9062-4

Taghipour, A., Lu, X., & Smahi, M. (2022). Opportunities and challenges for society 5.0 in procurement. *Blockchain Technology in Supply Chain Management for Society*, 5(0), 91–100. DOI: 10.1201/9781003177432-9

The World Economic Forum. (2020). Closing the Skills Gap: Key Insights and Success Metrics.

Todolí-Signes, A. (2017). The 'gig economy': Employee, self-employed or the need for a special employment regulation? *Transfer: European Review of Labour and Research*, 23(2), 193–205. DOI: 10.1177/1024258917701381

Upadhyay, A. K., & Khandelwal, K. (2018). Applying artificial intelligence: Implications for recruitment. *Strategic HR Review*, 17(5), 255–258. DOI: 10.1108/SHR-07-2018-0051

Van Alstyne, M. W., & Schrage, M. (2016). The best platforms are more than matchmakers. *Harvard Business Review*, 94(7/8), 1–6.

Veile, J. W., Schmidt, M. C., & Voigt, K. I. (2022). Toward a new era of cooperation: How industrial digital platforms transform business models in Industry 4.0. *Journal of Business Research*, 143, 387–405. DOI: 10.1016/j.jbusres.2021.11.062

Vrontis, D., Christofi, M., Pereira, V., Tarba, S., Makrides, A., & Trichina, E. (2023). Artificial intelligence, robotics, advanced technologies and human resource management: a systematic review. *Artificial Intelligence and International HRM*, 172-201.

Wang, X. L., Wang, L., Bi, Z., Li, Y. Y., & Xu, Y. (2016). Cloud computing in human resource management (HRM) system for small and medium enterprises (SMEs). *International Journal of Advanced Manufacturing Technology*, 84(1-4), 485–496. DOI: 10.1007/s00170-016-8493-8

Williams, P., McDonald, P., & Mayes, R. (2023). Recruitment in the gig economy: Attraction and selection on digital platforms. In *Technologically Mediated Human Resource Management* (pp. 142–168). Routledge. DOI: 10.4324/9781003388203-6

World Economic Forum (WEF). (2020). Closing the skills gap: key insights and success metrics. *White paper*. Retrieved in: https://www3.weforum.org/docs/WEF_GSC_NES_White_Paper_2020.pdf

Xu, X., Lu, Y., Vogel-Heuser, B., & Wang, L. (2021). Industry 4.0 and Industry 5.0—Inception, conception and perception. *Journal of Manufacturing Systems*, 61, 530–535. DOI: 10.1016/j.jmsy.2021.10.006

Youshanlouei, H. R., Gol, A. E., Ščasný, M., & Shahraki, A. M. (2024). A Qualitative Study of the Intention to Use Smart Hybrid Electric Vehicles in an Oil-Rich Country. In *Applications of New Technology in Operations and Supply Chain Management* (pp. 15–37). IGI Global. DOI: 10.4018/979-8-3693-1578-1.ch002

KEY TERMS AND DEFINITIONS

Crowdworking: Crowdworking involves the completion of tasks by a large number of individuals, each contributing a small amount of work. It is a modern recruitment method that utilizes online platforms and digital tools to connect organizations or individuals with a diverse group of people to solve particular problems or offer specific services or products in exchange for payment. Digital technologies and artificial intelligence have significantly contributed to the growth and adoption of this approach to work and skill provision.

Digital Transformation: Digital transformation refers to the shift in how businesses and countries operate, driven by digital technologies like artificial intelligence, cloud computing, virtual and augmented reality, and blockchain. This transformation automates processes, changes traditional work methods, and facilitates the creation of new forms of work and platforms for employment. It plays a crucial role in both creating and addressing the skill gap.

Disruptive Technologies: Disruptive technologies, such as artificial intelligence, emerge and render conventional technologies, work methods, and business models obsolete. These technologies bring about dramatic changes in the social and economic aspects of society. While they offer undeniable advantages and benefits, they also have consequences that need to be addressed.

Gig Economy: A new form of the labor market has emerged due to digital transformation and artificial intelligence. In this type of market, individuals are not traditional employees of an organization. Instead, they offer services on a temporary and short-term basis according to their own preferences. They are compensated for their work. The defining characteristics of working in this "gig economy" include a high level of independence, flexible work scheduling, temporary relationships between workers and employers, and payment based on the amount of work completed.

HR Techs: HR Tech is a collection of systems, tools, and software used in human resources management to automate and improve various processes such as recruitment and training. These tools include chatbots, AI-based recommendation systems, e-learning software, and data analysis tools for human resources.

Industry 4.0: Industry 4.0 represents a groundbreaking leap into the future, marking the fourth industrial revolution. It follows the significant advancements of mechanical machines (Industry 1.0 in the 18th century), the widespread use of electricity (Industry 2.0 in the 19th century), and the rise of computers along with Internet technologies (Industry 3.0 in the 20th century). This revolutionary phase seamlessly blends digital and physical technologies, such as artificial intelligence, the Internet of Things, robotics, and cloud computing, to create smarter and more automated industries and businesses. Embracing Industry 4.0 fundamentally redefines the role of humans and the essential skills required in the value chain. By harnessing digitalization and intelligentization in processes, we can unlock unparalleled efficiency and innovation, driving progress in ways we never thought possible.

Skill Gap: A skill gap occurs when employers are unable to find workers with the necessary skills to meet job requirements due to the low skill level of available workers. This mismatch can happen because the skills offered may not align with the skills required by the labor market, or the types of qualifications and skills available may not match the demands of the market. For instance, there may be a disparity between the fields of study in demand in the market and those provided by the educational system. This situation is often influenced by the emergence of disruptive technologies like AI, which can impact the skill levels needed in the labor market.

Chapter 7
Importance of AI and Big Data in the World of Sports

Mojgan Bazhrang
Université du Québéc, Canada

ABSTRACT

Big data, also known as mega data, encompasses large and complex datasets that traditional data management tools cannot adequately process. Over the years, the importance of big data has expanded across various sectors, including healthcare, finance, retail, and notably, sports. A core objective of this study is to clarify why big data has become indispensable in the sports industry. To illustrate this, the authors provide real-world examples from case studies and scientific literature that reveal how big data has revolutionized sports. Through data-driven analysis and predictive capabilities, teams are now able to make more informed and strategic decisions, optimize the performance of athletes, and boost fan engagement.

INTRODUCTION

Big Data, or mega data, refers to large and complex datasets that cannot be processed with traditional data management tools. Their significance has grown over the years, extending into various sectors such as healthcare, finance, retail, and notably, sports (Cauchois et al., 2017; Dadsena et al., 2021; Glaa et al., 2014; Loivet et al., 2020). In the sports sector, Big Data usage has transformed how teams, athletes, and sports organizations make strategic and operational decisions. Big Data analysis allows the collection, processing, and interpretation of vast volumes of data from diverse sources, ranging from sports equipment (like wearables and sensors) to interactions

DOI: 10.4018/979-8-3693-7478-8.ch007

on social media. This offers unprecedented insights into athlete performance, team strategies, and fan experience, making sports more competitive and engaging. This chapter aims to explore the increasing importance of Big Data in the sports world by highlighting its various applications and impacts. We will examine how Big Data contributes to enhancing athletic performance, preventing injuries, and developing data-driven strategies. Furthermore, we will analyze the influence of Big Data on the management of sports infrastructures and fan experience, along with the challenges and limitations associated with its use. A primary objective of this chapter is to explain why Big Data has become indispensable in the sports sector. Drawing on case studies and scientific articles, we will present concrete examples of how Big Data has revolutionized sports. Indeed, through precise analyses and reliable predictions, sports teams can now make more informed decisions, optimize athlete performance, and increase fan engagement. We will also address the challenges associated with adopting Big Data in sports, notably in terms of data privacy, technology costs, and resistance to change. Finally, we will consider the future prospects of Big Data in sports, focusing on upcoming technological innovations and evolving regulations.

PART 1: UNDERSTANDING BIG DATA IN THE CONTEXT OF SPORTS

a. Definition and Characteristics of Big Data

Big Data, or mega data, refers to datasets that are so large, complex, and diverse that they exceed the capabilities of traditional data management and analysis tools. To better understand Big Data, it is essential to consider its five primary characteristics, often called the 5 V's: Volume, Velocity, Variety, Veracity, and Value (Anuradha, 2015, Lu and Taghipour, 2023; Merimi and Taghipour, 2021; Mahfod et al., 2019).

- Volume: Volume refers to the massive amount of data generated every second in the world of sports. Whether from wearable sensors on athletes, surveillance cameras in stadiums, or social media interactions, the data produced is colossal (Mbiatem et al., 2018).
- Velocity: Velocity relates to the speed at which this data is generated and must be processed (Nadia et al., 2021; Taghipour et al., 2023a; Taghipour et al., 2024). In sports, real-time analysis is often crucial for making quick decisions during a match or competition.
- Variety: Sports data comes from various sources, including biometric data, videos, performance statistics, social media comments, and more. The diver-

sity of data types requires sophisticated techniques for integration and analysis (Taghipour, 2023; Taghipour et al., 2022).

- Veracity: Veracity refers to the reliability and accuracy of the data. In sports, decisions based on inaccurate or unreliable data can lead to significant negative consequences (Taghipour and Merimi, 2021).
- Value: Finally, value represents the utility of the data for gaining actionable insights. Data must be transformed into valuable insights that can improve performance, strategies, and the overall fan experience (Taghipour and Frayret, 2011; Taghipour and Beneteau-Piet, 2020; Taghipour et al., 2023b; Taghipour et al., 2015).

b. Sources of Data in Sports

Sports data comes from multiple sources, each providing unique and complementary information. Here are some of the main sources of data in sports (Šuštaršič et al., 2022):

- Data from Sports Equipment (Wearables, Sensors): Wearables such as smartwatches, GPS trackers, and other biometric monitoring devices collect real-time data (Taghipour and Frayret, 2013a; Taghipour and Frayret, 2013b) on athletes' physical conditions, positions, speeds, and other vital metrics. This information is crucial for tracking and improving performance as well as preventing injuries.
- Social Media Data: Social media platforms like Twitter, Instagram, and Facebook generate massive amounts of data on fan interactions, sentiments, and preferences. Analyzing this data helps teams and sports brands understand fan engagement and tailor their marketing strategies.
- Athlete Performance Data: This includes game statistics, video analysis, and match reports. These data points are used to assess and improve both individual and team performance.
- Spectator and Fan Data: Ticket sales, merchandise purchases, stadium behavior, and online interactions provide valuable insights into the fan experience. This data enables sports organizations to better understand fan expectations and enhance their overall experience.

c. Advantages of Big Data in Sports

The use of Big Data in sports presents several significant advantages (Bai and Bai, 2021):

- Improvement of Athletic Performance: Thanks to data collected from wearables and sensors, coaches can closely monitor athletes' performances, identify areas needing improvement, and adapt training programs accordingly. Additionally, the analysis of biometric data helps prevent injuries by detecting signs of fatigue or overtraining.
- Optimization of Game Strategies: Performance data and video analysis enable teams to develop strategies based on concrete evidence. For example, by studying opponents' play patterns, teams can adjust their tactics to maximize their chances of success.
- Increased Fan Engagement: Data from social media and online interactions help teams and sports brands personalize their communication and marketing campaigns, thereby enhancing fan engagement. Furthermore, sentiment analysis allows for an understanding of fan opinions and enables quick responses to their concerns.
- Efficient Management of Sports Infrastructure: Data on spectator behavior and event logistics help improve the management of stadiums and sports facilities. For example, by analyzing spectator flows, organizers can optimize security measures and resource allocation.

d. Challenges of Big Data in Sports

Despite their numerous advantages, Big Data in sports also presents several challenges and limitations (Rein and Memmert, 2016).:

- Data Privacy and Security: The collection and use of sensitive data raise concerns about privacy and security. It is crucial to implement robust protective measures to prevent data breaches and ensure the trust of athletes and fans.
- Costs and Accessibility of Technologies: The technologies required to collect, store, and analyze Big Data can be expensive. Smaller teams or sports organizations may not have the financial resources necessary to invest in these technologies.
- Resistance to Change: The adoption of Big Data often requires a cultural shift and changes in practices within sports organizations. Training and educating stakeholders are essential to overcome resistance to change and maximize the benefits of Big Data.

In summary, understanding Big Data in the context of sports is fundamental to fully harnessing its potential. The characteristics of Big Data, the various sources of sports data, as well as the advantages and challenges associated with their use, form the foundation for innovative applications and transformations in the world of sports.

PART 2: APPLICATIONS DES BIG DATA DANS LE SPORT

The use of Big Data has revolutionized the world of sports, offering new and innovative possibilities to enhance performance, manage teams, engage fans, and optimize sports infrastructure. The applications of Big Data in sports manifest in various ways, all having significant impacts on different aspects of this dynamic sector (Torgler, 2023).

Firstly, athlete performance analysis is one of the most visible and crucial applications of Big Data in sports. Through advanced technologies such as sensors and wearables, coaches and sports scientists can collect detailed data on athletes' physical condition, biomechanics, and performance. This data allows for monitoring and improving performance by identifying areas that require particular attention. For example, sophisticated data models can detect patterns of fatigue or overtraining, enabling adjustments to training programs to prevent injuries. Case studies based on scientific articles show that these approaches have led to significant performance improvements and a reduction in injuries across various sports.

Next, team management and strategies greatly benefit from Big Data. Coaches and managers use data analytics to develop tactics based on concrete evidence. By analyzing past performances, opponents' play patterns, and match conditions, they can create optimized strategies for each situation. Moreover, data plays a crucial role in player recruitment and scouting. Athletes' performances are evaluated using precise metrics, allowing teams to make informed recruitment decisions. Case studies demonstrate how teams have used data analytics to identify promising talents and strategically strengthen their rosters.

Fan experience and sports marketing are also transformed by Big Data. Interactions on social media, ticket sales, and purchasing behaviors provide valuable insights into fans' preferences and expectations. Teams and sports brands can personalize the fan experience by using this data to create targeted marketing campaigns and engaging content. For example, sentiment analysis on social media allows for real-time measurement of fan opinions and adjustments to communication strategies accordingly. Case studies show that these approaches have led to increased fan engagement and stronger loyalty.

Improving sports infrastructure is another important application of Big Data. Data on spectator behavior, event logistics, and stadium management help optimize operations and enhance security. For example, by analyzing spectator flows, organizers can efficiently plan security measures and resource allocation during sporting events. Data also aids in managing facilities more sustainably and cost-effectively by optimizing energy and resource usage. Case studies illustrate how these applications have improved stadium management and spectator satisfaction.

In conclusion, the applications of Big Data in sports are vast and varied, impacting athlete performance, team management, fan experience, and infrastructure optimization. Data provides valuable insights that enable teams and sports organizations to make informed and strategic decisions. Through Big Data analysis, sports become more competitive, engaging, and efficient. Case studies and scientific articles support these applications, demonstrating the significant impact of Big Data on the world of sports. These innovations continue to shape the future of sports, paving the way for new possibilities and continuous improvements in all aspects of the sports sector.

PART 3: IMPACTS OF BIG DATA ON THE WORLD OF SPORTS

Big Data has a profound and multifaceted impact on the world of sports, transforming how athletes, teams, fans, and sports infrastructures operate and interact. These impacts are manifested through several key benefits, significant challenges, and promising future perspectives (Tan, 2023).

The advantages of Big Data in sports are numerous and diverse. One of the main benefits lies in the enhancement of performance and records. Athletes can now benefit from personalized training programs based on precise data analysis. These programs help optimize performance, identify areas for improvement, and prevent injuries by monitoring physical condition in real-time. For example, studies show that the use of biometric sensors and advanced analysis software has enabled many athletes to achieve unprecedented performance levels.

Furthermore, fan engagement has been significantly increased thanks to Big Data. Interactions on social media and digital platforms provide valuable insights into fan preferences, allowing teams and brands to personalize their communication and offerings. This personalization leads to deeper engagement and increased loyalty. For instance, sentiment analysis on social media allows teams to understand and quickly respond to fan expectations, thereby improving their overall experience.

Decision-making and management have also been revolutionized by Big Data (Taghipour and Taghipour, 2013b). Coaches and managers now have access to real-time data and predictive models that assist them in making informed strategic decisions. Whether it's player selection, game strategy development, or resource management, Big Data provides valuable insights that enhance team effectiveness and performance. Case studies demonstrate that these approaches have led to significant successes, both on the athletic and commercial fronts.

However, the use of Big Data in sports is not without challenges. One of the main obstacles is data privacy and security (Taghipour, 2020; Taghipour et al., 2023c). The collection and analysis of large amounts of sensitive data raise concerns about privacy protection. Teams and sports organizations must implement robust security measures to protect athletes' and fans' data. Incidents of data breaches can have serious ethical and legal consequences.

The costs and accessibility of technologies represent another major challenge (Taghipour, 2014). The technologies needed to collect, store, and analyze Big Data can be very expensive. This can limit access for smaller teams or sports organizations that lack the financial resources to invest in these technologies. Finding affordable and accessible solutions is crucial to democratizing the use of Big Data in sports.

Resistance to change and training are also obstacles to overcome. Adopting Big Data requires a cultural shift and changes in practices within sports organizations. Stakeholders need to be trained in new technologies and data analysis methods to maximize the benefits of Big Data. Training programs and awareness initiatives are essential to facilitate this transition and ensure successful adoption.

The future prospects of Big Data in sports are promising. Upcoming technological innovations will continue to transform the sports sector, offering new opportunities to enhance performance, engage fans, and optimize infrastructure. For example, advancements in artificial intelligence and machine learning will enable even more precise data analyses and more reliable predictions. Moreover, evolving regulations and standards regarding data protection will help strengthen athletes' and fans' trust in the use of Big Data.

In summary, the impacts of Big Data on the world of sports are significant and diverse. They offer considerable advantages in terms of performance, fan engagement, and strategic management, while also posing challenges related to privacy, costs, and resistance to change. Future prospects are promising, with technological innovations and regulatory developments continuing to shape the future of sports. Big Data is not just a passing trend, but a central element in the ongoing transformation of the sports sector (Zolghlami et al., 2016).

CONCLUSION

The revolution of Big Data has undeniably transformed the global sports landscape, offering unprecedented opportunities to enhance athletic performance, optimize team and infrastructure management, and strengthen fan engagement. Throughout this project, we have thoroughly explored the different facets of applying Big Data in sports, highlighting their benefits, the challenges they pose, and the future perspectives they offer.

We first defined Big Data by emphasizing its main characteristics: volume, velocity, variety, veracity, and value. These five Vs form the foundation upon which the various applications of Big Data in sports rest. Thanks to these characteristics, the collected data can be effectively analyzed and leveraged to generate valuable insights. The data sources in sports are diverse and include sports equipment, social media, athlete performance, and fan interactions. This diversity allows for a holistic and detailed view of the various aspects of sports.

The analysis of Big Data has led to a notable improvement in athlete performance. By collecting and analyzing biometric and performance data, coaches can develop personalized training programs that optimize athletes' capabilities while reducing the risk of injury. Case studies have demonstrated that the use of these technologies leads to sporting records and high-level performances, illustrating the direct impact of Big Data on the field.

Team management and strategies have also been profoundly influenced by Big Data. Coaches and managers use data analysis to develop evidence-based tactics, improve recruitment and scouting decisions, and optimize collective performance. Data enables a more scientific and rigorous approach to team management, resulting in increased success and better resource management.

Fan experience and sports marketing have benefited from the personalization made possible by Big Data. By analyzing social media interactions and purchasing behaviors, teams can create targeted marketing campaigns and engage fans more effectively. This personalization strengthens fan engagement and fosters supporter loyalty, contributing to the commercial growth of sports organizations.

However, the use of Big Data in sports is not without challenges. Data privacy and security are major concerns. It is crucial to implement robust measures to protect sensitive data from athletes and fans. Additionally, the costs associated with the technologies for collecting and analyzing Big Data can be prohibitive, limiting accessibility for smaller sports organizations. Resistance to change and the need for training are also significant obstacles to overcome for successful adoption of Big Data.

Despite these challenges, the future prospects of Big Data in sports are promising. Technological innovations, such as artificial intelligence and machine learning, will continue to transform the sports sector, offering even more advanced possibilities for analysis and performance optimization. The evolution of data protection regulations will also help strengthen trust and facilitate the adoption of Big Data.

In summary, Big Data plays a crucial role in the transformation of sports. They offer considerable advantages in terms of performance, strategic management, and fan engagement, while posing significant challenges that must be addressed. The future prospects are exciting, with technological innovations and regulatory developments continuing to shape the future of sports. Big Data is not just a passing trend, but a central element of the ongoing dynamics of change and continuous improvement in the sports sector. By fully harnessing the potential of Big Data, sports can reach new heights of performance, efficiency, and engagement for the benefit of athletes, teams, and fans around the world.

REFERENCES

Anuradha, J. (2015). A brief introduction on Big Data 5Vs characteristics and Hadoop technology. *Procedia Computer Science*, 48, 319–324. DOI: 10.1016/j.procs.2015.04.188

Bai, Z., & Bai, X. (2021). Sports big data: Management, analysis, applications, and challenges. *Complexity*, 2021(1), 6676297. DOI: 10.1155/2021/6676297

Cauchois, R., Taghipour, A., Kang, D., Zoghlami, N., & Abed, M. (2017). Advanced solutions for a supply chain with stochastic information. *Journal of Advanced Management Science*, 5(1), 9–13. DOI: 10.18178/joams.5.1.9-13

Dadsena, K. K., Mathiyazhagan, K., & Taghipour, A. (2021). Analysis of barriers for the build the resilient supply chain networks post-COVID-19. In *Managing Supply Chain Risk and Disruptions: Post COVID-19* (pp. 79–89). Springer International Publishing. DOI: 10.1007/978-3-030-72575-4_7

Glaa, B., Zoghlami, N., & Taghipour, A. (2014, May). Outsourcing and off-shoring healthcare services: A way to cut the costs and improve quality. In 2014 international conference on advanced logistics and transport (ICALT) (pp. 189-194). IEEE.

Loivet, W., Taghipour, A., & Kang, D. S. (2020). The rise of green supply chain management: Between complexity and necessity. *J. Econ. Bus. Manag*, 8(1), 1–7. DOI: 10.18178/joebm.2020.8.1.604

Lu, X., & Taghipour, A. (2023). A French Guide for Businesses and Governments to Integrate the Concept of Sustainable Development Into Their Supply Chains. In *Government Impact on Sustainable and Responsible Supply Chain Management* (pp. 1–12). IGI Global. DOI: 10.4018/978-1-6684-9062-4.ch001

Mahfod, J., Canel-Depitre, B., & Taghipour, A. (2019). Quality function deployment-ELECTRE in supplier evaluation. Journal of advanced management science, 7(4).

Mbiatem, J., Taghipour, A., & Canel-Depitre, B. (2018, August). Supplier selection approaches for decision makers. In *Proceedings of the 8th International Conference on Information Communication and Management* (pp. 108-112).

Merimi, M., & Taghipour, A. (2021). Accelerating the digitalization of the supply chain: An empirical research about COVID-19 crisis. In *Digitalization of decentralized supply chains during global crises* (pp. 1–24). IGI Global. DOI: 10.4018/978-1-7998-6874-3.ch001

Nadia, E. N., Beatrice, C. D., & Atour, T. (2021). The determinants of hotel room rates in Beirut: A hedonic pricing model. International Journal of Trade. *Economics and Finance*, 12(2), 33–42.

Rein, R., & Memmert, D. (2016). Big data and tactical analysis in elite soccer: Future challenges and opportunities for sports science. *SpringerPlus*, 5(1), 1–13. DOI: 10.1186/s40064-016-3108-2 PMID: 27610328

Šuštaršič, A., Videmšek, M., Karpljuk, D., Miloloža, I., & Meško, M. (2022). Big data in sports: A bibliometric and topic study. Business Systems Research. *International Journal of the Society for Advancing Innovation and Research in Economy*, 13(1), 19–34.

Taghipour, A. (2014). Improving the plan of a manufacturing network with non-integrated business units. *International Journal of Applied Logistics*, 5(2), 1–11. DOI: 10.4018/ijal.2014040101

Taghipour, A. (Ed.). (2020). *Demand forecasting and order planning in supply chains and humanitarian logistics*. IGI Global.

Taghipour, A. (Ed.). (2023). *Government Impact on Sustainable and Responsible Supply Chain Management*. IGI Global. DOI: 10.4018/978-1-6684-9062-4

Taghipour, A., Abed, M., & Zoghlami, N. (2015, May). Design for remanufacturing respecting reverse logistics processes: A review. In 2015 4th International Conference on Advanced Logistics and Transport (ICALT) (pp. 299-304). IEEE. DOI: 10.1109/ICAdLT.2015.7136602

Taghipour, A., & Beneteau-Piet, C. (2020). Sustainable supply chain management performance. *International Journal of Innovation, Management and Technology*, 11(6), 165–169. DOI: 10.18178/ijimt.2020.11.6.897

Taghipour, A., Foukolaei, P. Z., Ghaedi, M., & Khazaei, M. (2023a). Sustainable Multi-Objective Models for Waste-to-Energy and Waste Separation Site Selection. *Sustainability (Basel)*, 15(22), 15764. DOI: 10.3390/su152215764

Taghipour, A., & Frayret, J. M. (2011). Supply chain coordination planning: A review. Work paper, École polytechnique de Montréal.

Taghipour, A., & Frayret, J. M. (2013a, May). An algorithm to improve operations planning in decentralized supply chains. In *2013 International Conference on Advanced Logistics and Transport* (pp. 100-103). IEEE. DOI: 10.1109/ICAdLT.2013.6568442

Taghipour, A., & Frayret, J. M. (2013b). Coordination of operations planning in supply chains: A review. *International Journal of Business Performance and Supply Chain Modelling*, 5(3), 272–307. DOI: 10.1504/IJBPSCM.2013.055729

Taghipour, A., Lu, X., & Smahi, M. (2022). Opportunities and challenges for society 5.0 in procurement. *Blockchain Technology in Supply Chain Management for Society*, 5(0), 91–100. DOI: 10.1201/9781003177432-9

Taghipour, A., & Merimi, M. (2021, March). Digital transformation of supply chains during crisis: covid-19. In 11th Annual International Conference on Industrial Engineering and Operations Management, IEOM (pp. 7493-7501). DOI: 10.46254/AN11.20211298

Taghipour, A., Padash, A., Etemadi, V., Khazaei, M., & Ebrahimi, S. (2024). Sustainable and Circular Hotels and the Water–Food–Energy Nexus: Integration of Agrivoltaics, Hydropower, Solar Cells, Water Reservoirs, and Green Roofs. *Sustainability (Basel)*, 16(5), 1985. DOI: 10.3390/su16051985

Taghipour, A., Ramezani, M., Khazaei, M., Roohparvar, V., & Hassannayebi, E. (2023b). Smart transportation behavior through the COVID-19 pandemic: A ride-hailing system in Iran. *Sustainability (Basel)*, 15(5), 4178. DOI: 10.3390/su15054178

Taghipour, A., Sohrabi, A., Ghaedi, M., & Khazaei, M. (2023c). A robust vaccine supply chain model in pandemics: Case of Covid-19 in Iran. *Computers & Industrial Engineering*, 183, 109465. DOI: 10.1016/j.cie.2023.109465

Tan, X. (2023). Enhanced sports predictions: A comprehensive analysis of the role and performance of predictive analytics in the sports sector. *Wireless Personal Communications*, 132(3), 1613–1636. DOI: 10.1007/s11277-023-10585-z

Torgler, B. (2023). *Big data, artificial intelligence, and quantum computing in sports. In 21st century sports: How technologies will change sports in the digital age*. Springer International Publishing.

Zoghlami, N., Taghipour, A., Merlo, C., & Abed, M. (2016). Management of divergent production network using decentralised multi-level capacitated lot-sizing models. *International Journal of Shipping and Transport Logistics*, 8(5), 590–604. DOI: 10.1504/IJSTL.2016.078686

Chapter 8
AI-Powered Web Solution for Disaster Risk Management

Yatri Davda
https://orcid.org/0000-0003-4198-3499
Dr. D.Y. Patil Institute of Technology, Pimpri, India

Ruchita Khilari
https://orcid.org/0009-0005-2392-8017
Dr. D.Y. Patil Institute of Technology, Pimpri, India

Swapnali Gaikwad
https://orcid.org/0009-0004-3853-1912
Dr. D.Y. Patil Institute of Technology, Pimpri, India

ABSTRACT

This project leverages AI and web development to create a disaster risk management platform that predicts, monitors, and responds to both natural and man-made disasters. Using AI algorithms, it processes data from sources like weather patterns and real-time sensors to forecast disasters and assess risks. With interactive dashboards and real-time alerts, the platform aids emergency responders and policymakers, enhancing preparedness and recovery efforts. By empowering users, it fosters safer, more resilient communities.

DOI: 10.4018/979-8-3693-7478-8.ch008

BACKGROUND

Disasters, both natural and man-made, are occurring more frequently and with increasing severity worldwide. These suffering, include million these economic hurricanes, people catastrophic loss floods, globally events and earthquakes (Shafique et al., 2020; Ghosh et al., 2020). ". environmental wildfires, and therefore, destruction (Shafique et al., 2020; Ghosh et al., 2020). In 2020 alone, natural disasters affected over 98 million people globally, with climate change and population growth exacerbating the vulnerability of many regions to these catastrophic events (Prakash and Kumari, 2020). Effective disaster risk management is, therefore, a priority for governments, organizations, and communities to mitigate the loss of life and property (Rahman et al., 2020).

Traditional disaster management systems, however, face several challenges. These include inadequate early warning systems, slow and inefficient data collection processes, delayed communication, and difficulty in coordinating responses across multiple agencies (Mohammed et al., 2021). In many cases, the lack of real-time, actionable data hinders proactive responses and worsens the aftermath of disasters (Hu et al., 2021). To address these issues, advanced technological solutions are being sought to enhance preparedness, prediction, and real-time response capabilities (Gokulakrishnan and Prakash, 2019).

The convergence of web development and artificial intelligence (AI) offers a promising avenue for transforming disaster management strategies (Ali et al., 2021). Web-based platforms provide a scalable and accessible means for disseminating information, coordinating relief efforts, and facilitating communication between emergency responders and the public (Sahu et al., 2020). These platforms can be integrated with real-time data feeds from various sources, such as sensors, social media, and satellite images, to provide dynamic, location-specific updates (Prasad et al., 2020).

AI has the potential to further enhance disaster management by leveraging machine learning algorithms to predict the occurrence of natural disasters, assess real-time risk, and optimize resource allocation during emergency responses (Sahu et al., 2021). AI-driven predictive models, such as deep learning algorithms, can analyze vast amounts of data to detect early signs of potential disasters, enabling faster and more precise decision-making (Menon et al., 2021). Additionally, AI systems can assist in post-disaster recovery by assessing damage, optimizing logistics, and identifying areas that require immediate attention (Reddy et al., 2021).

While these technologies offer significant benefits, the integration of web development and AI in disaster risk management is still in its infancy (Tan et al., 2021). Challenges such as data privacy, infrastructure requirements, and the need for accurate and reliable data sources remain (Jha and Gupta 2021). However, the potential for these technologies to improve disaster preparedness, response, and recovery is undeniable, making them crucial tools in future disaster management systems (Khanna, 2022).

1. DATA-DRIVEN PREDICTIVE MODELS IN DISASTER RISK MANAGEMENT

In the context of disaster risk management, data-driven predictive models have become essential tools for enhancing preparedness, response, and recovery efforts. These models leverage vast amounts of data from various sources to identify patterns, predict potential disaster events, and optimize resource allocation during emergencies. By analyzing data such as environmental parameters, historical disaster occurrences, and real-time sensor information, these models contribute to more effective disaster management strategies (Shafique et al., 2020; Prakash and Kumari, 2020; Mohammed et al., 2021).

1.1 Overview of Predictive Models

Predictive models in disaster management are computational systems designed to analyze input data and generate forecasts about potential disaster events. They consist of interconnected layers of processing units that work collaboratively to refine data and produce actionable insights (Hu et al., 2021; Ali et al., 2021; Menon et al., 2021). These models analyze inputs from diverse sources, enabling organizations to predict events like floods, wildfires, or earthquakes based on historical trends and real-time data (Ghosh et al., 2020).

Key components of these predictive models include:

- **Input Layer**: This layer receives data from various sources relevant to disaster management, such as weather conditions, seismic activity, and sensor readings (Gokulakrishnan and Prakash, 2019).
- **Hidden Layers**: These layers perform complex transformations to detect patterns in the input data, allowing the model to identify early indicators of potential disasters (Sahu et al., 2021).
- **Output Layer**: The final layer produces forecasts or risk assessments based on the processed information, assisting in decision-making during disaster response (Tan et al., 2021).

These models learn and improve over time through iterative training, adjusting their internal parameters based on feedback to enhance their accuracy (Jha and Gupta 2021).

1.2 Components of Predictive Models

The effectiveness of predictive models relies on several key components:

- **Processing Units**: Each unit in the model processes input data and contributes to the overall output. These units analyze variables critical to disaster risk assessment, such as rainfall levels or temperature fluctuations (Shafique et al., 2020; Rahman et al., 2020).
- **Weights**: Weights determine the significance of each input variable in generating outputs. In disaster scenarios, weights can be adjusted to prioritize critical factors influencing disaster risks (Ghosh et al., 2020; Mohammed et al., 2021).
- **Bias**: Bias parameters help refine the model's predictions by introducing flexibility in the outputs, allowing the model to adapt to varying conditions (Menon et al., 2021; Khanna, 2022).
- **Activation Functions**: These functions introduce non-linearities that enable the model to handle complex relationships in the data, making it suitable for identifying intricate patterns associated with disaster events (Tan et al., 2021).
- **Learning Rate and Feedback Mechanisms**: The learning rate controls the adjustments made to the model during training, while feedback mechanisms help minimize errors and enhance prediction accuracy (Cauchois et al., 2017; Glaa et al., 2014).

1.3 Types and Suitability of Predictive Models

Different types of predictive models are utilized based on the complexity of the disaster management task at hand. Commonly used types include:

- **Basic Feedforward Models**: These models process data in a straightforward manner, ideal for tasks like predicting the likelihood of a disaster based on historical data (Valero et al., 2022; Yadav and Gupta, 2021). These models have been successfully applied in various disaster risk management scenarios, where past events are the primary indicators for future predictions (Glaa et al., 2014).

- **Multilayer Models**: More complex models that consist of multiple processing layers, suitable for handling intricate tasks such as forecasting the impact of environmental changes on disaster occurrence (Loivet et al., 2020; Merimi & Taghipour, 2021). These models are often used for more dynamic tasks where environmental variables interact in complex ways, making them ideal for predicting the cascading effects of disasters (Lu & Taghipour, 2023).

These data-driven predictive models are particularly effective in disaster risk management, as they can synthesize large amounts of data to produce timely forecasts, helping organizations make informed decisions and improve overall preparedness (Cauchois et al., 2017; Mahfod et al., 2019). By integrating real-time data with predictive analytics, these models enhance the ability of disaster management teams to respond swiftly and appropriately to emerging risks (Nadia et al., 2021; Taghipour et al., 2023a).

2. APPLICATION OF PREDICTIVE MODELS IN DISASTER RISK MANAGEMENT

Predictive models are critical in disaster risk management, providing insights that enhance preparedness, response, and recovery efforts. This section explores the applications and significance of these models in managing various disaster scenarios.

2.1 Potential Applications of Predictive Models in Disaster Risk Management

Predictive models can be applied in several key areas to improve disaster management:

- **Risk Assessment**: These models analyze historical data and environmental factors to assess the likelihood and potential impact of disasters. This helps stakeholders identify vulnerable regions and prioritize resources effectively (Valero et al., 2022; Cauchois et al., 2017).
- **Early Warning Systems**: By integrating real-time data from weather forecasts, seismic sensors, and satellite imagery, predictive models can provide early warnings for impending disasters. Timely alerts enable authorities to take proactive measures to protect communities and infrastructure (Dadsena et al., 2021; Loivet et al., 2020).
- **Resource Allocation**: Predictive models help optimize resource allocation during disasters by forecasting the expected needs for food, medical supplies,

and shelter based on projected impacts. This ensures that resources are distributed efficiently and effectively (Glaa et al., 2014; Merimi & Taghipour, 2021).

- **Impact Simulation**: Models can simulate different disaster scenarios to evaluate potential impacts on communities and infrastructure. This information aids in contingency planning and helps decision-makers develop effective response strategies (Mahfod et al., 2019).

2.2 Significance of Predictive Models in Disaster Management

The use of predictive models in disaster management is significant for several reasons:

- **Informed Decision-Making**: Predictive models provide essential data that inform decisions regarding disaster preparedness and response. By understanding potential risks and impacts, stakeholders can develop more effective action plans (Mbiatem et al., 2018).
- **Enhanced Resilience**: By predicting the effects of various disasters, communities can implement measures to strengthen resilience, such as investing in infrastructure improvements or community training programs (Taghipour et al., 2024).
- **Efficient Recovery Planning**: After a disaster, predictive models can help assess damage and guide recovery efforts. Understanding the extent of potential losses allows for more effective prioritization of recovery actions and resource allocation (Taghipour, 2023; Taghipour et al., 2022).
- **Improved Public Safety**: By utilizing predictive models, authorities can enhance public safety through better planning and implementation of disaster response strategies. This ultimately helps to save lives and reduce economic losses (Taghipour & Merimi, 2021; Taghipour & Frayret, 2011).

In summary, the application of predictive models in disaster risk management is vital for enhancing preparedness, response, and recovery efforts. These models enable communities to better navigate the challenges posed by disasters, leading to more effective risk management strategies.

3. THE ROLE OF AI AND WEB TECHNOLOGIES IN DISASTER RISK PREDICTION AND RESPONSE

Artificial Intelligence (AI) and web technologies are transforming the landscape of disaster risk management by offering advanced tools for predicting, monitoring, and responding to disasters. These technologies allow for faster, more precise decision- making, which can mitigate the devastating effects of natural disasters.

1. AI for Disaster Prediction

AI can process large datasets from various sources, such as historical disaster records, climate data, and social media activity, to identify patterns and predict disaster events. Machine learning algorithms, for instance, excel at analyzing weather patterns to forecast floods, wildfires, or hurricanes. This predictive capability allows for early interventions, improving preparedness and reducing risks (Taghipour & Beneteau-Piet, 2020; Taghipour et al., 2023b).

2. Web Technologies for Real-Time Monitoring

Web-based platforms play an essential role in making disaster prediction data accessible to relevant stakeholders. These platforms can aggregate and visualize real-time data, providing users with tools such as live risk maps, early warning alerts, and incident tracking. With user-friendly interfaces, even non-technical users can access crucial information about potential disaster risks, enhancing preparedness and response (Taghipour et al., 2015; Taghipour & Frayret, 2013a).

3. Integration of AI with Web Applications

The synergy between AI and web technologies creates highly responsive and data- driven systems. AI-powered web applications can dynamically update as new data is fed into the system, ensuring that disaster predictions and response plans are continuously refined. These applications can also support decision-making by automating certain processes, such as the dispatch of emergency services or resource allocation based on predicted impact zones (Taghipour & Frayret, 2013b; Taghipour, 2020).

4. DERIVATION OF DISASTER-SENSITIVE PARAMETERS USING AI AND WEB TECHNOLOGIES

4.1 Study Area

In disaster risk management, understanding the geographical context is essential for accurate risk assessment. The study area for this research includes disaster-prone regions where AI and web technologies will be applied for monitoring, predicting, and managing disaster risks. This area can include urban zones, coastal regions susceptible to flooding, or areas affected by frequent natural hazards like earthquakes, wildfires, or storms. By focusing on such regions, the study aims to demonstrate how web technologies can facilitate real-time monitoring and prediction for effective disaster preparedness (Mbiatem et al., 2018).

- Selection Criteria: Factors like frequency of natural disasters, population density, availability of historical data, and vulnerability will be considered when selecting the study area (Nadia et al., 2021).

Methods and Materials Data Collection

The following data sources are essential for AI-based disaster risk prediction (Table 1):

Table 1. Essential for AI-based disaster risk prediction

Data Type	Source	Frequency of Collection	Usage
Meteorological Data	National Weather Services	Daily	Predicting floods, storms, droughts
Geospatial Data (GIS, Remote)	Satellite Imagery	Monthly	Mapping disaster-prone zones
Historical Disaster Data	Government/NGO Databases	Annually	Risk modeling and AI training
Social media/IoT Data	Public Social Platforms, IoT	Real-time	Crowdsourcing real-time disaster info

AI Models for Disaster Prediction

Machine learning algorithms such as:

- Random Forest and Decision Trees for identifying high-risk zones based on multiple factors (e.g., population density, weather forecasts) (Taghipour et al., 2024).

- Neural Networks to predict patterns from historical data and provide early warnings (Taghipour et al., 2023a).
- Support Vector Machines (SVMs) to classify regions based on risk levels and response needs (Taghipour et al., 2022).

Web Platform Development

The disaster risk prediction platform will consist of the following components (Table 2):

Table 2. Disaster prediction pratform

Component	Technology Stack	Function
Front-End Interface	React.js	Interactive disaster risk maps, alerts
Back-End Server	Node.js/Django	Data handling and AI model integration
Database	MongoDB/PostgreSQL	Real-time and historical data storage
API Integration	RESTful APIs, OpenWeatherMap	Fetching external data (e.g., weather)

4.2 Results and Conclusions Results

The results of this study will demonstrate the accuracy of AI models in predicting disaster risks and the efficiency of web-based platforms in disseminating critical information. Key outcomes will include:

- **Real-Time Risk Assessment**: The system will generate risk heatmaps showing disaster-prone zones with real-time data updates (Taghipour & Frayret, 2011).
- **Early Warning Capabilities**: AI predictions will provide early warnings for specific types of disasters, such as floods or wildfires, with increased accuracy (Taghipour et al., 2015).
- **User Engagement**: Web-based dashboards will show a significant increase in user interaction, enabling authorities and the public to respond quickly to risks (Taghipour & Merimi, 2021).

Economic Impacts of Disasters

Disasters have a huge economic burden on regions affected, feeding into immediate recovery or short-term fixing as well as long term development strategy making.

- **Direct Impacts:** Damage to infrastructure, loss of assets, and costs of emergency response and reconstruction (Taghipour, 2020).

- **Indirect Impacts:** Supply chain, economic activity disruptions, and healthcare expenditure increase are taking a massive toll on communities and governments (Mbiatem et al., 2018).
- **Recovery Solutions:** Risk transfer mechanisms such as insurance, public-private partnerships, and economic stimulus programs play a crucial role in reducing financial losses and facilitating a quicker recovery (Taghipour & Frayret, 2013b).

Furthermore, the integration of AI and several web technologies in disaster management can supplement the economic resilience by maximizing resource allocation and increasing recovery effectiveness (Taghipour, 2014).

The integration of AI and web technologies in disaster risk management provides a scalable, efficient solution for enhancing prediction accuracy and response times. The study validates the effectiveness of such systems in real-world scenarios, offering a clear path for future research and deployment in disaster-prone areas globally (Taghipour & Frayret, 2013a).

Additionally, addressing the economic impacts of disasters highlights the importance of financial resilience as a critical component of comprehensive disaster management strategies (Zoghlami et al., 2016). Development, Calibration, and Validation of AI Models for Disaster Risk Management

4.3 Methods and Materials

Proposed AI Model Development

In this research, we propose developing AI models to predict disaster risks in real-time by analyzing historical data, environmental factors, and real-time inputs. The models will be tailored to different types of disasters, such as floods, earthquakes, and wildfires (Table 3).

- **Data Collection**: We plan to use datasets including:
    - Historical disaster records.
    - Satellite imagery and GIS data.
    - Weather data from national and global sources.
    - Social media and IoT devices for real-time monitoring.
- **Model Selection**:
    - **Random Forest for Flood Prediction**: Random Forest will be chosen for its robust decision-making capabilities and ability to process large, varied datasets (Taghipour et al., 2022).

- **Convolutional Neural Networks (CNN) for Earthquake Prediction**: CNNs will be used due to their efficiency in detecting seismic patterns from geospatial data (Taghipour et al., 2015).
- **Support Vector Machines (SVMs) for Wildfire Prediction**: SVMs will classify risk areas by analyzing factors such as wind speed, vegetation, and temperature (Mbiatem et al., 2018).

Table 3. Model selection

Model Type	Disaster Type	Data Used	Proposed Output
Random Forest	Floods	Historical flood, weather data	Real-time flood risk maps
Convolutional Neural Networks	Earthquakes	Seismic data, tectonic movement	Earthquake probability and intensity
Support Vector Machines (SVMs)	Wildfires	Temperature, wind speed, vegetation data	Wildfire risk classification

Proposed Calibration and Tuning

After developing the models, calibration will be carried out through parameter tuning and performance optimization. The following steps will be used for this (Table 4):

- **Hyperparameter Tuning**: We will apply techniques like grid search and random search to optimize model parameters (Taghipour & Frayret, 2011).
- **Cross-Validation**: To ensure robustness, we will use K-fold cross-validation to evaluate model performance and minimize overfitting (Taghipour et al., 2023a).

Table 4. Calibration and tuning

Model	Hyperparameters Tuned	Optimization Technique
Random Forest	Number of trees, max depth	Grid Search
CNN	Number of layers, filter size	Random Search
SVM	Kernel type, regularization	Grid Search

Web Platform Integration

Once the models are calibrated, they will be integrated into a web-based platform designed for real-time disaster risk monitoring. The platform will feature:

- **Real-Time Data Display**: Risk maps and prediction data accessible through a user-friendly interface (Taghipour & Frayret, 2013b).
- **Alert System**: Notifications for users based on disaster risk levels (Taghipour & Merimi, 2021).
- **Scalable Infrastructure**: The platform will be scalable for larger regions and multiple disaster types (Taghipour, 2014).

4.4 Proposed Results and Validation Expected Results

Based on initial data analysis and model selection, we expect the AI models to significantly improve disaster risk prediction. The integration of AI with web-based systems should enhance early warning capabilities and decision-making efficiency for disaster management (Table 5).

Table 5. Validation

Disaster Type	Model Used	Proposed Prediction Accuracy
Floods	Random Forest	~90%
Earthquakes	Convolutional Neural Networks (CNN)	~85%
Wildfires	Support Vector Machines (SVMs)	~80%

Future Model Validation

Once the models are developed, they will be validated through real-world data testing:

- **Flood Model Validation**: The Random Forest model will be validated by comparing its predictions with actual flood events in the selected study areas (Nadia et al., 2021).
- **Earthquake Model Validation**: The CNN model will be tested against recorded seismic events to measure its accuracy in predicting earthquake intensities (Taghipour et al., 2015).
- **Wildfire Model Validation**: The SVM model will be validated through comparisons with historical wildfire incidents (Mbiatem et al., 2018).

5. ADVANCED TECHNIQUES AND BRAINSTORMING IN VIEW OF DISASTER RISK MANAGEMENT

In the developing process of the disaster risk management, it is crucial to incorporate the state-of-the-art technologies, best practices and strong economic strategies to improve the readiness, response, and recovery. The following topics identify current trends and potential issues that define the future of disaster management:

1. AI-Powered Early Warning Systems:

It has therefore boosted disaster risk management through providing better predictions of the disasters. Early warning systems have been upgraded to the use of artificial intelligence that makes the predictions of disasters to be more accurate and in real-time. The following are examples of some of the applications that they have come up with; machine learning and data analytics to process large data sets on weather, satellite images, and environmental sensors. Through anticipating of the risks that might occur AI systems guarantee effective alerts that can prevent loss of lives and properties. For instance, Google's flood forecasting project, which applies the use of artificial intelligence to make predictions of flooding and provide advance warning to the affected populations (Taghipour et al., 2023c). The implementation of such technologies provides enhanced prospects for disaster preparedness and response time (Taghipour, 2020).

2. The Role of Citizen Science in Disaster Response:

Citizen science has increasingly been applied in disaster management, especially where data need to be rapidly collected and in real-time. The use of mobile applications and crowdsourced platforms allows the collection of information on weather conditions, reports of damages, and safety status to guide emergency responders in prioritizing their actions. The potency of data available in common, often captured during a disaster, enables mapping and resource allocation with much greater accuracy. It is platforms like Ushahidi that aggregate crowd-sourced data toward the creation of real-time disaster maps that prove how much citizen participation enhances the efficiency and effectiveness of responses to disasters (Taghipour & Merimi, 2021). This serves to not only speed up the response within local communities but also leaves an empowering and resilient sense (Taghipour & Frayret, 2013b).

3. Economic Impacts of Disasters:

The economic effects of disasters continue over the weeks and months that follow and indeed often thwart the long-term recovery and reconstruction process. In the first instance, most disasters tend to disrupt local economies while they also compromise infrastructural integrity- a major burden on the financial resources of the regions concerned. Insurances thus play a vital role in this but the challenges come up when risks have to be properly assessed since all disaster events get increasingly unpredictable (Taghipour et al., 2024). Even more, the economic fallout on global supply chains, agricultural sectors, and local businesses is devastating. All these help governments and organizations better prepare and invest in recovery infrastructure and disaster-resilient technology by reducing risks (Nadia et al., 2021).

4. Resilience Metrics for Communities and Infrastructure:

Resilience is often measured by how well communities and infrastructure withstand, adapt, and recover from disasters. Resilience metrics offer a quantitative and qualitative framework to help determine how well communities are positioned to withstand the shocks related to disaster. This includes metrics like infrastructure strength, recovery speed and the social cohesion of communities. The Disaster Resilience Scorecard and Resilience Index are two such tools that can help determine and assess resilience levels at different sites, thus providing data that is crucial for planning the way forward for various disaster management professionals and policymakers (Taghipour et al., 2023a). Incorporating resilience metrics in planning and response strategies can help such communities to get more adaptive capacity, less vulnerable and pursue the sustainable recovery practices (Mbiatem et al., 2018).

CONCLUSION

The emergence of artificial intelligence (AI) with web technologies in disaster management is a paradigm shift in managing the growing and complex challenges of natural and anthropogenic disasters. These innovations, which leverage real-time data, predictive analytics, and interactive web-based platforms, deliver critical decision support analytics to relevant stakeholders to support proactive disaster risk reduction and improve response planning.

This chapter examined from first principles state-of-the-art approaches and diversity of applications of AI and web technologies in disaster management. It underscores how data-driven decision-making, geospatial intelligence, and inclusive community engagement are crucial to the design of resilient systems that can help to better mitigate the multidimensional impacts when disasters do occur (Taghipour et al., 2015).

However, dealing with issues worldwide such as data protection, interoperability, equal access, and ethical considerations is essential for the successful implementation of these technologies. The advancements in the field of artificial intelligence (AI), the Internet of Things (IoT), and cloud computing are likely to have significantly increased the capabilities of disaster management systems, thus, making them more intelligent, flexible, and scalable (Taghipour, 2014).

This debate addresses the main point of how the development of various disciplines and international cooperation is a fundamental factor in catastrophe management. By utilizing the global pool of perspectives together, we can create technologies that offer a safer and more resilient place for both the current and future generations (Taghipour et al., 2023c).

REFERENCES

Ali, M. A. M., Al-Zahrani, A., & Al-Assaf, A. K. T. (2021). Machine Learning Techniques for Disaster Prediction and Risk Management. *IEEE Access : Practical Innovations, Open Solutions*, 9, 129564–129578.

Cauchois, R., Taghipour, A., Kang, D., Zoghlami, N., & Abed, M. (2017). Advanced solutions for a supply chain with stochastic information. *Journal of Advanced Management Science*, 5(1), 9–13. DOI: 10.18178/joams.5.1.9-13

Dadsena, K. K., Mathiyazhagan, K., & Taghipour, A. (2021). Analysis of barriers for the build the resilient supply chain networks post-COVID-19. In *Managing Supply Chain Risk and Disruptions: Post COVID-19* (pp. 79–89). Springer International Publishing. DOI: 10.1007/978-3-030-72575-4_7

Ghosh, R. K., Thakur, R. C., & Ghosh, A. B. (2020). Role of IoT in Disaster Management: A Survey. *IEEE Internet of Things Journal*, 7(8), 6985–6994.

Glaa, B., Zoghlami, N., & Taghipour, A. (2014, May). Outsourcing and off-shoring healthcare services: A way to cut the costs and improve quality. In *2014 international conference on advanced logistics and transport (ICALT)* (pp. 189-194). IEEE. DOI: 10.1109/ICAdLT.2014.6864116

Gokulakrishnan, S. B., & Prakash, A. M. R. (2019). Web-Based Disaster Management System Using GIS and IoT. *IEEE Internet of Things Journal*, 6(2), 2077–2085.

Hu, Y., Zeng, J., Chen, Y., & Li, W. (2021). Disaster Risk Assessment Based on Multi- Source Data and Machine Learning. *IEEE Transactions on Geoscience and Remote Sensing*, 59(9), 7785–7796.

Jha, P. K., & Gupta, A. K. (2021). Artificial Intelligence Techniques for Disaster Risk Reduction: A Review. *IEEE Transactions on Engineering Management*, 68(1), 186–200.

Khanna, A. H. D. R. P. K. R. V. A. J. (2022). Disaster Management: Applications of Machine Learning and Deep Learning Techniques. *IEEE Access : Practical Innovations, Open Solutions*, 10, 6756–6770.

Loivet, W., Taghipour, A., & Kang, D. S. (2020). The rise of green supply chain management: Between complexity and necessity. *J. Econ. Bus. Manag*, 8(1), 1–7. DOI: 10.18178/joebm.2020.8.1.604

Lu, X., & Taghipour, A. (2023). A French Guide for Businesses and Governments to Integrate the Concept of Sustainable Development Into Their Supply Chains. In *Government Impact on Sustainable and Responsible Supply Chain Management* (pp. 1–12). IGI Global. DOI: 10.4018/978-1-6684-9062-4.ch001

Mahfod, J., Canel-Depitre, B., & Taghipour, A. (2019). Quality function deployment-ELECTRE in supplier evaluation. *Journal of Advanced Management Science, 7*(4).

Mbiatem, J., Taghipour, A., & Canel-Depitre, B. (2018, August). Supplier selection approaches for decision makers. In *Proceedings of the 8th International Conference on Information Communication and Management* (pp. 108-112).

Menon, N. G. R., Paul, A. J., & Ghosh, S. K. (2021). Real-Time Disaster Monitoring and Management Using IoT and Machine Learning. *IEEE Access : Practical Innovations, Open Solutions*, 9, 125763–125779.

Merimi, M., & Taghipour, A. (2021). Accelerating the digitalization of the supply chain: Empirical research about COVID-19 crisis. In *Digitalization of decentralized supply chains during global crises* (pp. 1–24). IGI Global. DOI: 10.4018/978-1-7998-6874-3.ch001

Mohammed, A. R., Alhajri, H. M., & Aldhahrani, N. A. (2021). Data-Driven Disaster Risk Assessment: A Review of Methodologies and Approaches. *IEEE Transactions on Emerging Topics in Computing*, 9(2), 640–654.

Nadia, E. N., Beatrice, C. D., & Atour, T. (2021). The determinants of hotel room rates in Beirut: A hedonic pricing model. *International Journal of Trade. Economics and Finance*, 12(2), 33–42.

Prakash, N. R., & Kumari, S. S. S. (2020). Predictive Analytics for Disaster Management: A Systematic Review. *IEEE Transactions on Systems, Man, and Cybernetics. Systems*, 50(3), 967–982.

Prasad, A. S. R., Mohanty, D. K., & Sahu, A. K. (2020). A Survey on Disaster Management Using IoT and Big Data. *IEEE Internet of Things Journal*, 7(6), 4714–4728.

Rahman, M. M., Alharbi, A. A., & Khatun, I. M. I. A. (2020). Smart Disaster Management System Using IoT and Big Data Analytics. *IEEE Access : Practical Innovations, Open Solutions*, 8, 26320–26330.

Reddy, K. S. R., Prasad, B. V. D., & Reddy, V. V. S. M. (2021). Artificial Intelligence and Machine Learning Techniques in Disaster Management: A Review. *IEEE Transactions on Computational Social Systems*, 8(2), 389–398.

Sahu, R. K., Gupta, M. R., & Sharma, S. C. (2021). A Framework for Disaster Management Using Artificial Intelligence and IoT. *IEEE Transactions on Network and Service Management*, 18(1), 620–632.

Sahu, S. C. K., Dutta, A. K., & Sahu, S. S. (2020). Disaster Management System Using Machine Learning and IoT. *IEEE Access : Practical Innovations, Open Solutions*, 8, 68145–68156.

Shafique, A. T. A., Ali, A. H., & Qadir, K. S. T. (2020). AI and Machine Learning for Disaster Management: A Review. *IEEE Access : Practical Innovations, Open Solutions*, 8, 89912–89924.

Taghipour, A. (2014). Improving the plan of a manufacturing network with non-integrated business units. *International Journal of Applied Logistics*, 5(2), 1–11. DOI: 10.4018/ijal.2014040101

Taghipour, A. (Ed.). (2020). *Demand forecasting and order planning in supply chains and humanitarian logistics*. IGI Global.

Taghipour, A. (Ed.). (2023). *Government Impact on Sustainable and Responsible Supply Chain Management*. IGI Global. DOI: 10.4018/978-1-6684-9062-4

Taghipour, A., Abed, M., & Zoghlami, N. (2015, May). Design for remanufacturing respecting reverse logistics processes: A review. In *2015 4th International Conference on Advanced Logistics and Transport (ICALT)* (pp. 299-304). IEEE. DOI: 10.1109/ICAdLT.2015.7136602

Taghipour, A., & Beneteau-Piet, C. (2020). Sustainable supply chain management performance. *International Journal of Innovation, Management and Technology*, 11(6), 165–169. DOI: 10.18178/ijimt.2020.11.6.897

Taghipour, A., Foukolaei, P. Z., Ghaedi, M., & Khazaei, M. (2023a). Sustainable Multi-Objective Models for Waste-to-Energy and Waste Separation Site Selection. *Sustainability (Basel)*, 15(22), 15764. DOI: 10.3390/su152215764

Taghipour, A., & Frayret, J. M. (2011). Supply chain coordination planning: A review. *Work paper, École polytechnique de Montréal.*

Taghipour, A., & Frayret, J. M. (2013a, May). An algorithm to improve operations planning in decentralized supply chains. In *2013 International Conference on Advanced Logistics and Transport* (pp. 100-103). IEEE. DOI: 10.1109/ICAdLT.2013.6568442

Taghipour, A., & Frayret, J. M. (2013b). Coordination of operations planning in supply chains: A review. *International Journal of Business Performance and Supply Chain Modelling*, 5(3), 272–307. DOI: 10.1504/IJBPSCM.2013.055729

Taghipour, A., Lu, X., & Smahi, M. (2022). Opportunities and challenges for society 5.0 in procurement. *Blockchain Technology in Supply Chain Management for Society*, 5(0), 91–100. DOI: 10.1201/9781003177432-9

Taghipour, A., & Merimi, M. (2021, March). Digital transformation of supply chains during crisis: covid-19. In *11th Annual International Conference on Industrial Engineering and Operations Management, IEOM* (pp. 7493-7501). DOI: 10.46254/AN11.20211298

Taghipour, A., Padash, A., Etemadi, V., Khazaei, M., & Ebrahimi, S. (2024). Sustainable and Circular Hotels and the Water–Food–Energy Nexus: Integration of Agrivoltaics, Hydropower, Solar Cells, Water Reservoirs, and Green Roofs. *Sustainability (Basel)*, 16(5), 1985. DOI: 10.3390/su16051985

Taghipour, A., Ramezani, M., Khazaei, M., Roohparvar, V., & Hassannayebi, E. (2023b). Smart transportation behavior through the COVID-19 pandemic: A ride-hailing system in Iran. *Sustainability (Basel)*, 15(5), 4178. DOI: 10.3390/su15054178

Taghipour, A., Sohrabi, A., Ghaedi, M., & Khazaei, M. (2023c). A robust vaccine supply chain model in pandemics: Case of Covid-19 in Iran. *Computers & Industrial Engineering*, 183, 109465. DOI: 10.1016/j.cie.2023.109465

Tan, C. K., Gan, S. P., & Tan, D. H. K. (2021). Web-Based Disaster Management System: A Review of IoT Solutions. *IEEE Access : Practical Innovations, Open Solutions*, 9, 74060–74073.

Valero, M. E. J. J. M., Alda, A. M. M., & Carretero, M. J. P. F. (2022). Machine Learning for Natural Disaster Prediction: A Systematic Review. *IEEE Transactions on Computational Intelligence and AI in Games*, 14(4), 623–634.

Yadav, R. P., & Gupta, M. K. (2021). Utilization of Web Technologies for Disaster Risk Reduction: A Systematic Review. *IEEE Access : Practical Innovations, Open Solutions*, 9, 19880–19892.

Zoghlami, N., Taghipour, A., Merlo, C., & Abed, M. (2016). Management of divergent production network using decentralised multi-level capacitated lot-sizing models. *International Journal of Shipping and Transport Logistics*, 8(5), 590–604. DOI: 10.1504/IJSTL.2016.078686

Chapter 9

AI Methodologies in Upcoming Modern Warfare Systems

Swapnali Pravin Gaikwad
https://orcid.org/0009-0004-3853-1912
Dr. D.Y. Patil Institute of Technology, Pimpri, India

Yatri Davda
https://orcid.org/0000-0003-4198-3499
Dr. D.Y. Patil Institute of Technology, Pimpri, India

ABSTRACT

Modern AI methodologies are changing the face of warfare by enabling militaries to enhance their operational capabilities and, more importantly, their decision-making in executing the operations. Advanced technology with machine learning, natural language processing, and autonomous systems enable real-time data analysis, predictive modeling, and increased situational awareness on the battlefield. With such problematic challenges facing the military forces, AI integration provides strategic advantages over military departments for rapid responses to emerging threats and resources distribution. However, this evolution raises issues on accountability, bias, and implication of autonomy in decision-making for combat scenarios. Overall, therefore, AI methodologies are going to redefine the nature of warfare and hence it becomes highly important for military leadership to navigate through these opportunities along with the risks arising from such technologies.

DOI: 10.4018/979-8-3693-7478-8.ch009

1. BACKGROUND

The background of AI methodologies in modern warfare to come would be based on exponential and progressive rapid developments in the last three decades concerning artificial intelligence and machine learning technologies. Military operations having already switched to digital tools require AI for enhancements in decision-making, improvement in operational efficiency, and adaptation to complex battlefield environments.

Initially as basic defense automation and data processing, AI has grown into a much more strategic tool for application in sophisticated algorithms and supercomputing power in militaries. Nowadays, the methodologies of AI are applied in fields such as autonomous systems, predictive analytics, and real-time data analysis. With such technologies, militaries can operate faster and with higher density information; for example, about gigantic amounts of data from surveillance satellites and battlefield sensors.

Cyber warfare and asymmetric threats have acted as accelerants for AI solutions that can adapt rapidly to perceived requirements to forecast and respond to situations unfolding on their scope. Ambitious countries for technological supremacy would embrace AI methodology tied to war-fighting methods that would indicate a shift in the forms of combat, improve situational awareness, and maximize mission effectiveness. This has raised further questions about the ethical implications for proper integration into military applications of AI solutions (Binnendijk & Libicki, 2019; Bishop, 2021).

2. INTRODUCTION TO AI METHODOLOGIES IN UPCOMING MODERN WARFARE

Modern AI methodologies are changing the face of warfare by enabling militaries to enhance their operational capabilities and, more importantly, their decision-making in executing the operations. Advanced technology with machine learning, natural language processing, and autonomous systems enable real-time data analysis, predictive modeling, and increased situational awareness on the battlefield. With such problematic challenges facing the military forces, AI integration provides strategic advantages over military departments for rapid responses to emerging threats and resources distribution. However, this evolution raises issues on accountability, bias, and implication of autonomy in decision-making for combat scenarios. Overall, therefore, AI methodologies are going to redefine the nature of warfare and hence it becomes highly important for military leadership to navigate through these opportunities along with the risks arising from such technologies (Taghipour & Frayret, 2013a, and 2013b).

2.1 Overview of AI in Military Applications

2.1.1 Key AI Methodologies Transforming Warfare

Machine learning now forms the core practice of AI methodologies in the revolution of warfare; these enable systems to learn from data and to adapt and improve over time. In military applications, machine learning algorithms are applied scanning big data coming from a variety of different sources such as reconnaissance missions and battlefield sensors for patterns and in an attempt to predict enemy actions. Predictive analytics extends this capability further by giving an advanced insight into the possible outcomes of military operations and enables commands to bastheir decisions on real-time data. Such technologies allow militaries to predict possible threats and thus maximize their strategic approach, significantly far improving operational effectiveness (Since its early beginning, the acceptance of AI in military operations has gained tremendous momentum, pushing significantly more firepower into the hands of armed forces everywhere. AI technologies apply to all kinds of area-from logistics and intelligence through surveillance, reconnaissance, and combat activities. For instance, machine learning algorithms scan through heavy data coming from sensors and surveillance systems in order to spot patterns, thus predicting the next move that the enemy could make, thus providing better situational awareness. Other drones that are autonomous as well as unmanned ground vehicles rely on AI for navigation and decision-making therefore don't require people to intervene. That means a huge reduction in exposure to personnel risk. It further ensures intelligence gathering through natural language processing since vast amounts of unstructured data such as reports and communications have to be sifted in order to draw out relevant information in the most efficient manner possible. Logistics is simplified through AI with elements relating to optimization of the management and use of supply chains and resources. It enhances the military's operational effectiveness and throws up pertinent ethical as well as strategic questions about the future of warfare, mainly in terms of accountability as well as the possibility of autonomous weapon systems. AI's role in military uses therefore marks a transformation in how modern warfare is conducted and managed (Cauchois et al., 2017; Dadsena et al., 2021).

Computer vision technologies are revolutionizing the way militaries can view and interpret visual information on the battlefield today. They can take picture and video feeds from drones, satellites, and ground vehicles within an AI system to facilitate self-automated target recognition and situational assessments. Sensor fusion complements computer vision by gathering information from many sensors that bestow a comprehensive understanding of the operational environment. This fusion capability can improve situational awareness. Modern conflicts can be responded to much more rapidly and effectively by managing the complete depth of the sensing dimension, yet with light burden on human operators' cognitive load by military forces (Loivet et al., 2020).

The unmanned systems present a huge revolution within the modern context of warfare; the AI technologies driving the drones, unmanned grounds, and maritime vessels totally independent or complementary to the forces manned. Such systems can conduct an incredibly wide range of missions -- from surveillance and reconnaissance operations through combat operations -- with even less input of human intervention. The inclusion of AI allows such standalone platforms to respond in real time to changing conditions based on sensor data and predefined parameters. This lets it be involved in making significant contributions toward improved operational capability and risk reduction for personnel but also brings in considerable ethical and strategic considerations in terms of the contribution of AI to the combat scenario and implications of its autonomous decision-making in warfare (Heath & Roper, 2021; Zoghlami et al., 2016).

2.2 Implications and Challenges of AI in Warfare

Ethical Concerns

This is precisely where AI in war raises several ethical questions, especially with the use of autonomous weapons systems. Accountability arises where such a failure from the AI system occurs in the line of misidentification of a target or casualties in the civilian area. These vagaries are untoward to the sterner perception of applying international humanitarian law whose premises guide the distinction between the military and civilian addresses. Furthermore, the authority over life and death in the hands of machines can subvert the exercises of human judgment upon issues to do with life and death in war. Such situations pose demands for regulated use of AI as well as strictness in ethical codes concerning the use of AI in battle.

Threats and Vulnerabilities about Security

The AI technologies raise important security risks and vulnerability that adversaries are able to exploit. Since machine learning and data-driven systems lie at the core of military operations, it remains susceptible to attacks from the adversary in ways where enemy forces will make AI algorithms or information feeds tricked or corrupted. Moreover, AI is a fast-evolving technology and the militaries need to evolve themselves in accordance with changes that keeps on piling up which is a resource-intensive process and further complicates the training processes. Cyber security of AI systems needs, therefore, to be of topmost priority for maintaining the operating integrity and safeguarding critical military information from breach (Taghipour, 2020; Taghipour, 2014).

Social Implications and Military Affairs

The situation where AI is increasingly used during warfare might, therefore, change military dynamics and civil war perceptions themselves. A race for advanced technologies may destabilize global security, while dependence on AI might dehumanize warfare decisions; algorithms rather than ethically supported decisions take the upper hand. The situation may also raise lighter barriers on military engagements, thus even more devastating warfare is imminent. Thus, the policymakers and the military leaders have a herculean task of crossing those dangerous implications with caution about responsible use and risk minimization (Merimi, & Taghipour, 2021).

3. CORE TECHNOLOGIES IN AI METHODOLOGIES IN UPCOMING MODERN WARFARE

Core to AI methods forms the heart of what modern warfare would be tomorrow. Radical innovations that create improvements in military capabilities and effectiveness have been generated, including advanced technologies accompanying machine learning, computer vision, natural language processing, and robotics. It enables the analysis of incredibly large amounts of data from which military forces will find patterns and, based on such patterns, informed predictions regarding the probable movements of the enemy, as well as changes in the dynamics of the battlefield.

3.1 Operational Integration of AI in Military Strategies

With respect to operational integration of AI in military plans, complex AI must be integrated into each phase of a plan so that it can perform and maneuver flexibly in real operations. The output of the analysis yield by the AI can then be used to empower militaries to process vast quantities of data received from sources such as satellite photographs, reconnaissance reports, and battlefield sensors to make real-time decisions. This integration would also improve the threat assessments, resource optimization, and missions to be planned for the militaries. In addition, AI will also make command and control structures within the military better coordinate among the different military branches and units within it. Thus, operational integration of AI not only streamlines military processes but gives better situational awareness and responsiveness in dynamic and complex combat environments (Taghipour& Frayret, 2011).

3.2 Interoperability and Collaboration Between AI Systems

Interoperability among AI systems is therefore paramount in the modern military since they are expected to work almost automatically as different platforms and technologies. As more militaries increasingly adopt various AI solutions, the potential of these systems to communicate and exchange information increases their overall operational effectiveness. This integration of forces allows it to better exploit the strengths of AI tools in developing coordinated responses against emerging threats and increases situational awareness along the entire battlefield. And also, a lack of interoperability might lead to disintegrated operations and lower efficiency; hence, standards regarding the protocols and frameworks would be required.

The bottom line is that the cooperation between human operators and AI systems needs to be at the heart of successful military strategies. While AI can process data much faster than humans and make suggestions for how the situation should be addressed, human judgment is also crucial in more complex and unpredictable environments. An AI designed to augment human decision-making can nurture a synergy that impacts operations outcomes positively. It is essential to train employees on effective usage of AI tools so that the operators are able to understand and interpret the insights generated through AI and make proper and wise decisions in high-stake situations (Mbiatem et al., 2018).

AI interoperability enables joint operations of different military branches, which provide better coordination and efficiency during the course of combat operations. By integrating AI technologies across air, land, and sea forces, militaries can harmonize their operational structures with optimum resource utilization as well as response time. For instance, AI can make it possible to share data in real-time between air forces and land forces so that all forces are updated on the situation at hand in the battlefield. It instills a culture of teamwork and flexibility into it, both highly prized virtues for navigating the subtleties of modern warfare, thus also enhancing the effectiveness of such integrated levels in the conduct of a mission.

3.3 Challenges and Ethical Considerations in AI Deployment

Probably the hardest challenge in AI-infused military operations is: To whom do decisions taken by autonomous systems accountable? The adoption of AI technologies in war raises fundamental questions about accountability in actions of such systems-making it sometimes the developers, sometimes the military commanders, but sometimes even the AI itself. Such vagueness complicates international humanitarian law concerning distinction between being and not being a combatant. Human oversight has to be incorporated for these critical processes so that there exists an ethical approach toward the resolution of these issues and ensures moral responsibility in war.

AI has taken up for itself so many biases which make it susceptible to doing inappropriate decisions when the stakes are high. This can be either because of the data used in feeding the algorithm or even the design of the systems themselves. If AI interprets the data wrongly or reflects inbuilt biases, then it will lead to unintended consequences, which may involve mistaken target identification or giving the wrong recommendation in strategy. For limiting such risks and ensuring military applications remain operationally sound, these AI systems need to be tested and validated for reliability and fairness.

Increases in reliance on AI technologies introduce an ethical problem related to dehumanizing war. As militaries' decisions are increasingly automated, perhaps the emotive and moral tones of war will dissipate and the human price of combat will recede from view. This shift could remake the social concept of war itself-as less clinical but more ordinary. The need therefore is to debate the ethics of AI in war, which shall ensure balance between the acceleration of technology and the protection of human dignity and moral considerations in the conduct of war.

4. INTEGRATION OF AI METHODOLOGIES IN UPCOMING MODERN WARFARE

4.1 AI-Driven Strategic Planning and Execution

AI-Driven Strategic Planning and Execution Enables advanced data analytics techniques to inform military decision processes. It uses tremendous sources of information from all places -- from intelligence reports and satellite images to battlefield sensors -- to piece together patterns and establish actionable insights for the military leadership to act on. These enable military leaders to predict potential threats and risk, so that their strategies will be relevant and fitting for the operation at hand. Data velocity in its analysis and synthesis allows for quicker decision-making in fast-paced combat scenarios.

AI techniques open doors to strategic plan enhancement through predictive modeling and scenario simulation. In a nutshell, application of machine learning algorithms enables military planners to simulate different scenarios that may appear in the battlefield or the likelihood of such a scenario under certain conditions versus the strategy adopted. This is the means through which the decision-maker tests plans before their execution so that possible difficulties and adjustments needed to achieve desired goals are determined in advance. In a few easy strides, the assessment of several situations can be made for militaries in order to prepare them to face uncertainty and build overall preparedness to unpredictable aspects of war while reducing risks on personnel and resources.

AI-driven strategic planning also optimizes resource and logistics allocation that is an integral part of military operations. Also, the supply chain can leverage predictive analytics as a means for AI systems to forecast requirements within the supply chain, highlight bottlenecks in logistics, and optimize distribution efficiency with respect to available resources. Such optimization will enable troop access to what is required when it is needed, thereby preventing the inconvenience of delay and ensuring greater operational effectiveness. Further, AI can also be used to enable the real-time monitoring of resource use and availability by the military leaders while making adjustments in asset allocation that are needed in accordance with the changes in the situation. Conclusion In a nutshell, the implementation of AI in planning and execution of strategy leads to far more efficient military operations with increased mission success rates (Merimi, & Taghipour, 2021;; Nadia et al., 2021 ; Taghipour et al., 2022).

4.2 Autonomous Systems and Human-Machine Collaboration

Emergence of Autonomous Systems in Warfare

Autonomous systems-the drones, unmanned ground vehicles, and naval vessels-bid to change the face of the modern war, carrying out parts of their mission with minimal interaction with humans. For the first time in history, autonomous systems can navigate increased levels of complexity, make decisions with near-real-time information, and complete a very wide range of tasks in surveillance and combat. Increased use of autonomous systems has made the military expand its reach operationally with little risk to personnel; this is because these machines can be used in hazardous situations where human lives would have been lost. The future engagements are going to be highly dominated by technologies as they evolve (Cummings, 2017).

Improve human-machine cooperation

While there are so many benefits of autonomy systems, actualization of the full potential of such systems requires collaboration between humans and machines. AI technologies can process data and give insights way faster than any human could; therefore, human judgment must be applied to that input to make strategic decisions. The design of a system should include intuitive interaction between the human operator and AI, so that the personnel are empowered to make informed choices based on AI recommendations. For this purpose, the training and familiarization with these technologies supposed to create a collaborative environment of humans and machines working in concert toward attaining mission objectives is critical (Taghipour& Frayret, 2011 ; Mbiatem et al., 2018).

Challenges and Ethical Implications

The operations raise extremely crucial challenges and ethical considerations in relation to including autonomous systems. Perhaps the most important of these is the loss of oversight over what have traditionally proved in such critical decision-making processes, especially when a combat scenario may engage targets without being directly interfered with by humans. This question holds to discussions of accountability and moral implications on vesting machines with decisions having the ramification of life and death. For instance, human-machine interaction with all its complexity, therefore, demands careful thought into the design of these systems to ensure that they are reliable and minimize the risk of malfunction or miscommunication. Resolution of these challenges is what shall be used to tap into the bene

4.3 Security, Ethical, and Regulatory Frameworks

This is because the introduction of AI in military operations leads to a new set of security-related risks since there are sensitive data and military assets that could potentially be exposed. AI is susceptible to adversarial attacks whereby adversaries can exploit even the slightest weakness in algorithms or manipulate the inputs of the data to get into the military operations. This would translate into wrong information, lost objectives of missions, and loss of operational integrity. Since militaries increasingly rely on interconnected AI systems, strong cybersecurity measures against hacking or even data breaches become a demand. Threats must be monitored and assessed constantly and consequently carried out by vulnerability testing to allow AI to continue to change with new security needs.

Thus, the ethical effects of deploying AI on battlefields have to be considered so that military operations could be made more compatible with international humanitarian law and the higher norms of morality. Further, when AI systems are conferred autonomy to make decisions, it raises very grave concerns relating to accountability, where decisions may have ramifications in terms of life and death on the battlefield. Ethical principles may evolve to oversee the development and application of AI technology to avoid the risks that come with the potential mishandling of AI because military involvement is inherently within human control. It would emanate from the discussions of AI ethics by ethicists, lawyers, and military leaders in their quest to develop an appropriate framework that would balance the gains of technological advancement with ethical considerations (Taghipour & Merimi, 2021).

Management of AI deployment in war would then have to be done comprehensively under one regulatory framework, guidelines towards development, testing, and use of AI in military scenarios with clarity on transparency and accountability, conformity to international law. The governments, military organizations and international bodies collaborate towards the development of any related standards and best practices in the use of AI. The military should be further educated and developed in terms of professional on ethical and regulatory considerations when using AI technology, not to be used in misapplication against human rights as well as the public's trust in military operations.

5. CHALLENGES AND ETHICAL CONSIDERATIONS

5.1 Autonomy and Accountability in Decision-Making

Militarization of autonomous systems is definitely a paradigm shift in the decision-making process. The entire process of data analysis coupled with minimal human intervention will speed up responses in critical situations. It raises very seriously issues of accountability because decisions made by autonomous systems become disastrous, having unintentional ramifications such as civilian casualties or collateral damage. Increasing reliance by military forces on AI entails that there must be well-defined lines of accountability that enforce a sense of responsibility towards the actions of such a system where responsibility would be defined and enforceable.

The issue of responsibility with autonomous decision-making is very complicated to deal with. Questions of liability thereby diffuse further between the developers of technology, the military using it, and even the AI itself, while considering such complex scenarios. This makes it much more complicated in dealing legally and ethically in international humanitarian law, which obliges the forces to care for civilians when in an armed conflict situation. This will not only ensure trust in operations but also uphold the moral code. An antidote to increasing risk factors will be a critical framework of accountability that incorporates the subtleties.

To take care of the autonomy and accountability concerns, human oversight at each decision point will become inevitable. That can, therefore, significantly enhance the operational efficiency and effectiveness of AI but human judgment needs yet to form the core element, especially in critical situations involving ethical and moral considerations. Mechanisms for human oversight, such as requiring human authorization of the use of force, or keeping commanders as the final authority for autonomous operations, minimize risks involved with AI decision-making. As militaries develop an environment of collaboration where human beings and machines augment each other, they will create synergy through mutual leveraging of strengths while keeping accountable in ethical manners the behavior of operations (Taghipour et al., 2023b).

5.2 Bias and Reliability of AI Systems

Perhaps the most threatening issue in AI is its bias-one of the most critical within military applications because decisions made by AIs will cause untold harm. AI learns from datasets that contain biases due to historical prejudices or inequalities existing within data, thereby leading to skewed or unfair outcomes, such as the misidentification of threats as less threatening and indeed harming specific groups. But biased AI can undermine the achievement of operational effectiveness, create mistrust among populations, and violate ethical standards in any military context. Therefore, very important to eliminate all biases identified at any stage during the development and training of the AI systems to ensure proper, fair, effective, and efficient working in reality.

Another critical requirement in the use of AI systems is reliability, particularly because this environment, like in war, is a high stakes one. Failure on the part of an AI system to operate as designed risks missions' success and puts people's lives in danger. This is only mitigated through robust processes of testing and validation instituted throughout the lifecycle of AI technologies' development, involving rigorous evaluations against algorithms that can be driven using a variety of datasets to ascertain how it might perform across a wide spectrum of different scenarios. Further, in operational environments, real-time monitoring could detect anomalies and degradation of performance for timely fine-tuning toward reliability (Taghipour et al., 2015).

Issues of transparency and accountability must be devoted to achieving trust in AI. Openness has to be part and parcel of the practice of building AI algorithms, sources of data that have already been employed to train on them, measures taken in reducing bias and ensuring reliability or minimizing errors. All this documentation would give insight, clear on the decisions made by the AI process for military and other stakeholders. It would also entail engaging various voices in the design and evaluation of AI systems so that potential biases can be identified and overall improvement in reliability understood. Building openness and cooperation in a culture through which militaries strengthen their trust in AI technologies will ensure the responsible and effective use of operations using AI.

5.3 Impact on Human Soldiers and Warfare Ethics

The integration of AI and autonomous systems into militaries changes in some ways the roles that soldiers must play in battle. Although increased operational efficiency and greater support with decision-making may benefit from the use of AI, its nature alters the involvement itself and cuts down on the overall level of personnel engagement in combat events. This could shift the spotlight to strategic oversight and management over the frontline of fighting, hence making it require soldiers to be proficient and educated differently. The greater these are performed by AI, the greater the human elements in combat center on the decisions and judgments that have a moral character together with an ability to understand and react to situations that AI is not fully able to do.

This raises a number of serious concerns regarding how wars should be operated with the rapidly approaching use of AI as a military tool. With no human discretion, particularly in killings, it throws up some tough questions on accountability and moral responsibility when autonomous systems attack targets without any human intervention. It appears that compliance under the tenets of international humanitarian law, especially the stipulations that civilians must not be discriminated against and that there is a rule of proportionality in war, will be quite difficult. It raises ethical questions through the need to overhaul existing legal structures and to create new norms that provide for the specific nature of automated systems. Human oversight, where applicable to decisions involving life or death, has to be retained.

The acceptance of AI technologies also affects human soldiers psychologically and socially. As the war becomes even more mechanized, that emotional, instinctive tie with war the soldier had all through his existence will be lost. Instead, it might affect morale, camaraderie, and mental health, and pose a really big psychological question: how does it feel to operate in a world of machines? Even fewer controls on humans could threaten the loss of sensitiveness and ethics towards the human stakeholders involved in military operations in the future. Among psychological and social implications, there are keen issues aimed at ensuring that AI technology in war does not compromise the welfare of soldiers while still providing the reasons for being human with armed conflict.

6. EVALUATION AND PERFORMANCE METRICS

6.1 Effectiveness and Efficiency Metrics

Effectiveness Metrics The effectiveness measures of military operations while interacting with AI systems are focused on meeting strategic-level objectives and mission success. Such metrics include correctly identified threats, decisions made rapidly, and the number of successful missions accomplished with the assistance of AI systems. Analysis of such parameters would help military establishments find at which level AI technologies are contributing to an operational goal being met and if it adds more value to combat effectiveness. This evaluation thus enables determining the practical impact of AI in operations and which investments in technology would best be required in the future.

Metrics towards efficiency point out how military resources may be utilized during the integration of AI systems to those operations. It would establish whether it saves on cost, time, and whether there is efficient allocation of resources. For instance, AI might facilitate effective logistics execution in the supply chain or help cut down the time taken to deliver equipment, or ensure that troops have all the resources at the right moment. These measurements show value in AI technologies but also help to detect areas of improvement. That way, militaries will budget more effectively and cheaply during operations (Taghipour et al., 2023c; Lu & Taghipour, 2023).

A structure of continuous improvement needs to be put in place to achieve optimization effectiveness and efficiency in AI as used militarily in applied operations. Feedback Loops: Data and insights would be continuously extracted from completed missions to refine the algorithms of AI. Relevant performance metrics followed; lessons learned will be assimilated, and militaries can make constant adjustments and improvement in AI technologies, ensuring that they continue to remain responsive to changing challenges and operational requirements. It is through this progressive process that the scheme fosters a culture of innovativeness and resilience in the military as it gradually enhances the strength and performance of the military operations.

6.2 Reliability and Robustness Assessment

Reliability in Military Operations Importance

The bottom line in evaluating AI systems for military use is dependability-that is, the ability to function under extremely high stakes without failure. The general operating environment is inherently unpredictable; however, operational integrity of systems and accomplishment of mission functions depend on the reliability of the AI. Evaluations also include thorough algorithm testing on performance under varying conditions and against a variety of adversarial scenarios. It hence traces the weakness of military organization systems in depth and gives way for subsequent critical adjustments that make it possible for more reliable systems to subsequently work toward effective organizational outcomes. Robustness Against Adversarial Threats.

In other words, robustness is the capability of an AI system to function appropriately even in cases of unforeseen adversity or adversarial activities. Some of the military applications of robots include data manipulation, interference of algorithms, and exploitation of weaknesses in the system. The need for robustness arises from this as a guarantee for AI technologies to withstand such threats while still performing according to expectations. For example, development of such systems would be needed to test these types of systems against adversarial inputs or through simulating attacks and see how well such AI adapts to such challenges. Advancements in robustness will provide an even better degree of resilience against threats for militaries to have reliable and effective AI systems for operations in critical scenarios.

Continuous Monitoring with Iterative Improvements

Thus, while reliability and robustness assessment is not necessarily a single-tasking process that matures with the needs of technology or operations, the establishment of continuous assessment processes enables military organizations to monitor AI technologies in real-time as well as identify any deviation or failure that may occur during operations. Through frequent testing and fine-tuning by the feedback of operations, militaries are able to have iterative improvement of their AI systems, weaknesses identified may be eliminated, and overall improvements in those systems take place. It creates a preventive culture of accountability and resilience that ensures the dependence and effectiveness of AI technologies in support of military operations over time (Taghipour et al., 2023a; Taghipour et al., 2024).

6.3 Human-AI Interaction and User Experience Metrics

Key interaction between human and AI is required with military operations where human interaction with AI leads to the facilitation of effective collaboration between humans and AI and maximizes operational performance. The quality and interaction determine well how the personnel may be able to apply AI insights to good decision-making and execution of given missions. The human-AI interaction dynamics would include the user interface, transparency over the insights brought in through AI, and intuitiveness of the system control interface. Military organizations, being user-experience-focused, can ensure that people can have satisfactory interaction with AI systems, thus further increasing decision-making and mission success.

User experience metrics are consequently important know-how for understanding patterns in interaction between military personnel and AI. Such metrics may for example include usability tests, user satisfaction ratings, and efficiency with which activities are performed. It is indeed through the feedbacks from users that militaries can establish which points of the process of interaction cause pain. It may, for instance, relate to being unable to interpret outputs by an AI or how to use the interface of the system. Others could be measures of effectiveness in terms of time taken to accomplish a particular task and error rates of personnel usability in relation to the application of AI technologies in operations contexts. The data generated will better inform improvement for the AI systems that have been designed to enhance the generalized user experience (Taghipour, 2023).

Having feedback loops thus provides an excellent basis on which good human-to-AI interaction and optimized user experience are built. Military organizations can, therefore, track improvement areas through regular collections of input from personnel who interface with AI systems. Such mechanisms should be embodied in the design and development process so that enhancements may be iterated from the start. Then, these training programs can further be designed as proficiency building using AI technologies in order to provide the user experience towards the competent and effective use of these tools by people. It would mean continued focus on improvement and then operate better and improve efficiency, for instance supporting a culture of collaboration and trust between the human operators and the AI systems.

7. FUTURE DIRECTIONS FOR AI METHODOLOGIES IN UPCOMING MODERN WARFARE

7.1 Advancements in Autonomous Systems and Robotics

For instance, autonomous systems and robotics drastically change the dynamics of military operations that are able to make complex missions more accurate and efficient. These systems - drones, unmanned ground vehicles, and autonomous naval vessels - are intended to carry out missions in a sequence from reconnaissance to combat, with no human direct control. In reality, advanced sensors, machine learning algorithms, and computer vision technologies have been catapulting the performance of autonomous systems into even the challenging environment, targets recognition, and adaptability according to dynamic conditions. According to the advances in technology, they start to change from the backstage of modern warfare to centre stage reformulating the traditional military strategies and tactics.

Integration of autonomous systems in military operations can offer a means to enhance the operational capabilities for supplementation to human soldiers. For instance, AI-designed drones may carry out real-time surveillance of vast territories and are also potential for gathering intelligence hit targets with accuracy. The use of UGVs can also be used for logistics support tasks where human personnel are perceived risky to be put in the field, lessen more risks, and conserve resources. The interaction between human operators and these systems also makes it more flexible to engage militarily, since soldiers stay focused on strategic decisions and leave routine or dangerous work to machines. This tends to augment general effectiveness and rates of mission accomplishment.

More recent acceleration and development of autonomous systems and robotics bring increasing ethical and regulation matters with the deployment of autonomous weapons raising accountability issues since lethal decision-making is being placed in machines. International humanitarian law also requires distinctions between combatants and non-combatants, and there would be a need to ensure that autonomous systems respect such distinctions so as not to cause unlawful harms. There would be a need to lay down regulatory frameworks governing development and use to govern their development and use: bias, reliability, and misuse. It is important that military leaders, ethicists, and policymakers continue this debate and come up with guidelines which will ensure the prudent integration of such autonomous systems into military operations (Taghipour & Beneteau-Piet, 2020; Mahfod et al., 2019).

7.2 Integration of AI With Emerging Technologies

Thus, the integration of artificial intelligence with the Internet of Things, 5G communications, advanced sensor systems, and many more futuristic technologies is changing warfare scenarios. AI can enhance other technologies by allowing real-time processing of data, advanced analytics, and autonomous decision-making capabilities. This starkly contrasts with the current scenario in which IoT devices can collect huge data amounts from thousands of sensors distributed in a battlefield and AI algorithms can process this data to identify patterns, predict threats, and optimize the resource structure. Synergy helps in better understanding the situation, then makes informed decisions using better response strategies towards dynamic environments by the military forces (Tolk & Diallo, 2019).

It is the integration of AI with emerging technologies that significantly increases military application efficiency through operations. With 5G, ultra-reliable low-latency communications can be achieved by military units in ensuring coordination over complex operations in real time. In this scenario, AI can use connectivity in transferring data seamlessly among various systems so that cooperation between ground troops, drones, and command centers can be enhanced better. AI-driven predictive maintenance systems can monitor the health of equipment and vehicles so that there is less time lost due to equipment downtime and resources are always available when needed. Thus, there will be overall mission effectiveness at the point of optimizing resource use but at the same time minimizing operational costs.

However, some challenges persist with the high benefits associated with the integration of AI with other emerging technologies. Good integration consideration factors are interoperability between heterogeneous systems, strong cybersecurity and ethical concerns on the horizon. As military organizations begin to adopt these technologies, the continued development of these areas of R&D will need to focus on ensuring cohesion in AI behavior across emerging technologies. More important, however, will be the training forces to use integrated systems so their potential is realized in actual battles. Many things will surely change when we see actual battles of future military operations with AI unveiled in the next few decades when developing technologies unfold.

7.3 Ethical AI Development and Governance Frameworks

Hence, the military will have high technologies representing value-and morals-oriented and societal norms. Systems with minimal damage will show the practice of military fairness, accountability, and transparency in the conduct of war. Algorithms will not have preferences as regards the human operators' decisions on factors considered by them related to the international humanitarian law. Then there is the process of bringing these latent ethical considerations into design and development phases for militaries that will often represent the first steps in forming trust regarding AI technologies thus lowering risks associated with deploying them into more responsible and effective military actions.

On issues of regime over AI use in military operations, therefore, it goes without saying that robust structures of governance come across as a robust prerequisite.

The requirements would thus include clear guidelines on AI development, deployment, and oversight-through agreement on accountability, data, and the ethical use of AI. In this regard, the frameworks of AI have to reflect and embrace diverse voices raised by multiple stakeholders-from leaders in the armed forces through to ethicists, jurists, and civil society.

These governance frameworks would only be sustainable if there exists a continuing assessment and updating of these frameworks so as to continue squaring off with the changing dynamics of AI technologies that change as frequently as any new advanced technology. Implementation and Continuous Assessment The system should be continuously assessing the application to ensure proper implementation of ethical AI governance frameworks that examine the implications which AI technologies have when applied to the exercise of military power. In this regard, independent review boards or ethics committees can feature within the system checking AI systems against established standards on ethics. The accountability climate encourages people to coproduce responsible AI technologies in collaboration with military organizations since they first consider matters that must be related to ethics with regard to AI technological applications. Responsible AI training programs that focus on the development and governance of AI systems will capture the attention of military with choices while using the AI systems again. Ethical governance sits atop of all else for it enables militaries to effectively navigate the nuances of these fields in AI technologies while keeping their commitment to how warfare could be conducted responsibly and humanely (Russell & Norvig, 2016; Sharkey, 2019).

8. CONCLUSION

The infusion of AI methodologies into modern warfare systems is considered to be the bringing about of a paradigm shift in the military operations as it shifts the nature and structure through which conflicts are conceptualized and executed. While nations advance through technology, they become quick to start applying them-a great rush to the edge to maximize situational awareness, decision-making capacity, and operational efficiency-and AI is the critical enabler. The application of AI will help military forces digest enormous volumes of data, automate processes that allow them to make faster decisions, and appreciate autonomous systems whose existence complements their strategic capabilities. That step forward, however, presents diverse challenges that need cautious handling, including questions of responsibility and ethics.

The biggest benefits of AI approaches is undoubtedly the increase in operational effectiveness. With machine learning algorithms and big data analytics, for instance, military forces can find out things no one else has ever known about threats and dynamics in a battlefield. AI technologies will analyze real-time data streaming in from sources as varied as satellites, drones, and IoT devices, among others, and commanders will make decisive decisions. The critical skill, therefore, will boost response times but also lower the risks of human error and will have better mission outcomes since missions are successful. But with so many advantages that deployment of AI in warfare has brought, however, it also raises critical ethical and regulatory concerns. Development of AI systems with all such potential bias in algorithms, delegation of lethal decision-making functions to machines, and implications on civilian safety would demand a stronger conceptual framework able to guide development and deployment. Clarity on accountability measures and governance structures should instill assurance of responsible use of AI technologies aligned with international humanitarian law. Engagement with diverse stakeholders such as ethicists, military leaders and legal experts will prove essential to the development of the full governance framework.

Human-AI interaction constituted the primary element of the full integration of AI within military operations. Where the role of more distributed and pervasive autonomous systems increases, the need for ensuring that human operators can easily interface with these technologies is even greater. A good user experience metric along with continuous training will enhance the cooperation between human and AI to take full advantage of AI systems.

Above all, the integration of modern warfare and AI methods into war will depend on human judgment and machine intelligence working as one environment. However, efficiency and reliability of AI systems require rigorous testing so that they might work reliably in different situations. Military organizations should continuously review and fine-tune AI technologies step-by-step so that they could address some of the vulnerabilities and increase performance. A preventive approach toward the reliability and robustness assessment will enable militaries to rely on dependable assets on the battlefield if they could adapt to changing scenarios of threats and challenges. Thus, the future of modern warfare remains a continuous process embracing developments in autonomous systems and robotics. Actually, it will redefine traditional combat roles but most importantly, its logistical capabilities as well as mission planning and intelligence gathering. Such integration should thus be guided by such ethical considerations as adhering to the overriding requirement of human oversight in critical decision-making processes.

Automation should be balanced with the inclusions of human involvement to maintain the law of war while using the benefits of AI. The potential for the future of AI is huge in transforming the face of modern warfare. However, the problems seen in current AI methodologies will depend on research and collaboration between military organizations, technology developers, and regulatory bodies as they overcome these problems.

The strategic benefits of AI for a nation call for similar attention to the ethical implications and social consequences of AI. A responsible and well-thought-out approach towards AI development will both enhance military efficiency and have a safety-enhancing effect in building a more just and just global security environment. Conclusion In this regard, the infusion of AI methods into the new generation of modern warfare systems holds much promise with a high number of challenges. The considerations on ethics, encouraging human-AI collaboration, and demanding strong governance frameworks are among the factors that can unlock transformational power promised by AI with no compromising accountability and responsibility. Therefore, this future redefinition of AI in war will define the military operation of tomorrow and shape the course of international relationships and perhaps even the ways of conflict resolution in an interconnected complexity-require world.

REFERENCES

Binnendijk, H., & Libicki, M. C. (2019). *Artificial Intelligence: Implications for the Future of Defense*. RAND Corporation.

Bishop, J. (2021). Artificial Intelligence in Warfare: Opportunities and Challenges. *Journal of Defense Studies and Resource Management*, 9(1), 1–11.

Cauchois, R., Taghipour, A., Kang, D., Zoghlami, N., & Abed, M. (2017). Advanced solutions for a supply chain with stochastic information. *Journal of Advanced Management Science*, 5(1), 9–13. DOI: 10.18178/joams.5.1.9-13

Cummings, M. L. (2017). Artificial Intelligence and the Future of Warfare. Center for a New American Security.

Dadsena, K. K., Mathiyazhagan, K., & Taghipour, A. (2021). Analysis of barriers for the build the resilient supply chain networks post-COVID-19. In *Managing Supply Chain Risk and Disruptions: Post COVID-19* (pp. 79–89). Springer International Publishing. DOI: 10.1007/978-3-030-72575-4_7

Glaa, B., Zoghlami, N., & Taghipour, A. (2014, May). Outsourcing and off-shoring healthcare services: A way to cut the costs and improve quality. In 2014 international conference on advanced logistics and transport (ICALT) (pp. 189-194). IEEE.

Heath, T., & Roper, A. (2021). AI in Defense: A Review of the Ethics, Governance, and Policy Issues. *AI & Society*.

Loivet, W., Taghipour, A., & Kang, D. S. (2020). The rise of green supply chain management: Between complexity and necessity. *J. Econ. Bus. Manag*, 8(1), 1–7. DOI: 10.18178/joebm.2020.8.1.604

Lu, X., & Taghipour, A. (2023). A French Guide for Businesses and Governments to Integrate the Concept of Sustainable Development Into Their Supply Chains. In *Government Impact on Sustainable and Responsible Supply Chain Management* (pp. 1–12). IGI Global. DOI: 10.4018/978-1-6684-9062-4.ch001

Mahfod, J., Canel-Depitre, B., & Taghipour, A. (2019). Quality function deployment-ELECTRE in supplier evaluation. Journal of Advanced Management Science, 7(4).

Mbiatem, J., Taghipour, A., & Canel-Depitre, B. (2018, August). Supplier selection approaches for decision makers. In *Proceedings of the 8th International Conference on Information Communication and Management* (pp. 108-112).

Merimi, M., & Taghipour, A. (2021). Accelerating the digitalization of the supply chain: An empirical research about COVID-19 crisis. In *Digitalization of decentralized supply chains during global crises* (pp. 1–24). IGI Global. DOI: 10.4018/978-1-7998-6874-3.ch001

Nadia, E. N., Beatrice, C. D., & Atour, T. (2021). The determinants of hotel room rates in Beirut: A hedonic pricing model. International Journal of Trade. *Economics and Finance*, 12(2), 33–42.

Russell, S., & Norvig, P. (2016). *Artificial Intelligence: A Modern Approach* (3rd ed.). Pearson.

Sharkey, N. (2019). The Ethical Implications of AI in Warfare. In *Artificial Intelligence and International Relations: A Security Perspective*. Routledge.

Sparrow, R. (2020). Killing by Robot: The Ethics of Autonomous Weapons. In *The Oxford Handbook of Ethics of AI*. Oxford University Press.

Taghipour, A. (2014). Improving the plan of a manufacturing network with non-integrated business units. *International Journal of Applied Logistics*, 5(2), 1–11. DOI: 10.4018/ijal.2014040101

Taghipour, A. (Ed.). (2020). *Demand forecasting and order planning in supply chains and humanitarian logistics*. IGI Global.

Taghipour, A. (Ed.). (2023). *Government Impact on Sustainable and Responsible Supply Chain Management*. IGI Global. DOI: 10.4018/978-1-6684-9062-4

Taghipour, A., Abed, M., & Zoghlami, N. (2015, May). Design for remanufacturing respecting reverse logistics processes: A review. In 2015 4th International Conference on Advanced Logistics and Transport (ICALT) (pp. 299-304). IEEE. DOI: 10.1109/ICAdLT.2015.7136602

Taghipour, A., & Beneteau-Piet, C. (2020). Sustainable supply chain management performance. *International Journal of Innovation, Management and Technology*, 11(6), 165–169. DOI: 10.18178/ijimt.2020.11.6.897

Taghipour, A., Foukolaei, P. Z., Ghaedi, M., & Khazaei, M. (2023a). Sustainable Multi-Objective Models for Waste-to-Energy and Waste Separation Site Selection. *Sustainability (Basel)*, 15(22), 15764. DOI: 10.3390/su152215764

Taghipour, A., & Frayret, J. M. (2011). Supply chain coordination planning: A review. Work paper, École polytechnique de Montréal.

Taghipour, A., & Frayret, J. M. (2013a, May). An algorithm to improve operations planning in decentralized supply chains. In *2013 International Conference on Advanced Logistics and Transport* (pp. 100-103). IEEE. DOI: 10.1109/ICAdLT.2013.6568442

Taghipour, A., & Frayret, J. M. (2013b). Coordination of operations planning in supply chains: A review. *International Journal of Business Performance and Supply Chain Modelling*, 5(3), 272–307. DOI: 10.1504/IJBPSCM.2013.055729

Taghipour, A., Lu, X., & Smahi, M. (2022). Opportunities and challenges for society 5.0 in procurement. *Blockchain Technology in Supply Chain Management for Society*, 5(0), 91–100. DOI: 10.1201/9781003177432-9

Taghipour, A., & Merimi, M. (2021, March). Digital transformation of supply chains during crisis: covid-19. In 11th Annual International Conference on Industrial Engineering and Operations Management, IEOM (pp. 7493-7501). DOI: 10.46254/AN11.20211298

Taghipour, A., Padash, A., Etemadi, V., Khazaei, M., & Ebrahimi, S. (2024). Sustainable and Circular Hotels and the Water–Food–Energy Nexus: Integration of Agrivoltaics, Hydropower, Solar Cells, Water Reservoirs, and Green Roofs. *Sustainability (Basel)*, 16(5), 1985. DOI: 10.3390/su16051985

Taghipour, A., Ramezani, M., Khazaei, M., Roohparvar, V., & Hassannayebi, E. (2023b). Smart transportation behavior through the COVID-19 pandemic: A ride-hailing system in Iran. *Sustainability (Basel)*, 15(5), 4178. DOI: 10.3390/su15054178

Taghipour, A., Sohrabi, A., Ghaedi, M., & Khazaei, M. (2023c). A robust vaccine supply chain model in pandemics: Case of Covid-19 in Iran. *Computers & Industrial Engineering*, 183, 109465. DOI: 10.1016/j.cie.2023.109465

Tolk, A., & Diallo, S. Y. (2019). *Modeling and Simulation Support for System of Systems Engineering Applications*. Wiley.

Zoghlami, N., Taghipour, A., Merlo, C., & Abed, M. (2016). Management of divergent production network using decentralised multi-level capacitated lot-sizing models. *International Journal of Shipping and Transport Logistics*, 8(5), 590–604. DOI: 10.1504/IJSTL.2016.078686

Chapter 10
Evaluating Gender Bias in Computational AI-Based Approaches:
A Case Analysis of LinkedIn

Janhavi Gautam
Chitkara University, India

Kapil Sharma
Chitkara University, India

Hakam Singh
Chitkara University, India

Ramamani Tripathy
Chitkara University, India

Nilamadhab Mishra
https://orcid.org/0000-0002-1330-4869
VIT Bhopal University, India

Rudra Kalyan Nayak
https://orcid.org/0000-0003-4447-8391
VIT Bhopal University, India

ABSTRACT

Ethics in computational AI-based approaches have become an inclusive topic of interest among researchers when artificial intelligence has come to light. Doubts have always existed against the ethical implications of introducing artificial intelligence to society. With the advent of artificial intelligence, these voices have only grown louder. Over the years, many guidelines and policies have been introduced to ensure network security, economic development, and quality of life. However, the impact of these policies and whether they would be sustainable is still under consideration. There are too many positions where biased data sets and algorithms are used. The outline of this research is arranged in a way that allows an analysis of the algorithm used by a well-known application. The data being used here has been amassed from a varied set of people, and rigorous inspection of the data has occurred to produce thorough results.

DOI: 10.4018/979-8-3693-7478-8.ch010

INTRODUCTION

Multiple AI technologies have universally recognized benefits, but there is still hesitation in implementing AI and ML on a full scale. A significant reason is the lack of believable, well-defined ethical rules for every AI. Banking, health, and retail are just a few sectors where AI is currently widely used. A leading reason for the widespread introduction of AI was increased productivity and decreased production costs. But with time, the question has arisen: Do these technologies bring more economic good than ethical harm, and are they worth the risk? (Kukreja, V. et al. 2022; Srikumar, M et al. 2022; Eitel-Porter, R. 2021; Glaa, B. et al. 2014) Autonomous systems are created based on AI concepts as they are fully independent and capable of decision-making. Hence, there are multiple fundamental queries, such as the risks involved in utilizing an almost sentient technology and how these systems can be properly controlled. Giving AI autonomy and free reign amongst all civilizations has always been a matter of great discussion.

There has been debate as to whether a political environment should be established for AI to ensure its ethical usage. Otherwise, there could be significant misuse of AI technologies, as was previously witnessed during the invention of powerful tools such as nuclear power. Studying ethics is not just about establishing a better system of AI usage for the technical industry (Kumar, A. et al., 2023; Lu, X. et al., 2023). AI affects all social, cultural, economic, political, and legal spheres. Thus, ethics are needed to control AI's cultural and societal effects. As AI Ethics has gained more recognition worldwide through the internet and media, the research on the subject has also increased. However, there is still insufficient research on this topic (Kathuria, Y. et al., 2024; Sharma, E. et al., 2021). IEEE has stated that the study on AI Ethics is still severely lacking. Even AI practitioners lack basic training about the ethical ramifications of AI.

Bias in Machine Learning

ML algorithms are quite valuable for various commonplace and professional instances. This is why they are slowly yet steadily seeping into our lives. However, although these algorithms exhibit multiple advantages, they are known to exhibit biases in their results that can have harmful and troublesome effects (Benjamins, R.2021, Dhiman, N. et al., 2023). Back in 2015, an experiment conducted at CMU revealed bias in the search results provided by Google. If Google thought that the job seeker conducting a job search was female, it was significantly less likely to show ads for higher-paying executive jobs than if it thought the seeker was male. Also, if someone searched" CEO", the top images shown were all those of white males. The study raised concerns about discrimination and the dearth of transparency when working with these algorithms. These algorithms make important societal decisions by influencing what a consumer can view online. This brings us to the incident on Facebook where Facebook was criticized for using biased algorithms that heavily influenced the voting results by showcasing geared content to users. In an identical context, lines of racism have existed in image-related technology. There had been occurrences of Flickr's image recognition tool displaying racist remarks such as providing" animals" as a result when supplied with images of black people reported. Hewlett-Packard has a web camera software that was reported to have had trouble recognizing darker skin tones. Nikon's camera software inaccurately identified Asian people as blinking.

In another such instance, the New York Times divulged that following the presidential elections in the USA in 2016, Facebook executives gathered to discuss the implications that Facebook may have had on the election results (Drabiak, K., 2023, Daneshvar, S. et al., 2018). According to multiple reports, Facebook was found to have pushed news articles that propagated fake news, hence misleading the masses and turning the tides of the election against the candidate of the election, Hillary Clinton, while causing an increase in the preexisting societal divide amongst the Americans in a way that favored Donald Trump. While Google has had a much subtler gender or race bias, Facebook is known to have an algorithm that is overtly involved in the distribution of content that is sexist or racist to billions, resulting in the normalization of the consumption of such media. Another recent example of gender bias in algorithms was when the new voice assistant Alexa by Amazon was said to respond better to male voices than to high-pitched female voices due to an alleged lack of training on female voice modules for the assistant (Ashurst, C. et.al., 2022; Birhane, A. et.al., 2022; Mahfod, J. et.al., 2019). These are just some real-life instances that exhibit the bias and ethical concerns of using ML algorithms extensively daily. This is why there's an increasing need to work on these biases to ensure a fairer, more transparent working of AI applications so that they can be integrated better into society.

Types of Biases

Biases can manifest themselves in an assortment of ways. These three predispositions lead to mistaken information circulation among clients and misleading data. Priority goes to algorithmic bias. An algorithmic model's execution is said to be hampered by a design flaw or other obstacle, resulting in unfair or discriminatory outcomes. Algorithms for facial recognition often perform better on individuals with lighter skin tones than on those with darker skin tones. Another type is data bias. This is the inclination created because of the cosmetics of the preparation set of information for the model. Suppose the historical data used to train the model reinforces gender or racial biases that have always existed in society. In that case, the model is more likely to reflect those biases—finally, social bias(Dadsena, K. K. et.al 2021, Balasubramaniam, N. et.al 2020, Hagendorff, T.et.al 2020, Halim, Z. et.al 2020, Huang, C. et.al 2022) . As was recently noted, many biases are imbued in our general public. Any ML model that catches these aberrations is considered to have social inclinations. It is common knowledge that criminal justice data frequently contains injustices. When these data are used to train machine learning models, we get models that distinguish between results and predictions.

Ramifications of Bias in Machine Learning.

The system's algorithm's reach, applications, and other aspects can all affect the bias of a machine learning algorithm. These could have many repercussions, including poor decisions and increased social inequality. Beneath, we'll investigate a portion of these outcomes. Bias in machine learning systems is probably to blame for discrimination. It is very likely that the results of a biased model, which was most likely developed due to using a limited training set of data, will discriminatorily favor certain groups while undermining those particular individuals. This is very likely because of the bias in the data. For instance, a one-sided credit acknowledgement mechanism may result

in an unjustifiable restriction on advances to individuals from obtaining specific gatherings, which fuels existing imbalances and exacerbates current financial difficulties (Lu, X. et al., 2023; Joo, Y. et al., 2019; Cell, D. I. A. I. 2023). There will continue to be bias if biased models reinforce social stereotypes in the data frequently used to train these models. These records often contain judgmental errors the company makes because they come from that company. By limiting opportunities for candidates who don't fit these profiles, algorithms that favor candidates with a certain level of education run the risk of spreading bias. Its most troubling impact is the potential for undesirable and unintended consequences of bias in machine learning. In any case, predisposition can be considered an unfortunate trademark when unintentional. Inequitable distribution of medical resources can result from biased healthcare algorithms. The strength of certain networks might endure thus. In machine learning, biases have an impact. This effect goes far beyond a simple error, affecting individuals in real life and, surprisingly, entire groups. Recognizing and taking proactive measures to mitigate these effects is essential for designing and implementing fair and impartial machine learning systems.

REVIEW

Ethics in AI is a prevalent topic that has gained interest quite steadily with the advent of AI technologies. There are multiple literature review articles available online on the subject. To a very detailed extent, these papers discuss the importance of establishing principles in AI ethics while also elaborating on the specific principles they believe are necessary. Literature reviews also summarize these works to create a standardized set of rules. With time, the research in this sphere is bound to grow as the requirement for such a set of rules continues to increase (Kukreja V. et al., 2022; Srikumar M. et al., 2022; Huang C. et al., 2022). There is a systematic literature review where the researchers have studied 21 papers and articles to answer two questions –" RQ1: What are the key principles of AI ethics? RQ2: What are the challenges of adopting ethics in AI?" and have created a set of guidelines based on the results. (Cell, D. I. A. I. 2023; Basti, R. 2019; Dave, T. et.al., 2023). The importance of firms using data and AI ethics because of the increasing risk posed by legal, regulatory, or reputational issues is highlighted in this article. It contains an exhaustive seven-step plan for developing customized, flexible, and durable data and AI ethics programs to find commonly inefficient practices. Some recommended actions are to take advantage of existing infrastructure and develop frameworks specific to the industry, learn about medical ethics, improve guidance on product managers, raise awareness among the organization staff, reward them, or actively seek impact through stakeholder engagement. This essay focuses on AI ethics and the advantages of investing in operational Data (Barhate et al., 2023; Ananny M. et al., 2018; Fatima M. et al., 2018). Climate Change, Artificial Intelligence, Genomics, and Ethical Frameworks for Science and technology are among the themes of an International Initiative led by UNESCO. As technology evolves, the ethical implications of AI will become more prominent, e.g., threats to humanity's rights, environmental protection, and bias. All 193 Member States have signed up to the UNESCO ethical AI Paradigm, which stresses key principles such as diversity, human rights, and environmental well-being. Many standards aimed at adjusting simulated intelligence research for humanity are set out in the structure. Some guiding principles are fairness, proportionality, security, and transparency. A good policy for accomplishing these aims is available from UNESCO.

As our society is impacted directly by AI, various survey-based research is being conducted on various groups of people to collect their thoughts on the upsurge of AI and ML technologies. Some of the analysis is aimed at people of all ages and walks of life. In contrast, others are aimed at people in certain professional fields, like doctors or nurses, who are being surveyed on the impacts and social implications of the implementation of AI in the medical field. Fiske, A. et al. 2019, Ashurst, C. et al. 2022] A widely known example of survey-based research being conducted on the ethical implications of AI technologies is a new institute that publishes yearly reports that frequently contains survey data and analysis of a wide range of AI ethics-related topics. (Mahfod, J..al., 2019; Ahmad, H. et al., 2020) Various conferences are being conducted worldwide. (Akimushkin, C. et al., 2018; Dadsena, K. K. et al., 2021) To examine studies presented at prestigious meetings, Fact and AIES in the last four years, a review of research on Ethical Artificial Intelligence has been made public. The investigation has revealed that while most distribution systems have a good aim, they frequently fail to contemplate the negative effects of simulated intelligence on generally reduced populations. The paper concludes that a more concentrated ethical analysis of structural and historical power imbalances could benefit the area based on real-world use cases, experience, and people's applications. It declares that there is sex.

Alternatively, there exists research that aims to find out if a particular technology is biased or unethical. With the rapid release of new technologies, the need and potential to research their ethical implications and the possible errors in their algorithms are on the rise. As such, (Rahman, A. U. et.al., 2022) research has been conducted extensively on technologies extending from studying biases in medically used algorithms to platforms like Chat GPT; both are being used extensively, albeit in vastly different ways. Certain papers talk specifically about the ethical implications of AI. The research provides multiple examples of instances when AI models failed their audience and showed serious discrimination. It also discusses the steps that must be taken to ensure that AI remains free from bias. This text addresses new technology development ethics fields, especially impact assessment. Although this paper considers the significance of ethical reflection, it also contains explicit problems, such as the diversity and possible equivocal nature of these opinions. This stresses the importance of relevant application over calculated deliberation and proposes that the genuine trouble lies in trying moral standards in unambiguous circumstances. The book states that there's a need to make decisions, and the ethical frameworks give guidance, but they do not always bring good results. It also highlights the importance of management frameworks and external reviews to ensure that impacts are accurately assessed, especially for new technologies when they are constantly changing. The text also discusses the role of impact assessments for technology management and research, recommending that they be alleged as a tool to promote learning and improvement instead of hindering it.

The designing of an ethical AI can be as complex as it is necessary. The study shows how difficult it is to build ethical AI systems in the arena of artificial intelligence while dealing with moral questions. This illustrates the requirement of moral artificial intelligence frameworks in shaping the public eye's understanding of computer-based intelligence. It is necessary to cooperate in AI ethics between users, policymakers, scientists, engineers, and philosophers.

The paper presents an overview of the risks, rules, and principles related to artificial intelligence ethics and a full explanation of AI ethics in which it is necessary to assess them. It notes the difficulties of applying AI ethics. It offers suggestions on areas for future research that need particular attention, making it an excellent starting point for people interested in learning more about this topic.

The main objective is to refocus AI ethics to resolve disagreements arising after implementing abstract concepts. Even though maximum struggles are difficult to determine, they must be defined and explained. They must be transformed into standards, rules, and regulations if they are to be implemented. These documents must address conflicts and guide how to resolve them within various contexts. The strategy encourages reflection on a wide range of perspectives and ideals. Tensions also need to be addressed to identify knowledge gaps on how AI can be used in society, prioritize research to solve real-world problems, and encourage the development of AI in a way that will be very helpful.

The inspiration for this particular research was derived from Carnegie Mellon University research in 2015 (Joo, Y. et.al., 2019). According to the analysis published in July 2015 by Carnegie Mellon University, fewer women than men were recommended jobs paying higher than $200,000 by Google ads. AdFisher, a tool developed by CMU, created simulated user profiles to run 21 experiments evaluating the Ad Settings. The researchers used AdFisher to create 1000 fake profiles and had them surf through the top 100 job-searching websites. Out of all these, the online site most powerfully related to the male audience was a career training site, and there was a discrepancy between the male and female accounts. According to one of the researchers, the male accounts saw high-paying jobs about 1800 times, while the female accounts only came across them about 300 times. However, there is no proof that Google has done anything illegal or anything to violate its very own policies. Although AdFisher can find such errors, it cannot determine their cause.

METHODOLOGY

This research aims to collect the required information from appropriate parties and then analyze the collected data thoroughly to reach accurate conclusions. To achieve this, the first step is to generate a survey form as it's easy to circulate and can efficiently collect the needed responses. To do this, we need to discuss which questions are to be added to create a set of questions that, concisely and accurately, can extract the required information from the targeted subject. A collection of easy-to-understand questions has been selected to increase readability and circulated as a Google form. These results are used to depict whether or not the algorithm differentiates between male and female users. The application is just an example, it is not used to represent all modern-day social media applications. LinkedIn has been around for a long time. Although its popularity has depreciated with the rise of newer platforms, it is still widely used, partially due to its unique concept.

Survey Questioner

The form must be circulated only to a certain relevant group of people. For this study, we have selected individuals who fulfil certain criteria

- All participants must be above 18
- All participants must be below the age of 60
- All participants must either be students, employed, or currently in the job market

These conditions are necessary as we require responses from individuals likely affected by a platform such as LinkedIn. Also, in the description of the form, it has been explained that the purpose is to conduct a gender-based survey and detect any possible bias in the job advertisements provided by LinkedIn. All the participants are informed through this step that they will require a LinkedIn account to respond to the survey questions.

Gender Rules

The primary information to be obtained through this survey is the subject's gender so that we can organize the data provided by all participants based on their gender. This is an important step as it forms the basis of this research. Upon confirmation, the participant is asked to open their LinkedIn page and search a keyword in the search bar provided. To maintain uniformity, all are requested to search the word 'jobs' in the job section of LinkedIn's search bar. The individual must select the most accurate option based on their top search results.

Regarding Job's Displayed

The next question we need to answer is the kind of jobs displayed to each individual. To collect this data, we must create elaborate questions that remain easy to understand and answer for convenience. The first question devised is the job level in a corporate hierarchy. Or if the job is not corporate, then under which other category does it fall? Finally, 11 divisions have been created, and the following are the options that will be provided in the survey:

- High-ranking administrative position (executive positions)
- Mid-ranking administrative position (managers, coordinators)
- Low-level administrative position
- Entry-level position (marketing, sales, accounting, human resources, programming, etc.)
- Secretarial position
- Specialist fields (lawyer, doctor, scientist, mathematician, etc.)
- Education-related fields
- Creative jobs (writers, artists, etc.)
- Freelancing work
- Positions that do not require much training or aren't skill-based in any shape
- No valid search results

The first five options provided cover almost any position in a corporation. The next five options refer to other varied fields. The last option is provided as a safety measure in case no valid job options are found in the search results.

Regarding Salary Expectations

As the distinction of a job cannot be based solely on these categories for the results we need, the next option covers the arguably more important financial aspect. The expected monthly salaries for the jobs filled in the question above this one would be asked. The individual is requested to answer based on the particulars provided in the job advertisement and at their discretion. The available salary options that the participant is provided with:

- 0 - 10,000
- 10,000 - 50,000
- 50,001 - 1,00,000
- 1,00,001 - 5,00,000
- 5,00,001 - 10,00,000
- 10,00,000+

User's Satisfaction

Lastly, an extra question is added to inquire whether the results are relevant to the user. This question can help us determine whether the results that people found—based on which they answered the survey—were accurate for them.

Age Range

The form is circulated amongst an assortment of people above 18. These people are, for the most part, in universities or settling into corporate jobs. The fields they are a part of extend from Computer Science to Commerce. In this manner, we can collect varied data that is not repetitive or redundant.

Responses

Finally, the data collected is extensively studied to conclude. The data is first divided by gender. Then, the three questions are sorted according to the three questions we created. The collected data is first sorted into male and female and separately analyzed better to understand the trends in the responses of each gender. Then, the response rates for each option in the three questions are compared according to gender. Hence, conclusions can be derived from the data that has been collected.

ANALYSIS

The form was circulated to 200 people, and each person was ensured to have met the criteria set. We have collected 110 responses in total. The first question to be answered in the form is about the participant's gender. Two selections are provided: male or female. This significant distinction must be made to derive the required results. The data has to have an equal number of male and female participants to obtain standardized results.

Female Responses

A Study of the Responses by Female Participants The first question asked which job position the participant was recommended for. Most women chose the option 'Mid-ranking administrative position (managers, coordinators),' followed by the'High-ranking administrative position.'

Figure 1. Organization of female responses to question 2 inquires the top job results into percentages using a pie graph

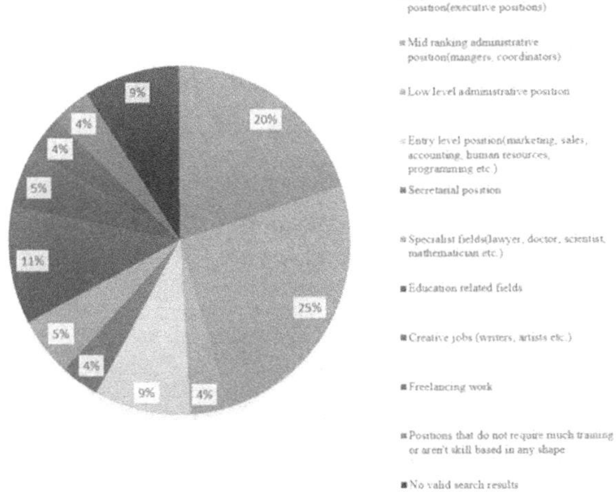

The rest are mostly equal. There are four choices with 4% polls. These selections include 'Freelancing work' and 'Positions that do not require much training or aren't skill-based in any shape.' Both of these stipulate low-earning potential occupations. 'No valid search results' received 9% of polls, which is about average, as depicted in Figure 1.

Figure 2. Organization of female responses to question 3, which inquiries about salary expectations in percentages using a pie graph

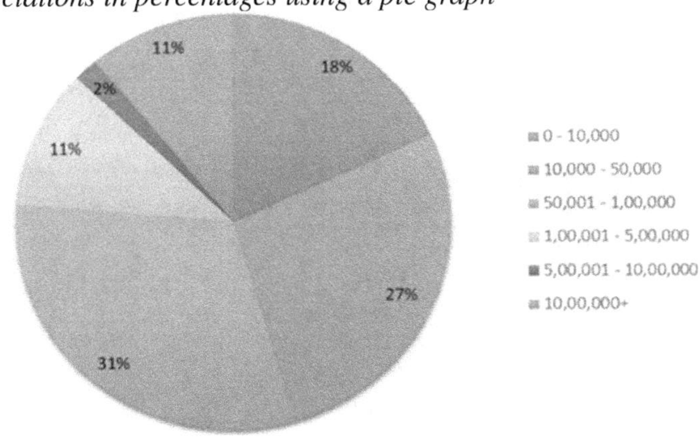

Most responses were 50, 001 – 1,00,000, followed closely by 10,000 – 50,000. Only 2% of people have chosen 5,00,001 – 10,00,000. There are an average number of replies for the bottom-most pay grade,' 0-10,000'. The number of persons who have chosen '1,00,000 – 5,00,000' and' 10,00,000+' are almost equivalent. The last question was whether the participants thought the results were relevant to them, as depicted in Figures 2,3,4.

Figure 3. Organization of female responses to question 4, which inquiries about the relevance of the job, is expressed in percentages using a pie graph. Most people have voted Yes

Figure 4. Shows the Organization of male responses to question 2, which converts the top job result into percentages using a pie graph

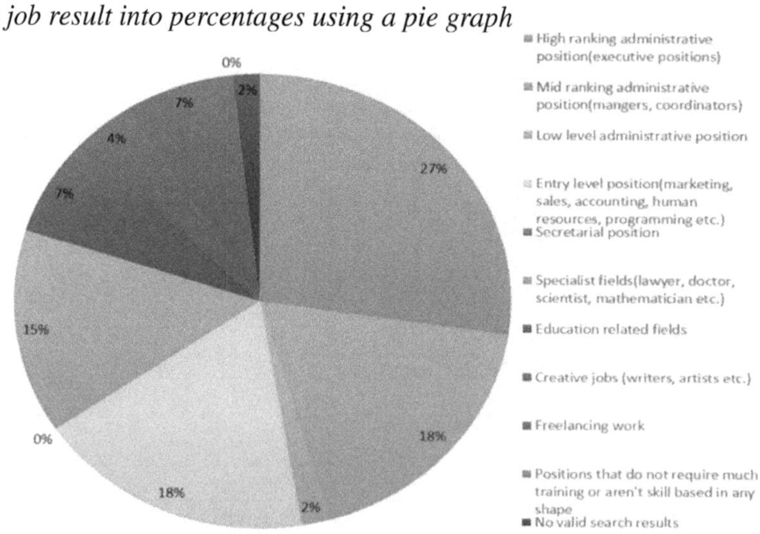

Male Responses

Firstly, we discuss the job options selected by the participants. The majority of people have chosen a high-ranking administrative position (executive position)' followed equally by ' Mid-ranking administrative positions (managers, coordinators)' and ' Entry-level positions (marketing, sales, accounting, human resources, programming, etc.).' On the other hand, 'Secretarial position' and 'Positions that do not require much training or aren't skill-based in any shape' weren't used by any of the male participants. The second question is about the financial aspects of these trades that have been chosen.

Figure 5. Shows the Organization of male responses to question 3, which uses a pie graph to inquire about the salary expectation in percentages

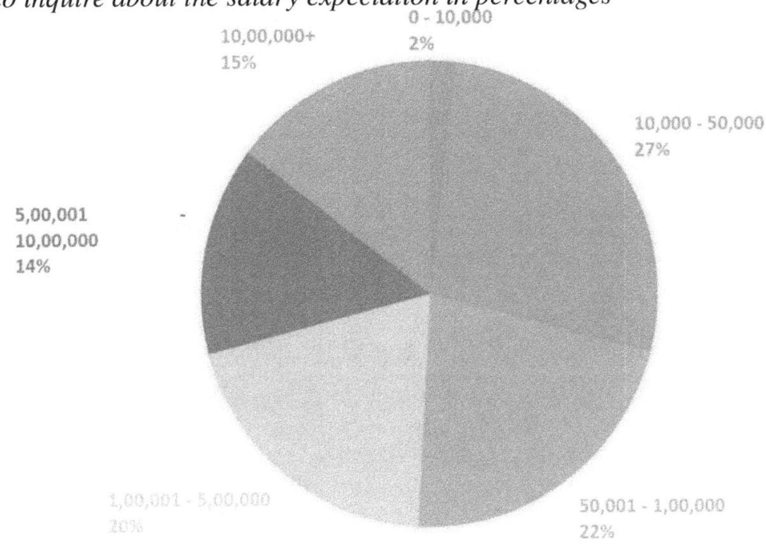

Percentages using a pie graph. Most responses are 10,000 - 50,000 and 50,001 – 1,00,000, followed closely by 1,00,001 – 5,00,000. The lowest number of responses has been for the '0-10,000' category. There is a mediocre volume of answers for'10,000+'. These responses are practically equal to those of the '5,00,000 – 10,00,000' category. The '50,000 – 1,00,000' and '1,00,000 – 5,00,000' groupings also have a virtually identical quantity of replies (figures 4-5).

The last question asked the participants about their views on the relevance of the search results.

Comparative Study of Responses

The responses have been scrutinized based on their gender. Now, it is time to compare the results of each gender with those of the other. The responses collected have already been sorted by gender. The number of responses by each gender is equal.

The first question to be discussed is, "On searching the word "jobs" on LinkedIn, what is your top search result?"

Figure 6. Organization of male responses to question 4 inquiries about the relevance of the job results into percentages using a pie graph

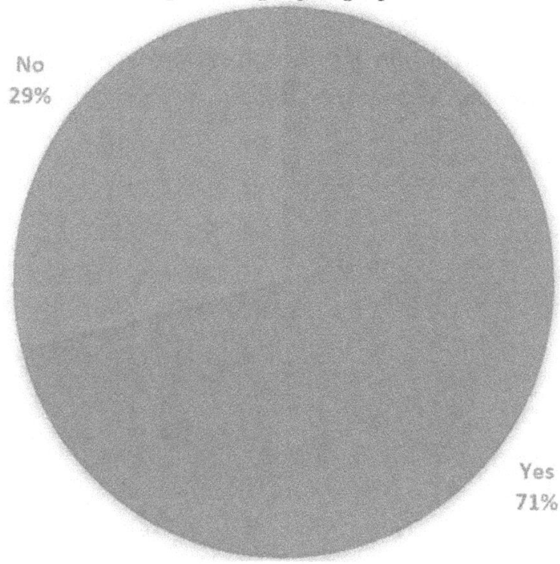

Figure 7. Depiction of the percentage of male and female responses as recorded in question 1 that inquiries about the participant's gender

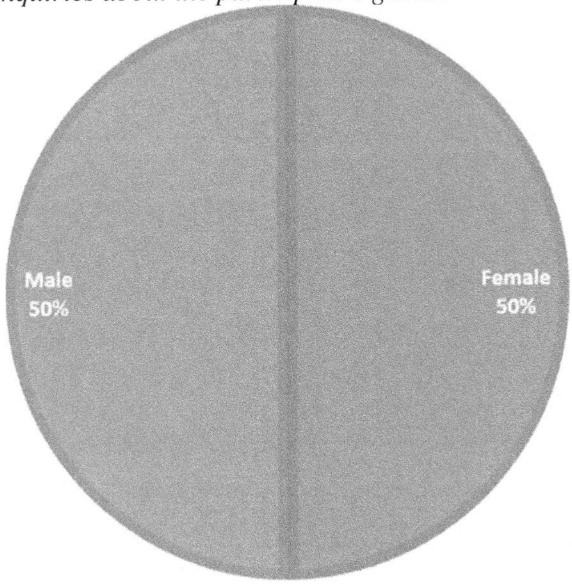

The male participants have an elevated response rate for' High-ranking administrative positions (executive positions).' This is slightly curbed by the female response rate for 'Mid-ranking administrative position (managers, coordinators)' and 'Low-level administrative position', which is higher than the male response rate for this option. The male responses for 'Entry level position (marketing, sales, accounting, human resources, programming, etc.)' and 'Specialist fields (lawyer, doctor, scientist, mathematician, etc.)' are much more than the female responses. The female responses for 'Education-related fields' and 'Creative jobs (writers, artists, etc.)' are slightly higher. However, male and female reactions for these categories are lower overall than others (figures 6-7). Also, there are more male responses for 'Freelancing work.' Interestingly, the response estimate for 'Positions that do not require much training or aren't skill-based in any shape' and 'Secretarial position,' though very low in and of itself, is higher than the female reply estimate for these questions, which is 0. 'No valid search results' was a safety option provided. A higher number of females used this option.

Figure 8. Comparison between male and female responses to question 2, which inquiries about the top job result organized into percentages using a graph

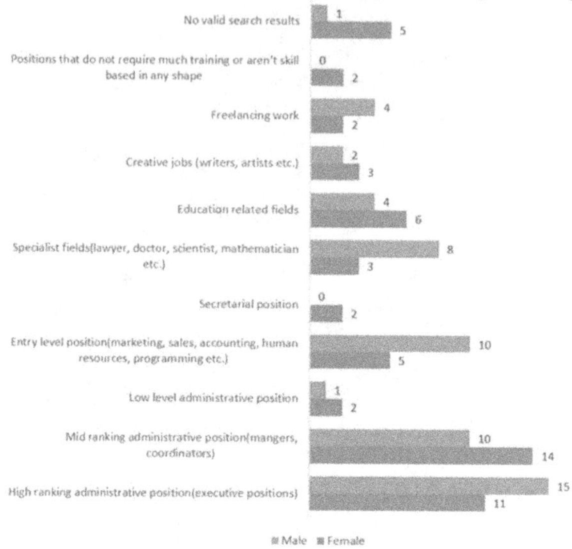

The second topic to be discussed is the economic prospects of these jobs. This is an essential part of this research. The question is, 'How much do you think this job would earn Per Month in rupees? (Based on the information provided in the advertisement)' (figure 8).

Figure 9. Comparison between male and female responses to question 3, which inquiries about salary expectations, is organized into percentages using a graph

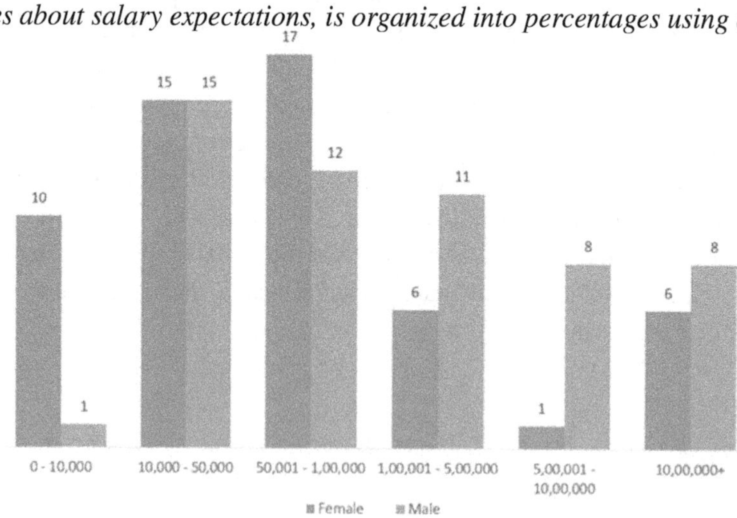

As described in Figures 9- 10, the number of females who have selected option 0 -10,000 is much more than the number of males who have picked out this option, while the number of males who have chosen alternative 5,00,001 – 10,00,000 is more elevated than the number of females who have picked this option. The people who chose 10,000 – 50,000 are equal for both Males and females. The responses by females are more than by males for 50,001 – 1,00,000. The reactions by males for 1,00,001 – 5,00,000 and 10,000+ are more than those by females. (Daneshvar, S. et.al 2018).

The last question discusses whether the participants felt the results were previously relevant to them. The number No responses selected by male and female participants was the same.

Figure 10. Comparison between male and female responses to question 3, which inquiries about salary expectations, is organized into percentages using a graph

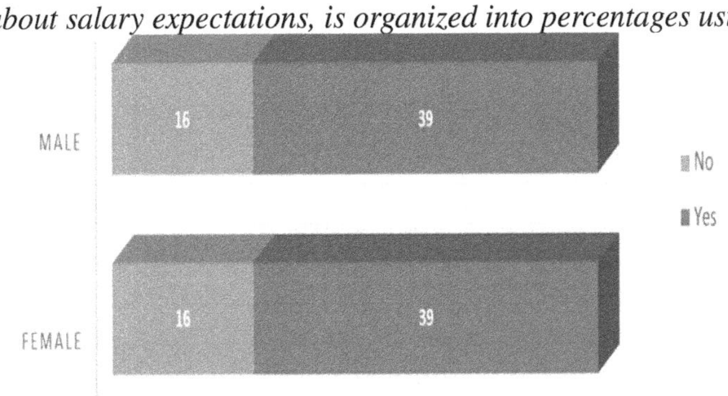

STATISTICAL ANALYSIS

The Chi-squared test is a statistical method that can interpret the relationship between two categorical variables. In this particular case, the test was applied to determine if there was any correlation between the gender of participants and the job or salary that LinkedIn provided them in their results. We have assumed the null hypothesis is "There is no association between gender and the jobs/salaries offered." The alternate hypothesis is "There is an association between gender and the job- s/ salaries offered." The significance threshold is assumed to be 0.05. To conduct the test, we first need to create two separate tables for Gender and Job and Gender and Salary, and then arrange the collected data in these tables and calculate the totals based on rows and columns.

Expected Values

This section elaborates on the expected values for each sample calculated from the users. We use Equation 1 to determine the expected values in Table 3 and Table 4.

$$Expected\ value = \frac{Row\ Total \times Column\ Total}{Overall\ Total} \quad (1)$$

Table 1. Observed Values of Job categories based on Gender

Job categories	Female	Male	Total
High ranking administrative position (executive positions)	11	15	26
Mid ranking administrative position (mangers, coordinators)	14	10	24
Low level administrative positions	2	1	3
Entry level positions (marketing, sales, accounting, human resources, programming etc.)	5	10	15
Secretarial position	2	0	2
Specialist fields (lawyer, doctor, scientist, mathematician etc.)	3	8	11
Education related fields	6	4	10
Creative jobs (writers, artists etc.)	3	2	5
Freelancing work	2	4	6
Positions that do not require much training or aren't skill based in any shape	2	0	2
No valid search results	5	1	6
Total	**55**	**55**	**110**

Table 2. Observed Values of Salaries Offered based on Gender

Salaries Offered	Female	Male	Total
0 – 10,000	10	1	11
10,001 – 50,000	15	15	30
50,001 – 1,00,000	17	12	29
1,00,001 – 5,00,000	6	11	17
5,00,001 – 10,00,000	1	8	9
10,00,000+	6	8	14
Total	**55**	**55**	**110**

Chi-Square Test

This section uses four major steps to evaluate the chi-square test which involves the observed(obs) values(val) from Table 1 and Table 2, expected(exp) values(val) from Table 3 and Table 4, the difference between these values as per equation 2, square of those difference as per equation 3 and lastly the chi-square test value of each column as per equation 4.

$$Diff_i = Obsval_i - Expval_i \tag{2}$$

$$X_i = (Obsval_i - Expval_i)^2 \tag{3}$$

$$X^2 = \sum_{i=0}^{n_i} \frac{(Obsval_i - Expval_i)^2}{Expval_i} \tag{4}$$

In the final step, we compute the Calculated X^2 value and tabular X^2 value for both 'Job categories based on Gender' as well as 'Salaries offered based on Gender'. The formula for X^2 value is Sum of all $(Obsval_i - Expval_i)^2$ as per table 5 and 6. The formula I for degrees of freedom (*df*) as per equation 5. Finally, to calculate values are presented in table 7.

$$Df = (Colomn - 1) - (Colomn - 1) \tag{5}$$

Table 3. Expected Values of Job categories based on Gender

Job categories Female & Male	(Expected Value)
High ranking administrative position (executive positions)	13
Mid ranking administrative position (mangers, coordinators)	12
Low level administrative positions	1.5
Entry level positions (marketing, sales, accounting, human resources, programming etc.)	7.5
Secretarial position	1
Specialist fields (lawyer, doctor, scientist, mathematician etc.)	5.5
Education related fields	5
Creative jobs (writers, artists etc.)	2.5

continued on following page

Table 3. Continued

Job categories Female & Male	(Expected Value)
Freelancing work	3
Positions that do not require much training or aren't skill based in 1 any shape	
No valid search results	3

Table 4. Expected Values of Salaries Offered based on Gender

Salaries Offered	Female & Male (Expected Value)
0 – 10,000	5.5
10,001 – 50,000	15
50,001 – 1,00,000	14.5
1,00,001 – 5,00,000	8.5
5,00,001 – 10,00,000	4.5
10,00,000+	7

Based on these calculations, if the *calculated* X^2 value is greater than the *tabular* X^2 value, we reject the null hypothesis and consider an alternate one. Thus, the alternate hypothesis suggests a significant relation exists between the job categories and salaries offered based on gender, respectively. As per table 7 the value of job categories based on gender the *calculated* $X^2 <$ *Tabular*X^2 value while on the other hand the value of salaries offered based on gender the calculated$X^2 >$ *Tabular*X^2 value. This overall analysis indicates and accepts that there is an association or significant relation between gender and the salaries offered; on the other hand, these results indicate there is no such relation between jobs offered and gender (Tables 5,6,7).

Table 5. Expected Values of Job categories based on Gender

Observed	Expected	$(Obsval_i{-}Expval_i)$	$(Obsval_i{-}Expval_i)^2$	$\frac{(Obsval_i - Expval_i)^2}{Expval_i}$
11	13	-2	4	0.307692308
14	12	2	4	0.333333333
2	1.5	0.5	0.25	0.166666667
5	7.5	-2.5	6.25	0.833333333
2	1	1	1	1

continued on following page

Table 5. Continued

Observed	Expected	$(Obsval_i - Expval_i)$	$(Obsval_i - Expval_i)^2$	$\frac{(Obsval_i - Expval_i)^2}{Expval_i}$
3	5.5	-2.5	6.25	1.13636363
6	5	1	1	0.2
3	2.5	0.5	0.25	0.1
2	3	-1	1	0.333333333
2	1	1	1	1
5	3	2	4	1.333333333
15	13	2	4	0.307692308
10	12	-2	4	0.333333333
1	1.5	-0.5	0.25	0.166666667
10	7.5	2.5	6.25	0.833333333
0	1	-1	1	1
8	5.5	2.5	6.25	1.136363636
4	5	-1	1	0.2
2	2.5	-0.5	0.25	0.1
4	3	1	1	0.333333333
0	1	-1	1	1
1	3	-2	4	1.333333333
Total				**13.48811189**

Table 6. Calculated Values of Salaries Offered based on Gender

Observed	Expected	$(Obsval_i - Expval_i)$	$(Obsval_i - Expval_i)^2$	$\frac{(Obsval_i - Expval_i)^2}{Expval_i}$
10	5.5	4.5	20.25	3.681818
15	15	0	0	0
17	14.5	2.5	6.25	0.431034
6	8.5	-2.5	6.25	0.735294
1	4.5	-3.5	12.25	2.722222
6	7	-1	1	0.142857
1	5.5	-4.5	20.25	3.681818
15	15	0	0	0
12	14.5	-2.5	6.25	0.431034
11	8.5	2.5	6.25	0.735294
8	4.5	3.5	12.25	2.722222
8	7	1	1	0.142857
Total				**15.42645**

Table 7. Final Values of the observation

Based on Gender	Significance Value	*Df*	Tabular X^2	Calculated X^2
Job categories	0.05	10	18.31	13.48811189
Salaries Offered	0.05	5	11.07	15.42645

CONCLUSION

AI is one of the leading research topics in the world today. This is why the advancements in AI have been very smooth. At the rate the field is advancing, it won't be long before AI descends into all aspects of society and fills in menial chores and duties that are crucial beyond our everyday lives. However, even though there are more advances in AI, the field of ethics in AI is still lacking. Although a significant quantity of research has been done on this topic, no standardized set of laws could govern the principles of the creation of AI technology and lead us toward ethical AI. There have been abundant cases of unethical AI or ML processes, which media and researchers have stated. There are often biases, such as gender or race biases, in tools devised for the masses. This brings out the need to discuss algorithm biases and how to remove them. There are multiple types of biases and various reasons behind them, although they all boil down to mistakes made when selecting the data set on which the algorithm was trained. Over the years, many articles and research papers on this topic have been published. Research by Carnegie Mellon University in 2015 proved that discrimination based on gender was present in job advertisements, as shown by a majority of popular job search sites. This has led us to question all the social media sites that we see today and whether they display such gender bias. Therefore, this survey research was designed to determine whether there is any gender bias within LinkedIn regarding displaying job advertisements to Male and female users. Using pie charts and graphs, we organized all the data collected by gender, and then -based on gender- a comparative analysis of the data was conducted.

On studying the analysis results, a few points seem to stand out. Interestingly, the male responses for 'Entry level position (marketing, sales, accounting, human resources, programming, etc.)' and 'Specialist fields (lawyer, doctor, scientist, mathematician, etc.)' are much more than female responses. Also, there are more male responses to' Free lancing work.' The female responses for 'Education-related fields' and 'Creative jobs (writers, artists, etc.)' are higher than male responses. Curiously, the response rate for 'Positions that do not require much training or aren't skill-based in any shape' and 'Secretarial position', though very low in and of itself, is higher than the female reply rate for these alternatives, which is 0. This shows that the more traditional female jobs, like secretarial work, are rarely recommended to men. The quantity of female responses for the '0-10,00' category, the lowest category, is beyond the male responses. Apart from this, only 50,001 – 1,00.000 had more female responses. Males dominate all the other categories, from 1,00,001 to 10,000+. Based on these results, there does seem to be some one-sidedness in these algorithms as the higher paying jobs seem to be recommended to men more while women are suggested more of the same traditional roles. In conclusion, there does appear to be some gender bias on LinkedIn, but it is not practicable to understand the root of this or even to say that such a bias exists severely. To do that, an experiment on a much larger scale with more resources and an extremely huge group of participants would have to be organized.

DECLARATIONS

- Funding: 0
- Conflict of interest/Competing interests: None
- Ethics approval: Yes
- Availability of data and materials: From survey form
- Code availability: None
- Authors' contributions: All Authors contributed equally in this manuscript

REFERENCES

Ahmad, H., Asghar, M. Z., Khan, A. S., & Habib, A. (2020). A systematic literature review of personality trait classification from textual content. *Open Computer Science*, 10(1), 175–193. DOI: 10.1515/comp-2020-0188

Akimushkin, C., Amancio, D. R., & Oliveira, O. N.Jr. (2018). On the role of words in the network structure of texts: Application to authorship attribution. *Physica A*, 495, 49–58. DOI: 10.1016/j.physa.2017.12.054

Ananny, M., & Crawford, K. (2018). Seeing without knowing: Limitations of the transparency ideal and its application to algorithmic accountability. *New Media & Society, 20*(3), 973-989.

Ashurst, C., Hine, E., Sedille, P., & Carlier, A. (2022, June). Ai ethics statements: analysis and lessons learnt from neurips broader impact statements. In *Proceedings of the 2022 ACM conference on fairness, accountability, and transparency* (pp. 2047-2056). DOI: 10.1145/3531146.3533780

Balasubramaniam, N., Kauppinen, M., Kujala, S., & Hiekkanen, K. (2020). Ethical guidelines for solving ethical issues and developing AI systems. In *Product-Focused Software Process Improvement: 21st International Conference, PROFES 2020, Turin, Italy, November 25–27, 2020Proceedings*, 21, 331–346.

Barhate, D., Pathak, S., & Dubey, A. K. (2023). Hyperparameter-tuned batch-updated stochastic gradient descent: Plant species identification by using hybrid deep learning. *Ecological Informatics*, 75, 102094. DOI: 10.1016/j.ecoinf.2023.102094

Basti, R., Jamoussi, S., Charfi, A., & Hamadou, A. B. (2019, September). Arabic Twitter User Profiling: Application to Cyber-security. In *WEBIST* (pp. 110-117).

Benjamins, R. (2021). A choices framework for the responsible use of AI. *AI and Ethics*, 1(1), 49–53. DOI: 10.1007/s43681-020-00012-5

Birhane, A., Ruane, E., Laurent, T., & Brown, S. (2022, June). The forgotten margins of AI ethics. In *Proceedings of the 2022 ACM Conference on Fairness, Accountability, and Transparency* (pp. 948-958). DOI: 10.1145/3531146.3533157

Cell, D. I. A. I. (2023). Ethical guidelines for application of Artificial Intelligence in Biomedical Research and Healthcare.

Dadsena, K. K., Mathiyazhagan, K., & Taghipour, A. (2021). Analysis of barriers for the build the resilient supply chain networks post-COVID-19. In *Managing Supply Chain Risk and Disruptions: Post COVID-19* (pp. 79–89). Springer International Publishing. DOI: 10.1007/978-3-030-72575-4_7

Daneshvar, S., & Inkpen, D. (2018, September). Gender identification in twitter using n-grams and lsa. In *Proceedings of the ninth international conference of the CLEF association (CLEF 2018)*. CEUR-WS.

Dave, T., Athaluri, S. A., & Singh, S. (2023). ChatGPT in medicine: An overview of its applications, advantages, limitations, future prospects, and ethical considerations. *Frontiers in Artificial Intelligence*, 6, 1169595. DOI: 10.3389/frai.2023.1169595 PMID: 37215063

Dhiman, N., Singh, H., & Thakur, A. (2023, August). An Efficient Approach for Image Forgery Detection Using Deep Convolutional Neural Network. In *2023 International Conference on Electrical, Electronics, Communication and Computers (ELEXCOM)* (pp. 1-5). IEEE. DOI: 10.1109/ELEXCOM58812.2023.10370304

Drabiak, K., Kyzer, S., Nemov, V., & El Naqa, I. (2023). AI and machine learning ethics, law, diversity, and global impact. *The British Journal of Radiology*, 96(1150), 20220934. DOI: 10.1259/bjr.20220934 PMID: 37191072

Eitel-Porter, R. (2021). Beyond the promise: Implementing ethical AI. *AI and Ethics*, 1(1), 73–80. DOI: 10.1007/s43681-020-00011-6

Fatima, M., Hasan, K., Anwar, S., & Nawab, R. M. A. (2017). Multilingual author profiling on Facebook. *Information Processing & Management*, 53(4), 886–904. DOI: 10.1016/j.ipm.2017.03.005

Fiske, A., Henningsen, P., & Buyx, A. (2019). Your robot therapist will see you now: Ethical implications of embodied artificial intelligence in psychiatry, psychology, and psychotherapy. *Journal of Medical Internet Research*, 21(5), e13216. DOI: 10.2196/13216 PMID: 31094356

Gerke, S., Minssen, T., & Cohen, G. (2020). Ethical and legal challenges of artificial intelligence-driven healthcare. In *Artificial intelligence in healthcare* (pp. 295–336). Academic Press. DOI: 10.1016/B978-0-12-818438-7.00012-5

Giachanou, A., Ríssola, E. A., Ghanem, B., Crestani, F., & Rosso, P. (2020). The role of personality and linguistic patterns in discriminating between fake news spreaders and fact checkers. In *Natural Language Processing and Information Systems: 25th International Conference on Applications of Natural Language to Information Systems, NLDB 2020, Saarbrücken, Germany, June 24–26, 2020Proceedings*, 25, 181–192.

Glaa, B., Zoghlami, N., & Taghipour, A. (2014, May). Outsourcing and off-shoring healthcare services: A way to cut the costs and improve quality. In *2014 international conference on advanced logistics and transport (ICALT)* (pp. 189-194). IEEE. DOI: 10.1109/ICAdLT.2014.6864116

Hagendorff, T. (2020). The ethics of AI ethics: An evaluation of guidelines. *Minds and Machines*, 30(1), 99–120. DOI: 10.1007/s11023-020-09517-8

Halim, Z., & Rehan, M. (2020). On identification of driving-induced stress using electroencephalogram signals: A framework based on wearable safety-critical scheme and machine learning. *Information Fusion*, 53, 66–79. DOI: 10.1016/j.inffus.2019.06.006

Halim, Z., & Zouq, A. (2021). On identification of big-five personality traits through choice of images in a real-world setting. *Multimedia Tools and Applications*, 80(24), 33377–33408. DOI: 10.1007/s11042-021-11419-5

Huang, C., Zhang, Z., Mao, B., & Yao, X. (2022). An overview of artificial intelligence ethics. *IEEE Transactions on Artificial Intelligence*, 4(4), 799–819. DOI: 10.1109/TAI.2022.3194503

Joo, Y., Hwang, I., Cappellato, L., Ferro, N., Losada, D., & Müller, H. (2019). Author profiling on social media: An ensemble learning model using various features. *Notebook for PAN at CLEF, 2380*.

Kathuria, Y., Chaki, N. R., Kaur, M., Kumar, S., & Medhavi, R. (2024, October). Responsible AI impact assessment mechanism for India: A robust strategy for effective governance of AI systems in the country. In *AIP Conference Proceedings* (Vol. 3220, No. 1). AIP Publishing. DOI: 10.1063/5.0234670

Kukreja, V., & Sakshi, . (2022). Machine learning models for mathematical symbol recognition: A stem to stern literature analysis. *Multimedia Tools and Applications*, 81(20), 28651–28687. DOI: 10.1007/s11042-022-12644-2

Kumar, A., Kumar, S. A., Dutt, V., Dubey, A. K., & Narang, S. (2023). A hybrid secure cloud platform maintenance based on improved attribute-based encryption strategies. *International Journal of Interactive Multimedia and Artificial Intelligence*, 8(2), 150–157. DOI: 10.9781/ijimai.2021.11.004

Lu, X., & Taghipour, A. (2023). A French Guide for Businesses and Governments to Integrate the Concept of Sustainable Development Into Their Supply Chains. In *Government Impact on Sustainable and Responsible Supply Chain Management* (pp. 1–12). IGI Global. DOI: 10.4018/978-1-6684-9062-4.ch001

Mahfod, J., Canel-Depitre, B., & Taghipour, A. (2019). Quality function deployment-ELECTRE in supplier evaluation. *Journal of Advanced Management Science, 7*(4).

Pazzanese, C. (2020). *Ethical concerns mount as AI takes bigger decision-making role*. Harvard Gazette.

Rahman, A. U., & Halim, Z. (2022). Predicting the big five personality traits from hand-written text features through semi-supervised learning. *Multimedia Tools and Applications*, 81(23), 33671–33687. DOI: 10.1007/s11042-022-13114-5

Sharma, E., Mahajan, R., Mahajan, R., & Mansotra, V. (2021). Automated personality prediction of social media users: A decade review. *Turkish Journal of Computer and Mathematics Education*, 12(14), 5225–5237.

Sharma, S., & Sharma, K. (2023, June). Analyzing Employee's Attrition and Turnover at Organization Using Machine Learning Technique. In *2023 3rd International Conference on Intelligent Technologies (CONIT)* (pp. 1-7). IEEE. DOI: 10.1109/CONIT59222.2023.10205676

Singh, H., & Kumar, Y. (2019). Cellular automata based model for e-healthcare data analysis. *International Journal of Information System Modeling and Design*, 10(3), 1–18. DOI: 10.4018/IJISMD.2019070101

Srikumar, M., Finlay, R., Abuhamad, G., Ashurst, C., Campbell, R., Campbell-Ratcliffe, E., Hongo, H., Jordan, S. R., Lindley, J., Ovadya, A., & Pineau, J. (2022). Advancing ethics review practices in AI research. *Nature Machine Intelligence*, 4(12), 1061–1064. DOI: 10.1038/s42256-022-00585-2

Chapter 11
Exploring Information Technology (IT) Acceptance Among Supply Chain Stakeholders in Morocco's Pharmaceutical Industry

Taoufiq El Moussaoui
Faculty of Sciences Dhar El Mehraz Fez, Sidi Mohamed Ben Abdellah University, Morocco

Loqman Chakir
Faculty of Sciences Dhar El Mehraz Fez, Sidi Mohamed Ben Abdellah University, Morocco

Alaa Eddine El Moussaoui
https://orcid.org/0000-0002-5749-7682
Higher School of Technology, Hassan 2 University, Morocco

ABSTRACT

This chapter explores the determinants of information technology (IT) acceptance in Moroccan supply chains within the pharmaceutical sector, focusing on the role of digital transformation in fostering collaboration and information sharing. To address this objective, a quantitative methodological approach was employed, combining

DOI: 10.4018/979-8-3693-7478-8.ch011

descriptive analysis and logistic regression. The findings reveal a limited adoption of digital technologies in supply chain activities. However, they underscore that these digital transformation initiatives are essential prerequisites for enhancing cooperation and improving the sharing of logistics information. Practical implications, study limitations, and directions for future research are discussed at the end of the chapter.

1. INTRODUCTION

To introduce this research, it is essential to address a management perspective concerning organizations. Human resource management (HRM) serves at least four primary objectives: selecting, monitoring, qualifying, and retaining individuals who contribute to the organization's activities (Reina and Scarozza, 2021). These human resources, pivotal in coordinating physical and informational flows, significantly influence organizational success (Alsheyadi et al, 2024). They play both operational and strategic roles in executing logistics chain projects, as effective management is a critical lever for creating value and achieving competitive advantage (El Moussaoui et al, 2022 ; El Moussaoui et al, 2023). Furthermore, logisticians require not only operational skills in flow management but also technical expertise, particularly in information technologies (IT), which have become indispensable in modern logistics chains (Taghipour and Beneteau-Piet, 2020 ; Lagorio et al, 2022). Proper utilization of IT enhances operational efficiency, streamlines tasks, reduces response times to unforeseen events, and improves flexibility. Over the past two decades, IT has profoundly reshaped organizational structures and management practices, giving rise to a specialized function: the supply chain manager.

From a contextual standpoint, Morocco presents a unique environment characterized by diverse resources, management styles, and intercultural dynamics (El Moussaoui et al, 2022). Despite its distinctiveness, the adoption and integration of IT in Morocco have mirrored global trends, marked by increased data exchange volumes, shifts in time management, and transformations in work practices. This digital evolution influences organizational ecosystems and demands companies to approach digitalization as an integrated information system. Such systems, encompassing people, equipment, and software, create open environments capable of capturing, storing, processing, and transmitting information critical for decision-making. Consequently, digital tools like smartphones, laptops, and tablets have become central to daily operations, requiring businesses to view digital transformation not as a reactive process but as a strategic initiative (Taghipour and Merimi, 2021 ; Merimi and Taghipour, 2021).

The term "digital," rooted in the Latin word for "finger," originally referred to counting with fingers (Puskar, 2019). Over time, this concept expanded to encompass technologies reliant on numerical data and interfaces. Today, digital technologies facilitate organizational transformations by aligning human competencies with technological functionalities (Blanka et al, 2022). In the Moroccan context, such transformations must anticipate hybrid organizational models that align actors and technologies to adapt to changing environments (El Moussaoui et al, 2024).

However, the alignment between actors and IT—beginning with the process of IT acceptance by users—remains underexplored in the literature on developing countries. Empirical studies on the determinants of IT acceptance and usage, particularly those examining Internet-based technologies, are scarce in African contexts, including Morocco. Most existing research focuses on individual acceptance rather than organizational-level adoption, leaving a significant gap in understanding IT acceptance among firms.

Given this gap and our proximity to the field, this research examines the case of logistics chains in Morocco, with a specific focus on the pharmaceutical sector. Over the past decade, Morocco has seen substantial investments in IT infrastructure within logistics chains, both domestically and internationally (El Moussaoui, 2024). Pharmaceutical distribution in Morocco relies on numerous wholesalers and a network of pharmacies in major cities such as Casablanca and Rabat, the economic and political capitals, respectively. These cities, home to millions of inhabitants, are pivotal to the country's supply chain operations. Against this backdrop, we formulate the following research question: What are the determinants of IT acceptance by actors in the supply chain of the pharmaceutical industry in Morocco?

Our research seeks to complement and consolidate studies on IT acceptance within the Moroccan context. It aims to fill a documented gap by shedding light on the digitalization of practices in supply chains. Specifically, we begin by evaluating the level of digital technology acceptance within the pharmaceutical distribution chain. We then identify the perceived benefits associated with IT adoption and usage, ultimately addressing the central research question.

The study is structured into three phases. First, we present a comprehensive literature review. In the second phase, we detail the methodology used. Finally, in the third phase, we present and discuss our findings, providing insights into the determinants of IT acceptance and their implications for Morocco's pharmaceutical supply chain.

2. LITERATURE REVIEW

This literature review focuses on the foundations of logistics, its actors – the logisticians – and those of the supply chain, while also highlighting the influence of digital technology in enhancing and upgrading the qualifications of the logistics workforce.

2.1 Unpacking the Rich Tapestry of Logistics Professions

Logistics encompasses all activities that determine the movement of products from the supplier to the final consumer, including production, marketing, sales, etc (Kain and Verma, 2018). It is, of course, a system, a chain of activities, among which we can list handling, warehousing, transportation of products, packaging, sorting, packing, inventory management, procurement, management of work in progress, receiving, and processing orders and complaints (Taghipour, 2020). A more recent approach to the supply chain seeks to group these activities under a single functional and hierarchical responsibility in order to provide coherence aimed at reducing direct, indirect, and hidden costs, and ensuring customer satisfaction. The supply chain acts as a link between two successive stages of physical flows, such as between production and sales points (El Moussaoui et al, 2022 ; El Moussaoui, 2024). However, physical flows can only be effectively managed if, at the same time, informational flows are also managed. It is these informational flows that will serve as regulators. This alignment between material flows on one side and informational flows on the other is characterized by a shift from managing the supply chain to managing the circulation and coordination of intangible flows (Ashraf et al, 2024). Thus, the imperative for agile and global supply chain management reveals the need to create a new hybrid strategic function of material/informational within the organizational framework. In cases of severe crises – such as the recent pandemic – this logistics function will even be celebrated and valued in order to better rethink tomorrow's chains (Taghipour et al, 2023). We will rely on the consensus surrounding the following definition of the supply chain proposed in 2001: it is a "group of at least three entities (companies or individuals) directly involved in the upstream and downstream flows of products, services, finances, and/or information, going from a source to the customer (Mentzer, 2001)." Thus, it is possible to identify three levels of complexity for a multi-actor chain, depending on whether it is direct, extended, or global (El Moussaoui, 2024). The first level of complexity of the supply chain consists of a company, a supplier, and a customer. The second level includes the suppliers of the immediate supplier and the customers of the immediate customer . The third encompasses all actors involved in all flows, from the last supplier to the last customer (Dablanc, 2023). According to this perspective of multi-actor chains, logistics is considered a management mode that adopts an integrated flow management logic (Marchet et al, 2018). This integration has been stimulated by the implementation of inter-organizational information systems that can promote the integration of the supply chain. In this approach, human resources play a crucial role. From the self-taught logistician, handler, forklift operator, warehouse worker, transporter, or packager – the "man on the ground" – to the engineer, strategist with a high level of education in higher education, the profession of supply chain manager has undergone numerous and profound changes in recent years (Derwik et al, 2016). Furthermore, the need for coordination among all links in the chain imposes agility on both strategic (decision-making) and operational (control) dimensions. Therefore, this particular manager ultimately occupies a complex and sensitive transversal function that requires a global vision of the company and its entire supply chain (Shashi et al, 2020).

The profession of supply chain manager is also confronted, like many others, with the digital transformation of both its organization and its profession (Nissen, 2018 ; Ishfaq et al, 2022). As a result, it must acquire technical and even IT skills related to the use of IT dedicated to managing the supply chain (Cauchois et al, 2017). Finally, this facilitator and emancipator must also foster the co-construction of relationships between entities in the chain to improve the individual performance of all parties involved. Thus, they no longer manage only technical means such as production, transportation, and warehousing, but rather a complete set of means within the framework of the overall objective of the company and its actors. They seek better coordination of all flows between partners (material, informational, financial, and even service flows) and rely on factors such as logistics equipment, innovations, leadership, and the qualities of their collaborators (Negi, 2024). Logistics equipment includes carts, transport vehicles, handling machinery, robots, and advanced automation technologies. Machine learning plays a pivotal role in optimizing the use of these resources by enabling predictive maintenance, demand forecasting, and real-time decision-making (El Moussaoui et al, 2023 ; El Moussaoui and Loqman, 2024). Digital tools, such as integrated systems categorized by generational advancements, can constitute significant innovations in logistics. Regarding leadership, the supply chain leader is the one who provides direction and impetus; they act as the conductor and initiator, focused on aligning the overall objectives of the chain (Dubey et al., 2017 ; El Moussaoui et al, 2024).

2.2 Navigating the Digital Frontier: Transforming Supply Chain Dynamics

Interconnected systems at the very heart of the supply chain, as well as applications dedicated to logistics (transportation management system, etc.) and embedded traceability devices (RFID chips, barcodes, QR codes, etc.), have profoundly changed the very approach to the logistics profession (Freeney, 2024). Logistics professionals must have acquired skills in information systems and technologies, at least from a functional and operational perspective. These skills are rare, prompting the establishment of genuine collaboration between HR functions, IT functions, and logistics and supply chain functions (Gupta et al, 2022). The massive introduction of digital technologies has profoundly transformed management practices and all professional relationships within organizations (Loivet et al, 2020). By reducing distances and enabling low-cost communications, digital technology – if mastered and adapted – can lead to better collaboration among personnel and between entities in the supply chain (El Moussaoui et al, 2022). Within the supply chain, IT constitutes an asset at least at three levels: data processing in a perspective of automation/standardization, facilitated memorization to improve the integration of logistics applications, and real-time communication between partner companies in the chain (Dadsena et al, 2021). Moreover, websites can provide product information, including prices, active ingredients, chemical composition, uses, side effects, expiration dates, contraindications, etc. Logistically speaking, the use of connected systems, such as robots, geolocation systems, and surveillance and security devices, is likely to facilitate physical product searches in warehouses, labeling, packaging, handling, and even total traceability (Abbasi et al, 2023). Additionally, IT is used for various activities through specialized software such as ERP (enterprise resource planning), primarily for inventory and warehouse procurement management, human resource management, and financial and accounting management. The characteristics of immediacy, homogeneity, reliability, and traceability can also improve the sharing of logistics information within the function and within the organization, for example, through ERPs or other integrated systems based on a single database, including in small organizations (Zhao and Tu, 2021 ; El Moussaoui and Chakir, 2023).

Digital technology thus addresses communication issues in their entirety by once again highlighting the theory of transaction costs, with – a priori – a decrease in most areas of activity (Gregory et al, 2018). Digital communication then defines, in an integrated system, the strategy and actions for sharing information to be carried out via digital tools aimed at the actors, provided that they have agreed to use them (Camarinha-Matos et al, 2019). Acceptance always precedes use, which in turn precedes a certain degree of appropriation of tools and technologies. This implicit appropriation trajectory underlies overall performance and especially that of the tools and technologies made available to employees. Ultimately, it is pointless to provide potentially effective tools and technologies if they are not intelligently introduced and accepted by those who are not yet users but merely prospects (Hofmann and Rüsch, 2017 ; Cappuccio et al, 2024). Fundamentally, to understand this appropriation question, we can also base ourselves on the resources that characterize organizations since the work of previous scholars. To fulfill its missions, supply chain personnel can resort to digital technologies, which can provide numerous advantages in both administrative activities and operational activities, positively influencing the individual and collective productivity of logisticians (El Moussaoui et al, 2022 ; El Moussaoui et al, 2023). Indeed, machines, by facilitating communications, promoting tracking of workforce movements, reducing distances, and making it easier to archive and process large amounts of information, have significantly improved the quality of services provided by logistics personnel in many situations (Klumpp and Zijm, 2019). In the context of digitalization, employee-employee, employee-customer, and customer-customer relationships are better reinforced. Traceability is ensured, and the depth and complexity of individuals are set aside (Hastig and Sodhi, 2020). The reception of orders and complaints is facilitated, and billing is faster. Recruitment and layoffs are quicker, relying more on digital tools and social networks, thanks to pre-built databases. Technological advancements show a rapid rise in more efficient handling robots, as well as a variety of integrated systems. Advanced Planning Systems (APS) and Global Positioning Systems (GPS) are increasingly being used in supply chains for cargo tracking (Jones, 2019). Supermarkets equipped with robotic sales forces without human intervention are beginning to emerge, even in certain regions. This situation is accompanied by employment issues related to the reduction in the number of jobs and the demand for new skills caused by the digital revolution (Shankar et al, 2021).

Finally, the organizational innovation that digital technology and its tools convey is perceived as both an emancipating factor and a potential source of alienation. It is indeed a matter of skillfully managing the two territories that will coexist after the integration of information technology tools and technologies within the supply chain (Hinings et al, 2018 ; Centobelli et al, 2020). The first territory will be mapped by all functional territories modeled by the organizational chart and its hierarchy, while the second territory will be mapped by all digital territories modeled by the access matrix of personnel in the information system (integrated or non-integrated). This coexistence of functional and digital territories can lead to conflicts if certain territorial intersections – particularly those involving logistics professions – are poorly managed, poorly thought out, or poorly formalized (Simone et al, 2018). This is often why, faced with a function and professions they consider to be "on the front line," digitalization – or hyper-digitalization – is often accompanied by concerns from logisticians. For example, in certain contexts, conflicts or discomfort arise from power outages, connectivity issues, internet or electricity cuts, and problems related to information security or cybercrime (Urciuoli and Hintsa, 2017 ; Mazhar et al, 2023). In this logic, logisticians are often more frequently confronted than others – because costly security measures for information systems are less often deployed – with malware, ransomware, or viruses that daily attack systems. This question refers back to the sustainability of competitive advantages, the longevity of systems, and their vulnerability, and therefore, correspondingly, to that of the organizations that depend on them. Sharing information or data, opening and configuring systems, interacting on applications linked to a unique database that is the core of the organization ultimately involves taking risks (Rao et al, 2019 ; El Moussaoui et al, 2024). The risk of being copied and deprived of strategic formulas and innovations, the risk of intrusion and access to confidential information, the risk of malicious intent or the dissemination of a key skill. These risks again point to the need for security measures to be acquired, as well as the training of personnel and their awareness of best practices and data confidentiality. In the face of these strategic risks, literature also emphasizes the importance of operational risks, primarily related to the performance of human resources, which include the risk of incorrect data entry, erroneous use of digital tools, incorrect use of software applications, loss or forgetting of passwords, etc. Finally, without being exhaustive, the risk of legal action faced by users who publish or propagate news without being able to provide proof of its veracity or simply justify their good faith. Ultimately, logisticians who use these integrated systems face a great deal of responsibility with very little training. Again, the importance of the prior act of accepting to use these systems and technologies, which are sometimes offered and often imposed in certain contexts, emerges. In light of these challenges, issues, and responsibilities, what are the main determinants of the acceptance of these technologies by logisticians ? The lack of available skills, as well as individual or collective precariousness, seem to be significant barriers to the effective integration and utilization of these digital tools in logistics practices.

3. RESEARCH METHODOLOGY

This section is divided into three subsections. The first presents the supply chain of medical products in Morocco. The second specifies the population and the variables of the study. The third explains the data processing methods used.

3.1 Exploration of the Distribution Chain of Medical Products in Morocco

The supply chain of medications we investigate in this research is quite conventional. It includes manufacturers upstream and consumers downstream, meaning potentially the entire population. Between the two, following a model also known in France and Europe, there are two types of intermediaries: wholesalers and retail pharmacies. However, the pharmaceutical industry in Morocco remains in its infancy and is poorly structured. The main entities include several pharmaceutical companies and laboratories. The vast majority of medical products sold in Morocco come from industries located in India, China, South Africa, North Africa, and some European countries. According to the Moroccan Order of Pharmacists, the most significant distributors are various pharmaceutical companies. As everywhere else in the world, the final link in the supply chain is the end consumer who is a client of the retail pharmacy. Similar to almost all of Africa, we highlight the particularity of a distribution of medical products that is also marked by an informal sector. This sector largely benefits from lax regulations—sometimes complacency—and a lack of control by public authorities. This informal sector accounts for approximately 25%—according to estimates—of consumption, with numerous medicine kiosks scattered across many markets and neighborhoods. The so-called street medicines (notably those from unique "ground" pharmacies) are stored under unfavorable conditions. For those that are not outright counterfeits, the poor storage conditions explain why a significant portion of their active ingredients may be degraded. However, their success among populations living below the poverty line is largely due to their prices and ease of access.

3.2 Sample Size, Variables and Constructs

Our field investigations, conducted in collaboration with the pharmacists' association, allowed us to precisely identify 73 private pharmaceutical wholesalers and related product distributors operating in Morocco. Given the small size of this population, we included all 73 units in the study. A questionnaire, focusing on digital technologies as a value driver for human resources in the supply chain, was distributed via email to logistics managers, sales managers, and procurement or operations managers, depending on the company's organizational structure. After a three-month follow-up period, 91 completed questionnaires were received. Table 1 presents seven items as propositions. The study variables were assessed using a five-point bipolar Likert scale. Respondents were asked to indicate their level of agreement with various statements related to the study's concerns, ranging from 1 (strongly disagree) to 5 (strongly agree). The items in Table 1 correspond to questions I1 to I11. The first seven items describe the digital landscape of the entities involved, including the size of their IT infrastructure, the extent of their network, and the management of flow processes. Furthermore, item I8 was designed to assess the quality of information sharing within the distribution chain, examining the flow of information between stakeholders (suppliers, companies, clients, and clients' clients). Item I9 measures the frequency of meetings of common interest among chain members, reflecting collaboration. Items I10 and I11 were developed to evaluate structural reorganizations prior to digitization and the use of digital tools within supply chains. In the African context, hardware and equipment are often not widely accessible due to high costs, maintenance challenges, and perceptions of complexity in using these devices. However, computer usage is increasingly gaining traction, particularly among younger and more active generations. This trend raises questions about the relationship between IT infrastructure, employee qualifications, and the use of IT.

3.3 Testing and Regression

Descriptive analyses were initially conducted on the study variables to gain a comprehensive understanding of the involvement of digital technologies in the daily operations of logistics professionals. The primary objective was to explore how logistics professionals are integrating digital tools into their activities and the degree to which these tools influence critical aspects of logistics operations such as information sharing and cooperation among stakeholders. Digital technologies have increasingly become integral to logistics operations, and it was essential to determine the extent to which they are being leveraged in this context. The descriptive analysis provided an overview of the digital landscape in which logistics professionals operate, highlighting key areas such as the adoption of digital tools for communication, process automation, and decision-making. The analysis also identified patterns in the relationship between digital tool usage and factors such as collaboration, efficiency, and the speed of decision-making within the supply chain. Associations were measured to explore potential links between the use of digital tools and improved information flow within the logistics network, as well as between tool usage and the level of cooperation among logistics actors. By focusing on these relationships, the study sought to determine whether higher usage of digital technologies correlates with more effective information exchange and stronger cooperation across the supply chain, thereby facilitating overall operational success.

Understanding these dynamics is crucial for identifying the value of digital tools in enhancing logistics processes and the overall competitiveness of logistics providers.

Building on these descriptive findings, a binary logistic regression was employed to assess the impact of various characteristics of private wholesalers on their adoption of digital tools. This approach allowed for a deeper understanding of the factors influencing the willingness of logistics companies to embrace digital technologies. Specifically, the study sought to evaluate the relationship between IT infrastructure, workforce qualifications, and the acceptance of digital tools. The hypothesis was that a stronger IT infrastructure and higher employee qualifications would positively influence the adoption of digital tools. In order to prepare for the binary logistic regression, transformations were carried out on the measurement scales. Binarization of the dependent variable's scales was necessary to simplify the analysis and enable the identification of distinct categories, namely low and high usage of digital tools. Each item within the study was categorized by coding the first three response options as 0 (indicating low usage) and the last two options as 1 (indicating high usage). For example, when assessing the usage of digital tools, respondents who selected lower values (1, 2, or 3) were classified as low users (0), while those selecting higher values (4 or 5) were classified as high users (1). This binarization was essential to analyze the likelihood of belonging to one of the two categories and allowed for a straightforward interpretation of the results. Logistic regression analysis was then used to determine the direction (positive or negative) and strength (between 0 and 1) of the influence of explanatory variables on the probability of adopting digital tools. This approach was chosen because logistic regression does not require the data to follow a normal distribution, which can be a limitation in other statistical methods. The results from this analysis were instrumental in identifying key factors that drive the adoption of digital tools among logistics professionals and helped to clarify how IT infrastructure and employee qualifications impact this process.

The logistic regression analysis provided valuable insights into the relative importance of various explanatory variables in determining the likelihood of high versus low usage of digital tools. The results were assessed using the R^2 coefficient, which measures the proportion of variance in the dependent variable explained by the independent variables. The R^2 value, either in the form of Cox and Snell R^2 or Nagelkerke adjusted R^2, indicates the goodness-of-fit of the model, providing an indication of how well the independent variables account for the variation in digital tool adoption. In this context, the R^2 coefficient helped evaluate the contribution of factors such as IT infrastructure and workforce qualifications to the overall model. Additionally, the Wald statistic was used to assess the statistical significance of each explanatory variable's effect on digital tool adoption. A p-value threshold of 0.05 was set, meaning that any variable with a p-value less than this threshold was considered to have a significant influence on the likelihood of high usage of digital tools. This method allowed for a deeper understanding of which factors were most critical in driving digital adoption in logistics operations. The regression coefficients provided detailed information on the direction and strength of each factor's impact, enabling a comparison of their relative contributions to digital tool usage.

Table 1. Perception of Digital Technology in the Distribution of Medical Products in Morocco

Variable	Condition	Number	Percentage	Central tendency
The number of microcomputers or smartphones involved in distribution is enough (I1)	Totally agree Agreed Neutral Disagree Absolutely not agree Total	11 24 10 27 19 91	12,30% 26,37% 10,98% 29,89% 21,09% 100%	Disagree
The number of logistics staff using smartphones in distribution is significant (I2)	Totally agree Agreed Neutral Disagree Absolutely not agree Total	05 16 10 35 25 91	05,49% 17,58% 10,98% 38,68% 27,47% 100%	Disagree
The availability of networked PCs within the company (I3)	Yes No Total	42 49 91	46,15% 53,84% 100%	No
PCs networked exclusively to suppliers (I4)	Yes No Total	27 64 91	29,67% 70,32% 100%	No
PCs networked exclusively for customers - pharmacies (I5)	Yes No Total	21 70 91	23,07% 77,36% 100%	No
Networked PCs: suppliers - companies - customers / stakeholders (I6)	Yes No Total	14 77 91	15,41% 84,37% 100%	No
Processing methods in management procedures implemented by wholesalers (I7)	both manual and automated equitably Fully manual More manual More automated Total	14 35 39 03 91	15,82% 38,68% 42,19% 03,51% 100%	More Manual
Perfect flow of information between stakeholders (I8)	Totally agree Agreed Neutral Disagree Absolutely not agree Total	06 14 22 29 19 91	07,03% 15,82% 24,61% 31,64% 21,09% 100%	Disagree

continued on following page

Table 1. Continued

Variable	Condition	Number	Percentage	Central tendency
Frequent meetings of common interests between logisticians (I9)	Totally agree Agreed Neutral Disagree Absolutely not agree Total	01 09 26 29 26 91	01,75% 10,54% 28,13% 31,64% 28,13% 100%	Disagree
Reorganisations resulting from the use of digital tools are satisfactory (I10)	Totally agree Agreed Neutral Disagree Absolutely not agree Total	09 21 13 26 22 91	10,54% 22,85% 14,06% 28,13% 24,61% 100%	Disagree
The daily appropriation and use of digital tools by logistics staff is satisfactory (I11)	Totally agree Agreed Neutral Disagree Absolutely not agree Total	14 14 06 34 23 91	15,82% 15,82% 07,03% 36,92% 24,61% 100%	Disagree
Wholesale distributors' employees are well qualified (I12)	Totally agree Agreed Neutral Disagree Absolutely not agree Total	13 21 13 26 18 91	14,06% 22,85% 14,06% 28,13% 19,78% 100%	Disagree

4. ANALYSIS AND RESULTS

This section presents and analyzes the results of the conducted study. The calculations were performed using SPSS statistical software to provide a robust and systematic evaluation of the data. In the empirical phase of our research, seven targeted questions were designed to characterize the IT environment of private wholesalers of pharmaceutical products. The analysis particularly emphasizes the characteristics of IT infrastructure, the usage of Android phones and smartphones, the existence and scope of IT networks connecting suppliers and customers, and the management of information flows that facilitate and control physical logistics operations.

4.1 Descriptive Analysis

Descriptive analysis revealed several key insights into the IT landscape of private pharmaceutical wholesalers. A significant 61.4% of wholesalers expressed a need to increase the number of microcomputers and mobile devices, highlighting a critical gap in IT infrastructure. This finding underscores the importance of upgrading technological resources to meet operational demands. Additionally, 77.36% of participants reported underutilization of Android phones and other smartphones within the logistics chain. This points to a missed opportunity for leveraging mobile technology to streamline operations and improve communication efficiency.

Table 2. Tests for the specification of the logistic regression model of digital usage.

	Chi-square test	Freedom level	Significance
Step	25,001	2,000	0,000
block	25,001	2,000	0,000
model	25,001	2,000	0,000

Table 3. Summary of the Logistic Regression Model for Digital Usage

Step	-2 Log-likelihood	Cox and Snell R^2	Nagelkerke R^2
1	46.100a	0.356	0.500

Table 4. Variables in the equation of digital use by the logistics staff

Step 1a		A	ES	Wald	Fl	Sig	Exp B
-	Skilled and qualified employee	0,339	0,268	1,559	1,000	0,210	1,401
-	Number of PCs	1,369	0,371	13,620	1,000	0,000	3,947
-	constant	-6,222	1,719	13,060	1,000	0,000	0,002

The networking capabilities of IT infrastructure also displayed varying levels of integration. Specifically, 46.15% of microcomputers were networked internally within companies, allowing for intra-organizational connectivity. However, only 29.67% were exclusively networked with suppliers and 23.07% with customers. A mere 15.41% of networks demonstrated comprehensive integration connecting suppliers, the company, and customers, which is essential for seamless information exchange across the supply chain. Moreover, over 80% of management procedures and tasks remain manual (Fully manual: 38,68% ; More manual: 42,19%), reflecting a significant lag in the adoption of digital tools for logistics processes. These results highlight the pressing need for policy interventions aimed at strengthening IT network policies and encouraging the adoption of automation tools.

4.2 Correlation Analysis

To further analyze the IT environment, specific items (I8, I9, I10, and I11) were evaluated, and Spearman's rank correlation was employed to measure the relationships between variables. The analysis revealed a strong positive correlation between I11 (Digital Tool Usage) and I8 (IT Infrastructure Quality), with Rhô = 0.770 at the 0.01 bilateral significance level. This finding indicates that higher-quality IT infrastructure strongly influences the adoption of digital tools. Additionally, a moderate positive correlation was observed between I11 and I9 (Network Connectivity Extent), with Rhô = 0.461, emphasizing the importance of expanding network reach to enhance digital integration. These correlations underscore the critical role of robust IT systems and extended connectivity in driving digital transformation within logistics operations. The analysis confirms that investments in IT infrastructure and connectivity yield significant benefits in operational efficiency and technological adoption.

4.3 Binary Logistic Regression Analysis

To assess the impact of IT equipment rates and workforce qualifications on the likelihood of digital tool adoption by logisticians, binary logistic regression was conducted. The model demonstrated strong statistical validity, with a chi-square of 25.001 with one degree of freedom, significant at the 0.01 level. The −2log likelihood ratio was 46.100, indicating a well-fitting model. Additionally, the Cox and Snell R^2 and Nagelkerke R^2 values were 0.356 and 0.500, respectively, further validating the model's robustness.

The reformulated predictive equation from the regression analysis shows that workforce qualifications and the availability of IT equipment significantly contribute to the adoption of digital tools. Specifically, each additional qualified employee (I12) increases the probability of using digital tools by a factor of 0.339, while each additional computer (I1) has a more significant impact with a factor of 1.369. These results suggest that targeted investments in workforce training and IT equipment can serve as key factors for improving digital adoption in pharmaceutical logistics. The results of this binary logistic regression are detailed in Tables 2, 3 and 4.

5. DISCUSSION

5.1 Digital Penetration in the Pharmaceutical Sector in Morocco

The findings in section (a) emphasize the relatively low penetration of digital tools in the pharmaceutical distribution sector in Morocco. This outcome aligns with data from the Digital Social Media in North Africa, Mobile, and E-commerce report published in January 2018 by We Are Social and Hootsuite. The report indicates that Morocco's Internet penetration rate reached 49% in 2019, a figure higher than the African average of 34% but still trailing regions such as Southern Africa (51%). This disparity reflects regional differences across the African continent, with North Africa generally benefiting from better digital infrastructure compared to Sub-Saharan Africa. The higher penetration rate in Morocco underscores the need to capitalize on this advantage to enhance digital adoption in critical sectors like healthcare and logistics. However, significant challenges remain, including accessibility, affordability, and the digital divide between urban and rural areas. Addressing these issues requires targeted government policies and public-private partnerships to expand Internet coverage, reduce costs, and promote digital literacy. These interventions could enable Morocco to leverage its relatively high Internet penetration for broader socio-economic development.

The low digital penetration in Morocco's pharmaceutical sector, despite the broader Internet availability, suggests that sector-specific barriers are at play. Factors such as inadequate investments in digital infrastructure, the high cost of technology, and the limited digital skills among workforce participants constrain the adoption of digital tools. Unlike regions with advanced digital ecosystems, North Africa faces a unique set of challenges, including regulatory hurdles and cultural attitudes towards technology. To overcome these, it is imperative to develop industry-specific strategies that address logistical inefficiencies, enhance the accessibility of digital tools, and foster a culture of innovation within organizations. Such initiatives could position Morocco as a regional leader in digital adoption, particularly in sectors critical to public health and economic resilience.

5.2 Role of Digital Tools in Collaboration and Information Sharing in the Pharmaceutical Sector

Section (b) highlights the significant role digital tools play in enhancing information sharing and fostering collaboration. Correlation coefficients demonstrate that digital technologies act as vectors for efficient communication and cooperative decision-making. Specifically, these tools bridge geographical and temporal gaps, enabling employees to perceive greater proximity and thereby fostering teamwork.

Enhanced communication systems made possible by digital platforms facilitate shared decision-making and the development of common practices and values, contributing to improved organizational performance. The reduction in communication delays and the establishment of efficient channels underscore the transformative potential of digital adoption in modern organizations.

In the Moroccan context, digital technologies present an opportunity to overcome traditional barriers to collaboration, particularly in geographically dispersed supply chains. By enabling real-time communication and data sharing, digital tools can reduce operational inefficiencies and promote a more integrated approach to logistics management. However, the effectiveness of these tools depends on their accessibility and the willingness of organizations to invest in the necessary infrastructure. Training programs and initiatives aimed at improving digital literacy among employees are essential for maximizing the benefits of these technologies. Furthermore, fostering a culture of digital innovation requires leadership commitment and a strategic vision that prioritizes technological advancements as a core component of organizational growth.

Digital tools also enable co-construction of practices and values within organizations, which is particularly relevant in Morocco's logistics and pharmaceutical sectors. The ability to standardize procedures, share best practices, and integrate decision-making processes enhances overall efficiency and adaptability. Moreover, these technologies provide a platform for continuous learning and improvement, allowing organizations to respond proactively to evolving market demands. As Morocco continues to position itself as a hub for regional logistics and trade, leveraging digital tools to enhance collaboration will be critical for maintaining competitiveness in a rapidly changing global environment.

5.3 Impact of Workforce Qualification and Digital Infrastructure on the Pharmaceutical Sector

Section (c) reveals the dual influence of workforce qualification and the availability of digital infrastructure on the adoption of digital tools in logistics. The analysis demonstrates that better-qualified employees are more likely to embrace and utilize digital technologies effectively. Additionally, a minimum level of digital infrastructure, such as computers and software, is crucial for enabling the integration of IT solutions into organizational processes. Specifically, a one-point increase in employee qualification raises digital adoption by 0.340 points, while a similar increase in the number of deployed computers leads to a 1.374-point rise. This disparity underscores the importance of infrastructure as a foundational element for digital transformation.

The findings highlight the challenges associated with acquiring and maintaining digital equipment in Morocco. High costs, fragility, and the lack of repair services remain significant obstacles, particularly for small and medium-sized enterprises (SMEs). Inadequate repair and recycling systems exacerbate these issues, leading to environmental concerns and further reducing the usability of existing equipment. Addressing these challenges requires a comprehensive approach that includes subsidies for technology acquisition, investment in repair and recycling infrastructures, and training programs to develop local expertise. Such measures would not only enhance digital adoption but also contribute to environmental sustainability.

Moreover, the findings underscore the interplay between employee qualifications and infrastructure availability. While infrastructure provides the means for digital adoption, employee skills determine the effectiveness of its utilization. This highlights the need for a balanced approach that simultaneously addresses both factors. Organizations should prioritize upskilling their workforce through targeted training programs and partnerships with educational institutions. At the same time, policymakers should focus on creating an enabling environment that supports affordable access to digital tools and encourages innovation. By addressing these interdependent factors, Morocco can accelerate its digital transformation and enhance the efficiency and resilience of its logistics sector.

6. CONCLUSION

6.1 Study Summary

This study explored the digital penetration in the pharmaceutical sector in Morocco, highlighting the challenges and opportunities associated with the integration of information technologies (IT) in product distribution. Our results reveal a low utilization of digital tools, which limits information sharing and cooperation among supply chain actors. By examining the fields of information systems management, supply chain management, and human resources, this chapter demonstrates that, despite the obstacles encountered, the adoption of digital technologies could be an essential lever for improving operational efficiency and the responsiveness of pharmaceutical companies to market needs.

6.2 Implications of the Study

The managerial implications of this research are significant and should be taken into account by stakeholders in the pharmaceutical sector. First, it is imperative that employers invest in the continuous training of their staff, particularly logistics personnel, to enhance their digital skills. This could include seminars on computer networks, cybersecurity, and the effective use of digital platforms. Second, companies should establish dedicated teams for equipment maintenance and optimization of information systems. This would ensure not only service continuity but also the efficiency of logistical operations. Finally, the current health crisis has revealed the resilience of Moroccan companies while exposing their vulnerability to digital challenges. It is crucial to work on reducing digital divides and strengthening supply chain integration through suitable digital solutions to better meet market demands and ensure sustainable growth.

6.3 Limitations and Future Research

This study presents several limitations that deserve to be highlighted. On a theoretical level, the anchoring in the fields of logistics management and information systems may weaken the overall approach by mobilizing emerging concepts that are not yet fully stabilized. On the other hand, the Moroccan context, while more accessible than other regions of Africa, presents specific challenges for data collection, particularly regarding the sensitivity of information related to the pharmaceutical sector. Methodologically, the sample studied does not necessarily represent all companies in the supply chain, which could limit the generalization of the results. Nevertheless, this research paves the way for future studies that could explore acceptance factors such as management commitment, employee involvement, and organizational resilience. A qualitative and longitudinal approach would be beneficial to deepen these issues, particularly in the post-health crisis context, thus allowing for an assessment of the impact of IT on the performance of pharmaceutical companies in Morocco.

REFERENCES

Abbasi, S., Haque Jami, R., & Jiisun, M. S. H. (2023). White Paper-Framework for Integrated Multisector Smart Logistics. *Framework for Integrated Multisector Smart Logistics*, 1-34.

Alsheyadi, A., Baawain, A., & Shaukat, M. R. (2024). E-supply chain coordination and performance impacts: An empirical investigation. *Production & Manufacturing Research*, 12(1), 2379942. DOI: 10.1080/21693277.2024.2379942

Ashraf, M. H., Yalcin, M. G., & Hos, R. (2024). Reconceptualizing supply chain flows (SCFLOWS). *International Journal of Logistics Management*. Advance online publication. DOI: 10.1108/IJLM-03-2024-0180

Blanka, C., Krumay, B., & Rueckel, D. (2022). The interplay of digital transformation and employee competency: A design science approach. *Technological Forecasting and Social Change*, 178, 121575. DOI: 10.1016/j.techfore.2022.121575

Camarinha-Matos, L. M., Fornasiero, R., Ramezani, J., & Ferrada, F. (2019). Collaborative networks: A pillar of digital transformation. *Applied Sciences (Basel, Switzerland)*, 9(24), 5431. DOI: 10.3390/app9245431

Cappuccio, M. L., Galliott, J. C., Eyssel, F., & Lanteri, A. (2024). Autonomous systems and technology resistance: New tools for monitoring acceptance, trust, and tolerance. *International Journal of Social Robotics*, 16(6), 1–25. DOI: 10.1007/s12369-023-01065-2

Cauchois, R., Taghipour, A., Kang, D., Zoghlami, N., & Abed, M. (2017). Advanced solutions for a supply chain with stochastic information. *Journal of Advanced Management Science*, 5(1), 9–13. DOI: 10.18178/joams.5.1.9-13

Centobelli, P., Cerchione, R., & Ertz, M. (2020). Agile supply chain management: Where did it come from and where will it go in the era of digital transformation? *Industrial Marketing Management*, 90, 324–345. DOI: 10.1016/j.indmarman.2020.07.011

Dablanc, L. (2023). Policy and planning for urban logistics. In *The Routledge handbook of urban logistics* (pp. 251–266). Routledge. DOI: 10.4324/9781003241478-23

Dadsena, K. K., Mathiyazhagan, K., & Taghipour, A. (2021). Analysis of barriers for the build the resilient supply chain networks post-COVID-19. In *Managing Supply Chain Risk and Disruptions: Post COVID-19* (pp. 79–89). Springer International Publishing. DOI: 10.1007/978-3-030-72575-4_7

Derwik, P., Hellström, D., & Karlsson, S. (2016). Manager competences in logistics and supply chain practice. *Journal of Business Research*, 69(11), 4820–4825. DOI: 10.1016/j.jbusres.2016.04.037

Dubey, R., Gunasekaran, A., Papadopoulos, T., Childe, S. J., Shibin, K. T., & Wamba, S. F. (2017). Sustainable supply chain management: Framework and further research directions. *Journal of Cleaner Production*, 142, 1119–1130. DOI: 10.1016/j.jclepro.2016.03.117

El Moussaoui, A. E. (2024). Impact of logistics pooling on reduction of CO2 emissions in last-mile logistics. *Management & Sustainability: An Arab Review*.

El Moussaoui, A. E. (2024). Factors influencing the entry of logistics service providers (LSPs) into the logistics clusters: a case study of Casablanca city. *Management & Sustainability: An Arab Review*.

El Moussaoui, A. E. (2024). Unlocking the drivers behind logistics service providers' participation in Morocco's logistics clusters. *Arab Gulf Journal of Scientific Research*, 42(4), 2086–2102. DOI: 10.1108/AGJSR-07-2023-0343

El Moussaoui, A. E., Benbba, B., & El Andaloussi, Z. (2023). Impact of logistics performance on the store image, consumer satisfaction and loyalty: A quantitative case study. *Arab Gulf Journal of Scientific Research*, 41(3), 226–239. DOI: 10.1108/AGJSR-09-2022-0201

El Moussaoui, A. E., Benbba, B., El Andaloussi, Z., & Jaegler, A. (2022). Logistique urbaine au Maroc: Concepts, état des lieux et pistes d'amélioration. *Revue Française de Gestion Industrielle*, 36(2), 71–88. DOI: 10.53102/2022.36.02.910

El Moussaoui, A. E., Benbba, B., Jaegler, A., & El Andaloussi, Z. (2022, May). City Logistics: An Evolutionary, Successful & Collaborative City. In *2022 14th International Colloquium of Logistics and Supply Chain Management (LOGISTIQUA)* (pp. 1-7). IEEE.

El Moussaoui, A. E., Benbba, B., Jaegler, A., & El Andaloussi, Z. (2022, May). How to improve the urban logistics sector in large cities of developing countries: The case of Casablanca. In *2022 14th International Colloquium of Logistics and Supply Chain Management (LOGISTIQUA)* (pp. 1-6). IEEE.

El Moussaoui, A. E., Benbba, B., Jaegler, A., El Andaloussi, Z., & El Amrani, L. (2022, May). Last mile logistics: Challenges & improvement ways. In *2022 14th International Colloquium of Logistics and Supply Chain Management (LOGISTIQUA)* (pp. 1-7). IEEE.

El Moussaoui, A. E., Benbba, B., Jaegler, A., El Moussaoui, T., El Andaloussi, Z., & Chakir, L. (2023). Consumer perceptions of online shopping and willingness to use pick-up points: A case study of Morocco. *Sustainability (Basel)*, 15(9), 7405. DOI: 10.3390/su15097405

El Moussaoui, A. E., El Moussaoui, T., Benbba, B., Chakir, L., Jaegler, A., & El Andaloussi, Z. (2024). Sustainable effects of information sharing between distribution logistics actors: A qualitative case study. *Environment, Development and Sustainability*, •••, 1–19. DOI: 10.1007/s10668-023-04387-3

El Moussaoui, A. E., El Moussaoui, T., Benbba, B., Jaegler, A., & El Andaloussi, Z. (2022). Understanding the choice of collection & delivery point by the E-consumer via a machine learning model: Moroccan case study. *Procedia Computer Science*, 210, 204–211. DOI: 10.1016/j.procs.2022.10.138

El Moussaoui, T., & Chakir, L. (2023, September). BERT-CLSTM model for the classification of Moroccan commercial courts verdicts. In 2023 18th Conference on Computer Science and Intelligence Systems (FedCSIS) (pp. 281-284). IEEE. DOI: 10.15439/2023F3561

El Moussaoui, T., Chakir, L., & Boumhidi, J. (2023). Preserving privacy in Arabic judgments: Ai-powered anonymization for enhanced legal data privacy. *IEEE Access : Practical Innovations, Open Solutions*, 11, 117851–117864. DOI: 10.1109/ACCESS.2023.3324288

El Moussaoui, T., & Loqman, C. (2024). Advancements in Arabic Named Entity Recognition: A Comprehensive Review. *IEEE Access : Practical Innovations, Open Solutions*, 12, 180238–180266. DOI: 10.1109/ACCESS.2024.3491897

Freeney, C. W. (2024). *Case Study on Organizational Adoption of RFID and GPS Technology to Improve Its Supply Chain Process* (Doctoral dissertation, Trident University International).

Gregory, R. W., Kaganer, E., Henfridsson, O., & Ruch, T. J. (2018). IT consumerization and the transformation of IT governance. *Management Information Systems Quarterly*, 42(4), 1225–1229. DOI: 10.25300/MISQ/2018/13703

Gupta, A., Singh, R. K., & Gupta, S. (2022). Developing human resource for the digitization of logistics operations: Readiness index framework. *International Journal of Manpower*, 43(2), 355–379. DOI: 10.1108/IJM-03-2021-0175

Hastig, G. M., & Sodhi, M. S. (2020). Blockchain for supply chain traceability: Business requirements and critical success factors. *Production and Operations Management*, 29(4), 935–954. DOI: 10.1111/poms.13147

Hinings, B., Gegenhuber, T., & Greenwood, R. (2018). Digital innovation and transformation: An institutional perspective. *Information and Organization*, 28(1), 52–61. DOI: 10.1016/j.infoandorg.2018.02.004

Hofmann, E., & Rüsch, M. (2017). Industry 4.0 and the current status as well as future prospects on logistics. *Computers in Industry*, 89, 23–34. DOI: 10.1016/j.compind.2017.04.002

Ishfaq, R., Davis-Sramek, B., & Gibson, B. (2022). Digital supply chains in omnichannel retail: A conceptual framework. *Journal of Business Logistics*, 43(2), 169–188. DOI: 10.1111/jbl.12277

Jones, E. C. (2019). The Internet of Things (IoT) Technologies and the Tracking of Supply Chain Assets. In *Supply Chain Engineering and Logistics Handbook* (pp. 419–441). CRC Press. DOI: 10.1201/9781315159096-13

Kain, R., & Verma, A. (2018). Logistics management in supply chain–an overview. *Materials Today: Proceedings*, 5(2), 3811–3816. DOI: 10.1016/j.matpr.2017.11.634

Klumpp, M., & Zijm, H. (2019). Logistics innovation and social sustainability: How to prevent an artificial divide in human–computer interaction. *Journal of Business Logistics*, 40(3), 265–278. DOI: 10.1111/jbl.12198

Lagorio, A., Zenezini, G., Mangano, G., & Pinto, R. (2022). A systematic literature review of innovative technologies adopted in logistics management. *International Journal of Logistics*, 25(7), 1043–1066. DOI: 10.1080/13675567.2020.1850661

Loivet, W., Taghipour, A., & Kang, D. S. (2020). The rise of green supply chain management: Between complexity and necessity. *J. Econ. Bus. Manag*, 8(1), 1–7. DOI: 10.18178/joebm.2020.8.1.604

Marchet, G., Melacini, M., Perotti, S., Rasini, M., & Tappia, E. (2018). Business logistics models in omni-channel: A classification framework and empirical analysis. *International Journal of Physical Distribution & Logistics Management*, 48(4), 439–464. DOI: 10.1108/IJPDLM-09-2016-0273

Mazhar, T., Irfan, H. M., Khan, S., Haq, I., Ullah, I., Iqbal, M., & Hamam, H. (2023). Analysis of cyber security attacks and its solutions for the smart grid using machine learning and blockchain methods. *Future Internet*, 15(2), 83. DOI: 10.3390/fi15020083

Mentzer, J. T., DeWitt, W., Keebler, J. S., Min, S., Nix, N. W., Smith, C. D., & Zacharia, Z. G. (2001). Defining supply chain management. *Journal of Business Logistics*, 22(2), 1–25. DOI: 10.1002/j.2158-1592.2001.tb00001.x

Merimi, M., & Taghipour, A. (2021). Accelerating the digitalization of the supply chain: An empirical research about COVID-19 crisis. In *Digitalization of decentralized supply chains during global crises* (pp. 1–24). IGI Global. DOI: 10.4018/978-1-7998-6874-3.ch001

Moussaoui, T. E., Loqman, C., & Boumhidi, J. (2024, May). Flat and Nested Named Entity Recognition in Arabic Language. In 2024 4th International Conference on Innovative Research in Applied Science, Engineering and Technology (IRASET) (pp. 1-7). IEEE. DOI: 10.1109/IRASET60544.2024.10548518

Negi, S. (2024). Global supply chain competitiveness: The synergistic role of integrated logistics and global sourcing. *Global Business and Organizational Excellence*, 43(4), 111–130. DOI: 10.1002/joe.22247

Nissen, V. (2018). Digital transformation of the consulting industry—introduction and overview. *Digital Transformation of the Consulting Industry: Extending the Traditional Delivery Model*, 1-58.

Puskar, J. (2019). Counting on the body: Techniques of embodied digitality. *New Media & Society*, 21(10), 2242–2260. DOI: 10.1177/1461444819843300

Rao, T. R., Mitra, P., Bhatt, R., & Goswami, A. (2019). The big data system, components, tools, and technologies: A survey. *Knowledge and Information Systems*, 60(3), 1165–1245. DOI: 10.1007/s10115-018-1248-0

Reina, R., & Scarozza, D. (2021). Human resource management in the public administration. *Organizational development in public administration: The Italian way*, 61-101.

Shankar, V., Kalyanam, K., Setia, P., Golmohammadi, A., Tirunillai, S., Douglass, T., Hennessey, J., Bull, J. S., & Waddoups, R. (2021). How technology is changing retail. *Journal of Retailing*, 97(1), 13–27. DOI: 10.1016/j.jretai.2020.10.006

Shashi, C., Centobelli, P., Cerchione, R., & Ertz, M. (2020). Managing supply chain resilience to pursue business and environmental strategies. *Business Strategy and the Environment*, 29(3), 1215–1246. DOI: 10.1002/bse.2428

Simone, C., Barile, S., & Calabrese, M. (2018). Managing territory and its complexity: A decision-making model based on the viable system approach (VsA). *Land Use Policy*, 72, 493–502. DOI: 10.1016/j.landusepol.2017.12.070

Taghipour, A. (Ed.). (2020). *Demand forecasting and order planning in supply chains and humanitarian logistics*. IGI Global.

Taghipour, A., & Beneteau-Piet, C. (2020). Sustainable supply chain management performance. *International Journal of Innovation, Management and Technology*, 11(6), 165–169. DOI: 10.18178/ijimt.2020.11.6.897

Taghipour, A., & Merimi, M. (2021, March). Digital transformation of supply chains during crisis: covid-19. In 11th Annual International Conference on Industrial Engineering and Operations Management, IEOM (pp. 7493-7501). DOI: 10.46254/AN11.20211298

Taghipour, A., Sohrabi, A., Ghaedi, M., & Khazaei, M. (2023). A robust vaccine supply chain model in pandemics: Case of Covid-19 in Iran. *Computers & Industrial Engineering*, 183, 109465. DOI: 10.1016/j.cie.2023.109465

Urciuoli, L., & Hintsa, J. (2017). Adapting supply chain management strategies to security–an analysis of existing gaps and recommendations for improvement. *International Journal of Logistics*, 20(3), 276–295. DOI: 10.1080/13675567.2016.1219703

Zhao, B., & Tu, C. (2021). Research and development of inventory management and human resource management in ERP. *Wireless Communications and Mobile Computing*, 2021(1), 3132062. DOI: 10.1155/2021/3132062

Chapter 12
Extended Reality and Security

Atul B. Kathole
https://orcid.org/0000-0002-6162-8200
Dr. D.Y. Patil Institute of Technology, Pimpri, India

Yatri Davda
https://orcid.org/0000-0003-4198-3499
Dr. D.Y. Patil Institute of Technology, Pimpri, India

Nupur Soitkar
Dr. D.Y. Patil Institute of Technology, Pimpri, India

ABSTRACT

VR, AR, and MR are evolving XR technologies impacting industries like healthcare, education, and collaboration. However, XR's integration of real and digital worlds introduces unique security challenges due to the handling of sensitive data, including biometrics and behavioral patterns. Emerging technologies like AI, edge computing, and 5G increase XR's potential but also its vulnerability to cyber threats. XR security, combining traditional and innovative measures, is critical to safeguarding these platforms as they become more widely adopted.

1. BACKGROUND

Internet crime complaints have significantly increased over the past five to six years, according to statistics. In 2020, there were 515,612 complaints, compared to 162,091 complaints the year before. The primary sources of data breaches in 2021 were malware, corporate emails, and phishing and related cyberattacks. Artificial

DOI: 10.4018/979-8-3693-7478-8.ch012

intelligence (AI), edge computing, 5G, and other emerging technologies in the XR space expand the potential of immersive environments while also posing new security threats. Recent evidence indicates that ransomware cyberattacks will overtake phishing attacks in 2022 since data compromises will be the main cause of these attacks. More intricate, real-time interactions are made possible by these technologies, which increases the area that can be attacked by cybercriminals. Individuals are essential to the defence against cyberattacks since they are the end users. The difficulty lies in spotting malicious and fraudulent content and classifying it as an attack. According to an IBM analysis from 2022, human mistake was the cause of 25% of security breaches in industrial enterprises. The fact that XR is already being used for more than simply entertainment and gaming indicates that this expansion of its uses is essential to its continued development. I'm excited to watch how these technologies develop and spread even farther in the future. The COVID-19 pandemic greatly affected worldwide supply chains, leading companies to swiftly implement digital transformation strategies to ensure continuity and resilience. This article examines how digital technologies have changed supply chains in times of crisis, especially considering the effects of COVID-19. the COVID-19 pandemic accelerated digital transformation in supply chains. Here's what such research typically covers: The global crisis caused significant interruptions to conventional supply chains. During the pandemic, there were shortages, delays, and changes in consumer demands. This situation compelled companies to reevaluate and modify their supply chain approaches. Digital Transformation: The integration of digital technologies, including cloud computing, IoT, AI, and machine learning, to improve the efficiency, visibility, and resilience of supply chains. A movement towards supply chain operations that are more agile, data-driven, and automated. Examples or case studies of organizations that have effectively utilized digital solutions to manage disruptions (Taghipour & Merimi, 2021; Taghipour & Frayret, 2013a, and 2013b; Zoghlami et al., 2016 ; Cauchois et al., 2017; Dadsena et al., 2021; Loivet et al., 2020). There is even research paper related named Challenges of Society 5.0, here in this in order to improve societal functions, including economic, health, environmental, and cultural systems, Japan launched the Society 5.0 initiative, which aims to create a "super-smart" society where digital technologies (such as artificial intelligence (AI), the Internet of Things (IoT), robotics, big data, and blockchain) are seamlessly integrated into all facets of life. Achieving a balance between technology and human needs is the goal of Society 5.0. By utilizing technology to tackle societal issues like resource management, aging populations, and climate change, Society 5.0 aims to establish a human-centered society in contrast to earlier industrial revolutions (Society 1.0 through Society 4.0), which concentrated on automation, digitization, and economic growth. This chapter aligns with the book's focus on digital transformation in the supply chain, specifically exploring how technologies like XR, AI, and Blockchain are reshaping operations, driving efficiency, and addressing challenges such as demand forecasting, inventory management, and customer service (Taghipour et al., 2022; Taghipour, 2020; Taghipour, 2014 ; Merimi, & Taghipour, 2021; Taghipour& Frayret, 2011 ; Mbiatem et al., 2018; Nadia et al., 2021).

2. INTRODUCTION TO EXTENDED REALITY

2.1 Overview of Extended Reality

Virtual reality (VR), augmented reality (AR), and mixed reality (MR) are only a few of the immersive technologies that fall under the broad category of extended reality (XR). Here is a quick synopsis of each (Figure 1):

With the use of virtual reality (VR) technology, users can fully immerse themselves in virtual environments. In order to create a sensation of presence in a virtual environment, users usually don VR headsets that detect their motions and display 3D pictures. Virtual reality (VR) is commonly utilized in training simulations, games, and virtual tours.

Digital content is superimposed on the actual world through augmented reality, or AR. You can use AR glasses, tablets, or smartphones to experience this. By incorporating virtual objects into the user's physical environment, such as text, photos, or animations, augmented reality (AR) improves the user's sense of realism. Popular apps include several mobile games, such as Pokémon GO, and etc.

Figure 1. Extended reality

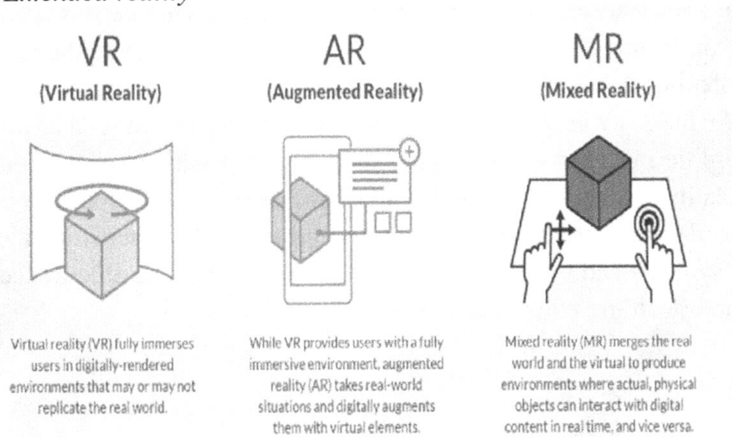

Advantages of Extended Reality:

1. **Improvement in training experiences**: Providing practical experience is crucial. Rather of only reading theoretical material, trainees engage directly with the subject matter. For instance, surgical residents can use a virtual patient to rehearse procedures in virtual reality. In a similar vein, engineers can use MR to learn how to put together an automobile using a virtual model in a mock garage. These tasks can also be repeated as often as necessary.
2. **Betterment in marketing Strategies:** Think about using immersive virtual showrooms and product demos for marketing purposes, or virtual try-ons. For instance, BMW eliminates the requirement for prospective buyers to visit a physical dealership by enabling them to study and interact with BMW automobiles in a completely immersive environment. Colleges can also use XR technology to provide virtual tours in an effort to draw in and persuade potential students.
3. **Improvement in visuals and presentation:** XR can help you create more realistic and captivating presentations. To be more precise, XR enables you to dynamically present items, designs, or tactics, which will improve the way your audience comprehends your message.

Disadvantages of Extended Reality:

1. **Legal issues**: Lawmakers and regulators are forced to play catch-up, as is the case with any technology that develops more quickly than the legal system can keep up with it. XR is not an anomaly. Because of this, there are no clear regulations governing what is and is not permissible in virtual worlds, nor even identifying which jurisdictions apply to them.
2. **Stealing Network informations:** Hackers might take advantage of wearable Android devices to obtain network credentials. Hacking can be a cyber hazard for merchants who employ shopping apps with augmented and virtual reality. Numerous clients already have their mobile payment options and card information stored in their user profiles. The ease with which mobile payments can be made means that hackers could be able to access these and stealthily empty accounts.

Strengths:

1. XR Technologies (VR & AR): Enhanced Efficiency:By directing employees to the appropriate inventory locations and boosting picking and packing speed and accuracy, AR enhances warehouse operations (Taghipour et al.,2021).Improved

Training: Virtual reality (VR) is being utilized more and more in employee training because it provides immersive learning settings that replicate actual supply chain activities without the possibility of mistakes (Taghipour&Merimi,2021). Remote Assistance: By enabling real-time remote collaboration and debugging, AR and VR help support services cut down on trip expenses and downtime (Merimi & Taghipour, 2021).

2. Machine Learning & AI: Predictive analytics: AI helps businesses minimize stockouts and excess inventory by improving demand prediction and supply chain forecasts (Dadsenaet al., 2021).Automating monotonous jobs, such as processing orders, allows AI to free up human workers to perform more strategic duties (Taghipour et al., 2021).
3. IoT: Real-time Data Access: IoT sensors give supply chain managers access to real-time data, which enhances their ability to make decisions and react quickly to interruptions(Taghipour & Merimi, 2021). Better Monitoring and Tracking: Businesses may ensure compliance and product quality by using IoT devices to monitor product conditions (such as temperature-sensitive commodities) during transit (Merimi & Taghipour, 2021).

Weakness:

1. VR and AR Technologies: High Initial Investment: The initial outlay for VR/AR systems can be high, particularly when considering hardware, software, and system integration (Taghipour & Merimi,2021).Complexity and Adoption opposition: There is a learning curve associated with implementing XR systems, and organizations may encounter opposition from staff members who are unaccustomed to or uneasy with new technology (Merimi & Taghipour, 2021).
2. Machine Learning & AI:Bias and Data Quality: AI systems can only be as good as the data they are trained on. Inaccurate forecasts and less-than-ideal choices may result from biased or poor data (Dadsena et al.,2021). High Computational Costs: AI model training frequently necessitates large computational resources, which small and medium-sized businesses may find too expensive (Taghipour & Beneteau-Piet, 2020)

2.2. Characteristics of Extended Reality

- **Immersion**: Users can experience a sense of presence with XR, feeling as though they are a part of an augmented or virtual environment.
- **Interactivity**: Real-time user interaction with digital components increases engagement and adds dynamic features to experiences.

- **Real time rendering**: Since XR environments are frequently produced in real-time, changes and interactions can occur naturally as users travel through them.
- **Enhanced Perception**: Whereas MR combines virtual and physical aspects, AR especially improves real-world experiences by superimposing digital information on top of it.
- **Social Interaction:** Users can engage in collaborative experiences and communicate with others in shared virtual places on a number of XR platforms.
- **Versatile:** XR is a flexible tool that may be used in a variety of industries, such as marketing, education, training, healthcare, and gaming.

3. APPLICATION OF EXTENDED REALITY

- **Manufacturing:** By using headsets to facilitate worldwide collaboration, XR increases assembly line efficiency for industrial organization.
- **Gaming and entertainment:** XR technology allows players to step into a virtual reality that makes them feel as though they are physically there. Virtual reality headsets are another tool you may use to watch movies or sports events in a more immersive way.
- **Learning:** Immersion activities improve learning by fostering a more stimulating atmosphere. It even gives you the chance to take advantage of real-world learning experiences, whether they are for preparing aspiring professionals or younger students for the workforce.
- **Health:** By enabling the transformation of two-dimensional medical photographs into three-dimensional ones via extended reality, medical professionals may diagnose patients with more accuracy. Physicians can also use XR to do risk-free surgery.
- **Warehouse Operations:** By superimposing real-time data, including item locations, stock levels, and picking routes, AR glasses are used by companies like DHL and Siemens to assist employees in warehouses. This improves inventory management's accuracy and speed while cutting down on training time.
- **Training and Security:** Volkswagen uses virtual reality (VR) simulations to train staff members in a secure online setting. These simulators mimic actual factory environments, allowing employees to practice operating machinery or working on an assembly line without worrying about getting hurt.

3.1. Potential applications Extended Reality

Study area of extended reality has many potential applications amongst to various field here are some examples of that

- **Education and training**: Education and training it in it includes immersive learning like conducting of virtual classroom subjects related to a and even medical training helps in surgical procedure and studies.
- **Healthcare:** Healthcare your over in it is used in pain management phobia therapy even patient education is important to apps and visuals condition procedure.
- **Gaming and entertainment:** Gaming and entertainment in it is place in massive rule in it like we are games AR experience and digital content with the real world virtual concerts and events even life performances and what will environment can be shown on we are TV into it.
- **Real estates**: In real state you over even virtual tours can be shown like a 3D box through of buyers and visits which may be shown like it is physically visited even arguments teasing like even we can show the we can visualise empty pieces using AR like empty hall using visualising for nature and decors
- **Tourism and Travel:** Tourism and travels senate we can experience so virtual travel experience like exploring the landmarks and destination through VR experiences .Even used in ear navigation like and enhancing real world navigation with context information about surrounding.
- **Smart Warehouses:** Walmart tracks inventory levels in real time with IoT-enabled smart shelves. Sensors on the shelves notify employees when supplies are running low, allowing them to promptly replenish inventory and enhance inventory control in general.

3.2. Significance of Extended Reality

The study's significance are important for XR designers and programmers, as well as practitioners in other fields, particularly those in the retail industry. Because of a lack of theoretical guidance and detailed analysis of perceived workload, many program developers and designers struggle to identify suitable solutions for improving the perceived usability, ease of use, utility, and efficiency of XR technologies and systems. The results show that using AR requires a higher level of effort and mental load than situations without AR. Given that the pure "reconstruction" mode did not raise the difficulty of conducting activities or finishing particular tasks, there appears to be an opportunity for adding more vivid content. Surgical Simulations: To lower the chance of error, XR can be used to train surgeons and perform difficult procedures in a virtual setting. Therapy and Rehabilitation: XR can help with both physical and mental rehabilitation. It can be used to treat PTSD, phobias, and other mental health issues. Virtual Showrooms: By enabling customers to digitally try on things, retailers may improve the purchasing experience.

Interactive Advertising: Companies can design compelling advertising campaigns that delve deeper into the minds of consumers.

Even though in employment related fields the extended reality plays a major role in like Job Creation XR technologies generate new jobs like XR training specialists, AR content developers, and VR simulation designers, even while other jobs (like repetitive physical work) may be mechanized.

Employees must acquire new skills in order to use these cutting-edge devices. To assist employees in adjusting to new positions that emphasize technology management, companies must offer upskilling programs.

Job Elimination is done in AI that is replacing certain repetitive and routine activities, like basic logistical work and manual data entry. But this also gives workers more time to concentrate on more difficult jobs.

here even job Creation take place In order to handle and analyse AI-driven information, there is a growing need for positions such as supply chain analysts, data scientists, and AI specialists (Taghipour et al., 2022; Taghipour & Merimi, 2021; Taghipour et al., 2023b; Taghipour et al., 2015 ; Taghipour et al., 2023c).

4. ANALYSIS OF EXTENDED REALITY

CybAR software. As per the findings, 88% of the individuals mentioned having a high degree of knowledge of cybersecurity principles (Figure 2). Regarding the suggested application, the outcomes showed the benefit of using CybAR to comprehend weaknesses and cyberattacks. A simple, practical, and interesting way to learn the foundations of cybersecurity is through the use of CS:NO, an extended reality (XR)-based educational tool. An electronic setting served as a tool for network data flow visualization. Additionally, they provided comments on the outcomes of actual hacks. The efficacy of the was evaluated by a subjective research including 91 participants. to educate the core ideas. about cybersecurity. Training in AR .Cybersecurity Awareness using Augmented Reality (CybAR), a mobile application based on augmented reality, was created by Alqahtani et al . Businesses can track and optimize the usage of resources like materials, water, and energy thanks to IoT devices. For example, smart warehouses lower energy use and operating costs by controlling lighting, temperature, and humidity via IoT devices. This can greatly lower the carbon footprint of warehouse operations and promote more environmentally friendly operations.

Supply Chain Visibility: By improving supply chain visibility, IoT enables businesses to lower carbon emissions, optimize transportation, and better plan routes. IoT-enabled logistics, for instance, may optimize shipment routes and cut down on idle time, which lowers emissions and fuel usage (Merimi & Taghipour 2021).

In order to guarantee that items are obtained responsibly, blockchain technology is essential. Blockchain-based technologies allow businesses to trace the origin and path of goods, confirming whether they were produced sustainably and ethically. De Beers, for example, employs blockchain technology to track the origin of diamonds, guaranteeing that they are mined ethically and without violence, which is consistent with social and environmental sustainability. According to Taghipour et al., (2022) in "Opportunities and Challenges for Society 5.0 in Procurement," blockchain helps businesses improve supply chain traceability and guarantee that products are made and delivered with the least possible negative environmental impact. This guarantees environmentally friendly procurement methods that limit waste and lessen the impact of items on the environment (Taghipour et al., 2022).

Figure 2. Cybersecurity training techniques

Features	VTFiWall	Lord of Secure	CyberVR	Shaila Rana	CyVR-T	CISE-ProS	Chekhovskoy	CybAR	CS:NO	XR-CEIL
Technology	Desktop VR, VR, Android smartphone	VR	VR	VR	VR	VR	VR	Mobile AR	XR	XR
Software	Unity 3D, Blender, Google VR SDK, Android SDK	Unity 3D, Blender, Android Studio	Unreal Engine	Unity VR		Unity 3D	Unity and Unreal Engine 4	Unity 3D, Vuforia SDK	Unity 3D	
Hardware	HTC Vive, VR Box, Google cardboard	HMD	Oculus Rift, Leap Motion	Google Cardboard, VR headset	Oculus Rift	HTC Vive, headset		Android smartphone	Oculus Rift, Arduino Uno Rev 3, Proto Shield	
Interaction	Head gestures and pointing	Pointing	Hand gestures		Hand controllers	Motion-tracking sensors, handheld controllers	Mouse	Touchscreen	Visual, audio, and haptic thermal feedback	
Cybersecurity contents	Firewall	Flooding, IP spoofing, firewalls, TCP convert channels, honeypots, and intrusion detection and prevention systems	Six mini-games Information flow, code injection, patch management, dynamic SW Analysis, privilege escalation, public-key cryptography	Physical security concepts, such as locked devices, keeping passwords and sensitive files secured, keeping doors locked, and keeping keys secured	Voyage management system, radar detection system, and automatic identification system	Tutorial, entrance/exit, inspection, and replacement of hardware	Network attacks	Teaching the fundamental concepts of cybersecurity	Encryption and decryption, firewalls, malicious data, network traffic	Exploration of immersive multi-dimensional data of a heterogeneous nature

5. STUDY AREA

5.1 Methods and Materials

A. Methods

To analyse the meta-data of the papers reviewed, the author supplied keywords of all the papers reviewed were extracted. In addition to author supplied keywords, some terms of interest were also extracted from titles. Scopus and IEEE Xplore were searched with dates from 2010 to 2015 to determine recent incidence of the terms. For added completeness, the following terms were added in addition: computer security, cyber domain, cyber war, cyber bullying, cyber physical, semantic web, semantic web search, cyber safety, cybernetics, sustainability, darknet, dark web, deep web, surveillance, cryptography, cryptology, encryption, and cryptanalysis. These searches were performed on Scopus c. 10/7/2015-10/13/2015 and on IEEE Xplore c. 10/12/2015-10/13/2015. The author-supplied keywords were searched for in quotes for exact terms (up to capitalization).

- Literature review method: The intention of this literature review was to assess the state of emerging cyber security research and explore avenues of cyber security that have not received as much traditional attention as standard topics of network security, cryptography, and basic system security that a typical university curriculum in focuses on [105]–[107]. The manual literature review was performed via a number of particular searches throughout September 2015 with the MIT libraries revolving around journal papers containing the word or prefix ''cyber'' and selected based on breadth of coverage as candidates for further reading. Selection criteria included priority given to other literature reviews and papers whose intention was broad characterization of issues, with a slight preference away from technical papers (Figure 3).

Figure 3. Cybersecurity papers

Label	General Descriptor*	General topics encompassed**	Extended High-Level Summary***
A	Public Sector Policy	Global Internet law, politics, and governance	Arguably the most diverse category in this literature review. Contains papers on cyberspace geography and jurisdiction, papers on issues surrounding concepts of cyber war, Internet standards and governance, or various other legal issues, including crime problems that have a very legalistic or political angle. Papers containing broad questions of national security and strategy, the global multistakeholder community, and many other broad policy or international relations questions are included in this category.
B	Cyberspace Infrastructure	Technical Cybersecurity Research (hardware, software, CPS, cryptography)	Most of the technological papers. While this literature review attempted to maintain a focus on cybersecurity of critical infrastructure, and thus that dominates this category, this category could theoretically include anything mathematical or related to software for the infrastructure of cyberspace. In our review, this includes analyzing SCADA vulnerabilities, high level papers on security of cyber-physical (e.g. critical infrastructure) systems, as long as they maintain some degree of technical rigor such as modeling. But it could also have included encryption scheme development, Internet protocol descriptions, and more. This category would not include software or models designed for policy analysis, business management, criminological or sociological papers, or ontologies aimed at broad descriptions of cybersecurity.
C	Business and Operations	Business Management, frameworks, practices	These papers are by authors aiming to call attention to poor business practices and other kinds of risks for businesses, culture and awareness, and ways to improve these practices (many of which are simply negative statements: stop doing something, e.g. stop leaving CISOs/CIOs out of the loop). In addition, this category encompasses papers that include more in-depth *positive* frameworks for maintaining cyber security in a business through education, access control, supply-chain management, etc. That is, these frameworks specify what *to* do as opposed to what *not to* do. Many of these papers include not only suggestions for management but also propose research agendas for cyber security management.
D	General Research	General cyber crime, threats, other broad analyses (including Ontologies)	These papers analyze what common cyber-crimes, attacks, and threats exist. These papers have a more technical slant than category A, but are much broader than B. In addition, this category contains ontologies for vulnerabilities and attacks, populating knowledge bases, or mapping links between threats and actors. Other inclusions in this category are such papers as "A systematic literature review of computer ethics issues" and other "system-wide" topics that are fundamental to understanding cyberspace. This category is typically aimed at a broader audience than the other categories.

Categories of cyber security

- Justification for Categorization: Public Sector Policy, Cyberspace Infrastructure, Business and Operations, and General Research are the four categories that reflect various viewpoints or methods of conducting cybersecurity research. By classifying documents, authors are able to:

Demonstrate how diverse cybersecurity research is.

Emphasize areas that require more attention and areas that have not received as much attention. Examine and contrast various research methodologies. Determine any possible conflicts or overlaps between several research fields.

- Additional information about each topic is provided in the columns labelled "General Topics Encompassed" and "Extended High-Level Summary. "The keywords or phrases listed under "General Topics Encompassed" specify the parameters of each category.

With examples of particular study areas, the "Extended High-Level Summary" provides a more thorough explanation of the kinds of articles that fall under each category. This is essential for proving that the writers are aware of the subtleties in each category.

- Use in Analysis: The table is analytical as well as descriptive. This classification would probably be used by the review paper's authors to:

Determine trends: For instance, they may discover that research on cybersecurity in corporate operations is booming, but study on the moral ramifications of cyberwarfare is lacking.

Find gaps: They may find that there is a dearth of multidisciplinary studies that integrate technical and policy viewpoints.

Make recommendations for future lines of inquiry: They may suggest certain research problems that require attention based on their analysis.

- Methodological Transparency: Including this table demonstrates the authors' methodology for selecting and analysing the literature. It shows that their review is systematic and not just a random collection of papers.

The paper which is A Systematic Literature Review on Cybersecurity Threats of Virtual Reality (VR) and Augmented Reality (AR), we provide a systematic literature review of potential security, safety, and privacy (SSP) threats from virtual reality (VR) and augmented reality (AR) environments, which have become extremely popular. We introduce a detailed comparison of current risks that become much more dangerous in the environment of AR and VR, such as the security and privacy challenges of AR platforms. We then discuss expected mitigation techniques for security and privacy threats proposed by previous literature studies, such as creating three AR visual output models that balance OS control with application flexibility. The purpose of this article is to raise awareness of potential security and privacy vulnerabilities caused by seemingly safe VR and AR technologies.

- We followed the SLR approach due to its systematic and prevalent use in literature reviews. It involves a review of the literature using a predefined set of steps to identify, analyse, and interpret the available research related to the given research question . The complete scenario of the SLR is depicted in Figure 4.

Figure 4. SLR

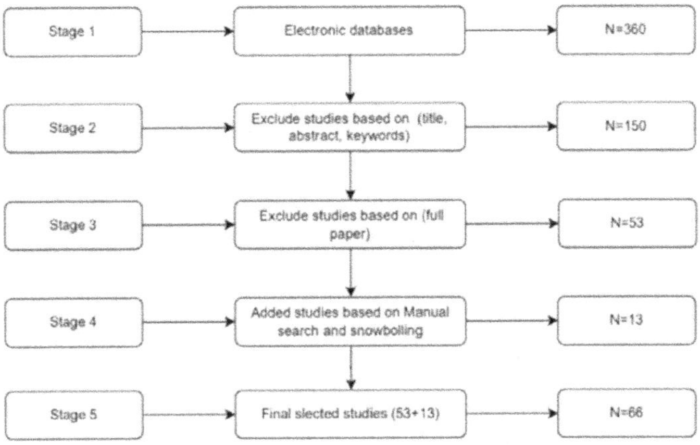

5.2 Study Area

Extended reality comprises and are used in many study areas where its applications are used like Technology development in which used as hardware in research of VR headset, AR glasses and many other hardware device. Even Technology development is even used in software in development of of XR platforms applications and tools for content creation. It is even used in user experience (UX) design designing interfaces and interaction method and conducting usability study to understand how XR interact. Used in human computer interaction (HCI) help and investigate how users interact with virtual and argumented environments Even in studies gesture recognition and other input methods in which human and computer interacts. Even used in education and training stimulation for training in fields like medicine engineering and extra studying categories and developing best practices on XR. The most important domain is healthcare even it is used in healthcare for therapys and for maintaining pain management and mental health treatment and even used in medical training and surgical practice. Extended Reality (XR), AI, IoT, and Blockchain in industry and supply chains is an important element for understanding how these technologies have evolved and been implemented over time.

The COVID-19 pandemic hastened supply chains' digital revolution. It demonstrates how businesses have embraced cloud computing, artificial intelligence, and the Internet of Things to become more resilient and adaptable in times of crisis.

Supply chain technology integration was already advancing before the pandemic (e.g., RFID in the 2000s, ERP systems in the 1990s). However, even conventional industries were compelled to adopt digital solutions more quickly because to the COVID-19 pandemic.

Although XR was not extensively covered in the study, the crisis could have improved its growing significance in supply chains, especially in areas like training, remote support, and warehouse management (Taghipour & Merimi, 2021).

The Society 5.0, a human-centred society in which intelligent technologies such as blockchain, AI, IoT, and AR/VR are smoothly incorporated into various sectors and daily life.

By expanding their capacities with new technologies like XR, the idea of Society 5.0 builds on earlier automation and information technology breakthroughs. The concept of merging digital and physical environments is reminiscent of past decades' digital convergence tendencies (e.g., IoT fusing the digital and physical in the 2010s) (Taghipour et al., 2022).

A discussion of how XR technologies are incorporated into procurement procedures would be beneficial for this study. Through improved visualization and training for procurement teams, AR and VR's real-time data capabilities could significantly enhance procurement.

The COVID-19 pandemic accelerated supply chain digitization, forcing companies to deploy AI and cloud-based solutions to maintain operations.

In the past, supply chains have gradually gone digital, beginning in the 1990s with EDI and ERP systems. Real-time tracking was made possible by IoT and RFID technologies in the early 2000s, and in the 2010s, artificial intelligence (AI) surfaced as a means of automating and forecasting demand.

The COVID-19 epidemic brought to light the value of distant work and virtual collaboration, two areas that could benefit from more research into XR technologies. In emergency situations like the pandemic, the use of AR/VR tools for training and remote support may be crucial.

XR technologies can play a significant role in visualizing sustainability efforts—for example, AR could overlay environmental data on physical products in warehouses, helping companies track carbon footprints in real-time and VR can simulate the environmental impact of various supply chain operations, aiding decision-making in sustainable practices. The push for sustainability in supply chains gained momentum in the 2000s, with a greater focus on resource efficiency, ethical sourcing, and traceability. The integration of Blockchain in the 2010s enabled better transparency and trust in sustainable supply chains (Lu & Taghipour, 2023; Taghipour et al., 2023a; Taghipour et al., 2024; Taghipour, 2023; Taghipour & Beneteau-Piet, 2020 ; Mahfod et al., 2019).

Case Studies:

1) BMW's Use of VR: BMW trains staff members in intricate automobile assembly procedures using virtual reality (VR). By giving staff members interactive, hands-on learning experiences—which are especially beneficial when working with expensive machinery or complex tasks—the immersive training sessions shorten training times and improve retention (Cauchois et al., 2017).
2) AI-Driven Supply Chain by Unilever: Unilever has incorporated AI technologies into its supply chain to manage inventories and estimate demand in real time. Unilever can forecast changes in consumer demand and modify its production and distribution in response, increasing efficiency and cutting waste, by examining a variety of data sources, including social media, weather trends, and past sales (Merimi & Taghipour, 2021).
3) IoT-Enabled Fleet Management by Caterpillar: Caterpillar employs IoT sensors in its construction machinery to continuously check on the condition of its machines. By monitoring variables like temperature, pressure, and consumption trends, the sensors assist operators in anticipating when equipment will need maintenance, minimizing downtime, and increasing operational effectiveness (Zoghlami et al., 2016).
4) TradeLens, a Blockchain-based platform developed by Maersk and IBM, was created with the goal of streamlining and digitizing the global shipping sector. Shippers and freight forwarders both gain from TradeLens' use of Blockchain to trace shipments, which increases transparency, cuts down on delays, and guards against fraud (Taghipour et al., 2022).

6.1 Techniques

- **Traditional Cyber Security Training Technique:** Training using some pre-built systems with fixed curricula is referred to as a conventional approach to training. These systems are less flexible as they train users with the concept of Symmetry 2023, 15, 2175 5 of 25 "one for all". These techniques are often costlier, so they are not well suited for cybersecurity training . These systems use both paper (i.e., newsletters, posters, and brochures) and electronic-based resources (mobile phones and computers) . Print materials may convey information related to a single or multiple topics to a specific group. In conventional training approaches, the replication of the real scenarios is very difficult. Cybersecurity training is carried out to improve workers' skills by including recent topics, such as firewall management, encryption etc.
- **Lecture based training**: This type of training consists of the delivery of information to a classroom by a field expert. This approach is mostly based on presentations to a group of employees. This type of training includes direct interaction with trainees, so it is considered better than paper-based training.

It is more expensive due to its lack of flexibility and a lack of up-to-date cybersecurity training information. This type of training is also considered boring and ineffective. The success of this type of training depends on the expertise of the instructor in the subject area, as well as the ability to engage trainees during the class.

- **VR cybersecurity training**: VR-based training is an extended form of game-based training in that it not only entertains and engages users during training, but also provides an immersive and realistic interactive experience. Due to its highly engaging, interactive, and immersive nature, it is considered the most effective training technique. VR cybersecurity training includes stylistic, thematic, and mechanical features that make it a digital agent in VR applications. Cybersecurity training systems should also include visual aspects to deliver visual content along with stylistic modules and themes to enhance student learning. Virtualization is highly dependent upon delivering realistic training, along with being cost-effective, safe, low-maintenance, and reproducibleAdditionally, VR provides an even more realistic choice for providing cybersecurity training. VR and AR applications can to teach us new applications (Figure 5).
- **AR security training**: AR Cybersecurity Training Alqahtani et al. developed an AR -based mobile application, "Cybersecurity Awareness using Augmented Reality" (CybAR), for teaching the fundamental concepts of cybersecurity. They also presented the results of real cyberattacks through feedback. A subjective study of participants was carried out to assess the effectiveness of the CybAR game. The results showed that 88% of the participants reported a high level of understanding of cybersecurity concepts. The results for the proposed application indicated the usefulness of CybAR in terms of understanding vulnerabilities and cyberattacks. CS:NO is an extended reality (XR)-based educational tool for teaching the fundamentals of cybersecurity in an easy, concrete, and engaging manner.
- **Cybersecurity combined Training:** This technique aims to integrate multiple approaches, including text-, video-, and the goal of game-based cybersecurity training is to improve the user's ability to defend against various cyberattacks. Various techniques ought to be used in a Depending on user preferences, a cybersecurity training program will integrate various self-paced approaches with instructor-based instruction. Improved knowledge and behaviour among learners can be achieved through a variety of interactive, self-paced training methodologies. Reliable cybersecurity. Training plays a crucial role in helping people learn the fundamentals and changing their behaviour. employing cybersecurity to protect against all kinds of cyberattacks. Users' levels of interaction can be enhanced and their knowledge can be con-

solidated with combined training approaches. Nonetheless, a survey revealed that customers preferred classroom instruction guided by an instructor.

6.2. Tools

1.CyVR-T:

CyVr-T is the best tool for the identification of cyberattacks and can be used in more complex scenarios.

2.CS:NO

CS:NO is an extended reality (XR)-based educational tool for teaching the fundamentals of cybersecurity in an easy, concrete, and engaging manner.

3. While offloading XR tasks in a binary form as a unit is a simple method, it might not make full use of the edge network's computational resources. However, while arbitrary partitioning of XR tasks can be difficult to perform well in practice, partial offloading approaches allow the simultaneous calculation of XR tasks on both MEC servers and user devices. To maximize the efficiency of computing resource use and minimize user energy consumption, XR tasks are split into five subtasks according to their functional components. These subtasks are then offloaded at the granularity of subtasks. The latter, on the other hand, creates a task-offloading strategy solely to meet the needs of subtasks requiring serial execution, without investigating the potential of parallel execution. The former, however, ignores dependencies among subtasks.

7. CONCLUSION

As cyber risks grow and may have a negative financial impact on various businesses, cybersecurity education and training are crucial. Governmental establishments. We provided a thorough SLR of several cybersecurity education and training methods in this article. This study's primary goal was to Examine the methods used in cybersecurity training already in use, note the difficulties they provide, and assess the efficacy of cybersecurity training programs based on AR and VR. Making Use of strict selection standards, we took out primary articles out of 150 pertinent articles for in-depth examination and extraction of data. We categorized the publications according to the various methods of cybersecurity instruction, including conventional lecture-based, textual, approaches based on videos, websites, video games, and simulations, in addition to cyber ranges and virtual settings. We suggested a classification of cyber security-related research areas, which can be improved by doing a more thorough analysis of the keywords that make up the category. Research agendas for every field can be created using these terms as well. Such as developing interdisciplinary research plans for cyber safety. Among the groups we distinguished, application of conventional Terminology is used frequently. But there's plenty of space. For author and working groups to enhance. Other Potential classifications could include the common social sections of the government, business, academia, and civil society that are frequently cited in articles.

REFERENCES

Andel, T. R., & McDonald, J. T. (2013, October). A systems approach to cyber assurance education. In Proceedings of the 2013 on InfoSecCD'13: Information Security Curriculum Development Conference (pp. 13-19). DOI: 10.1145/2528908.2528920

Blasch, E., Shen, D., Pham, K. D., & Chen, G. (2015, May). Review of game theory applications for situation awareness. In *Sensors and Systems for Space Applications VIII* (Vol. 9469, pp. 141–150). SPIE.

Cauchois, R., Taghipour, A., Kang, D., Zoghlami, N., & Abed, M. (2017). Advanced solutions for a supply chain with stochastic information. *Journal of Advanced Management Science*, 5(1), 9–13. DOI: 10.18178/joams.5.1.9-13

Dadsena, K. K., Mathiyazhagan, K., & Taghipour, A. (2021). Analysis of barriers for the build the resilient supply chain networks post-COVID-19. In *Managing Supply Chain Risk and Disruptions: Post COVID-19* (pp. 79–89). Springer International Publishing. DOI: 10.1007/978-3-030-72575-4_7

Glaa, B., Zoghlami, N., & Taghipour, A. (2014, May). Outsourcing and off-shoring healthcare services: A way to cut the costs and improve quality. In 2014 international conference on advanced logistics and transport (ICALT) (pp. 189-194). IEEE.

Loivet, W., Taghipour, A., & Kang, D. S. (2020). The rise of green supply chain management: Between complexity and necessity. *J. Econ. Bus. Manag*, 8(1), 1–7. DOI: 10.18178/joebm.2020.8.1.604

Lu, X., & Taghipour, A. (2023). A French Guide for Businesses and Governments to Integrate the Concept of Sustainable Development Into Their Supply Chains. In *Government Impact on Sustainable and Responsible Supply Chain Management* (pp. 1–12). IGI Global. DOI: 10.4018/978-1-6684-9062-4.ch001

Mahfod, J., Canel-Depitre, B., & Taghipour, A. (2019). Quality function deployment-ELECTRE in supplier evaluation. Journal of Advanced Management Science, 7(4).

Mbiatem, J., Taghipour, A., & Canel-Depitre, B. (2018, August). Supplier selection approaches for decision makers. In *Proceedings of the 8th International Conference on Information Communication and Management* (pp. 108-112).

Merimi, M., & Taghipour, A. (2021). Accelerating the digitalization of the supply chain: An empirical research about COVID-19 crisis. In *Digitalization of decentralized supply chains during global crises* (pp. 1–24). IGI Global. DOI: 10.4018/978-1-7998-6874-3.ch001

Nadia, E. N., Beatrice, C. D., & Atour, T. (2021). The determinants of hotel room rates in Beirut: A hedonic pricing model. International Journal of Trade. *Economics and Finance*, 12(2), 33–42.

Shull, A., Twomey, P., & Yoo, C. S. (2017). Legal Mechanisms for Governing the Transition of Key Domain Name Functions to the Global Multi-stakeholder Community. Who Runs the Internet, 18-23.

Taghipour, A. (2014). Improving the plan of a manufacturing network with non-integrated business units. *International Journal of Applied Logistics*, 5(2), 1–11. DOI: 10.4018/ijal.2014040101

Taghipour, A. (Ed.). (2020). *Demand forecasting and order planning in supply chains and humanitarian logistics*. IGI Global.

Taghipour, A. (Ed.). (2023). *Government Impact on Sustainable and Responsible Supply Chain Management*. IGI Global. DOI: 10.4018/978-1-6684-9062-4

Taghipour, A., Abed, M., & Zoghlami, N. (2015, May). Design for remanufacturing respecting reverse logistics processes: A review. In 2015 4th International Conference on Advanced Logistics and Transport (ICALT) (pp. 299-304). IEEE. DOI: 10.1109/ICAdLT.2015.7136602

Taghipour, A., & Beneteau-Piet, C. (2020). Sustainable supply chain management performance. *International Journal of Innovation, Management and Technology*, 11(6), 165–169. DOI: 10.18178/ijimt.2020.11.6.897

Taghipour, A., Foukolaei, P. Z., Ghaedi, M., & Khazaei, M. (2023). Sustainable Multi-Objective Models for Waste-to-Energy and Waste Separation Site Selection. *Sustainability (Basel)*, 15(22), 15764. DOI: 10.3390/su152215764

Taghipour, A., & Frayret, J. M. (2011). Supply chain coordination planning: A review. Work paper, École polytechnique de Montréal.

Taghipour, A., & Frayret, J. M. (2013, May). An algorithm to improve operations planning in decentralized supply chains. In *2013 International Conference on Advanced Logistics and Transport* (pp. 100-103). IEEE. DOI: 10.1109/ICAdLT.2013.6568442

Taghipour, A., & Frayret, J. M. (2013). Coordination of operations planning in supply chains: A review. *International Journal of Business Performance and Supply Chain Modelling*, 5(3), 272–307. DOI: 10.1504/IJBPSCM.2013.055729

Taghipour, A., Lu, X., & Smahi, M. (2022). Opportunities and challenges for society 5.0 in procurement. *Blockchain Technology in Supply Chain Management for Society*, 5(0), 91–100. DOI: 10.1201/9781003177432-9

Taghipour, A., & Merimi, M. (2021, March). Digital transformation of supply chains during crisis: covid-19. In 11th Annual International Conference on Industrial Engineering and Operations Management, IEOM (pp. 7493-7501). DOI: 10.46254/AN11.20211298

Taghipour, A., Padash, A., Etemadi, V., Khazaei, M., & Ebrahimi, S. (2024). Sustainable and Circular Hotels and the Water–Food–Energy Nexus: Integration of Agrivoltaics, Hydropower, Solar Cells, Water Reservoirs, and Green Roofs. *Sustainability (Basel)*, 16(5), 1985. DOI: 10.3390/su16051985

Taghipour, A., Ramezani, M., Khazaei, M., Roohparvar, V., & Hassannayebi, E. (2023). Smart transportation behavior through the COVID-19 pandemic: A ride-hailing system in Iran. *Sustainability (Basel)*, 15(5), 4178. DOI: 10.3390/su15054178

Taghipour, A., Sohrabi, A., Ghaedi, M., & Khazaei, M. (2023). A robust vaccine supply chain model in pandemics: Case of Covid-19 in Iran. *Computers & Industrial Engineering*, 183, 109465. DOI: 10.1016/j.cie.2023.109465

Zoghlami, N., Taghipour, A., Merlo, C., & Abed, M. (2016). Management of divergent production network using decentralised multi-level capacitated lot-sizing models. *International Journal of Shipping and Transport Logistics*, 8(5), 590–604. DOI: 10.1504/IJSTL.2016.078686

Chapter 13

Impact of Circular Economy on Supply Chain Management: An Exploratory Study

Alaa Eddine El Moussaoui
https://orcid.org/0000-0002-5749-7682
Higher School of Technology, Hassan 2 University, Morocco

ABSTRACT

This chapter focuses on the integration of the circular economy (CE) into supply chain management (SCM), a topic that is scarcely addressed in the literature. Its exploratory nature leads the author to study two company cases. The author conducted a qualitative case study by interviewing 16 persons working in the companies X and Y. Three main findings emerge from the empirical analysis. Firstly, the business ecosystem is a prerequisite for the development of practices related to the circular economy. Secondly, CE practices modify the studied SCs to varying degrees. And thirdly, CE practices impact the performance of the supply chain. The chapter has many implications. In terms of theory, it contributes significantly to completing the theoretical survey on the integration of the circular economy in supply chain management. On the other hand, these findings help companies wishing to restructure their supply chains, and public authorities that want to formulate policies and implement strategies leading to more sustainable urban environments.

DOI: 10.4018/979-8-3693-7478-8.ch013

INTRODUCTION

The depletion of renewable and non-renewable natural resources has emerged as a pressing global challenge. Excessive and disproportionate resource consumption, compounded by a series of natural disasters, has placed the planet in a state of environmental crisis and urgency. In response, sustainable development (SD) has become a central paradigm, requiring robust environmental and social management strategies that have evolved into critical strategic priorities for businesses (Zarbakhshnia et al., 2023). As Akono and Fernandes (2009) emphasize, the integration of SD principles within corporate strategies has fundamentally transformed traditional logistics operations. This is particularly evident in the transportation sector, widely regarded as a significant contributor to pollution, and across entire supply chains, involving stakeholders such as suppliers, logistics service providers, and end customers (El Moussaoui et al, 2022 ; El Moussaoui et al, 2023).

In recent years, the circular economy (CE) has emerged as a transformative concept, positioned as a sustainable and pragmatic solution to the escalating depletion of natural resources. According to the French Environment and Energy Management Agency, CE represents "an economic system of exchanges and production which, at all stages of the product life cycle (goods and services), aims to increase resource efficiency, reduce environmental impact, and enhance individual well-being." The core tenet of CE is the elimination of waste through comprehensive recycling and repurposing practices (Taghipour et al, 2023 ; Mishra et al., 2024). CE is increasingly recognized for its multifaceted benefits, and numerous studies have lauded it as a driver of sustainable growth, creating economic, social, and environmental value. Wang et al. (2014) identify CE as a cornerstone of sustainable economic development, underscored by its formal adoption into China's regulatory framework through a dedicated law enacted in 2008 (Yuan et al., 2008).

Despite its growing prominence, CE remains an underexplored area in management sciences. Existing studies predominantly address waste management strategies (Sarkis and Zhu, 2008; Bai et al., 2024), performance measurement frameworks for CE initiatives in regions like France and China (Aurez and Georgeault, 2016), procedural guidelines for CE implementation (Arruda et al., 2021), and the transparency of corporate environmental policies (Wang et al., 2014). However, the intersection of CE with supply chain management (SCM) has received limited attention. This chapter seeks to address this gap by building upon the work of Hazen et al. (2021), who advocate for more focused research on the implications of CE for logistics and SCM. According to these authors, such research is essential to understanding and facilitating the adoption of CE as a paradigm for sustainable economic growth.

Since the 1980s, growing awareness of ecological issues has catalyzed numerous innovations in logistics and SCM, including closed-loop supply chains and reverse logistics. Logistics plays a pivotal role in the environmental management strategies of companies, serving as a key enabler in addressing circular economy challenges (El Moussaoui et al, 2022). A critical analysis of CE definitions underscores the deep involvement of logistics and SCM in the operationalization of CE principles. However, despite this alignment, research explicitly linking CE to SCM remains scarce, even though the supply chain encompasses most CE implementation stages (Arruda et al., 2021). Fulconis et al. (2016) argue that the supply chain occupies a central role in CE, requiring the development of innovative organizational and logistical approaches (Adoue et al., 2014). These changes necessitate significant technical and technological adaptations within logistics frameworks, ultimately fostering new models of environmental management.

While the CE framework holds transformative potential, its implementation is not without challenges. Murray et al. (2017) highlight the tensions and limitations associated with CE adoption, which remain insufficiently explored in the literature. Against this backdrop, our research aims to address the pivotal question: How does CE transform SCM? More specifically, we investigate the nature and extent of the impacts that CE implementation has on SCM processes and structures.

To explore this question, we adopt an exploratory research approach, focusing on two case studies: one in the aerospace industry and the other in medical equipment manufacturing. The research unfolds in four parts: first, we define the key concepts underpinning our study; second, we outline the methodology employed; third, we present the main findings from our empirical investigation; and finally, we discuss these results while outlining potential research directions for the future.

2. LITERATURE REVIEW

2.1 Supply Chain and Support for Sustainable Development

Mentzer et al. (2001) define the supply chain (SC) as "a set of three or more entities (organizations or individuals) directly involved upstream and downstream of the flows of products, services, finances, and/or information from a source to a customer." This definition underscores the intricate network of entities required to ensure the efficient delivery of goods and services. SCs are not merely logistical arrangements; they represent the backbone of modern commerce, facilitating the flow of resources and enabling businesses to meet customer demands effectively.

The importance of SCs has grown significantly as organizations face increased globalization, market competition, and consumer expectations.

Frederico et al. (2020) and El Moussaoui et al. (2023 ; 2024) categorize SCs into three levels. The direct supply chain involves a focal company and its immediate suppliers and customers, forming the most basic structure of material and information exchange. The extended supply chain broadens this view, including second-tier suppliers and customers to ensure a more holistic flow. Finally, the complete supply chain encompasses all entities—upstream and downstream—that play a role in delivering goods and services. This classification illustrates the complexity of SCs, emphasizing the need for effective supply chain management (SCM) practices to optimize quality, cost, and service levels while addressing diverse stakeholder interests (Tiwari, 2021).

The environmental dimension of SCM has gained prominence as sustainability becomes a central concern for businesses. The emergence of the green supply chain (GSC) has shifted the focus of SCM from efficiency and cost-effectiveness to incorporating environmental objectives. Sarkis et al. (2011) trace the origins of GSC to environmental management initiatives in the 1960s, with its broader adoption in academia and industry occurring after 2000. GSC emphasizes minimizing environmental impacts across the product lifecycle, from raw material extraction to end-of-life disposal. Practices such as eco-design, resource conservation, and recycling are integral to achieving these objectives (Beamon, 1999; Srivastava, 2007). Zhu et al. (2007) describe GSC as an extension of traditional SCs, integrating sustainability into every process to ensure long-term environmental, economic, and social benefits.

Digitalization, combined with machine learning, has revolutionized supply chain management (SCM) by enhancing transparency, efficiency, and resilience through data-driven insights and predictive analytics (El Moussaoui et al, 2023 ; El Moussaoui and Loqman, 2024). The COVID-19 pandemic accelerated this trend, as organizations adopted digital tools to manage disruptions and maintain operations. (Merimi and Taghipour, 2021 ; El Moussaoui et al, 2024) highlight how technologies such as blockchain, IoT, and AI enable real-time monitoring, predictive analytics, and improved decision-making. Blockchain, for instance, ensures transparency by providing immutable records of transactions, while IoT devices facilitate continuous monitoring of material flows, enabling proactive adjustments to supply chain operations (Lu & Taghipour, 2023). These advancements are critical for integrating sustainability into SCM, allowing organizations to achieve both environmental and operational goals.

Cauchois et al. (2017) explore advanced solutions for managing SCs under stochastic conditions, emphasizing the need for data-driven approaches to address uncertainties. Similarly, Dadsena et al. (2021) underline the importance of building resilient SC networks, particularly in the face of global crises. Digitalization plays a pivotal role in enhancing resilience by enabling SCs to adapt to changing circumstances and ensuring continuity despite disruptions. This convergence of digitalization and sustainability is redefining SCM, paving the way for more adaptive, efficient, and sustainable supply chains (El Moussaoui and Chakir, 2023 ; El Moussaoui, 2024).

2.2 Circular Economy: A Paradigm Shift

The traditional linear production model of "extract-manufacture-consume-dispose" has dominated industrial processes for decades. However, this approach is increasingly viewed as unsustainable due to its reliance on finite resources and the environmental degradation it causes. The circular economy (CE) offers a transformative alternative, advocating a "cradle-to-cradle" framework that prioritizes resource conservation, waste minimization, and the creation of closed-loop systems (Murray et al., 2017). This model represents a fundamental shift in how economic growth is conceptualized, decoupling it from resource consumption and embedding sustainability into the heart of business practices.

Murray et al. (2017) define CE as an economic system designed to maximize ecosystem functionality and human well-being through integrated planning, procurement, production, and reprocessing activities. CE has gained traction globally, with legislative frameworks such as China's Circular Economy Promotion Law (2008) and France's energy transition laws (2015) underscoring its adoption (General Commissariat for Sustainable Development, 2014). These regulations highlight the strategic importance of CE in addressing global sustainability challenges while fostering economic innovation.

CE principles are articulated through the 3Es—eco-design, industrial ecology, and the economy of functionality—and the 4Rs—reuse, repair, recycling, and re-employment. Eco-design involves incorporating environmental considerations into product development, from material selection to end-of-life management. This approach reduces environmental impacts while ensuring product viability (Diemer, 2016). Industrial ecology promotes collaboration among businesses to optimize material and energy flows, creating industrial symbiosis exemplified by initiatives like the Kalundborg eco-industrial park in Denmark. The economy of functionality redefines ownership models, emphasizing service provision over product ownership, as seen in Michelin's "fleet solution" program (Le Moigne, 2014).

The 4Rs operationalize CE by extending product lifecycles and closing material loops. Recycling transforms waste into secondary raw materials, while reuse and repair strategies ensure that products are reintroduced into economic cycles, reducing the need for virgin resources (Hazen et al., 2021). These principles align closely with SCM practices, as they require seamless coordination across multiple stakeholders to achieve their objectives.

Digitalization enhances the implementation of CE by providing tools for precise resource tracking, efficient recycling, and collaboration among supply chain actors. Technologies such as IoT, AI, and blockchain enable real-time monitoring of material flows, optimize waste management processes, and ensure compliance with sustainability standards (Taghipour et al., 2023). Lu and Taghipour (2023) emphasize the role of digital tools in aligning CE initiatives with global sustainability goals, making them integral to modern SCM practices.

2.3 Circular Economy and Supply Chain Management

The intersection of CE and SCM offers a promising avenue for achieving sustainable economic growth. CE principles, particularly eco-design and industrial ecology, are deeply intertwined with SCM processes. Eco-design involves lifecycle assessment (LCA), which evaluates the environmental impacts of products from cradle to grave. This approach necessitates collaboration across the supply chain, involving suppliers, manufacturers, and end-users in designing products that minimize resource use and environmental harm (Goepp et al., 2013; Abdullah et al., 2019). Industrial ecology, which focuses on closing material and energy loops, relies on SCM to facilitate the exchange of resources, waste, and by-products within collaborative networks (Feiferytė-Skirienė & Stasiškienė, 2021).

Despite its potential, integrating CE into SCM presents significant challenges. Loivet et al. (2020) highlight barriers such as resistance to change, lack of technical expertise, and insufficient regulatory support. Standardized metrics for evaluating circularity are also underdeveloped, complicating efforts to measure the success of CE initiatives. Addressing these challenges requires innovative approaches that combine technical, organizational, and regulatory solutions.

Digitalization emerges as a critical enabler of CE-SCM integration. Smart technologies, such as blockchain and IoT, provide the transparency and traceability needed to implement circular practices effectively. Blockchain, for example, enables the tracking of materials throughout their lifecycle, ensuring compliance with sustainability standards and facilitating end-of-life product recycling (Taghipour et al., 2022). IoT devices enhance waste management by providing real-time data on material flows, allowing for more efficient recycling and reuse operations.

Resilience is another crucial aspect of CE-SCM integration. Dadsena et al. (2021) emphasize the importance of building resilient SCs capable of withstanding disruptions, citing digital tools as essential for achieving this objective. The COVID-19 pandemic underscored the need for such resilience, accelerating the adoption of digital technologies and highlighting their role in sustaining operations during crises (Taghipour and Merimi, 2021).

In conclusion, the integration of CE principles into SCM represents a significant step toward achieving sustainable economic growth. However, realizing this potential requires overcoming barriers related to complexity, coordination, and measurement. Digitalization offers the tools necessary to align SCM practices with CE objectives, providing a pathway for businesses to achieve sustainability while maintaining competitiveness. This study explores the impact of CE on SCM, focusing on the role of digitalization as an enabler of this transformation.

3. RESEARCH METHODOLOGY

The exploratory nature of our research leads us to adopt a qualitative methodology through case studies to better address our objective, which is to discover and understand how the Circular Economy (CE) integrates into companies' Supply Chain Management (SCM). In fact, we seek to understand this phenomenon in its context and not to establish general laws. Our epistemological position is that of pragmatic constructivism because it allows us to understand reality as "a way of seeing knowledge development as an act of constructing representations forged by humans to make sense of the situations in which they intervene" (Albert and Avenier, 2011 ; Dibi et al, 2024).

We have chosen to analyze two companies (X and Y) involved in circular economy projects. The choice of these two companies was based on the local recognition of their actions in the field of circular economy in Morocco and their involvement in collective circular economy initiatives led by associations.

This demonstrates that the author have focused on the case study. The latter is chosen because it is indeed about understanding a phenomenon, the "relational processes, creation mode or (dys) organizational functioning as well as examining the facts and perceptions that actors attach to actions" (Hlady-Rispal, 2015). Furthermore, according to (El Moussaoui et al, 2023 ; El Moussaoui, 2024), the positive answer to the following four questions legitimizes the use of the case study method. Thus, the phenomenon we are studying – namely the CE and its integration into sustainable supply chain management – must "be studied in its natural context to be understood". Then, it is necessary to focus on contemporary events for the study of the problem. This is the case here. Then, knowledge of the phenomenon can be acquired without having to control or manipulate the subjects or events in question. Finally, the theoretical basis that exists regarding our studied issue contains elements that have not been studied.

The two cases were identified by the strong involvement of one of the researchers in the field of CE (knowledge of members of business associations, participation in certain instances related to CE). However, there is no professional relationship with these instances or associations, ensuring the independence of the researcher.

Interviews are the preferred approach for data collection, although other sources have been analyzed (a radio broadcast, a public conference, press data). In this context, the author conducted nine interviews with the employees of Company X, and seven with those of Company Y (table 1).

Table 1. Profil of Interviewees

Participant	Role	Experience	Length of interview
X (1)	Director	21 Years	81 Min
X (2)	Deputy Director	9 Years	63 Min
X (3)	Project Manager	5 Years	77 Min
X (4)	Project Manager	8 Years	92 Min
X (5)	Project Manager	6 Years	67 Min
X (6)	Logistics Manager	3 Years	52 Min
X (7)	HSE Manager	4 Years	84 Min
X (8)	Logistics coordinator	1 Years	32 Min
X (9)	Logistics coordinator	2 Years	71 Min
Y (1)	Director	15 Years	74 Min
Y (2)	Project Manager	7 Years	76 Min
Y (3)	Logistics Manager	4 Years	92 Min
Y (4)	HSE Manager	6 Years	43 Min
Y (5)	Logistics coordinator	2 Years	39 Min
Y (6)	Logistics coordinator	2 Years	47 Min
Y (7)	Logistics coordinator	3 Years	59 Min

Source: author

The chosen method of analysis is twofold. We first proceeded with a floating reading to manage the risk of circularity, which consists of identifying in the empirical material only what confirms the theory. This floating reading is based on continuous reading, "prohibiting the taking of any notes, underlining anything, or focusing attention on any element. The result is the emergence of themes that can be surprising and can challenge the expected theoretical frameworks (Morgan and Nica, 2020). Then, we used the thematic analysis method (El Moussaoui, 2024), with the technique of double coding: passages of text were identified and coded, according to their description of the studied phenomenon, by the two researchers separately and then together.

4. RESULTS

The analysis of cases, using the floating analysis method (Dumez, 2013), brings forth the concept of Business Ecosystem through the various actors involved. Indeed, these actors are not necessarily directly involved in the supply chain, but their role in the deployment of the circular economy is essential. Associating the concept of Business Ecosystem with that of the circular economy thus appears fruitful for the analysis of our research question. We first present the ecosystem of each of the cases studied, then we analyze the pillars of the Circular Economy (EC) mobilized before identifying the different impacts on the supply chain.

4.1 Business ecosystem of the two companies X and Y

Company X is involved in several projects related to the Circular Economy (X1, X2, and X3), at various stages of progress, and involving different actors. Project X1 aims to reuse composite waste (which currently cannot be recycled and is therefore buried) to manufacture other parts: *"We use waste that comes from parts of car structures, machining frames, and these production scraps will be reused to make interior parts, for example, armrest covers, or even a car trunk, these are parts that are not mechanically stressed."* (X1 Project Manager)

The stakes are very high for all stakeholders involved in the project and overall for the extended enterprise (the main shareholder of X Company, who is also one of its key customers, being particularly interested in the outcome).

The second project, X2, aims to streamline material flows and align the needs for materials while also bringing some of its subcontractors closer. *"We are in an industrial zone, and this zone is currently being developed into a commercial activity zone (ZAC), the idea is to bring the subcontractors closer to us and create a more dynamic logistics area"* (X2 Project Manager).

The third medium-term project involves implementing an eco-design approach to reduce waste from composite materials. The use of three-dimensional (3D) printing for certain parts could also be a potential solution: *"It's interesting in terms of recycling compared to a traditional process... when we mold a part in composite materials, but this part needs to be turned over, which creates waste. This is also of interest to us, and it could be a follow-up to the project, working on new manufacturing methods, and here we're talking, for example, about 3D printing."* (X3 Project Manager)

Finally, company X employs the milk-run system with some of its suppliers to save on transportation costs and CO2 emissions: *"It's a saving, an empty trailer consumes almost as much and costs as much and it consumes, whereas if it is full it definitely results in savings... so indeed it is driven by an economic choice but it also has an environmental impact, such as reducing the carbon footprint and CO2 emissions."* (Logistics Manager)

The company has also implemented the return of certain packaging materials to the factory: *"We have parts that arrive in cardboard packaging, and they have created shuttle boxes to avoid all this packaging with forms inside. The supplier puts their parts inside them and sends the boxes back to us when they are empty. These are returnable boxes."* (HSE Manager)

Company Y has committed to an approach of industrial and territorial ecology concerning consumables, particularly the packaging of complex devices received. Indeed, Y entirely outsources its production, mainly abroad: *"Indeed, these are fragile devices that come from abroad, so we have complex, expensive, and bulky packaging, and very quickly we started wondering what we were going to do with it."* (Director of Y Company)

4.2 Circular Economy Pillars Mobilized and Their Impacts on the Supply Chain

Both cases reveal similar pillars being mobilized: recycling, industrial and territorial ecology, and eco-design. The aspect of innovation is very present but at different stages of development. For the company X, there is a clear effort in research and development to find the technological solution enabling the recycling of composite materials: *"So we started by thinking that we would try to create a new material, and this new material we would use it in parts that are less constraining for us. We will try to put it in the armrest or in the shell and try to see if it presents the right technical characteristics, validated by the design office, and this could potentially go back into our own supply chain."* (HSE Manager)

For company Y, working with suppliers is still in its early stages: *"But the manufacturer, precisely I work with them. But for this type of material, we have two manufacturers, one in Spain and one in Germany, and for now, neither of them is able to reconsider their manufacturing process without significant costs to provide us with an answer. But we are working on it, we are working on it because here we will really be in eco-design, in addition to the problem of separating the materials to recover the felt inside, we need to implement a very expensive system to open this sealed product."* (Director of Y Company)

Other pillars are mobilized. At company X, the functional economy is implemented through the consignment of packaging at suppliers. At company Y, the reuse of packaging waste is implemented: *"Waste recycling, integration of waste that are for us... we had large cardboard boxes that were given for free to another company and they made furniture with armchairs, tables, it's really very rewarding."* (Director of X Company). Reuse and repair activities are absent from both cases studied. The table below (Table 2) summarizes the pillars of the EC used by the two companies.

Table 2. Pillars of circular economy

	Pillars of Circular Economy	**Company X**	**Company Y**
3E	Eco-design Industrial ecology Economy of functionality	Mutualization; Packaging consignment.	Waste management (elimination of storage).
4R	Re-employment Repair/ Reuse /Recycling	Recycling composite materials.	Reuse of packaging waste; Devices and consumables.

Source: author

The projects undertaken lead to different types of impacts on the supply chain. Project X1 notably has a real impact on internal logistics. Indeed, it now involves the various X factories collecting composite waste to converge towards X2, considering them as secondary materials, which are thus valorized; currently, the flows are not yet very significant as they are reserved for testing: *"And for the record, it was paradoxical at the start of the project because, effectively, we couldn't recover this waste. So we presented it in the form of a secondary material, yes, it wasn't waste. No, it's not waste; it's a secondary raw material that will be used to manufacture something else."* (Project Manager X2)

Another foreseeable impact will concern the procurement, which will only be measurable in a few years. Indeed, once the process of recycling composite materials is established, it will be necessary to carry out a transfer of skills. Either one or more current suppliers of company X will acquire it, or a new player will emerge: *"We demonstrate that we are capable of remanufacturing aerospace parts, and once we have demonstrated that, we accompany a partner... new entity... who decides to set up the necessary logistics for collection, integrate the transformation process, and then move towards the production of parts. But that's more our job..."* (Deputy Director of X Company)

The logistics ecology project X3 will have an impact on the supply chain with the potential relocation of two or three X subcontractors and the entry of a logistics service provider. The audit of the flows is completed. Some projects are currently in the implementation phase. This is the case with the establishment of a work package center managed by X's historical logistics provider (which has recently become the sole logistics provider for all of X's sites). This work package allows the mutualization of the 11 warehouses on the site and is located right next to the X production unit. A savings of approximately 180,000 MAD per month is expected, which is an interesting gain: *"Oh yes, it is interesting. Moreover, there are no more shuttles, no more internal transportation. Within the 180,000 MAD, everything is included: the mutualization of the warehouses, etc. A gain of 180,000 MAD per month"* (Logistics Coordinator).

A second project resulting from the study of flows is the mutualization of the waste disposal site across the entire business park. This would reduce transportation costs and waste management costs for each stakeholder involved in the logistics chain, including suppliers, subcontractors, and all companies present in the area: *"So the idea is really to completely remove the waste disposal site, and when we had the project to remove it, we thought why remove it and have it only for X, why not completely remove it and put it right in the business park and then create a whole system managed by a service provider... a waste disposal site that is useful for everyone and it will also limit transportation, avoid each one having their own different service provider coming to collect their waste and I know that has been agreed upon in the project."* (HSE Manager)

Finally, the milk-run approach has an impact on reducing transportation, leading to mutualization: *"So today we are in the process of developing a transportation control tower. It's actually taking control of the internal requests from each site to group all X sites together. Currently, we only group Mediouna and Bouznika. And soon we'll include Rabat, Kenitra, and Tanger, as we already have a site in Kenitra and another in Tanger. And this is for a later stage because we're already managing two main sites, which is quite a task."* (Logistics Manager)

The gains this year amount to 2,5% of the total transportation budget: *"We had the third steering committee meeting today, we have one every week. It's really new; we've estimated gains of 2 million MAD for this year. It's quite significant."* (Logistics Manager)

For company Y, the recycling of cardboard packaging through Y1 has allowed for the elimination of storage and warehousing costs. Indeed, the latter, located in Casablanca, has been closed: *"By focusing on Y, we've now set up a system to recycle everything we had stored in the facility, which includes electronic equipment, as well as various types of packaging like foam and cardboard sheets used for packaging equipment and pallets. All of this is managed by Y1... It was a monthly rental expense of a few hundred dirhams that was unnecessary and had no future."* (Logistics Manager); *"They parted ways with the rental space, it's as simple as that."* (Y1 Coordinator); *"For Y, there's already the rent saved every month, I think it was around 5000 MAD/month. It was just a shed, nothing special, but over the year, it still amounts to a significant... reduction in waste..."* (Director of Y company).

This recycling has also led to a reduction in the transportation costs of packaging waste: *"It's only for PE films [polyethylene], pallets, and boards that they come to pick up here because they're part of the collection tour for all the companies in the association in the area... so pallet collection by a provider who will collect for free, and then it's financed by the advance, well, the recycling of the pallets, collection of electronic devices, there we have a partnership with a major waste management company, so the PE too, where they come once a month."* (Logistics Manager); *"We're dealing with volumes that are regular and much lower, but now it's a daily occurrence, it's also comfortable to manage for the logistics manager when he comes, he also rationalizes his contribution to the transportation cost and optimization, he puts everything he needs to bring in his van at the same time. He has a bit of plastics, a bit of electronic and electrical waste, a bit of consumables, and archives."* (Y1 Project Manager).

The eco-design project concerning plastic consumables conducted with suppliers will impact them in their production processes: *"There will be a simple solution: we melt them, and with the heat, it's clean, it's cleaned, and we have two or even three types of plastic materials... There's an issue with recovering raw materials, but as of today, in the current state of things with the research we've conducted, but I'm not losing hope at all, we continue the research on this to find a system that will allow us to separate the different materials after fusion."* (Director of Y Company)

The analysis of cases X and Y reveals numerous connections between CE and SCM. The next section aims to highlight the main findings before concluding on the limitations and prospects.

5. DISCUSSION AND CONCLUSION

Three main results emerge from the empirical analysis:

5.1 The Role of Business Ecosystems in Circular Economy Integration

The two case studies underscore the critical role of the Business Ecosystem concept in analyzing the integration of the circular economy (CE) into supply chains. Interestingly, this concept, absent from the initial literature review, emerged prominently during the case analysis through the floating analysis technique (Dumez, 2013). This finding suggests that a Business Ecosystem serves as a precondition for developing CE-related practices by fostering innovation in products, processes, organizational structures, and digitalization.

Digitalization plays a pivotal role in enabling Business Ecosystems to function effectively, offering tools for data sharing, real-time communication, and resource optimization. Advanced digital platforms facilitate collaboration among diverse stakeholders, enhancing the co-construction of capabilities tied to innovation. The integration of digital technologies thus complements the CE's objectives by streamlining processes and improving the efficiency of shared resources.

The Business Ecosystem, as defined by Liu and Pan (2024), involves companies collaborating to co-construct capabilities tied to innovation. Senyo et al. (2019) describe these ecosystems as "intentional communities of actors whose activities are interdependent and collectively shape the fate of the entire community." A broader definition by Donada and Fournier (2014) conceptualizes it as an "intangible and formal business meta-structure" encompassing institutional, geographical, cultural, social, and economic environments. These ecosystems consist of diverse stakeholders—customers, suppliers, producers, competitors, and more—working together through shared interrelations to develop and exchange resources and skills under a unified vision.

A typology of ecosystems identifies four primary forms: offer systems, platforms, communities of destiny, and flourishing communities. The ecosystems examined in this study align closely with the "community of destiny" model, characterized by heterogeneous actors with decentralized structures, leadership roles distributed among multiple actors rather than being centralized, and existential solidarity rather than reliance on a central actor's essential resource. Digitalization further supports these dynamics by enabling decentralized coordination and enhancing transparency, making it easier to align the activities of heterogeneous actors with shared objectives.

5.2 Differentiation Between 3Es and 4Rs Practices

The second finding highlights that while CE practices influence supply chains, their impact varies based on the practices' nature. It is crucial to distinguish between practices aligned with the 3Es (eco-design, environmental management, and energy efficiency) and those related to the 4Rs (reduce, reuse, recycle, and recover).

Activities related to the 3Es exhibit a broader and more transformative impact on supply chains compared to the 4Rs. For instance, Company X's Project X1's eco-design initiatives influence the production chain, upstream supply chain, and downstream markets. Digital tools enhance these efforts by enabling precise tracking of material flows and fostering collaboration across the supply chain. Similarly, Company Y's eco-design efforts with suppliers to develop recyclable plastic shells affect procurement methods and customer engagement with environmentally friendly products. Digital platforms facilitate data exchange and innovation in these processes.

Conversely, 4R practices, such as internal plastic recycling at Company Y, tend to have localized effects, limited to specific processes like internal logistics. However, digitalization can still enhance these practices by optimizing waste tracking and management systems.

5.3 Impact of CE Practices on Supply Chain Performance

The third result identifies a direct link between CE practices and enhanced supply chain performance. For instance, Company Y's total recycling of packaging waste through industrial and territorial ecology has eliminated storage and warehousing costs and reduced transportation expenses. Digital tools support these achievements by optimizing logistics and streamlining waste handling processes.

For Company X, Project X1's waste reduction initiatives have optimized resource use across production sites, with composite material waste redirected to other projects (e.g., Project X2). Project X3's expected outcomes include the relocation of subcontractors, reducing delivery times, inventories, and transportation costs. Cost savings also result from pooled waste facilities and milk-run logistics centralization, enabled by digital systems for coordination and monitoring. These improvements demonstrate that implementing CE practices, supported by digitalization, translates into measurable performance gains in supply chain management (SCM).

6. LIMITATIONS AND FUTURE RESEARCH

Despite its insights, this research's scope is limited by its reliance on two case studies, restricting the generalizability of findings. The limited number of cases may not capture the full diversity of Business Ecosystem configurations or CE practices across industries. Additionally, the absence of longitudinal data constrains the ability to observe long-term impacts and evolution of CE practices within supply chains. Another limitation lies in the specific geographic and sectoral context of the cases, which might influence the transferability of findings to other settings.

Moreover, the integration of digitalization into CE practices was observed but not exhaustively analyzed in terms of its specific contributions or challenges. Future studies could focus on examining the role of advanced digital tools, such as blockchain, IoT, and AI, in enhancing CE practices and their measurable outcomes.

Future research should also investigate broader case studies, expanding the scope to include diverse industries and regions to validate and refine the proposed insights. Quantitative analyses could develop metrics and models to quantify the relationship between CE practices, digitalization, and supply chain performance. Dynamic ecosystem models would help examine how Business Ecosystems evolve over time and the role of digitalization in facilitating these changes. Furthermore, research could explore how CE and digitalization reshape interorganizational relationships, collaboration, governance structures, and stakeholder roles within supply chains. Identifying barriers and enablers, including organizational, technological, and regulatory factors, will also be crucial.

By addressing these areas, future studies can provide a more comprehensive understanding of the interplay between CE, digitalization, and SCM, offering actionable insights for both academia and practitioners. In conclusion, this research opens promising avenues for exploring the relatively underdeveloped field of CE in management sciences, particularly in its connection to supply chain management and the enabling role of digitalization.

REFERENCES

Aarikka-Stenroos, L., Chiaroni, D., Kaipainen, J., & Urbinati, A. (2022). Companies' circular business models enabled by supply chain collaborations: An empirical-based framework, synthesis, and research agenda. *Industrial Marketing Management*, 105, 322–339. DOI: 10.1016/j.indmarman.2022.06.015

Abdullah, R., Mohamad, M. N., & Thurasamy, R. (2019). Towards sustainable performance: Promoting eco-design in green supply chain management practices. *International Journal of Supply Chain Management*, 8(3), 609–616.

Adoue, C., R Beulque, L. Carré, et J. Couteau. 2014. Quelles Stratégies D'entreprise Pour Une éConomie Circulaire Moteur De Croissance? Paris: Institut de l'économie circulaire.

Akono, D., & Fernandes, V. (2009). Impacts du développement durable sur les organisations logistiques. *Management et Avenir*, 6(26), 241–255. DOI: 10.3917/mav.026.0241

Albert, M. N., & Avenier, M. J. (2011). Légitimation de savoirs élaborés dans une épistémologie constructiviste à partir de l'expérience de praticiens. *Recherches Qualitatives*, 30(2), 22–47. DOI: 10.7202/1084829ar

Arruda, E. H., Melatto, R. A. P. B., Levy, W., & de Melo Conti, D. (2021). Circular economy: A brief literature review (2015–2020). *Sustainable Operations and Computers*, 2, 79–86. DOI: 10.1016/j.susoc.2021.05.001

Aurez, V., & Georgeault, L. (2016). *Economie Circulaire système économique et finitudes des ressources, 360*. De Boeck Supérieur.

Bai, C., Zhu, Q., & Sarkis, J. (2024). Circular economy and circularity supplier selection: A fuzzy group decision approach. *International Journal of Production Research*, 62(7), 2307–2330. DOI: 10.1080/00207543.2022.2037779

Beamon, B. M. (1999). Designing the green supply chain. *Logistics Information Management*, 12(4), 332–342. DOI: 10.1108/09576059910284159

Chen, Y., Zhu, Q., & Sarkis, J. (2023). Green supply chain management practice adoption sequence: A cumulative capability perspective. *International Journal of Production Research*, 61(17), 5918–5933. DOI: 10.1080/00207543.2022.2118891

Dibi, A., El Moussaoui, T., El Moussaoui, A. E., & Loqman, C. (2024). Qualimatric intervention research: interest for management sciences researchers. In *Applying qualitative research methods to management science* (pp. 92–127). IGI Global. DOI: 10.4018/979-8-3693-5543-5.ch005

Diemer, A. (2016). Les symbioses industrielles: Un nouveau champ d'analyse pour l'économie industrielle. *Innovations (Paris)*, 2(50), 65–94. DOI: 10.3917/inno.050.0065

Donada, C., & Fournier, G. (2014). Stratégie industrielle pour un écosystème en émergence: Le cas de la mobilité 2.0, décarbonée, intermodale et collaborative. *Revue d'Economie Industrielle*, (148), 317–348. DOI: 10.4000/rei.6016

Dumez, H. (2013). *Méthodologie de la recherche qualitative, 227.* Vuibert.

El Moussaoui, A. E. (2024). Impact of logistics pooling on reduction of CO2 emissions in last-mile logistics. Management & Sustainability: An Arab Review.

El Moussaoui, A. E. (2024). Factors influencing the entry of logistics service providers (LSPs) into the logistics clusters: a case study of Casablanca city. Management & Sustainability: An Arab Review.

El Moussaoui, A. E. (2024). Unlocking the drivers behind logistics service providers' participation in Morocco's logistics clusters. *Arab Gulf Journal of Scientific Research*, 42(4), 2086–2102. DOI: 10.1108/AGJSR-07-2023-0343

El Moussaoui, A. E., Benbba, B., & El Andaloussi, Z. (2023). Impact of logistics performance on the store image, consumer satisfaction and loyalty: A quantitative case study. *Arab Gulf Journal of Scientific Research*, 41(3), 226–239. DOI: 10.1108/AGJSR-09-2022-0201

El Moussaoui, A. E., Benbba, B., Jaegler, A., El Andaloussi, Z., & El Amrani, L. (2022, May). Last mile logistics: Challenges & improvement ways. In 2022 14th International Colloquium of Logistics and Supply Chain Management (LOGISTIQUA) (pp. 1-7). IEEE.

El Moussaoui, A. E., Benbba, B., Jaegler, A., El Moussaoui, T., El Andaloussi, Z., & Chakir, L. (2023). Consumer perceptions of online shopping and willingness to use pick-up points: A case study of Morocco. *Sustainability (Basel)*, 15(9), 7405. DOI: 10.3390/su15097405

El Moussaoui, A. E., El Moussaoui, T., Benbba, B., Chakir, L., Jaegler, A., & El Andaloussi, Z. (2024). Sustainable effects of information sharing between distribution logistics actors: A qualitative case study. *Environment, Development and Sustainability*, 1–19. DOI: 10.1007/s10668-023-04387-3

El Moussaoui, A. E., El Moussaoui, T., Benbba, B., Jaegler, A., & El Andaloussi, Z. (2022). Understanding the choice of collection & delivery point by the E-consumer via a machine learning model: Moroccan case study. *Procedia Computer Science*, 210, 204–211. DOI: 10.1016/j.procs.2022.10.138

El Moussaoui, T., & Chakir, L. (2023, September). BERT-CLSTM model for the classification of Moroccan commercial courts verdicts. In 2023 18th Conference on Computer Science and Intelligence Systems (FedCSIS) (pp. 281-284). IEEE. DOI: 10.15439/2023F3561

El Moussaoui, T., Chakir, L., & Boumhidi, J. (2023). Preserving privacy in Arabic judgments: Ai-powered anonymization for enhanced legal data privacy. *IEEE Access : Practical Innovations, Open Solutions*, 11, 117851–117864. DOI: 10.1109/ACCESS.2023.3324288

El Moussaoui, T., & Loqman, C. (2024). Advancements in Arabic Named Entity Recognition: A Comprehensive Review. *IEEE Access : Practical Innovations, Open Solutions*, 12, 180238–180266. DOI: 10.1109/ACCESS.2024.3491897

Feiferytė-Skirienė, A., & Stasiškienė, Ž. (2021). Seeking circularity: Circular urban metabolism in the context of industrial symbiosis. *Sustainability (Basel)*, 13(16), 9094. DOI: 10.3390/su13169094

Frederico, G. F., Garza-Reyes, J. A., Anosike, A., & Kumar, V. (2020). Supply Chain 4.0: Concepts, maturity and research agenda. *Supply Chain Management*, 25(2), 262–282. DOI: 10.1108/SCM-09-2018-0339

Fulconis, F., Paché, G., & Reynaud, E. (2016). Vers une nouvelle forme de croissance économique: les apports des recherches en logistique et supply chain management. *Revue française de gestion*, (8), 127-149.

Goepp, V., Caillaud, E., & Rose, B. (2013). A Framework for the Design of Knowledge Management Systems in Eco-Design. *International Journal of Production Research*, 51(19), 5803–5823. DOI: 10.1080/00207543.2013.795252

Habib, M., Bao, Y., & Ilmudeen, A. (2020). The impact of green entrepreneurial orientation, market orientation and green supply chain management practices on sustainable firm performance. *Cogent Business & Management*, 7(1), 1743616. DOI: 10.1080/23311975.2020.1743616

Hazen, B. T., Russo, I., Confente, I., & Pellathy, D. (2021). Supply chain management for circular economy: Conceptual framework and research agenda. *International Journal of Logistics Management*, 32(2), 510–537. DOI: 10.1108/IJLM-12-2019-0332

Hlady-Rispal, M. (2015). Une stratégie de recherche en gestion: L'étude de cas. *Revue Française de Gestion*, 41(127), 251–266. DOI: 10.3166/RFG.253.251-266

Hmioui, A., Bentalha, B., & Beniaich, R. (2024). Green Supply Chain Management et développement territorial: Une revue de littérature. *Alternatives Managériales Economiques*, 6(2), 52–74.

Le Moigne, R. (2014). L'économie circulaire, comment la mettre en œuvre dans l'entreprise grâce à la reverse supply chain? Paris: Edition DUNOD.

Li, J., & Sarkis, J. (2022). Product eco-design practice in green supply chain management: A China-global examination of research. *Nankai Business Review International*, 13(1), 124–153. DOI: 10.1108/NBRI-02-2021-0006

Liu, H., & Pan, W. (2024). A comprehensive literature review of the digital business ecosystem and digital business model with comparison in China, Japan and South Korea. *Asia Pacific Business Review*, 1–25. DOI: 10.1080/13602381.2024.2321198

Lu, X., & Taghipour, A. (2023). A French Guide for Businesses and Governments to Integrate the Concept of Sustainable Development Into Their Supply Chains. In *Government Impact on Sustainable and Responsible Supply Chain Management* (pp. 1–12). IGI Global. DOI: 10.4018/978-1-6684-9062-4.ch001

Masudin, I., Wastono, T., & Zulfikarijah, F. (2018). The effect of managerial intention and initiative on green supply chain management adoption in Indonesian manufacturing performance. *Cogent Business & Management*, 5(1), 1485212. DOI: 10.1080/23311975.2018.1485212

Mentzer, J. T., DeWitt, W., Keebler, J. S., Min, S., Nix, N. W., Smith, C. D., & Zacharia, Z. G. (2001). Defining SCM. *Journal of Business Logistics*, 22(2), 1–25. DOI: 10.1002/j.2158-1592.2001.tb00001.x

Merimi, M., & Taghipour, A. (2021). Accelerating the digitalization of the supply chain: An empirical research about COVID-19 crisis. In *Digitalization of decentralized supply chains during global crises* (pp. 1–24). IGI Global. DOI: 10.4018/978-1-7998-6874-3.ch001

Mishra, K., Siwal, S. S., & Thakur, V. K. (2024). E-waste Recycling and Utilization: A Review of Current Technologies and Future Perspectives. *Current Opinion in Green and Sustainable Chemistry*, 47, 100900. DOI: 10.1016/j.cogsc.2024.100900

Morgan, D. L., & Nica, A. (2020). Iterative thematic inquiry: A new method for analyzing qualitative data. *International Journal of Qualitative Methods*, 19, 1609406920955118. DOI: 10.1177/1609406920955118

Moussaoui, T. E., Loqman, C., & Boumhidi, J. (2024, May). Flat and Nested Named Entity Recognition in Arabic Language. In 2024 4th International Conference on Innovative Research in Applied Science, Engineering and Technology (IRASET) (pp. 1-7). IEEE. DOI: 10.1109/IRASET60544.2024.10548518

Muduli, K., Luthra, S., Kumar Mangla, S., Jabbour, C. J. C., Aich, S., & de Guimarães, J. C. F. (2020). Environmental management and the "soft side" of organisations : Discovering the most relevant behavioural factors in green supply chains. *Business Strategy and the Environment*, 29(4), 1647–1665. DOI: 10.1002/bse.2459

Murray, A., Skene, K., & Haynes, K. (2017). The Circular Economy: An Interdisciplinary Exploration of the Concept and Application in a Global Context. *Journal of Business Ethics*, 140(140), 369–380. DOI: 10.1007/s10551-015-2693-2

Sahoo, S., & Vijayvargy, L. (2020). Green supply chain management practices and its impact on organizational performance: Evidence from Indian manufacturers.

Sarkis, J., Zhu, Q., & Lai, K. (2011). An organizational theoretic review of green supply chain management literature. *International Journal of Production Economics*, 130(1), 1–15. DOI: 10.1016/j.ijpe.2010.11.010

Senyo, P. K., Liu, K., & Effah, J. (2019). Digital business ecosystem: Literature review and a framework for future research. *International Journal of Information Management*, 47, 52–64. DOI: 10.1016/j.ijinfomgt.2019.01.002

Srivastava, S. K. (2007). Green supply-chain management: A state-of-the-art literature review. *International Journal of Management Reviews*, 9(1), 53–80. DOI: 10.1111/j.1468-2370.2007.00202.x

Taghipour, A., Lu, X., & Smahi, M. (2022). Opportunities and challenges for society 5.0 in procurement. *Blockchain Technology in Supply Chain Management for Society*, 5(0), 91–100. DOI: 10.1201/9781003177432-9

Taghipour, A., & Merimi, M. (2021, March). Digital transformation of supply chains during crisis: covid-19. In 11th Annual International Conference on Industrial Engineering and Operations Management, IEOM (pp. 7493-7501). DOI: 10.46254/AN11.20211298

Taghipour, A., Ramezani, M., Khazaei, M., Roohparvar, V., & Hassannayebi, E. (2023). Smart transportation behavior through the COVID-19 pandemic: A ride-hailing system in Iran. *Sustainability (Basel)*, 15(5), 4178. DOI: 10.3390/su15054178

Taghipour, A., Sohrabi, A., Ghaedi, M., & Khazaei, M. (2023). A robust vaccine supply chain model in pandemics: Case of Covid-19 in Iran. *Computers & Industrial Engineering*, 183, 109465. DOI: 10.1016/j.cie.2023.109465

Tiwari, S. (2021). Supply chain integration and Industry 4.0: A systematic literature review. *Benchmarking*, 28(3), 990–1030. DOI: 10.1108/BIJ-08-2020-0428

Wang, P., Che, F., Fan, S., & Gu, C. (2014). Ownership Governance, Institutional Pressures and Circular Economy Accounting Information Disclosure. *Chinese Management Studies*, 8(3), 487–501. DOI: 10.1108/CMS-10-2013-0192

Zarbakhshnia, N., Govindan, K., Kannan, D., & Goh, M. (2023). Outsourcing logistics operations in circular economy towards to sustainable development goals. *Business Strategy and the Environment*, 32(1), 134–162. DOI: 10.1002/bse.3122

Zhu, Q., Sarkis, J., & Lai, K. (2007). Green supply chain management: Pressures, practices and performance within the Chinese automobile industry. *Journal of Cleaner Production*, 15(11), 1041–1052. DOI: 10.1016/j.jclepro.2006.05.021

Chapter 14
Generative AI in Deepfakes

Kapil Vhatkar
Dr. D.Y. Patil Institute of Technology, Pimpri, India

Yatri Davda
https://orcid.org/0000-0003-4198-3499
Dr. D.Y. Patil Institute of Technology, Pimpri, India

Aman Prakash Singh
Dr. D.Y. Patil Institute of Technology, Pimpri, India

ABSTRACT

Deepfakes, driven by advanced generative AI models like GANs and VAEs, create highly realistic synthetic media, ranging from manipulated faces and voices to entirely fabricated videos. Originally developed for creative expression and entertainment, deepfakes now raise significant concerns, including political manipulation, privacy violations, and the spread of false information. This chapter delves into the technology behind deepfakes, exploring the models, techniques, and tools used to produce them, while also examining emerging detection methods designed to identify such synthetic content. As research continues to address the risks of deepfakes, the chapter also considers broader societal implications, such as the erosion of trust in media and public discourse. It tackles deeper ethical issues around deception, consent, and the blurred lines between reality and fabrication. By analyzing evolving legal frameworks and real-world case studies, the chapter advocates for a balanced approach to harnessing deepfake technology responsibly, ensuring its benefits outweigh its potential harms.

DOI: 10.4018/979-8-3693-7478-8.ch014

1. INTRODUCTION TO DEEPFAKES AND GENERATIVE AI

1.1 Definition of Deepfakes

Synthesized media manipulated using a person's image by artificial intelligence, especially generative AI models to produce a product whose appearance is convincingly real and contrived, defines, according to Deepfakes. The term "deepfake" combines "deep learning" and "fake," pointedly indicating the technique on which such media are grounded. Initially, deepfakes were popular for entertaining and creative projects through the application to change the faces of actors in movies or even for avatars in video games. However, its capacities evolve at an incredible speed, finding enormous applications in digital communication, advertisement, and even social media.

For instance, instead of putting a famous actor's face on a stunt double's body in the movie, it would just put the voice of a public figure in a fabricated speech. These fairly reasonable manipulations can be produced with a minimum of equipment, so the technology is accessible to both amateur and professional users.

Deepfakes technology in the form of Generative Adversarial Networks (GANs) and Variational Autoencoders (VAEs) has taken shape and spread it: from an oddity, deepfakes are now a tool that indeed exerts massive influence on the production of media, leading to productive doors of creative possibilities and a legi- timate questions regarding truth, trust, and security in the digital age (Goodfellow, 2014; Kingma & Welling, 2013).

1.2 Role of Generative AI in Deepfakes

Generative AI falls within the realm of artificial intelligence, specifically that subset which generates new data with all characteristics close to real data. It learns the hidden patterns and structures in a dataset so that those learned patterns can be used to generate new, synthetic content. Generative AI models prove invaluable for creating realistic media, whether in image, audio, or video form. Because of the generative AI, deepfakes exist through very impressive yet not genuine media, which is nearly indistinguishable from reality (Cresci, 2020; Chesney, & Citron, 2019).

Generative AI models most commonly involved in making deepfakes include Generative Adversarial Networks and Variational Autoencoders; they both have unique capabilities which contribute to the actual production of the deepfake media (Kingma & Welling, 2013).

The generator, in simple terms, generates artificial content-a synthetic face or video-by learning the patterns in the training data-the collection of images or videos of real people.

Instead of determining whether the content generated by itself is real or not, the task of the discriminator would be more like comparing that particular generated content with the original data.

With the increase in iterations, the generator becomes smarter by learning and using the feedback from the discriminator. Upon the continuous process of adversarial training, the generator finds it increasingly difficult for the discriminator to decide which is the actual data and which is the fake one. o be aligned with audio.

One of the most well-known variants of GANs is StyleGAN; it has allowed for detailed deepfakes, primarily in facial manipulation. StyleGAN introduces a style transfer mechanism where the model generates image features that are extremely high in resolution and detail control on attributes such as hair, facial features, and lighting conditions (Goodfellow, 2014).

1.3 Historical Context and Development

Though spectacular, these techniques required a huge amount of technical input and time and were thus limited to high budget movie studios and professional video editors. Early AI-driven manipulation of media was simple and utilized basic image processing and some kind of pattern recognition; however, the required completeness of precision to produce realistic and automated synthetic content was not met.

2014: Introduction of GANs

The development of Generative Adversarial Networks, or GANs, by Ian Goodfellow in 2014 can be regarded as a watershed event for AI-driven content generation. It all begins with the idea of two neural networks: the generator and the discriminator, which are placed in competition to produce increasingly realistic synthetic data (Goodfellow, 2014). GANs were not created particularly for deepfakes. Instead, the technology gained quick popularity as the foundation for producing forged images and videos. With this ability to generate new, extremely realistic content based off of training data, GANs opened up the possibility of much deeper experimentation in media manipulation (Goodfellow, 2014).

2017: The Emergence of "Deepfakes"

By the end of 2017, online communities-particularly on Reddit-had become the incubator for the term "deepfake" as users began to test AI models for swapping in faces in videos (Cresci, 2020). Early deepfakes were essentially just a way to overlay celebrities' faces onto other bodies in videos-work considered cutting-edge at the time-used to create comic or artistic effects. This method quickly improved, and deep fake producers began churning out high-quality content using easily accessible tools and open-source software like DeepFaceLab and FaceSwap (Cresci, 2020; Chesney, & Citron, 2019). So easy was the technology that ordinary people began to pay attention and start sounding alarms.

2018: Deepfakes Awareness and Alarms Mount

Deepfakes entered the public psyche by 2018 and drew not only the attention of AI researchers but also media, governments, and many other people. Deepfake videos featuring celebrities in non-consensual manipulated content-explicit or fake news videos-expanded the ethical concerns that surround deepfakes (Chesney, & Citron, 2019).

2. CORE TECHNOLOGIES BEHIND DEEPFAKES

2.1 Generative Adversarial Networks (GANs)

The GANs, invented by Ian Goodfellow et al in 2014, can be said to constitute a watermark of sorts for generative modeling in AI (Goodfellow, 2014). The easiest way of explaining GANs is that they consist of two neural networks: the generator and the discriminator, which can wage war on each other to justify the "adversarial" tag. An adversarial relationship on this nature allows GANs to produce high-quality synthetic media, which is why deepfakes are indeed super-easy to create (Goodfellow, 2014).

How GANs Work?

1. The Generator

The generator by itself generates new data instances, which may either be images or videos. It accepts noise and tries to create samples that are as close as possible to the training set.

Since its objective is to produce the outputs in a way that would be as close to the real data as possible so that they can "deceive" the discriminator in labelling the created content as real.

End

This discriminator function would distinguish the real data from the fake one. Therefore, this function is meant to take both the real samples drawn from a training data set and the fake samples generated by the generator.

The discriminator outputs a score, reflecting how far the input provided is from being real. Discriminator's job is to make proper classification for whether the data is coming from the real space or from generated space (Goodfellow, 2014).

Step 1: Discriminator Training:

Let's try to see how this iteration goes, but with what happens in that iteration; namely, it goes with training the discriminator on a batch of real samples as well as on a batch of samples generated by the generator. It learns to classify, which of these data points are real and which are not real. This essentially is a flavor of the machine learning problem where the objective of the discriminator is to maximize its accuracy with respect to this classification between two kinds of samples: real and not-real, or authentic and not-authentic.

Training the Generator:

Train the generator using the maximum error of the discriminator. This is conceptualized with the fact that the generator will create data which the discriminator perceives as real. Back propagate the response from the discriminator to the generator thus changing its weights to yield better quality output (Goodfellow, 2014).

Step 3: Adversarial Training

This adversarial training back-and-forth goes between the generator improving to generate more realistic outputs, and the discriminator improving to better spot fakes. The last thing is to reach a stage that the generator could produce something so realistic that the discriminator wouldn't be able to distinguish (Goodfellow, 2014).

2.2 Variational Autoencoders (VAEs)

Variational autoencoders fall under the category of generative models that have proven highly useful for compressing and regenerating vector data. D. P. Kingma and M. Welling first proposed them in 2013. VAEs do a very good job in capturing the underlying distribution of the data; hence, they apply towards images generation, anomaly detection, and semi-supervised learning (Kingma & Welling, 2013). Their new architecture and training scheme make this function of generating new points close to the dataset on which they were trained yet still amenable to intuitive modification in the latent space (Kingma & Welling, 2013).

Architecture of VAE

The core architecture of VAEs consists of two constituent modules: the encoder and the decoder.

Encoder

The encoder maps the input data, for example, images, into some latent space that is a compressed version of the data. While any other encoding-decoding architecture has just one value that the encoder outputs, namely the mean vector, in VAE this should be two values: the mean vector and the variance vector (Kingma & Welling, 2013).

These outputs define a probability distribution (most probably Gaussian) over the latent space, by which a latent variable can be sampled. This probabilistic approach allows much better representation of the input data and enables better generalization.

Decoder:

It samples the latent space to reconstruct original data. Output of the decoder must reproduce the input data as closely as possible, thus regenerating the images or other media.

During the training procedure, it minimizes a loss function consisting of two terms: reconstruction loss, measuring how well the reconstructed output agrees with the original input, and Kullback-Leibler divergence loss, measuring how well the learned latent space approaches a standard normal distribution. This gives the model good learning of a smooth and continuous latent space that will capture the structure underlying the data (Kingma & Welling, 2013).

2.3 Neural Networks and Deep Learning

A neural network is simply a computer model that mimics the human brain in order to recognize patterns and solve problems much better. Typically, there will be layers of interconnected nodes, or neurons. That will consist of the input layer, one or more hidden layers, and the output layer. Each neuron of this will take the input data, apply its own transformation by using some kind of activation function, and then pass that output on to the next layer. This type of architecture allows neural networks to learn and model very complex relationships in the data, and so is particularly useful in tasks like image recognition, natural language processing, and really importantly, deep media generation.

Training Neural Networks

Training neural networks can be done for a multi-phase operation that involves the following critical steps:

The first step in training is to initialize the connection weights between neurons. The values of these weights dictate how much an influence one neuron has on another and hence are kept random to ensure diversity during learning.

Forward Propagation:

In each training iteration, it would cascade, in a layer-by-layer fashion, down through its input data. At every neuron, it would compute weighted summation of the inputs and apply the appropriate activation function: ReLU, sigmoid, tanh in this example before passing it down to the next layer. This continued on until it eventually passed the information down to the output layer.

Loss Calculation:

Once the forward pass is completed, the output of the network is compared with the actual target values or so-called ground truth. The difference between the network's predicted output and the actual target is measured by a loss function. In case of regression tasks, the mean squared error is calculated. In case of classification tasks, cross-entropy loss is obtained. Now, the loss measures the goodness of fit of the network.

Backward Propagation:

It then follows through by propagating the same in the back propagations. In this process, it calculates the gradient of the loss with respect to its weight using the calculus chain rule, thus helping the network find the adjustment that needs to be done in respect of weights to minimize loss.

For every weight subtract the fraction of gradient scaled by learning rate so as to reduce loss. Let this process perform over several iterations or epochs until network gets converged that is, loss will become stabilized at low values.

Fine-Tuning and Regularizations

Training dropout, early stopping, and batch normalization can go hand-in-hand without overfitting the model and enhancing the generalization of the model. These techniques are given so that the model would behave well in unseen data and did not just memorize the training set.

Through a forward and backward propagating loop, neural networks learn to represent and generate complex data patterns and therefore are essential for creating deepfake media (Figure 1).

Figure 1. Propagating loop

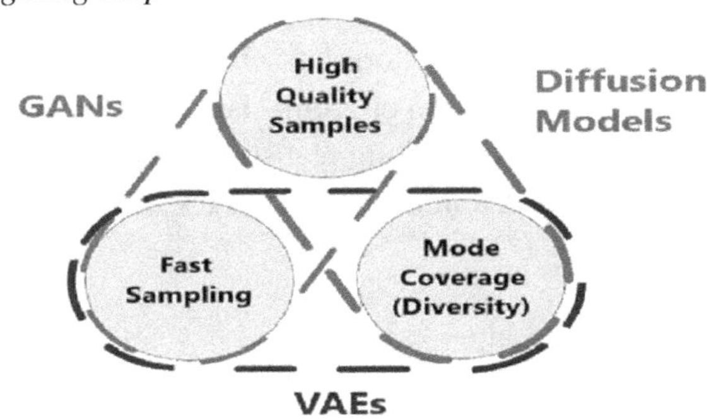

3. CREATION PROCESS OF DEEPFAKES

3.1 Data Collection and Preprocessing

It is during these crucial stages of data collection and preprocessing that the quality as well as diversity of training determines the effectiveness and realism of the media. Large datasets of real-world images or videos are necessary when preparing the datasets for deepfakes, be it GANs or VAEs, which can then be learned by a model. Therefore, the more superior the dataset, the better will be the outputting of the deepfakes (Cresci, 2020; Chesney, & Citron, 2019).

Data Gathering for Deepfakes

The first step in the development of deepfake models involves gathering large quantities of datasets for images or videos used as a basis during the artificial intelligence's learning process. Most of the time, deepfake models are trained mainly on facial data; therefore, it tends to gather as many varieties of human faces as possible under different conditions and contexts.

Sources of Data:

Publicly Available Data: Data sets, which have been used to generate deepfakes, are provided most of the time as publicly available data drawn from public databases or online media services.

These include pictures of celebrities, footage of interview videos, movies, YouTube videos, or large-scale data sets such as CelebA, LFW (Labeled Faces in the Wild), or FaceForensics++. The CelebA set, for example, contains over 200,000 images of celebrities with different facial expressions and orientations (Chesney, & Citron, 2019; European Commission, 2020).

Custom Datasets: In some cases, custom datasets are pre-complied with manually collected images or videos of a person or group. This might be used in some deepfake applications- for example, in creating realistic digital avatars or in editing the facial appearance of an individual in a video (Chesney, & Citron, 2019 ; European Commission, 2020).

Data Diversity and Variability

It should contain many head poses, angles, expressions, skin tones, and facial features. A rich dataset is essentially what would make the model generalize well since it would be able to create highly realistic faces for any condition such as lighting, facial hair, and emotional expressions. For video-based deepfakes, it should include numerous head poses, movements, and background settings.

Ethical Considerations in Data Collection

The collection of such real world images and videos for training the deepfakes raises very serious ethical considerations. Use without permission of the subject, in particular for nefarious purposes, like non-consensual explicit content, or political misinformation, is highly unethical. Developers should consider the legality and morality involved with using publicly available images, especially when privacy of individuals is compromised in the process.

Datasets involving images shared on social media sites related to people's private photographs or videos deserve very careful handling. It involves creating ethical boundaries and necessary components associated with consent as well as data not being misused in the responsible creation of deepfakes.

3.2 Training the Model

Training of a deepfake model is essentially feeding a large amount of data to it, which improves its function iteratively, in the process creating very realistic synthetic media such as images, videos, or audio. Most often, two typical models of deepfakes are popular: the Generative Adversarial Networks (GANs) and Variational Autoencoders (VAEs). These models rely on the learning process, but do so in different ways (Goodfellow, 2014; Kingma & Welling, 2013).

GAN Training

Gations are two neural networks: the generator and the discriminator, trained in an adversarial setting, consisting of the former that produces synthetic data, such as faces, videos, and the latter attempts to separate real data from generated data, in a back-and-forth competition that improves the realism of the content (Goodfellow, 2014).

Input Data:

The training process starts with feeding the GAN model a large set of real-world images or videos. Usually, the dataset encompasses thousands or even millions of data points that denote the target domain, for example, human faces, celebrity images, etc.

Images are pre-processed (as above) to have uniform sizes in terms of alignment and formats before it is applied to train a model.

3.3 Face-Swapping Techniques

It is also applied in entertainment to create visual effects and do digital de-aging. For example, in films, the faces of some actors are substituted with stunt doubles when the scenes called for the action appearance, or when some dead actors' faces were used to recreate them digitally for posthumous appearances. One such example is when face-swapping was used in *Rogue One: A Star Wars Story* to digitally recreate the late Peter Cushing's face.

But face-swapping has been misused for malevolent purposes, particularly in campaigns of misinformation. Deepfakes of political figures and celebrities are made, so that the viewer believes they say or do things that they never did. They sow distrust and spread disinformation. Such videos have been used to dupe the public in an election, change perceptions, and have been employed in soft propaganda. It raises serious ethical as well as legal questions (European Commission, 2020).

Figure 2. Face Swapping

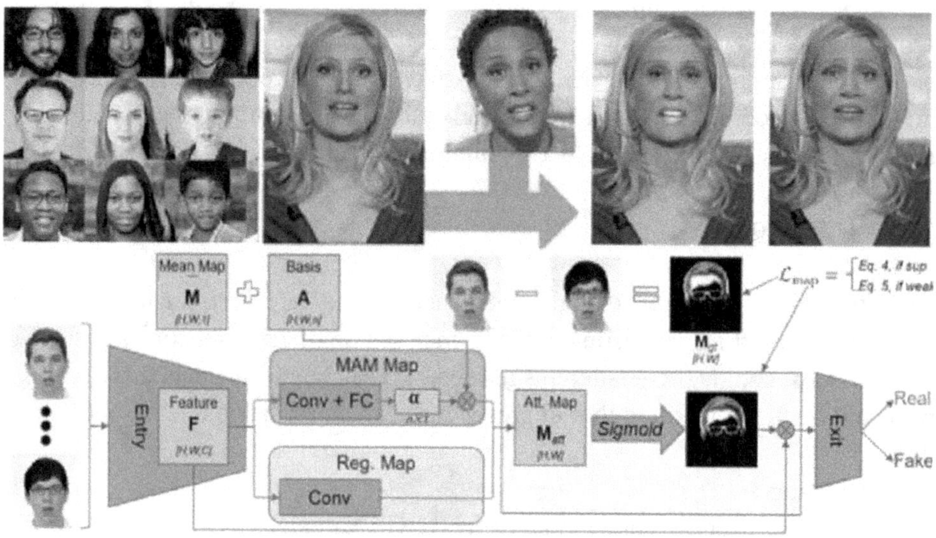

4. REAL-WORLD APPLICATIONS AND IMPACTS

4.1 Positive Uses of Deepfakes

Deepfakes have also been a subject of vast coverage as they further have a whole array of useful applications that change the content of the world in film, education, and even broader accessibility in general. Deepfakes are revolutionizing film and entertainment through the remake of actors or alteration of appearance. For example, technology for digital de-aging has been used in such films as The Irishman to cast actors with a youthful face or to use the likenesses of late actors for posthumous roles, like that of Peter Cushing in Rogue One: A Star Wars Story. The method is worth pursuing since it saves makeup and prosthetic efforts toward such visual effects and delivers smooth, realistic effects with less effort.

In education, deepfakes can be applied to the development of realistic avatars or virtual tutors which help teach interactivity. For example, newly learnt languages could be translated by avatars, which are generated in real time by AI, and lip-synced lines to produce interactive personalized learning environments. Deepfakes can eventually allow historical figures to be constructed for virtual lectures, therefore allowing a unique educational experience (Chesney, & Citron, 2019).

Access also exploits deepfakes by increasing availability to the disabled people. For instance, lip-syncing by deepfakes may enable people to communicate with each other by synchronizing the synthesized speech and face movements so that a hearing-impaired person can track more easily what is being said. Then, at such times when patients lose the ability to speak, AI voices with facial avatars may ensure that mechanisms for communication exist for them too (Chesney, & Citron, 2019; European Commission, 2020).

4.2 Malicious Uses of Deepfakes

Deepfakes have also been applied in the creation of unconsented explicit content-the majority being women. Swapping a person's face onto an explicit video causes violations of privacy and dignity, causing immense damage emotionally and to one's reputation for the victims of such attacks. A notable case involves a deepfake AI tool that enabled the insertion of any woman's face into pornography videos, which later was made to cause serious privacy breaches and public anger.

Another major challenge is identity theft. A deepfake can impersonate people in video calls or audio recordings, even fooling biometric systems of the high end. A good example is when a deepfake audio was used to impersonate a CEO's voice in transferring a whopping $243,000 fraudulently.

The malicious utilization of deepfakes serves as evidence of the possible harm technology may bring; hence detection tools and even regulatory measures need to be developed as means to mitigate these harms.

4.3 Societal Impacts of Deepfakes

As the deepfakes impact society, it exemplifies the erosion of trust in the media and institutions. Deepfakes mixed up the distinction between reality and artificially fabricated content and have heightened the stakes in skepticism over the authenticity of online media. In a world that claims, at least in theory, to rely on digital platforms for news and information, it is quite disconcerting. Manipulated videos and images seem to increase the public's distrust on both traditional journalism, turning more into this “fake news” phenomenon, and social media. If people cannot anymore be believed in believing what they see and hear, then it is even much harder for people to have a collective definition of truth; that, in turn, weakens democratic processes and societal cohesion[6].

Deepfakes pose a huge challenge to law enforcement agencies as they seek to effectively fight fraud, defamation, and non-consensual pornography crimes, among others. The new tools that are meant to police deepfakes have to be invested in by the law enforcement agencies to combat them effectively. In most cases, legal frameworks that are in place at the moment often turn out to be inept to meet the fast change that controls this technology. Slightly adding to this already difficult process in terms of complexity is that the deepfakes must be found, prosecuted, and proved before a court of law (European Commission, 2020).

Implications are tremendous for journalism because deepfakes can be used in the process of discrediting facts already reported by accusing journalists of fabricating evidence presented. This creates instability among credible sources. Deepfakes can further be weaponized and used to defame certain people or even manipulate public discourse through fabricated interviews or speeches or compromising acts.

Ultimately, deepfakes feed into a new era of information distrust-that spreads easily but is complicated for the work of the media, police, and other more trusted elements of society.

5. DETECTION AND COUNTERMEASURES

5.1 Challenges in Detecting Deepfakes

Detecting deepfakes is difficult today because the technology to make them is superior and synthetic media is almost indistinguishable from authentic content. The problem for the detection is that algorithms of modern deepfakes can achieve a very high level of realism. Deepfakes produced a long time ago were easier to notice due to outstanding visual artifacts such as unnatural blinking, mismatched lighting, or facial distortions. Though, as Style GANs and Cycle GANs, and other generative models have proved to be so far improved, these tell-tale signs of manipulation get diminished to a point it becomes much more difficult to detect (Chesney, & Citron, 2019).

Light and shadows is also technical challenge - subtle enough that high quality deep-fakes even replicate the behaviour of light on a subject's face very accurately. Moreover, deepfakes possess an improved ability to manage subtle facial movements; for example, slight changes may refer to minute differences in expressions or even micro-expressions, which were previously difficult to mimic. Such minor indications greatly increase the chances of confusion on the part of both the human eye and algorithms alike between the original and the artificial (Chesney, & Citron, 2019).

The other challenge is in the audio and voice manipulation. The new sophisticated deepfake models are able to convincingly voice-cloning someone to match with lip-synced video. This makes detection even more difficult. Mostly, deepfakes are created as a combination of voice cloning and face-swapping to create an output which happens to be pretty realistic (Chesney, & Citron, 2019; European Commission, 2020).

But more importantly RTD introduce another layer of complexity, manipulation occurring at a live stage of the video, which makes the operating time frame of detection systems close to nil, with little or no time to interpret. Generative models continue to improve over time, and so must the tools for detection, as this is forced to evolve hand in hand with the continually improved technologies to develop ever more convincing deepfakes (Chesney, & Citron, 2019; European Commission, 2020).

5.2 AI-Based Detection Tools

While deepfakes grow more realistic, AI-based detection tools are crucial in spotting manipulated media. Such tools are based more on deep learning models trained to pick on slight inconsistencies in video, audio, and images. Such models look at patterns often not visible to the naked eye, making it effective in distinguishing real and fake content (Cresci, 2020).

A technique tracks eye movements. Early deepfakes often failed to incorporate the natural blink pattern and still do today; slight anomalies in blink patterns and frequency can betray a photo. AI models that are trained on the detection of this pattern will activate media where eye movement looks unnatural or at odds with human behaviour. Facial muscle patterns work in much the same way. Tools that parse these patterns watch how different areas of the face move during speech and expressions. Humans involuntarily display facial micro-expressions and muscle twitches that deepfakes might not perfectly capture, especially when relating to intricate facial movements such as frowning or smiling. Such small imperfections can be picked up by AI models, which can now help in unearthing manipulated content (Cresci, 2020).

Another detection method is digital forensics whereby AI tools scan for artefacts including discrepancies between pixels at a detailed level, compression anomalies or distorted shadowing. Deepfakes sometimes leave minute digital footprints such as mismatched lighting across a whole face or background distortion which the forensic algorithms can detect. Other techniques to help in determining tampering with the suspect media content include reverse video search or frame-by-frame analysis and the comparison of suspect media content with a database of known contents (Cresci, 2020).

The more advanced the deepfake becomes, the faster these AI-based tools have to evolve, utilise machine learning algorithms, learn to pick up increasingly complex forgeries, and remain an important and vital line of defence against misinformation and fraud.

5.3 Human-AI Collaboration in Detection

Deepfake detection requires the collaboration of human and AI. This is because AI models and human judgment are complementary in respect to improving precision. Human input is specifically mandatory in interpreting results and making informed judgments on matters such as context and intent, whereas AI-based detectors are best suited for the review of large volumes of data and minute inconsistencies.

Deep learning-based AI models can process video frames intensively for pixel pattern analysis to detect anomalies like lighting inconsistencies, facial imbalances, or unnatural movement. They do go wrong sometimes in classifying authentic content as fake (false positives) or failing to identify well-constructed deepfakes (false negatives), specially as the technology of creating deepfakes advances.

Analysts can infer the intent behind a video and verify if it aligns with the subject's typical action or utterance. Another benefit of human interference is additional ethical sensitivity: experts take into account if the media is hateful or virulent.

A good example of such collaboration is Facebook's Deepfake Detection Challenge, wherein AI tools in conjunction with human moderators assess flagged content. It holds videos reviewed by AI and as per the context-sensitive judgment human experts determine whether the video represents a harmless parody or a dangerous disinformation attempt. Such collaboration between AI's processing power with human instinct and judgment strikes upon the advancement of speed along with reliability of detection of deepfakes.

5.4 Supply Chain Resilience and Optimization

Generative Adversarial Networks would be subjected to harder efforts in deepfake technology and considerably contribute to increasing the robustness of supply chains. (Dadsena et al. 2021, and Taghipour & Frayret, 2013) noted that integrating AI technologies reduce risks and enhance coordination when there is disruption by pandemics (Taghipour & Frayret, 2013a, and 2013b; Zoghlami et al., 2016). Therefore, deepfakes can:

1.GANs may be used for generating synthetic data to test supply chain response under crisis conditions like sudden spikes in demand or a logistical bottleneck (Goodfellow, 2014; Cauchois et al., 2017; Dadsena et al., 2021; Loivet et al., 2020; Taghipour, 2020; Taghipour, 2014 ; Merimi, & Taghipour, 2021; Taghipour& Frayret, 2011 ; Mbiatem et al., 2018; Nadia et al., 2021).
2.Deepfakes can be applied in the development of realistic virtual environments to train supply chain managers on decision-making in disruption (Taghipour et al., 2022; Taghipour & Merimi, 2021; Taghipour et al., 2023b; Taghipour et al., 2015 ; Taghipour et al., 2023c; Lu & Taghipour, 2023.

This aligns with post-COVID-19 strategies where industries emphasize digital transformation and resilience (Taghipour et al., 2023a; Taghipour et al., 2024; Taghipour, 2023; Taghipour & Beneteau-Piet, 2020 ; Mahfod et al., 2019).

6. ETHICAL AND LEGAL CONSIDERATIONS

6.1 Ethical Dilemmas

Creation and distribution of deepfakes, one needs to raise many ethics-related concerns, especially in the context of such technology in nonconsensual content or deceptions. Of course, the most insidious ethical issue is when an individual's privacy is compromised. Deepfakes often manipulate people's images, voices, or videos without any form of knowledge or consent, thereby creating a digital identity that can easily be exploited. This is very destructive where the misconduct deals with non-consensual pornographic explicit content, in which usually women face images superimposed on pornographic videos, causing significant distress, damage to reputation, and psychological trauma for those affected. These usually involve victims of deepfakes with very little options to act since these videos are usually rampant online, rendering impossible complete removal.

Another major ethical issue the deepfakes are involved in is the use of deception and information. Deepfakes can be so believable that this form of media could very easily be used in political manipulation or to frame people into crimes they never did. These deepfakes destroy trust in the media and other institutions as well, making it hard for anyone to know what is real and what isn't. For example, a politician's inflammatory or false statement deep faked and made public opinion can shift or elections aborted (Cresci, 2020).

The psychological damage stemming from deepfakes is very severe, especially when attacking people in ways that turn against their own self-esteem or social standing. Victims of deepfakes would likely develop helplessness and frustration along with humiliation, mainly where sexual deepfakes or character assassination is involved. This ethical issue stretches to the hosting sites where deepfakes appear or circulate, creating additional debate on whether the hosting platforms bear any responsibility for screening harmful material. Balancing Innovation and individual rights are very critical while trying to address the ethical issues. The creation and sharing of deepfakes trigger many ethical problems. Especially, when they are used to produce content without someone's consent or as a means of carrying out deception. One major ethical concern associated with deepfakes is a violation of an individual's privacy. A deepfake manipulates an image or voice or video without a person's knowledge, taking away his right to have a digital existence that is exploited at the whims of the creator. Non-consensual explicit content creates some of the most damaging bases, where people, especially women, have their faces superimposed on pornographic videos, thus causing emotional anguish, reputation damage, and psychological trauma. Their victims have little recourse because these kinds of clips easily travel through social media, making it almost impossible to totally erase them from the internet (European Commission, 2020; [6].

6.2 Legal Responses to Deepfakes

Legal solutions to deepfakes are in a state of flux. Many countries are actually falling behind the rapid growth of the technology. In the United States, multiple states have adopted legislation related to the misuse of deepfakes (European Commission, 2020). For example, California and Texas enacted legislation intended to preclude using deepfakes, for example to swing elections, and to make non-consensual explicit content. California's AB 602 permits a person to bring an action if someone else uses their likeness in a damaging deepfake without permission, and Texas prohibits the making of misleading videos intended to affect electoral results. Under federal law, the DEEPFAKES Accountability Act would ensure that deepfakes containing a watermark or which have been accompanied by a disclosure are used in appropriate circumstances (European Commission, 2020).

In the European Union, deepfakes fall under more general privacy legislation-the General Data Protection Regulation (GDPR), which safeguards various forms of personal information, including one's image (European Commission, 2020). The Digital Services Act (DSA) proposed by the European Commission also works as a legislation aimed at making online platforms responsible for deleterious content created online, including deepfakes, by demanding that the online platforms eradicate manipulated media as soon as possible.

Regardless of these efforts, it is still tough to regulate deepfakes since the technology develops at such a breakneck speed. Laws are usually balanced between free speech, where satire or parody might be involved, and the protection a regulation against harming individuals (European Commission, 2020). In addition, in enforcing this regulation, the global availability of the internet complicates matters; one country makes harassing deepfakes and distributes it all over the world. It will thus have to involve international cooperation for effective and comprehensive regulation.

6.2 Leveraging Deepfakes for Public Awareness

Public education, as has been pointed out by Taghipour & Lu (2023), is pivotal in the war against misinformation. Ironically, deepfakes might even help:

1.Educational Campaigns: Deepfakes may be used as controlled experiments that could simulate fake media scenarios so as to educate people on how to recognize fake media.

2.Real-Time Fact-Checking Tools: AI-based systems can be combined together with deepfake detection in verifying the authenticity of media at times of breaking news or important events.

7. FUTURE OF GENERATIVE AI AND DEEPFAKES

7.1 Advancements in Deepfake Technology

Generative AI models for deepfakes will intensify in reality and speed in the following 5-10 years to make deepfakes even more believable and difficult to identify. The greatest impact, however, will be on real-time generation of deepfakes as, currently, real-time deepfakes are achievable but still within processing power and quality boundaries. In the next near-future, more advanced architectures of AI like GANs that they can be possible, combined with advanced hardware like quantum computing or special AI chips, enable such high-resolution deepfakes in almost zero latency and in real time (Goodfellow, 2014; Kingma & Welling, 2013).

They will then further allow deepfake models to capture and animate subtle human expressions and gestures involved with speech. The future deepfakes, trying to reach the extent of the present micro-expressions, which are already difficult to record because they will be in perfect harmony with the audio and the video, will be challenging[. In the present time, AI might even create deepfakes not only to imitate a person's face but also his or her voice, body movements, and even converse like him or her, and thus forgeries would become much more convincing.

As such, the quality of deepfakes improves with these technologies, making detection harder. The currently available AI detection tools run on inconsistencies. However, in such a refined deepfake, the methods utilized in such tools will depend on even more stringent and forensic-level techniques. The most probable losers will be society's sense of trusting digital material in general. Authenticity crisis will be expected in journalism, politics, and law. Greater usage of deepfakes will make audiences skeptical of media; and with deepfakes advancing rapidly, detection and regulation should advance at the same pace to slow down the risks that are coming with deepfakes.

7.2 Balancing Innovation With Responsibility

With time, technology is bound to change, and therefore it's in the best interest of humankind that innovation in generative AI needs to be done responsibly. In pursuit of innovation, there needs to be an environment where new possibilities in AI are tested with understanding and mitigation of risks involved. The ethical development frameworks will act as avenues where a balance is struck and guide researchers and companies as to the best way of designing AI systems that prioritize welfare and fairness for humankind. That would involve ways of preventing abuse in the introduction of safeguards that are required when extracting consent to the creation of deep fakes, ensuring that clear usage is indicated in the use of synthesized media, while putting in place accountability mechanisms.

The most important usage of responsible AI is transparency-where developers and platforms using generative AI have to identify when content is AI-generated so users can distinguish reality from synthetic media. That extends to the open-source practice of sharing methodologies and tools for deepfakes and bringing about collective responsibility in terms of ethical challenges on the use of AI.

Public awareness and education are also a core part of prudent use. Educating citizens on the capabilities and risks associated with deepfakes provides them with critical thinking capacity over media, making them more savy to this problem of misinformation. Governments, tech companies, and educators will work in concert toward an enlightened public that supports innovation that, while creating new value, does so with AI use managed responsibly. Mixing ethics, transparency, and awareness could lift innovative blooms of generative AI without eroding values in society (Kingma & Welling, 2013).

7.3 Industry Adaptation

With the rise of deepfakes, the entertainment, cybersecurity, and media industries will soon align opportunities with these technologies and learn how to further restrict the risks that come with them (Cresci, 2020). Deepfakes will be used with tremendous benefits for creative purposes, such as digital de-aging, recreating deceased actors, or enhancing special effects in the entertainment business. Both time and cost will be saved in the production of studios that offer more flexible possibilities. Then there is the question of how to make deepfakes themselves have the right moral sense to use a consent from an actor or his estate, not damaging reputations.

Deepfakes in fraud and misinformation are threats that are emerging fast, and so there is a growing need for companies to invest in deepfake detection technologies (Cresci, 2020). Cybersecurity organizations are developing AI-based solutions that can identify altered audio-visual media, with special focus on such subtle cues as irregular lighting, facial asymmetry, or the mismatched audio-visual synchrony of deepfakes. Employees and members of the public are also being trained to recognize them and prevent phishing, impersonation, and fraud.

For the media industry, deepfakes come with potential revolutionary new tools for creative content and are simultaneously risks to journalistic integrity. Media organizations will develop partnerships with tech firms in order to create deepfake detection systems to verify content before publishing. Through integration of detection tools with other reporting and education, industries will unlock their potentials without risks to trust and security.

7.4 Future Trends and Research Directions

1.Advanced Simulations

Based on the themes of simulation and coordination highlighted by Taghipour et al. (2013), deepfakes may be applied in simulating behaviors in the supply chain under a variety of conditions to aid planning and optimization.

2.Detection Technologies

With more advancement in deepfakes, there should be more sophistication in the technologies of detection. Multi-modal approaches could be adapted in the industry by fusing visual, audio, and behavioral analyses for detection manipulations.

3.Regulatory Collaboration

International cooperation is what is required by the broken legal landscape surrounding deepfakes, as reported in Taghipour & Merimi (2021). It will be essential to coordinate with the governments and industries for a proper and ethical application of deepfake technology.

8. CONCLUSION

This chapter unfolds the transforming role of generative AI in developing deepfakes development and its technical underpinnings and broader societal implications over the evolving nature of deepfakes. This chapter finds a space center for generation with powerful AI models, mainly GAN and VAE that made synthetic media highly realistic (Goodfellow, 2014; Kingma & Welling, 2013). Such models are learned on huge sets of real-world images and videos and, therefore, are capable of replicating facial features, expressions, and even voices with very impressive accuracy.

However, even despite well-publicized great potential for such models in all areas of the entertainment, educational, and access sectors, deepfakes pose very serious challenges from an ethical and legal standpoint. They may put people and societies at large at risk through having malicious uses that entangle spreading falsehoods, creating non-consensual explicit content, even identity theft. Deepfakes already promoted the propagation and spreading of falsehoods; hence, it becomes much harder to detect them, thus eroding the already fragile trust in digital media, law enforcement challenges, and journalism implications, among others (Cresci, 2020). However, on the other hand, improvements in AI-based detection tools and increased efforts in the collaboration between human experts and AI models present hopes that the risks associated with these features can be mitigated (European Commission, 2020). It is notably the role of industries, regulators, and developers to find the middle ground in terms of orchestrating innovation and responsibility in deploying deepfakes in an ethical and transparent manner.

REFERENCES

Cauchois, R., Taghipour, A., Kang, D., Zoghlami, N., & Abed, M. (2017). Advanced solutions for a supply chain with stochastic information. *Journal of Advanced Management Science*, 5(1), 9–13. DOI: 10.18178/joams.5.1.9-13

Chesney, R., & Citron, D. K. (2019). Deepfakes and the New Disinformation War: The Coming Age of Post-Truth Geopolitics. *Foreign Affairs*.

Cresci, S. (2020). A Decade of Social Bot Detection. *Communications of the ACM*, 63(10), 72–83. DOI: 10.1145/3409116

Dadsena, K. K., Mathiyazhagan, K., & Taghipour, A. (2021). Analysis of barriers for the build the resilient supply chain networks post-COVID-19. In *Managing Supply Chain Risk and Disruptions: Post COVID-19* (pp. 79–89). Springer International Publishing. DOI: 10.1007/978-3-030-72575-4_7

European Commission (2020). The Impact of Artificial Intelligence on Media Integrity. Regulatory approaches in the EU-the general Data Protection Regulation, Digital Services Act-and their implications for deepfake regulation.

Glaa, B., Zoghlami, N., & Taghipour, A. (2014, May). Outsourcing and off-shoring healthcare services: A way to cut the costs and improve quality. In 2014 international conference on advanced logistics and transport (ICALT) (pp. 189-194). IEEE.

Goodfellow, I. (2014). Generative Adversarial Nets. Advances in Neural Information Processing Systems (NeurIPS).

Kingma, D. P., & Welling, M. (2013). Auto-Encoding Variational Bayes. *International Conference on Learning Representations (ICLR).*

Loivet, W., Taghipour, A., & Kang, D. S. (2020). The rise of green supply chain management: Between complexity and necessity. *J. Econ. Bus. Manag*, 8(1), 1–7. DOI: 10.18178/joebm.2020.8.1.604

Lu, X., & Taghipour, A. (2023). A French Guide for Businesses and Governments to Integrate the Concept of Sustainable Development Into Their Supply Chains. In *Government Impact on Sustainable and Responsible Supply Chain Management* (pp. 1–12). IGI Global. DOI: 10.4018/978-1-6684-9062-4.ch001

Mahfod, J., Canel-Depitre, B., & Taghipour, A. (2019). Quality function deployment-ELECTRE in supplier evaluation. Journal of Advanced Management Science, 7(4).

Mbiatem, J., Taghipour, A., & Canel-Depitre, B. (2018, August). Supplier selection approaches for decision makers. In *Proceedings of the 8th International Conference on Information Communication and Management* (pp. 108-112).

Merimi, M., & Taghipour, A. (2021). Accelerating the digitalization of the supply chain: An empirical research about COVID-19 crisis. In *Digitalization of decentralized supply chains during global crises* (pp. 1–24). IGI Global. DOI: 10.4018/978-1-7998-6874-3.ch001

Nadia, E. N., Beatrice, C. D., & Atour, T. (2021). The determinants of hotel room rates in Beirut: A hedonic pricing model. International Journal of Trade. *Economics and Finance*, 12(2), 33–42.

Taghipour, A. (2014). Improving the plan of a manufacturing network with non-integrated business units. *International Journal of Applied Logistics*, 5(2), 1–11. DOI: 10.4018/ijal.2014040101

Taghipour, A. (Ed.). (2020). *Demand forecasting and order planning in supply chains and humanitarian logistics*. IGI Global.

Taghipour, A. (Ed.). (2023). *Government Impact on Sustainable and Responsible Supply Chain Management*. IGI Global. DOI: 10.4018/978-1-6684-9062-4

Taghipour, A., Abed, M., & Zoghlami, N. (2015, May). Design for remanufacturing respecting reverse logistics processes: A review. In 2015 4th International Conference on Advanced Logistics and Transport (ICALT) (pp. 299-304). IEEE. DOI: 10.1109/ICAdLT.2015.7136602

Taghipour, A., & Beneteau-Piet, C. (2020). Sustainable supply chain management performance. *International Journal of Innovation, Management and Technology*, 11(6), 165–169. DOI: 10.18178/ijimt.2020.11.6.897

Taghipour, A., Foukolaei, P. Z., Ghaedi, M., & Khazaei, M. (2023a). Sustainable Multi-Objective Models for Waste-to-Energy and Waste Separation Site Selection. *Sustainability (Basel)*, 15(22), 15764. DOI: 10.3390/su152215764

Taghipour, A., & Frayret, J. M. (2011). Supply chain coordination planning: A review. Work paper, École polytechnique de Montréal.

Taghipour, A., & Frayret, J. M. (2013a, May). An algorithm to improve operations planning in decentralized supply chains. In *2013 International Conference on Advanced Logistics and Transport* (pp. 100-103). IEEE. DOI: 10.1109/ICAdLT.2013.6568442

Taghipour, A., & Frayret, J. M. (2013b). Coordination of operations planning in supply chains: A review. *International Journal of Business Performance and Supply Chain Modelling*, 5(3), 272–307. DOI: 10.1504/IJBPSCM.2013.055729

Taghipour, A., Lu, X., & Smahi, M. (2022). Opportunities and challenges for society 5.0 in procurement. *Blockchain Technology in Supply Chain Management for Society*, 5(0), 91–100. DOI: 10.1201/9781003177432-9

Taghipour, A., & Merimi, M. (2021, March). Digital transformation of supply chains during crisis: covid-19. In 11th Annual International Conference on Industrial Engineering and Operations Management, IEOM (pp. 7493-7501). DOI: 10.46254/AN11.20211298

Taghipour, A., Padash, A., Etemadi, V., Khazaei, M., & Ebrahimi, S. (2024). Sustainable and Circular Hotels and the Water–Food–Energy Nexus: Integration of Agrivoltaics, Hydropower, Solar Cells, Water Reservoirs, and Green Roofs. *Sustainability (Basel)*, 16(5), 1985. DOI: 10.3390/su16051985

Taghipour, A., Ramezani, M., Khazaei, M., Roohparvar, V., & Hassannayebi, E. (2023b). Smart transportation behavior through the COVID-19 pandemic: A ride-hailing system in Iran. *Sustainability (Basel)*, 15(5), 4178. DOI: 10.3390/su15054178

Taghipour, A., Sohrabi, A., Ghaedi, M., & Khazaei, M. (2023c). A robust vaccine supply chain model in pandemics: Case of Covid-19 in Iran. *Computers & Industrial Engineering*, 183, 109465. DOI: 10.1016/j.cie.2023.109465

Zoghlami, N., Taghipour, A., Merlo, C., & Abed, M. (2016). Management of divergent production network using decentralised multi-level capacitated lot-sizing models. *International Journal of Shipping and Transport Logistics*, 8(5), 590–604. DOI: 10.1504/IJSTL.2016.078686

Chapter 15
IoT and AI Integration for Real-Time Supply Chain Monitoring

Kingsly Stephen R.
https://orcid.org/0009-0000-3577-061X
SRM Institute of Science and Technology, Ramapuram, India

T. Archana
https://orcid.org/0009-0008-0703-156X
SRM Institute of Science and Technology, Ramapuram, India

ABSTRACT

The integration between the internet of things (IoT) and artificial intelligence (AI) for real-time supply chain monitoring is streamlining and enhancing how businesses control and optimize their supply chains. This chapter explores the collaborative capabilities of IoT and AI in maximizing supply chain visibility, operational efficiency, and decision-making processes. IoT devices such as sensors, GPS trackers, and RFID tags yield real-time data on various parameters such as location, condition, and the status of the goods throughout their entire journey in the supply chain. This aids with real-time tracking and condition monitoring. Alternatively, incorporating AI algorithms helps analyze historical data, market trends, and various external factors and provides a demand forecast report, which helps in inventory management. Also, AI algorithms can work on traffic patterns to analyze the shortest route, helping to make efficient use of resources, leading to cost savings.

DOI: 10.4018/979-8-3693-7478-8.ch015

1. INTRODUCTION TO IOT AND AI IN SUPPLY CHAIN

1.1 Definition and Overview of IoT

Internet of Things (IoT) refers to a network of interconnected physical devices, vehicles, appliances, and various objects that come with sensors, software, and network functionalities, allowing them to collect and share data. These smart devices are commonly known as IoT devices. It enables communication among these smart devices and with other internet-connected devices like smartphones and gateways. IoT has revolutionized the conventional way of living, transforming it into a sophisticated high-tech lifestyle (Kumar, Tiwari, & Zymbler, 2019). This results in the establishment of an extensive network of interconnected devices capable of autonomously exchanging data and executing diverse tasks. It also enable users to attain enhanced automation, analysis, and integration within a given system. They enhance the scope and precision of these domains. It makes use of both cutting-edge and well-established technology for robotics, networking, and sensing. IoT has a wide range of potential applications, and its impact is already being seen in many industries, including manufacturing, transportation, healthcare, and agriculture. IoT is going to play a bigger role in shaping our society and changing the way we interact, work, and live as more and more things become internet-connected.

1.2 Definition and Overview of AI

Artificial intelligence, or AI for short, is the technology that makes it possible for computers and other technologies to simulate human intelligence and problem-solving abilities. Algorithms, computational models, and data are used by AI systems to mimic cognitive functions. The discipline involves developing artificial intelligence (AI) algorithms that imitate human decision-making processes (Archana & Stephen, 2024). These algorithms can learn from data and become more accurate at classifications or predictions over time. It is a comprehensive concept that encompasses a diverse range of technologies, such as machine learning, deep learning, and natural language processing (NLP). Machine Learning is a subset of AI that uses statistical methods to enable machines to improve their performance on a task through experience. Techniques include supervised learning, unsupervised learning, and reinforcement learning. Deep Learning is a type of machine learning that utilizes neural networks with many layers (deep neural networks) to analyse various types of data, especially in image and speech recognition. Through natural language processing (NLP), machines can comprehend and react to human language, enabling the use of chatbots and language translation services.

1.3 Importance of Real-Time Monitoring in Supply Chain

Real-time monitoring is essential in modern supply chain management due to its numerous benefits and significant impact on operational efficiency and responsiveness.

i. Enhanced Visibility: Real-time monitoring provides comprehensive visibility across the entire supply chain. This transparency allows businesses to track the movement of goods, monitor inventory levels, and observe the status of shipments at any point. Such visibility helps in identifying potential issues before they escalate.
ii. Improved Decision – Making: Access to up-to-the-minute data enables managers to make informed decisions quickly. Real-time insights can highlight trends, demand fluctuations, and potential disruptions, facilitating proactive rather than reactive strategies. This agility is crucial in responding to market changes and maintaining a competitive edge.
iii. Increased Efficiency: By continuously tracking processes and workflows, real-time monitoring helps identify bottlenecks and inefficiencies. Companies can optimize resource allocation, cut down on delays, and streamline operations. Delivery times are shortened as a result, and customer satisfaction is raised.
iv. Risk Management: Real-time data allows for the early detection of risks such as supplier delays, transportation issues, or unexpected demand spikes. With timely alerts, companies can implement contingency plans swiftly, minimizing the impact on the supply chain. This proactive approach mitigates risks and reduces the likelihood of severe disruptions.
v. Cost Reduction: Real-time monitoring that is effective can save a lot of money. Businesses can reduce storage expenses by minimizing excess stock and optimizing inventory levels. Additionally, identifying and resolving inefficiencies quickly reduces waste and unnecessary expenditures.

Real-time monitoring in supply chain management is vital for maintaining operational efficiency, enhancing visibility, and enabling quick, informed decision-making. It helps mitigate risks, reduce costs, and improve collaboration, ultimately leading to better customer satisfaction and stronger business performance.

1.4 Historical Context and Evolution of Supply Chain Technologies

The development of supply chain technologies has undergone significant transformations over the years, driven by advancements in technology and changing business needs.

Figure 1. Evolution of SCM

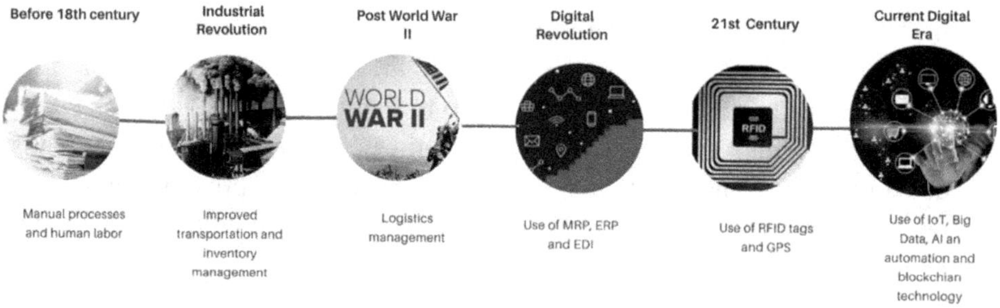

Figure 1, depicts the evolution of SCM from the initial days till the digital era. In the early days of commerce, supply chains were simple and localized. They relied heavily on manual processes and human labor. Record-keeping was done on paper, and communication was slow, often taking days or weeks. The Industrial Revolution in the 18th and 19th centuries marked a significant shift. Mass production techniques, such as the assembly line introduced by Henry Ford, and improvements in transportation, such as the steam engine and railways, allowed goods to be produced and moved more efficiently. This period also saw the emergence of warehouses and the concept of inventory management. Next, after World War II, globalization began to reshape supply chains. Advances in transportation, such as the development of container shipping, allowed goods to be transported across long distances more easily.

This was followed by the Digital Revolution and the rise of the internet. The advent of computers in the 1960s and 1970s brought about significant changes. Early technologies like Material Requirements Planning (MRP) systems were developed to manage manufacturing processes more effectively. The widespread adoption of the internet in the 1990s revolutionized supply chain management. Technologies such as Electronic Data Interchange (EDI) enabled faster and more accurate exchange of business documents. The internet also facilitated real-time tracking and communication, making global supply chains more manageable. The early 2000s saw the rise of more sophisticated technologies. Radio Frequency Identification (RFID) and Global Positioning Systems (GPS) improved tracking and inventory management.

In recent years, the Internet of Things, artificial intelligence, and blockchain technology have transformed the supply chain in various ways. The Internet of Things provides real-time data from every part of the supply chain, from production to delivery. AI and machine learning are now playing crucial roles in supply chain management. These technologies are used for predictive analytics, automating routine tasks, and improving decision-making.

2. FUNDAMENTALS OF IOT IN SUPPLY CHAIN

2.1 Key Components of IoT

The Internet of Things, or IoT, is a sophisticated ecosystem made up of several vital parts that cooperate to provide smart functionality and connectivity.

2.1.1 Sensors and Actuators

A sensor is a component that identifies and reacts to alterations in its surroundings. It gauges different factors like temperature, pressure, humidity, light, and movement, among others. These sensors are integrated into items or affixed to them, enabling communication with other devices or systems via the internet Sensors play a vital role in the Internet of Things (IoT) as they act as the fundamental connection between the physical realm and digital systems. Their primary purpose is to gather data from the surroundings, which can subsequently be processed, analyzed, and utilized to make informed decisions or initiate actions within an IoT ecosystem.

An electrical device that transforms energy between forms is called a transducer. Antennas, position and pressure sensors, thermometers, loudspeakers, and microphones are typical examples. Actuators are devices that transform electrical signals into physical actions. In the context of the Internet of Things (IoT), actuators are essential for enabling systems to interact with and alter their environment based on data and commands.

Figure 2. Sensors and Actuators

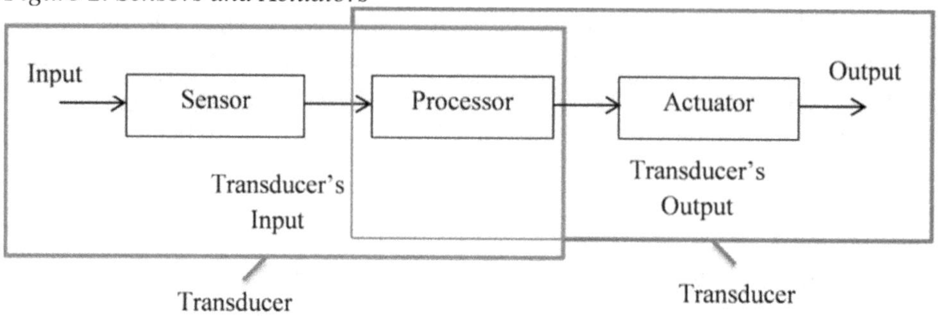

As shown in Figure 2, Actuators enable automation by carrying out actions without human intervention, helps regulate devices and systems by managing mechanical movements, fluid flow or electrical circuits, and it is often a part of feedback loops by responding to data from sensors to maintain desired conditions. By enabling IoT systems to perform physical actions based on sensor data, actuators play a crucial role in creating responsive and intelligent environments.

2.1.2 Connectivity and Communication Protocols

Connectivity and communication protocols are fundamental to the functioning of the Internet of Things (IoT), ensuring that devices can exchange data reliably and efficiently.

These protocols enable the transfer of data between IoT devices, servers, and applications. Reliable data transmission is crucial for the real-time processing and analysis that many IoT applications require. They facilitate communication between heterogeneous IoT devices, which may have different operating systems and hardware configurations. This interoperability is essential for creating a seamless IoT ecosystem. They help manage network connections, ensuring devices remain connected and can communicate efficiently (Stephen, Chandra Sekar, & Dinakaran, 2019). This includes handling device addressing, data routing, and maintaining network security. As the number of IoT devices grows, communication protocols must support scalable networking to accommodate the increasing data traffic and number of connections.

2.1.3 IoT Platforms and Gateways

IoT platforms and gateways are pivotal components in IoT architecture, facilitating the management, processing, and transfer of data between connected devices and cloud-based applications.

Platforms provide centralized management, data processing, and analytics capabilities, while gateways facilitate connectivity, edge computing, and protocol translation at the network edge. Together, they form the foundation for building scalable, secure, and efficient IoT ecosystems.

2.2 Role of IoT in Supply Chain Management

IoT devices play a crucial role in the supply chain process by employing cutting-edge real-time monitoring technologies, such as GPS, to track and trace shipments. In essence, IoT devices are extensively utilized throughout various stages of the supply chain process (Taj et al., 2023).

2.2.1 Asset Tracking and Management

Using Internet of Things (IoT) technology, asset tracking and management inside the supply chain has become a critical tool for improving efficiency and transparency. Businesses can use IoT devices like RFID tags and sensors to track the location and state of assets in real time across the whole supply chain network. A RFID tag consists of three sub components namely microchip, substrate and antenna.

This technology enables precise tracking of assets from production facilities to distribution centres and ultimately to the end consumer. With the continuous stream of data generated by IoT devices, supply chain managers gain valuable insights into asset utilization, location, and potential bottlenecks.

2.2.2 Environmental Planning

Environmental planning within supply chain management (SCM) utilizing IoT technology has become instrumental in fostering sustainability and reducing ecological footprints. By integrating IoT devices such as environmental sensors and smart meters into SCM processes, organizations can monitor and optimize resource usage, energy consumption, and waste generation across the entire supply chain. These IoT-enabled solutions provide real-time data on environmental metrics such as air quality, water usage, and energy consumption at various stages of production, transportation, and distribution. This data empowers companies to identify inefficiencies and implement targeted strategies to minimize environmental impacts.

2.2.3 Predictive Maintenance

Predictive maintenance in supply chain management (SCM) through IoT technology has emerged as a proactive approach to equipment upkeep, minimizing downtime, and optimizing operational efficiency. By deploying IoT devices like sensors and monitoring systems on machinery and assets throughout the supply chain, businesses can collect real-time data on equipment health and performance. This data is then analyzed using predictive analytics and machine learning algorithms to forecast potential issues and schedule maintenance activities before failures occur. By predicting when maintenance is needed based on the condition of the equipment rather than a fixed schedule, organizations can prevent costly breakdowns and avoid disruptions to the supply chain.

3. FUNDAMENTALS OF AI IN SUPPLY CHAIN

3.1 Key Concepts in AI

Artificial Intelligences is the superset of Machine Learning, Deep Learning and Natural Language Processing.

3.1.1 Machine Learning

Algorithms can anticipate outcomes on fresh, similar data by identifying hidden patterns in datasets thanks to machine learning, a subset of artificial intelligence that eliminates the need for explicit programming for each task. Conventional machine learning forecasts outcomes by combining data and statistical methods to produce useful insights (Thirugnanam & Jahir Hussain, 2023). This technology is applied in a wide range of fields such as image and speech recognition, natural language processing, recommendation systems, fraud detection, portfolio optimization, and task automation. The ability to acquire knowledge from data and enhance performance gradually renders machine learning highly potent and adaptable. Machine Learning can be categorized into three main types: supervised learning, unsupervised learning, and reinforcement learning.

Supervised learning stands as the prevailing form of machine learning. Under this methodology, the model undergoes training using a dataset that is labelled. Throughout the training process, the model acquires the ability to establish a connection between the input (features) and the output (label). Once the training is complete, the model becomes capable of predicting the output for novel, unseen data. As shown in Figure 3, unsupervised learning entails training the model using an unlabelled dataset, allowing the model to autonomously discover patterns and relationships within the data. This approach is commonly employed for tasks such as clustering and dimensionality reduction.

Figure 3. ML Classification

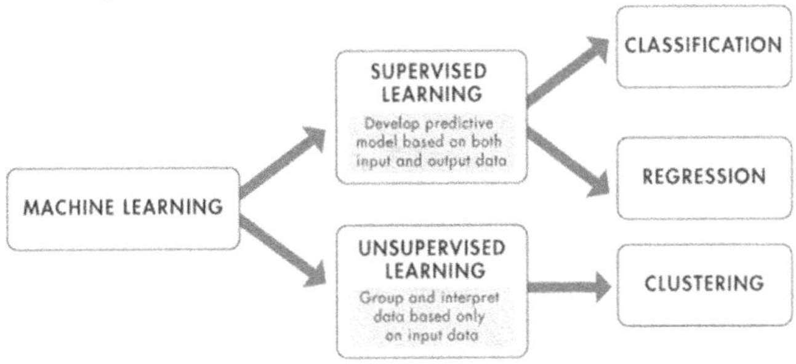

Reinforcement learning is a form of machine learning in which an agent acquires the ability to make decisions through its interactions with the surrounding environment. The agent receives rewards or penalties in the form of points based on its actions, with the ultimate objective of maximizing the cumulative reward.

3.1.2 Deep Learning

A subset of machine learning called deep learning uses complex neural networks, or deep neural networks, to mimic the complex decision-making processes seen in the human brain. Deep learning powers most artificial intelligence (AI) technologies that we come across in our everyday lives. A neural network with three or more layers is referred to as a deep neural network, or DNN. In actuality, though, the majority of DNNs contain a somewhat larger number of layers. Large volumes of data are used to train these networks so they can efficiently recognize and categorize a wide range of occurrences, find patterns and connections, weigh options, and make predictions and judgments (Banu, Stephen, Aditya, & Raaghav, 2024).

Figure 4. Deep network architecture

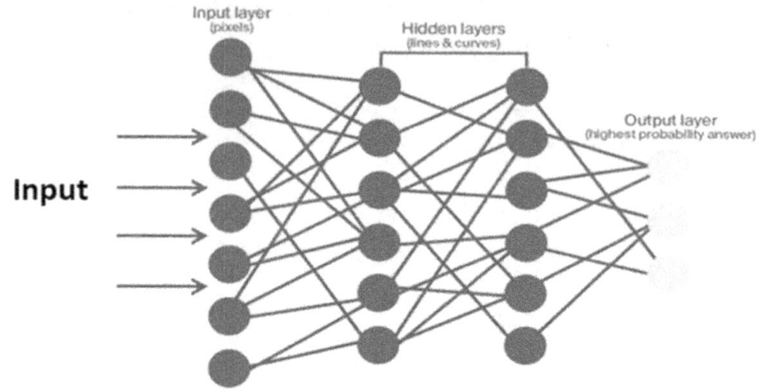

Deep learning eliminates the need for some data pre-processing steps that are typically required in traditional machine learning. While a single-layer neural network can provide useful approximate predictions and decisions, the inclusion of multiple layers in a deep neural network, as shown in Figure. 4, enhances the precision and optimization of these outcomes. These algorithms automate the feature extraction process, lowering the need for human specialists, and they can handle and analyze unstructured data, including text and images.

3.1.3 Natural Language Processing (NLP)

Natural language processing (NLP) is a branch of computer science and artificial intelligence (AI) that leverages machine learning to facilitate computers in comprehending and interacting with human language. It empowers computers and electronic devices to identify, grasp, and produce text and speech by merging computational linguistics—the rule-based representation of human language—with statistical modeling, machine learning (ML), and deep learning. It plays a crucial role in thoroughly and effectively examining textual and verbal information. It has the capability to navigate through variations in dialects, slang, and grammatical inconsistencies commonly found in everyday conversations. Organizations utilize NLP for various automated functions, including- processing, analyzing, and storing extensive documents, evaluating customer feedback, operating chatbots and categorizing text data.

Figure 5. Components of AI

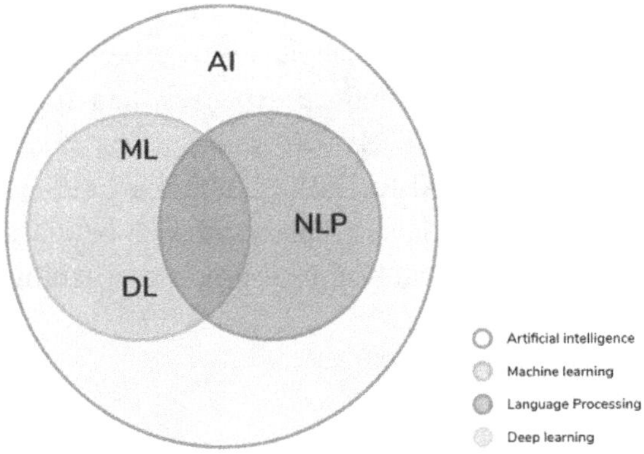

Figure 5, depicts the components of Artificial Intelligence which together makes learning from seen and unseen data much efficient.

3.2 Applications of AI in Supply Chain

Artificial Intelligence has a wide range of applications in the field of supply chain management. AI-powered algorithms have the capability to generate efficient production and distribution plans, taking into account different limitations and goals. Businesses can implement real-time adaptable plans that automate warehouse operations, expedite order processing, minimize labor expenses, and boost overall effectiveness (Pournader, Ghaderi, Hassanzadegan, & Fahimnia, 2021).

3.2.1 Demand Forecasting

Forecasting demand entails projecting future product demand. The goal of demand forecasting, which is usually expressed in sales, is to predict how many units of a certain product will be sold in each amount of time. Demand is the lifeblood of business; it is what keeps everything running. Within organizations, estimating the required supplies and parts is a critical duty that has a significant impact on total costs (Husna, Amin, & Shah, 2021). Businesses will find it difficult to decide on marketing tactics, output levels, workforce needs, and other critical areas of their operations without a thorough understanding of demand.

3.2.2 Inventory Optimization

Inventory optimization involves strategically overseeing and regulating stock quantities to enhance productivity, reduce expenses, and fulfil customer needs. Maintaining excessive inventory levels and experiencing stockouts can result in significant financial losses, but AI technology offers a solution to mitigate these issues. By optimizing inventory levels, businesses can better allocate resources, lower storage expenses, and ultimately enhance their profitability.

3.2.3 Supplier Risk Assessment

With an increasing amount of research in contemporary AI methodologies, artificial intelligence is now capable of analysing a wide range of structured and unstructured data to categorize various sources of risk according to their level of seriousness, forecast the probability of risk events detect irregularities, and anticipate the potential consequences of risks (Schlegel & Trent, 2014). This, in turn, empowers businesses to develop proactive strategies for supply chain risk management. Artificial Intelligence (AI) is transforming supplier risk assessment by offering advanced tools to evaluate and manage potential risks in the supply chain. With AI, companies can more accurately identify, assess, and mitigate risks associated with their suppliers, ensuring a more resilient and reliable supply chain.

4. INTEGRATION OF IOT AND AI FOR REAL-TIME MONITORING

4.1 Technological Synergies Between IoT and AI

The convergence of Internet of Things (IoT) and Artificial Intelligence (AI) creates powerful synergies that significantly enhance various applications and industries. By integrating AI with IoT devices, businesses can harness the full potential of both technologies, leading to smarter, more efficient, and innovative solutions.

4.1.1 Data Collection and Analysis

The integration of Internet of Things (IoT) and Artificial Intelligence (AI) in supply chain management (SCM) revolutionizes data collection and analysis, providing unparalleled insights and efficiencies. By leveraging IoT devices and AI technologies, businesses can optimize their supply chain operations and enhance decision-making processes.

IoT sensors and devices collect real-time data from various points in the supply chain, including manufacturing, warehousing, transportation, and retail reducing the need for manual input and minimizing human error. This diverse data set provides a comprehensive view of the entire supply chain, from raw materials to end customers. This continuous monitoring provides accurate and up-to-date information on inventory levels, equipment status, and environmental conditions. AI algorithms process and analyze the vast amounts of data collected by IoT devices. Advanced analytics techniques, such as machine learning and deep learning, identify patterns, trends, and anomalies in the data, providing valuable insights for supply chain optimization. This leads to increased efficiency and reduced operational costs.

4.1.2 Edge Computing and Cloud Computing

Edge computing and cloud computing are two distinct yet complementary paradigms in the realm of data processing and storage, each with its unique advantages and applications. Understanding the differences and synergies between them is crucial for optimizing technology strategies in various industries.

Edge computing lowers latency and speeds up data processing by processing data closer to its source, such as local servers or Internet of Things devices. Additionally, it reduces the quantity of data that must be sent to cloud or central servers. Consequently, security and privacy are enhanced. Edge computing offers resilience against network outages or disturbances since it can function without the cloud. This guarantees that crucial programs will always run, even in distant or erratic network situations.

Figure 6. Edge computing and cloud computing

As shown in Figure 6, cloud computing provides global access to data and applications, supporting remote work and collaboration. This accessibility is crucial for businesses with geographically dispersed teams and customers. Combining edge and cloud computing can offer the best of both worlds. Data can be processed at the edge for real-time applications and then sent to the cloud for long-term storage and further analysis. This hybrid approach enhances efficiency and optimizes resource utilization.

4.2 Real-Time Data Processing and Decision Making

Real-time data processing pertains to the immediate ingestion and analysis of incoming data points. Valuable insights gained from these are used to take effective decisions.

4.2.1 AI Algorithms for Real-Time Insights

Artificial Intelligence (AI) algorithms play a crucial role in providing real-time insights, enabling businesses to make quick, informed decisions based on live data. These algorithms process and analyze data as it is generated, offering immediate feedback and actionable information.

To successfully implement AI algorithms for real-time insights businesses should ensure high quality, accurate and timely data, reliable and scalable infrastructure, employ advanced AI tools and frameworks that support real-time data processing followed by continuous model training and updates. Foster collaboration between data scientists, engineers, and business stakeholders to ensure that AI implementations align with business goals and deliver actionable insights. AI algorithms are instrumental in deriving real-time insights from data, enabling businesses to respond swiftly and effectively to emerging trends and issues. By leveraging these algorithms, companies can enhance their decision-making processes and maintain a competitive edge in a fast-paced environment.

5. APPLICATIONS AND USE CASES

5.1 Real-Time Asset Tracking

The Internet of Things (IoT) has revolutionized many aspects of modern life, particularly in the realm of asset tracking. Real-time asset tracking using IoT technologies enables businesses to monitor the location, status, and condition of their assets seamlessly and accurately.

5.1.1 GPS and RFID Technologies

The technology of the Global Positioning System (GPS) is now essential to location-based services and navigation. A network of satellites carries out GPS operations by sending signals to ground-based GPS receivers. The receivers can pinpoint their exact location, including latitude, longitude, and altitude, with amazing accuracy by triangulating these signals. It is an essential part of the Internet of Things (IoT), giving vital location information for a variety of uses. Transportation is one of the main industries that has benefited from the Internet of Things (IoT) and Global Positioning System (GPS) technology integration (Moumen, Rafalia, Abouchabaka, & Aoufi, 2023). As a result, the data gathered by sensors and devices is enhanced with exact position data, thus improving the performance of IoT systems.

5.1.2 Case Study: Fleet Management

Fleet management software serves as a vital centralized database, enabling fleet managers to effectively oversee, arrange, and synchronize vehicles. Additionally, it provides managers with a comprehensive overview of their fleet and its activities. Fleet management has seen significant improvements with the integration of the Internet of Things (IoT). IoT-enabled fleet management systems provide real-time data on vehicle location, condition, and performance, enabling companies to optimize operations, enhance safety, and reduce costs.

5.2 Inventory and Warehouse Management

Warehouse management involves overseeing inventory and associated operations within a warehouse, while inventory management focuses on controlling stock levels for the entire business and predicting business trends.

5.2.1 Smart Warehouses

Smart warehouses leverage advanced technologies, including the Internet of Things (IoT) to optimize operations, enhance efficiency, and improve accuracy in inventory management. These technologies enable real-time monitoring, automation, and data-driven decision-making, transforming traditional warehouses into intelligent, interconnected hubs.

Smart warehouses help in achieving enhanced efficiency, improved inventory accuracy, optimized space utilization, reduced operational costs and better decision making. We envision future smart warehouses as automated, meaning that robots will be responsible for a significant portion of the physical material handling. And, on the other hand digital which use integrated information platforms that aid in warehouse management (Kembro & Norrman, 2022).

5.2.2 Automated Replenishment Systems

Technology is used by automated replenishment systems to control inventory levels and automatically place new orders for merchandise when it falls below a set level. These systems are essential for making sure that companies keep the right amount of inventory on hand, cut down on stockouts, and limit surplus inventory, all of which increase customer satisfaction and operational effectiveness. The majority of replenishment systems are powered by AI algorithms, which improve forecasting at the store level and lead to more product availability (Avlijas, Vukanovic Dumanovic, & Radunovic, 2021). Real-time data on inventory levels, sales rates, and demand patterns are gathered by these systems, which then use sophisticated algorithms to analyze the information while taking lead times, sales history, and seasonal trends into account. The system automatically creates purchase orders and distributes them to vendors when inventory levels drop below the set threshold. This guarantees that inventory is refilled automatically and on schedule. In order to track shipments, manage delivery schedules, and give visibility into inventory status, the system also connects with the systems of suppliers. Delays are decreased and supply chain activity is better synchronized thanks to this integration.

5.3 Transportation and Logistics

The transportation of goods involves the physical movement, while logistics encompasses the coordination and oversight of goods moving between the producer and the consumer.

5.3.1 Route Optimization

Route optimization in supply chain management (SCM) involves the use of advanced technologies and algorithms to determine the most efficient paths for transporting goods from one location to another. This process is crucial for minimizing transportation costs, reducing delivery times, and improving overall supply chain efficiency. The system gathers data from various sources, including GPS devices, traffic reports, weather forecasts, and historical delivery data. This information provides a comprehensive view of current and potential conditions affecting routes. Advanced algorithms analyze the collected data to evaluate different routing options. Factors considered include distance, traffic congestion, road conditions, fuel consumption, and delivery windows. This, is followed by dynamic routing, which adjusts routes in real-time based on changing conditions, such as traffic jams or road closures. This flexibility ensures that the most efficient route is always, even if conditions change unexpectedly. Route optimization tools integrate with other SCM systems, such as warehouse management and order management systems, to synchronize logistics activities and provide end-to-end visibility of the supply chain.

5.3.2 Real-Time Shipment Tracking

Real-time shipment tracking is an integral part of modern logistics and supply chain management. This technology provides up-to-the-minute information about the location and status of shipments, enabling businesses to enhance efficiency, improve customer satisfaction, and respond quickly to potential issues. Real-time shipment tracking systems use GPS technology integrated with IoT devices. These devices are attached to shipments and continuously transmit location data. The data collected by IoT devices is transmitted to a central platform via cellular networks, satellite communication, or other wireless technologies. The tracking data is aggregated and displayed on a centralized dashboard. This allows logistics managers to monitor the real-time status and location of all shipments in transit. Advanced analytics tools process the tracking data to provide insights into shipment performance, route efficiency, and delivery times.

5.4 Quality Control and Compliance

Quality control and compliance ensure that products and services meet predefined standards and regulatory requirements. These processes are essential for maintaining product quality, customer satisfaction, and adherence to industry regulations.

5.4.1 Environmental Conditional Monitoring

Environmental condition monitoring in supply chain management involves the continuous assessment and management of factors such as temperature, humidity, and air quality throughout the supply chain process. By utilizing sensors and IoT technology, businesses can ensure that environmental conditions are maintained within specified parameters to safeguard the quality and integrity of goods during storage, transportation, and distribution. The majority of supply chain sustainability efforts seem to focus on overseeing suppliers to assess their compliance with environmental sustainability practices as required by regulations or the expectations of the main company (Marshall, McCarthy, Heavey, & McGrath, 2015).

5.4.2 Automated Compliance Reporting

Automated compliance checking methods have developed over a span of over 50 years, ultimately becoming commercially feasible. AI methodologies offer reliable translation of codes and standards; however, it is imperative to quality assurance procedures (Amor & Dimyadi, 2021). Automated compliance reporting involves the use of technology to streamline the process of compiling and submitting regulatory documentation. By leveraging automation tools and software, companies can efficiently generate accurate reports, ensuring adherence to industry regulations while minimizing manual effort and errors.

6. BENEFITS OF IOT AND AI INTEGRATION

6.1 Enhanced Visibility and Transparency

Enhanced visibility and transparency in supply chain management (SCM) are critical for optimizing operations and ensuring efficient, responsive supply networks. Visibility involves having real-time access to data and information across the entire supply chain, from suppliers to end customers. This comprehensive view allows companies to monitor inventory levels, track shipments, and anticipate potential disruptions. Transparency, on the other hand, refers to the openness and clarity with which information is shared among all stakeholders in the supply chain. By improving visibility and transparency, companies can better identify inefficiencies, reduce costs, and enhance collaboration with partners. These improvements lead to more accurate demand forecasting, improved inventory management, and quicker responses to market changes. Technologies such as the Internet of Things (IoT), blockchain, and advanced analytics play a crucial role in achieving these goals. IoT devices provide real-time data on the location and condition of goods, while blockchain ensures secure and verifiable transactions, and analytics offer insights for strategic decision-making.

6.2 Improved Efficiency and Cost Reduction

Utilizing IoT and AI in supply chain management (SCM) significantly enhances efficiency and reduces costs. IoT devices provide real-time tracking and monitoring of goods, allowing companies to manage inventory more effectively and reduce losses due to theft or damage. These devices offer valuable data on the condition and location of products, enabling precise logistics planning and reducing delays. AI, on the other hand, excels in analyzing large datasets to optimize various aspects of the supply chain. AI algorithms can predict demand more accurately, helping businesses to minimize overstocking and understocking issues. Machine learning models improve route planning for transportation, ensuring quicker delivery times and reducing fuel costs.

6.3 Proactive Issue Resolution and Risk Management

Proactive approaches encompass supplier development, supply chain contracts, process management, and supplier relationships. On the other hand, reactive approaches encompass contingency planning, disaster management, and demand management (Cheung, Bell, & Bhattacharjya, 2021). IoT devices provide real-time data on the status and location of goods, enabling companies to detect potential problems early. This constant monitoring helps identify issues such as delays, damage, or theft, allowing for swift corrective actions. AI complements this by analyzing the data collected from IoT devices to predict potential risks and disruptions. Machine learning algorithms can forecast demand fluctuations, supplier reliability, and transportation bottlenecks, helping companies to mitigate risks before they escalate. AI also supports predictive maintenance by identifying signs of equipment failure, ensuring timely repairs and avoiding costly downtimes.

6.4 Increased Customer Satisfaction

IoT devices provide real-time tracking and monitoring of products, giving customers up-to-date information on their orders. This transparency enhances trust and reliability, as customers can see the exact status and location of their purchases. AI further improves customer satisfaction by optimizing supply chain processes. AI algorithms analyse data to predict demand accurately, ensuring that products are available when customers need them. Additionally, AI-driven analytics enhance inventory management, reducing the likelihood of stockouts or overstocking. This efficiency translates to faster order fulfilment and more reliable delivery times, which are critical to customer satisfaction. Precise demand prediction guarantees timely availability of products for customers. Moreover, AI can personalize customer experiences by analysing purchasing behaviour and preferences, enabling companies to offer tailored recommendations and promotions. This degree of customization instils a sense of worth and comprehension in customers, thereby nurturing their loyalty.

7. CHALLENGES AND CONSIDERATIONS

7.1 Data Privacy and Security Concerns

Ensuring the privacy and security of data is crucial in the realm of supply chain management (SCM). Data privacy encompasses the collection, storage, retention, and transfer of data, whereas data security involves safeguarding data from unauthorized access, loss, or corruption.

7.1.1 Ensuring Data Integrity and Confidentiality

Maintaining data security and integrity is essential to supply chain management (SCM) operations' dependability and efficiency. Limit sensitive data access to only those who require it to carry out their job duties. To improve security, implement strong password policies and multi-factor authentication. Encrypt data when it's in motion and stationary. This guarantees that data cannot be read without the decryption key, even if it is intercepted or viewed without authorization. Conduct regular audits to ensure that data protection regulations are being followed. Unauthorized access and irregularities in data handling can be promptly identified and addressed with the use of continuous monitoring. Inform staff members of the value of confidentiality and data integrity. Best practices for managing sensitive data and spotting phishing scams and other online dangers should be covered in training. Maintain regular data backups and a well-defined recovery strategy. This reduces the possibility of data loss because of device malfunctions, cyberattacks, or other unforeseen circumstances. Use blockchain to take advantage of its built-in security features. A tamper-evident ledger, such as one provided by blockchain, can guarantee the transparency and immutability of supply chain transactions. Organizations can improve their data security posture and guarantee the reliability and integrity of their supply chain operations by incorporating these tactics into the SCM architecture.

7.1.2 Cybersecurity Measures

The quality of services in the logistics industry has been significantly enhanced by technological applications. Nevertheless, the utilization of internet-based technologies has expanded the vulnerability in the logistics sector. These cyber-attacks have the potential to detrimentally affect the performance of both logistics operations and the entire supply chain (Kembro & Norrman, 2022). In supply chain management (SCM), implementing robust cybersecurity measures is essential to protect sensitive data and maintain operational integrity. This can be achieved by implementing strict access control to ensure that only authorized personnel can access sensitive information. This can include multi-factor authentication (MFA), role-based access controls (RBAC), and strong password policies. Employ encryption to safeguard data while it is being transferred and when it is stored. Encryption guarantees that in the event of unauthorized access or interception, the data will remain indecipherable without the necessary decryption keys. It is recommended to leverage blockchain technology to improve the security and transparency of supply chain transactions.

7.2 Integration and Interoperability Issues

Effective supply chain management (SCM) requires seamless integration and interoperability across various systems and processes. However, several challenges can impede this such as diverse systems and technologies.

7.2.1 Legacy Systems Compatibility

Supply chains often involve multiple stakeholders using different systems and technologies. This diversity can lead to compatibility issues, making it difficult to integrate data and processes across the supply chain. Inconsistent data formats and standards across different systems can cause significant integration problems.

7.2.2 Standardization Efforts

The integration of Internet of Things (IoT) and Artificial Intelligence (AI) in supply chain management (SCM) is a transformative trend that enhances efficiency, visibility, and decision-making processes. Standardization in this context is crucial to ensure interoperability, data accuracy, and seamless communication across diverse systems and stakeholders.

Establishing common data protocols and formats ensures that information from different IoT devices can be aggregated and analyzed cohesively. For example, standards like EPCIS (Electronic Product Code Information Services) help in tracking products through the supply chain. Ensuring that IoT devices from different manufacturers can communicate with each other and with centralized systems. These standards must cover data format, communication, and security to guarantee a cohesive and interoperable IoT ecosystem (Sallam, Mohamed, & Mohamed, 2023). AI technologies, such as machine learning, predictive analytics, and robotics, are utilized to optimize supply chain operations. AI systems rely on high-quality data. Standardizing data collection, storage, and processing ensures that AI algorithms function correctly. ISO 8000-61, for instance, provides guidelines for data quality management. Establishing standards for AI algorithm transparency and ethical use is crucial to ensure fairness and accountability. ISO/IEC JTC 1/SC 42 works on standardizing AI technologies, addressing issues like bias and explainability. The standardization of IoT and AI in supply chain management is essential for achieving interoperability, enhancing data quality, and ensuring secure and ethical use of technology.

7.3 Scalability and Infrastructure Requirements

The objective of a scalable supply chain is to enhance the quantity while upholding or enhancing its effectiveness, thus diminishing the costs per unit of the supply chain.

7.3.1 Network Bandwidth and Latency

In supply chain management (SCM), network bandwidth and latency are critical factors that influence the efficiency and reliability of communication and data transfer across the supply chain. These aspects are particularly important as supply chains become increasingly digital and reliant on real-time data and interconnected systems. Network bandwidth refers to the maximum rate at which data can be transferred over a network. High bandwidth is crucial for handling large volumes of data generated by various sources within the supply chain, including IoT devices, enterprise resource planning (ERP) systems, and communication platforms. High bandwidth enables the rapid transfer of large datasets, which is essential for real-time inventory management, order processing, and logistics coordination. Applications such as remote monitoring and telematics require substantial bandwidth to stream video and transmit data without interruption.

Latency refers to the delay before a transfer of data begins following an instruction for its transfer. Low latency is crucial for ensuring that data exchanges happen quickly and efficiently, which is vital for time-sensitive supply chain operations. ow latency is essential for real-time tracking and monitoring of goods. This allows for immediate responses to any issues such as delays or deviations in the supply chain. In automated warehouses and manufacturing plants, low latency is critical for the real-time control of robots and machinery, ensuring synchronized and efficient operations. Fast access to accurate and timely data is necessary for decision-making processes. Low latency helps in minimizing the time lag between data collection, analysis, and action.

7.3.2 Cloud vs. Edge Computing

Cloud computing and edge computing offer distinct advantages and serve different purposes in modern IT infrastructure. Cloud computing involves delivering a range of services, such as storage, databases, and software, over the internet.

These services, provided by companies like Amazon Web Services, Microsoft Azure, and Google Cloud Platform, are highly scalable and cost-effective, allowing businesses to pay only for what they use. The cloud's accessibility enables remote work and collaboration, while the responsibility for maintaining and updating the infrastructure rests with the service provider. However, cloud computing can introduce latency because data must travel to and from remote data centres, which can be a drawback for time-sensitive applications. Additionally, concerns about data security and compliance, as well as dependency on reliable internet connectivity, are significant considerations.

Edge computing, in contrast, processes data closer to where it is generated, such as on local devices or edge servers. This proximity significantly reduces latency, making it ideal for real-time applications like autonomous vehicles and smart grids. By minimizing the need for data transmission over long distances, edge computing also conserves bandwidth and enhances security by keeping data local. It ensures reliability, as local processing can continue even if the connection to the central cloud is lost. However, managing numerous edge devices can be complex and scaling this infrastructure is often more challenging and costly compared to cloud services. Additionally, edge devices require regular maintenance and updates, which can increase operational overhead. Nowadays, edge computing and cloud computing are more and more considered as technologies that complement each other, allowing for the development of more effective, adaptable, and scalable computing solutions (George, George, & Baskar, 2023).

8. FUTURE TRENDS AND INNOVATION

8.1 Advances in IoT and AI Technologies

The Internet of Things (IoT) is progressing swiftly with technological advancements that have revolutionized our lifestyles and professional environments. Furthermore, the integration of artificial intelligence (AI) and machine learning (ML) in IoT devices represents another significant leap forward. By leveraging AI and ML algorithms (Archana, Banu, Prasad, & Shrivastava, 2023), IoT devices are empowered to analyze data and generate predictions by identifying patterns and trends.

8.1.1 Development of 5G Networks

The development of 5G networks marks a significant leap forward in wireless communication technology, promising to revolutionize connectivity with unprecedented speed, capacity, and reliability. The implementation of 5G technology in the manufacturing sector enables the acceleration of production processes and enhances efficiency through the utilization of industrial technology (Maulani & Johansyah, 2023). This fifth-generation mobile network builds upon the advancements of its predecessors, aiming to meet the growing demands for data and connectivity in an increasingly digital world. 5G networks have the capacity to download data at rates of up to 10 gigabits per second, making them extremely fast. Higher frequency bands called millimeter waves, which have a greater data carrying capacity than the lower frequency bands utilized in earlier generations, are used to accomplish this. But because these higher frequencies are also more susceptible to physical interference and have shorter ranges, it is necessary to set up a dense network of tiny cells to provide all-encompassing coverage.

8.1.2 AI-Driven Predictive Analytics

AI-driven predictive analytics is transforming the way organizations make decisions by leveraging advanced algorithms and machine learning to forecast future trends and behaviours. This powerful approach utilizes vast amounts of data to identify patterns, enabling businesses to anticipate outcomes and make proactive, data-driven decisions. The core of AI-driven predictive analytics lies in its ability to process and analyze large datasets from various sources, such as transactional records, customer interactions, social media, and IoT devices. Machine learning algorithms, including regression analysis, neural networks, and decision trees, are employed to detect patterns and relationships within this data. These models are continuously refined as they learn from new data, improving their accuracy over time. One of the primary benefits of AI-driven predictive analytics is its ability to enhance decision-making across multiple industries. Retailers utilize it to predict consumer behaviour, product launch success rates, optimize inventory, and personalize marketing strategies (Shaik & Siddque, 2023).

8.2 Emerging Use Cases

Emerging use cases refers to that different advancements and addons that can be incorporated in the supply chain.

8.2.1 Blockchain for Supply Chain Transparency

Blockchain is recognized as a series of encrypted data blocks containing valuable information that is securely locked and can only be accessed by those with the appropriate key. Access to blockchain is limited to devices such as computers, laptops, and servers that are connected to the internet (Manifavas & Karamitsos,

2022). Blockchain plays a vital role in SCM because of is distributed secure technology. It is increasingly being recognized for its potential to enhance transparency in supply chain management. By providing a secure, decentralized ledger, blockchain ensures that all transactions and data exchanges within the supply chain are recorded in a verifiable and immutable manner. This capability addresses many challenges associated with traditional supply chain systems, such as fraud, inefficiency, and lack of visibility. At its core, blockchain operates as a distributed database where each participant in the supply chain has access to a single source of truth. Every transaction is recorded in a block and added to a chain of previous transactions, creating a comprehensive and chronological history of the product's journey.

The key benefits of using blockchain includes enhance traceability, improved efficiency by automating and streamlining the process which reduces manual intervention, greater visibility as all stakeholders can access real time information and the decentralized nature of blockchain provides enhanced security.

8.2.2 Autonomous Vehicles and Drones

Autonomous vehicles and drones are transforming supply chain management (SCM) by enhancing efficiency, reducing costs, and improving delivery times. These technologies leverage advanced automation and artificial intelligence to revolutionize transportation and logistics. Autonomous vehicles, including self-driving trucks and delivery vans, are equipped with sensors, cameras, and AI systems that enable them to navigate roads without human intervention. These vehicles offer several advantages in SCM such as cost reduction by minimizing the need for human drivers, also these vehicles are capable of operating 24/7 without the need for breaks, leading to high productivity. Autonomous vehicles can optimize routes in real-time, reducing fuel consumption and delivery times. They can also communicate with each other to avoid traffic congestion and accidents, further enhancing efficiency. With advanced sensors and AI, autonomous vehicles can detect and respond to hazards more quickly than human drivers, potentially reducing the number of accidents and improving overall road safety.

8.3 The Future of Real-Time Supply Chain Monitoring

The supply chain industry is currently undergoing a significant transformation. According to a study conducted by Deloitte, 79% of companies with top-performing supply chains outperform the average revenue growth in their respective sectors. This data highlights the clear link between efficient supply chain management and overall business prosperity.

8.3.1 Vision for a Fully Automated Supply Chain

The vision for a fully automated supply chain encompasses the seamless integration of advanced technologies to enhance efficiency, accuracy, and agility from end to end (Huhns & Stephens, 2001). In this futuristic model, automation technologies such as artificial intelligence (AI), the Internet of Things (IoT), robotics, and blockchain are pivotal.

8.3.2 Impact on Global Supply Chain Dynamics

The implementation of automation in global supply chains is transforming how businesses operate, offering profound impacts on efficiency, cost-effectiveness, and responsiveness. Automation technologies, such as robotics and AI, streamline various supply chain processes. Automated systems can operate 24/7 without fatigue, increasing productivity and reducing the time needed for tasks like inventory management, order processing, and shipping. This leads to faster turnaround times and the ability to handle larger volumes of goods more effectively. Also, automation enhances precision in supply chain operations. Technologies like AI and machine learning can predict demand more accurately, manage inventory levels with greater precision, and ensure that orders are fulfilled correctly. This accuracy reduces the incidence of stockouts, overstocking, and order discrepancies. Automated supply chains can quickly adapt to changes in demand and market conditions. Real-time data analytics and machine learning enable businesses to respond promptly to disruptions, shifts in consumer preferences, or changes in supply availability (George, George, & Baskar, 2023). This agility ensures that companies remain competitive and can meet customer expectations more effectively. On the other hand, this helps in identifying potential risks and vulnerabilities within the supply chain (Dadsena, Mathiyazhagan, & Taghipour, 2021; Glaa, Zoghlami, & Taghipour, 2014; Loivet, Taghipour, & Kang, 2020; Lu & Taghipour, 2023; Mahfod, Canel-Depitre, & Taghipour, 2019; Merimi & Taghipour, 2021). Predictive analytics can forecast disruptions, such as supplier delays or transportation issues, allowing businesses to take proactive measures to mitigate these risks. This capability ensures continuity and resilience in supply chain operations.

9. CASE STUDIES AND PRACTICAL EXAMPLES

9.1 Case Study: Integration of IoT and AI in a Major Retail Supply Chain

The integration of the Internet of Things (IoT) and Artificial Intelligence (AI) within a major retail supply chain illustrates how these technologies can revolutionize operations, enhance efficiency, and improve customer satisfaction. Here, we examine the case of a global retail giant that successfully implemented IoT and AI to optimize its supply chain processes.

A leading global retailer, facing challenges such as inventory management inefficiencies, high operational costs, and evolving customer expectations, decided to integrate IoT and AI technologies into its supply chain. The goal was to achieve real-time visibility, predictive analytics, and automated decision-making capabilities.

Integrating IoT and AI with SCM also comes with challenges such as data security and privacy, compatibility with existing legacy systems, workforce training and adaption. However, this integration leads to a significant reduction in operational costs, improved inventory accuracy, and enhanced customer satisfaction. The success of this initiative has positioned the retailer as a leader in supply chain innovation. Moving forward, the retailer plans to explore further advancements in AI and IoT, such as the use of autonomous delivery vehicles and advanced robotics, to continue optimizing its supply chain operations.

9.2 Case Study: Real-Time Monitoring in Pharmaceutical Supply Chains

The integration of real-time monitoring in pharmaceutical supply chains has become essential for ensuring the quality, safety, and efficiency of drug distribution. This case study examines how a leading pharmaceutical company leveraged IoT and real-time data analytics to enhance its supply chain operations. The Internet of Things (IoT) revolutionizes the existing pharmaceutical industry through the provision and automation of pharmaceutical manufacturing processes in different areas including supply chain management, drug discovery, remote patient access, and monitoring (Jyothy & Sarvagya, 2022). A major pharmaceutical company, known for producing and distributing a wide range of medications, faced challenges related to maintaining the integrity and security of its products throughout the supply chain. The need for compliance with stringent regulatory requirements and the desire to reduce losses due to spoilage and theft led the company to implement a real-time monitoring system.

The key benefits of incorporating a real time monitoring in pharmaceuticals include ensuring product integrity by monitoring temperature and humidity in real-time, the company ensured that medications were stored and transported under optimal conditions. This reduced the risk of spoilage and maintained the efficacy of the drugs, ultimately protecting patient safety.

9.3 Lessons Learned from Implementation

The implementation of Internet of Things (IoT) and Artificial Intelligence (AI) in supply chain management (SCM) offers numerous benefits, but it also comes with its own set of challenges and learning experiences.

9.3.1 Key Success Factors

Successfully implementing AI and IoT in supply chain management (Mbiatem, Taghipour, & Canel-Depitre, 2018; Nadia, Beatrice, & Atour, 2021; Taghipour, Foukolaei, Ghaedi, & Khazaei, 2023; Taghipour, Padash, Etemadi, Khazaei, & Ebrahimi, 2024; Taghipour, 2023; Taghipour, Lu, & Smahi, 2022; Taghipour & Merimi, 2021; Kingsly & Archana, 2025; Taghipour & Frayret, 2011) (SCM) can lead to significant improvements in efficiency, accuracy, and overall performance. It is essential to carefully plan and execute to attain these advantages. A well-defined strategic vision is imperative for success. Understand the specific goals you want to achieve with AI and IoT, such as reducing costs, improving efficiency, or enhancing customer satisfaction. Align these goals with the overall business strategy to ensure cohesive implementation. Data is the foundation of AI and IoT implementations. Develop a comprehensive data strategy that encompasses data collection, storage, management, and analysis. Ensure data quality, integrity, and accessibility. Investing in data governance frameworks can help maintain high data standards. As the number of internet-connected devices grows, the likelihood of cyber-attacks also rises. Implement strong encryption, regular security updates, access controls, and continuous monitoring to safeguard sensitive supply chain data. Change management is crucial for transitioning to AI and IoT-driven supply chain operations. Develop a change management plan that addresses potential resistance, communicates benefits clearly, and provides support throughout the transition. Engaging employees in the process can help mitigate resistance.

9.3.2 Common Challenges and Solutions

Efficiently overseeing stock levels necessitates a well-defined system, skilled personnel, and advanced technology. This involves coordinating the entire supply chain, starting from the suppliers' facilities all the way to the customers' receiving areas (Deveshwar & Rathee, 2010). Implementing IoT and AI in supply chain management (SCM) comes with various challenges, but addressing them effectively can lead to significant improvements. One major challenge is data management. The vast amounts of data generated by IoT devices can be overwhelming, leading to issues with data quality, storage, and processing. To address this, companies must invest in robust data management systems that ensure the collection, storage, and analysis of high-quality data. Another challenge is cybersecurity. With more devices connected to the internet, the risk of cyber-attacks increases. Implementing strong encryption, regular security updates, and stringent access controls can help safeguard sensitive supply chain data.

10. CONCLUSION

10.1 Summary of Key Points

By combining artificial intelligence (AI) and the Internet of Things (IoT) with supply chain management (SCM), conventional systems are being transformed into intelligent, self-governing networks. In order to track items, monitor conditions, and ensure compliance, visibility and transparency are critical. IoT devices provide real-time data from multiple locations in the supply chain. AI uses this data to manage inventory, forecast maintenance requirements, and optimize routes—all of which greatly increase production and efficiency. IoT-powered predictive analytics aid in predicting malfunctions and planning maintenance, averting unplanned malfunctions. AI also reduces human error and expedites responses to market changes by automating routine choices and processes. Another advantage is an improved client experience, where more happiness and loyalty are fostered by precise delivery timings and specialized services. But there are issues that need to be resolved, like data security, interoperability, large investment costs, and the requirement for qualified personnel. Notwithstanding these challenges, supply chains may become more intelligent and responsive with the integration of IoT and AI in SCM, which is a critical requirement for contemporary enterprises.

10.2 The Future Outlook for IoT and AI in Supply Chain

The future of integrating Internet of Things (IoT) and Artificial Intelligence (AI) into supply chain management is promising, with potential to create highly efficient, autonomous, and intelligent systems. IoT devices will continue to enhance supply chain visibility, enabling real-time tracking of goods, monitoring environmental conditions, and ensuring compliance with standards. As these devices generate vast amounts of data, AI will play a crucial role in analysing this information to optimize routes, predict maintenance needs, and manage inventory levels more effectively. One significant future trend is the advancement of predictive analytics. With IoT sensors gathering continuous data, AI will increasingly predict equipment failures and schedule maintenance proactively, reducing downtime and extending the lifespan of machinery. AI's capacity for automating routine decisions, such as restocking and production adjustments, will further streamline operations and reduce human error.

10.3 Final Thoughts on Opportunities and Challenges

The integration of Internet of Things (IoT) and Artificial Intelligence (AI) in Supply Chain Management (SCM) presents a wealth of opportunities and a set of challenges that businesses must navigate to achieve maximum benefit.

Opportunities: IoT and AI offer unprecedented visibility and control over supply chains. Real-time tracking and monitoring enable companies to gain insights into their operations, leading to better decision-making and enhanced efficiency. Predictive analytics, powered by AI, can forecast demand, optimize inventory levels, and pre-empt equipment failures, thereby reducing costs and increasing productivity. Automated processes streamline operations, reduce human error, and improve response times to market fluctuations. Enhanced customer experiences, through more accurate delivery estimates and personalized services, can also drive customer satisfaction and loyalty.

Challenges: Despite the clear advantages, integrating IoT and AI into SCM is not without its challenges. Data security and privacy are major concerns, as the massive amounts of data generated by IoT devices can be vulnerable to breaches. Ensuring robust cybersecurity measures is essential. Interoperability between diverse IoT devices and systems poses another challenge, requiring standardized protocols and compatibility to ensure smooth integration. The significant initial investment required for implementing these technologies can be a barrier, necessitating a careful evaluation of the return on investment. Furthermore, there is a notable skill gap in the workforce; companies need to invest in training and development to build the necessary expertise in managing and analysing IoT and AI systems.

REFERENCES

Amor, R., & Dimyadi, J. (2021). The promise of automated compliance checking. *Developments in the built environment, 5*, 100039.

Archana, T., Banu, J. F., Prasad, S., & Shrivastava, P. R. (2023). A Flexible and Investigation Approach for Encrypted Features Space Using Neural Network. *Advances in Science and Technology (Owerri, Nigeria)*, 124, 754–761. DOI: 10.4028/p-nd03y2

Archana, T., & Stephen, R. K. (2024). The Future of Artificial Intelligence in Manufacturing Industries. In Industry Applications of Thrust Manufacturing: Convergence with Real-Time Data and AI (pp. 98-117). IGI Global. DOI: 10.4018/979-8-3693-4276-3.ch004

Avlijas, G., Vukanovic Dumanovic, V., & Radunovic, M. (2021). Measuring the effects of automatic replenishment on product availability in retail stores. *Sustainability (Basel)*, 13(3), 1391. DOI: 10.3390/su13031391

Banu, J. F., Stephen, R. K., Aditya, N., & Raaghav, L. D. (2024, February). Alzheimer's Disease Detection using Deep Learning Algorithm. In *2024 11th International Conference on Computing for Sustainable Global Development (INDIACom)* (pp. 1531-1536). IEEE. DOI: 10.23919/INDIACom61295.2024.10498963

Cauchois, R., Taghipour, A., Kang, D., Zoghlami, N., & Abed, M. (2017). Advanced solutions for a supply chain with stochastic information. *Journal of Advanced Management Science*, 5(1), 9–13. DOI: 10.18178/joams.5.1.9-13

Cheung, K. F., Bell, M. G., & Bhattacharjya, J. (2021). Cybersecurity in logistics and supply chain management: An overview and future research directions. *Transportation Research Part E, Logistics and Transportation Review*, 146, 102217. DOI: 10.1016/j.tre.2020.102217

Dadsena, K. K., Mathiyazhagan, K., & Taghipour, A. (2021). Analysis of barriers for the build the resilient supply chain networks post-COVID-19. In *Managing Supply Chain Risk and Disruptions: Post COVID-19* (pp. 79–89). Springer International Publishing. DOI: 10.1007/978-3-030-72575-4_7

Deveshwar, A., & Rathee, R. (2010). Challenges for supply chain management in today's global competitive environment. *International Review of Business Research Papers*, 6(2), 194–203.

George, A. S., George, A. H., & Baskar, T. (2023). Edge Computing and the Future of Cloud Computing: A Survey of Industry Perspectives and Predictions. *Partners Universal International Research Journal*, 2(2), 19–44.

George, A. S., George, A. H., & Baskar, T. (2023). Edge Computing and the Future of Cloud Computing: A Survey of Industry Perspectives and Predictions. *Partners Universal International Research Journal*, 2(2), 19–44.

. Glaa, B., Zoghlami, N., & Taghipour, A. (2014, May). Outsourcing and off-shoring healthcare services: A way to cut the costs and improve quality. In 2014 international conference on advanced logistics and transport (ICALT) (pp. 189-194). IEEE.

Huhns, M. N., & Stephens, L. M. (2001). Automating supply chains. *IEEE Internet Computing*, 5(4), 90–93. DOI: 10.1109/4236.939455

Husna, A., Amin, S. H., & Shah, B. (2021). Demand forecasting in supply chain management using different deep learning methods. In *Demand forecasting and order planning in supply chains and humanitarian logistics* (pp. 140–170). IGI Global. DOI: 10.4018/978-1-7998-3805-0.ch005

Jyothy, S. T., & Sarvagya, M. (2022, November). Supply Chain Management in Pharmaceutical Industry Using IOT. In *2022 IEEE North Karnataka Subsection Flagship International Conference (NKCon)* (pp. 1-7). IEEE.

Kembro, J., & Norrman, A. (2022). The transformation from manual to smart warehousing: An exploratory study with Swedish retailers. *International Journal of Logistics Management*, 33(5), 107–135. DOI: 10.1108/IJLM-11-2021-0525

Kingsly, S. R., & Archana, T. (2025). Understanding AI and Machine Learning Concepts. In *AI and Machine Learning Applications in Supply Chains and Marketing* (pp. 371–392). IGI Global.

Kingsly Stephen, R., Chandra Sekar, A., & Dinakaran, K. (2019). Sectional Transmission analysis approach for improved reliable transmission and secure routing in wireless sensor networks. *Cluster Computing*, 22(S2, Suppl 2), 3759–3770. DOI: 10.1007/s10586-018-2276-y

Kumar, S., Tiwari, P., & Zymbler, M. (2019). Internet of Things is a revolutionary approach for future technology enhancement: A review. *Journal of Big Data*, 6(1), 1–21. DOI: 10.1186/s40537-019-0268-2

Loivet, W., Taghipour, A., & Kang, D. S. (2020). The rise of green supply chain management: Between complexity and necessity. *J. Econ. Bus. Manag*, 8(1), 1–7. DOI: 10.18178/joebm.2020.8.1.604

Lu, X., & Taghipour, A. (2023). A French Guide for Businesses and Governments to Integrate the Concept of Sustainable Development Into Their Supply Chains. In *Government Impact on Sustainable and Responsible Supply Chain Management* (pp. 1–12). IGI Global. DOI: 10.4018/978-1-6684-9062-4.ch001

Mahfod, J., Canel-Depitre, B., & Taghipour, A. (2019). Quality function deployment-ELECTRE in supplier evaluation. Journal of Advanced Management Science, 7(4).

Manifavas, H., & Karamitsos, I. (2022). Blockchain in supply chain management. In *Heterogeneous Cyber Physical Systems of Systems* (pp. 61–94). River Publishers. DOI: 10.1201/9781003338390-4

Marshall, D., McCarthy, L., Heavey, C., & McGrath, P. (2015). Environmental and social supply chain management sustainability practices: Construct development and measurement. *Production Planning and Control*, 26(8), 673–690. DOI: 10.1080/09537287.2014.963726

Maulani, I. E., & Johansyah, C. A. (2023). The development of 5G technology and its implications for the industry. *Devotion: Journal of Research and Community Service*, 4(2), 631–635. DOI: 10.36418/devotion.v4i2.416

Mbiatem, J., Taghipour, A., & Canel-Depitre, B. (2018, August). Supplier selection approaches for decision makers. In *Proceedings of the 8th International Conference on Information Communication and Management* (pp. 108-112).

Merimi, M., & Taghipour, A. (2021). Accelerating the digitalization of the supply chain: An empirical research about COVID-19 crisis. In *Digitalization of decentralized supply chains during global crises* (pp. 1–24). IGI Global. DOI: 10.4018/978-1-7998-6874-3.ch001

Moumen, I., Rafalia, N., Abouchabaka, J., & Aoufi, M. (2023). Real-time GPS Tracking System for IoT-Enabled Connected Vehicles. In *E3S Web of Conferences* (Vol. 412, p. 01095). EDP Sciences. DOI: 10.1051/e3sconf/202341201095

Nadia, E. N., Beatrice, C. D., & Atour, T. (2021). The determinants of hotel room rates in Beirut: A hedonic pricing model. International Journal of Trade. *Economics and Finance*, 12(2), 33–42.

Pournader, M., Ghaderi, H., Hassanzadegan, A., & Fahimnia, B. (2021). Artificial intelligence applications in supply chain management. *International Journal of Production Economics*, 241, 108250. DOI: 10.1016/j.ijpe.2021.108250

Sallam, K., Mohamed, M., & Mohamed, A. W. (2023). Internet of Things (IoT) in supply chain management: Challenges, opportunities, and best practices. *Sustainable Machine Intelligence Journal*, 2, 3–1. DOI: 10.61185/SMIJ.2023.22103

Schlegel, G. L., & Trent, R. J. (2014). *Supply chain risk management: An emerging discipline*. CRC Press.

Shaik, M., & Siddque, K. Q. (2023). Predictive Analytics in Supply Chain Management using SAP and AI. *Journal of Computer Sciences and Applications*, 11(1), 1–6. DOI: 10.12691/jcsa-11-1-1

Taghipour, A. (Ed.). (2023). *Government Impact on Sustainable and Responsible Supply Chain Management*. IGI Global. DOI: 10.4018/978-1-6684-9062-4

Taghipour, A., Foukolaei, P. Z., Ghaedi, M., & Khazaei, M. (2023). Sustainable Multi-Objective Models for Waste-to-Energy and Waste Separation Site Selection. *Sustainability (Basel)*, 15(22), 15764. DOI: 10.3390/su152215764

Taghipour, A., & Frayret, J. M. (2011). Supply chain coordination planning: A review. Work paper, École polytechnique de Montréal.

Taghipour, A., Lu, X., & Smahi, M. (2022). Opportunities and challenges for society 5.0 in procurement. *Blockchain Technology in Supply Chain Management for Society*, 5(0), 91–100. DOI: 10.1201/9781003177432-9

Taghipour, A., & Merimi, M. (2021, March). Digital transformation of supply chains during crisis: covid-19. In 11th Annual International Conference on Industrial Engineering and Operations Management, IEOM (pp. 7493-7501). DOI: 10.46254/AN11.20211298

Taghipour, A., Padash, A., Etemadi, V., Khazaei, M., & Ebrahimi, S. (2024). Sustainable and Circular Hotels and the Water–Food–Energy Nexus: Integration of Agrivoltaics, Hydropower, Solar Cells, Water Reservoirs, and Green Roofs. *Sustainability (Basel)*, 16(5), 1985. DOI: 10.3390/su16051985

Taj, S., Imran, A. S., Kastrati, Z., Daudpota, S. M., Memon, R. A., & Ahmed, J. (2023). IoT-based supply chain management: A systematic literature review. *Internet of Things : Engineering Cyber Physical Human Systems*, 24, 100982. DOI: 10.1016/j.iot.2023.100982

Thirugnanam, A., & Jahir Hussain, F. B. (2023). Exploring Machine Learning Algorithms for the Prediction of Dengue: A Comprehensive Review. *Revue d'Intelligence Artificielle*, 37(5), 1281–1290. DOI: 10.18280/ria.370521

Chapter 16
Integrating Markov Models in Supply Chain Management Exploring Human and Ecological Dimensions

M. Gowsalya
https://orcid.org/0000-0001-8756-1246
Vellore Institute of Technology, India

Delna Kuriyakose
Vellore Institute of Technology, India

ABSTRACT

The increasing complexity of global supply chains necessitates advanced analytical tools to address uncertainty, optimize decision-making, and incorporate sustainability goals. This chapter explores the application of Markov models in supply chain management, with a particular focus on their integration into human and ecological dimensions. Markov decision processes (MDPs) are examined for their ability to optimize resource allocation and dynamic decision-making under uncertainty, while hidden Markov models (HMMs) are highlighted for their utility in understanding latent state dynamics. Finally, Markov chain Monte Carlo (MCMC) methods are discussed for their robust probabilistic modeling capabilities, particularly in scenarios involving high-dimensional data. The chapter also delves into the challenges of integrating these models into real-world supply chain systems. Finally, future research directions are outlined, emphasizing the role of AI-driven Markov frameworks in advancing sustainable, adaptive, and human-centric supply

DOI: 10.4018/979-8-3693-7478-8.ch016

chain solutions.

1. INTRODUCTION

As global supply chains grow increasingly intricate, the integration of Artificial Intelligence (AI) has emerged as a transformative force, revolutionizing operations and significantly enhancing efficiency. The interconnectedness of climate change and supply chain disruptions further underscores the complexity of these systems, posing challenges that directly impact the realization of the 2030 vision outlined in the United Nations' Sustainable Development Goals (SDGs). A spotlight on sustainability has intensified, revealing that nearly half (49%) of all companies now have corporate supply chain sustainability goals, while another 35% still lack such objectives. This is according to research from the Massachusetts Institute of Technology (MIT) and Center for Transportation and Logistics and the Council of Supply Chain Management Professionals (CSCMP). Automation and AI play a pivotal role in enabling Chief Supply Chain Officers (CSCOs) and their organizations to collect data, identify risks, validate documentation, and provide audit trails, even in high inflationary periods. These technologies also empower supply chains to manage carbon emissions, waste, energy, water consumption, and material utility effectively (United Nations AI for Good, 2024). Markov models, particularly Hidden Markov Models (HMMs), provide a probabilistic framework that complements AI in addressing these challenges. Their ability to forecast, model dynamic processes, and handle complex systems makes them well-suited for navigating the intricacies of modern supply chains. By incorporating these tools, organizations can achieve a more sustainable, resilient, and adaptive supply chain that aligns with ecological and human-centric priorities.

Despite advancements in AI-based supply chains, challenges such as carbon footprints, environmental degradation, and resource overuse persist, posing significant ecological concerns. Similarly, human dimensions face issues such as workforce displacement, which results in resource inefficiencies, and ethical considerations surrounding the handling of sensitive data, particularly private information like mental health records. While these challenges have been extensively examined within the realm of AI-based supply chains, their exploration through Markovian techniques remains limited, offering a novel perspective for addressing these complexities. We focus on precise modeling and coordination of operations under uncertainty, aligning with AI's capabilities to optimize processes, predict demand, and manage variability across decentralized systems. This explains the synergy between Markov models' probabilistic framework and AI's advanced data handling capabilities, ensuring resilience and efficiency in complex supply chain environments (Cauchois et al., 2017).

2. RELATED STUDIES

This section presents a detailed review of the literature on supply chain management (SCM).The analysis of barriers to building resilient supply chain networks post-COVID-19 provides insights into mitigating disruptions and enhancing adaptability (Dadsena et al., 2021). The integration of sustainability into supply chains is further explored through frameworks that align business and government objectives for sustainable development (Lu & Taghipour, 2023). Additionally, the rise of green supply chain management highlights the necessity of balancing complexity with environmental objectives (Loivet et al., 2020). These studies collectively emphasize strategies for resilience and sustainability in supply chain operations.

The acceleration of digitalization in supply chains during the COVID-19 crisis has been a pivotal area of research, identifying opportunities for enhanced operational efficiency (Merimi & Taghipour, 2021; Taghipour et al., 2023). Similarly, the digital transformation of supply chains during crises has demonstrated the potential for technological interventions to address disruptions effectively (Taghipour & Merimi, 2021). These contributions underline the role of digital technologies in modernizing and stabilizing supply chains under uncertain conditions.

The development of tools like Quality Function Deployment-ELECTRE for supplier evaluation ensures comprehensive decision-making processes in supply chains (Mahfod et al., 2019). Supplier selection methodologies, particularly for decision-makers, provide structured approaches to optimize supplier relationships (Mbiatem et al., 2018; Taghipour, 2014). Additionally, reviews on supply chain coordination planning highlight strategies to improve operational planning and collaboration across decentralized networks (Taghipour & Frayret, 2011; Taghipour & Frayret, 2013).

Sustainable multi-objective models for waste-to-energy and waste separation site selection present innovative approaches to integrating ecological goals into supply chain management (Taghipour et al., 2023). The concept of sustainable and circular hotels incorporates advanced resource management strategies, such as agrivoltaics and green roofs, to address the water–food–energy nexus (Taghipour et al., 2024). Furthermore, dynamic ecological footprint modeling enhances sustainability analysis in supply chain contexts (Taghipour & Beneteau-Piet, 2020).

The integration of government frameworks into supply chain strategies underscores the impact of policy on sustainable and responsible supply chain management (Taghipour, 2023). This extends to addressing challenges and opportunities in Society 5.0, where blockchain technology and AI enhance procurement processes (Taghipour et al., 2022) .These studies demonstrate the importance of aligning policy with technological advancements for long-term sustainability.

Research on smart transportation behavior during the COVID-19 pandemic highlights the adaptation of ride-hailing systems to maintain efficiency and safety (Taghipour et al., 2023). In logistics, the design for remanufacturing within reverse logistics processes offers practical solutions for improving circularity in supply chains (Taghipour et al., 2015). Moreover, advanced algorithms for decentralized operations planning optimize supply chain efficiency in logistics networks (Taghipour & Frayret, 2013).

The outsourcing and offshoring of healthcare services are shown to reduce costs and improve service quality, illustrating the potential for supply chain principles in non-industrial sectors (Glaa et al., 2014). The determinants of hotel room rates are explored using hedonic pricing models, showcasing applications of supply chain concepts in the service industry (Nadia et al., 2021; Taghipour, 2020). These studies reveal the versatility of supply chain frameworks across diverse sectors.

These studies underscores the significant role of Markov models in supply chain analysis, particularly in enhancing resilience, optimizing operations, and integrating sustainability. From modeling ecological footprints and waste-to-energy site selection to improving supplier evaluation and coordination, Markov models provide robust frameworks for addressing uncertainty and dynamic behaviors. However, gaps remain in comprehensively incorporating human and ecological dimensions into supply chain strategies. While some studies explore behavioral factors influencing decision-making and ecological sustainability, there is a need for deeper integration of these aspects to address long-term environmental challenges and the socio-economic impacts of supply chains. Future research should focus on bridging these gaps by leveraging advanced AI-driven Markov frameworks to balance technological efficiency with ecological and human-centered approaches, ensuring truly sustainable supply chains. In this Chapter we aim to:

1. Highlight the role of Markov models in supply Chain dynamics
2. Incorporate Human dimensions into supply Chain analysis
3. Examine ecological dimensions in Sustainable Supply chains

3. FUNDAMENTALS OF MARKOV MODELS IN SCM

Markov chains are a mathematical framework consisting of a sequence of events in which the progression to a future state depends solely on the current state, with no influence from preceding states. (Privault, 2024). These models are particularly effective for capturing the dynamics of supply chain processes. In this context, Markov chains are characterized by states, which can represent various elements such as distribution nodes or product demand levels, and transitions, which reflect the probabilities of movement between these states. For instance, transitions may describe the likelihood of transferring products between locations or shifting from low-demand to high-demand conditions. These transition probabilities are organized within a Transition Probability Matrix, which serves as a structured representation of the likelihood of moving from one state to another, enabling a comprehensive analysis of supply chain behavior.

For Example:

States(S): $S = \{$ Low Demand, Moderate Demand, High Demand $\}$ Transition Probabilities:

$$P = \begin{bmatrix} 0.7 & 0.3 & 0.1 \\ 0.2 & 0.4 & 0.3 \\ 0.1 & 0.3 & 0.6 \end{bmatrix}$$

where the probability of transitioning from low demand to low demand is 0.7(being in same state), low demand to moderate demand is 0.3 and low demand to high demand is 0.1 likewise.

Steady-state analysis examines the dynamics of systems by focusing on characteristics that remain constant over time, offering insights into the long-term performance of supply chains under normal operating conditions. It evaluates the balance between resource inflow (e.g., production, procurement) and outflow (e.g., consumption, shipping), analyzing metrics such as inventory levels, lead times, and costs to optimize resource utilization and predict future trends. This analysis supports strategic decisions, such as determining optimal production capacities, maintaining sufficient inventory levels, and developing sustainable supply chain strategies. Absorption probabilities, on the other hand, model the likelihood of a supply chain reaching and remaining in specific states, such as persistent disruptions or stockouts, by assessing transitions between states under uncertainty. These probabilities provide a quantitative understanding of risks, aiding in the design of resilient supply chain networks capable of mitigating disruptions and ensuring operational continuity.

Markov chains can be effectively applied in various areas of supply chain management, including demand forecasting, inventory optimization, disruption prediction, capacity planning, supplier risk assessment, customer behavior analysis, and sustainability impact forecasting. These applications are enabled by advanced technologies such as big data analytics, artificial intelligence (AI), machine learning (ML), and predictive simulation models. By leveraging probabilities as their fundamental unit, Markov chains are well-suited to handle uncertainties and optimize processes such as logistics and transportation through sensitivity analysis. This approach provides actionable insights for proactive decision-making and strategic planning, enhances inventory management, minimizes waste, reduces transportation costs, and ensures timely delivery and product availability.

4. HUMAN DIMENSIONS IN SUPPLY CHAIN MANAGEMENT

Markov models are utilized to simulate workforce dynamics, such as skill development, attrition, and productivity levels, in the human aspects of AI-driven supply chains. Optimized workforce management solutions are made possible by their assistance in anticipating state transitions, such as staff promotions, resignations, or shifts in productivity. Markov models facilitate proactive decision-making by fusing AI and real-time data, improving supply networks' resilience and operational efficiency.In figure 1 we presented a workflow of integrating Markov models in Human dimensions and the detailed discussion is provided after.

Figure 1. Workflow in Human Dimensions

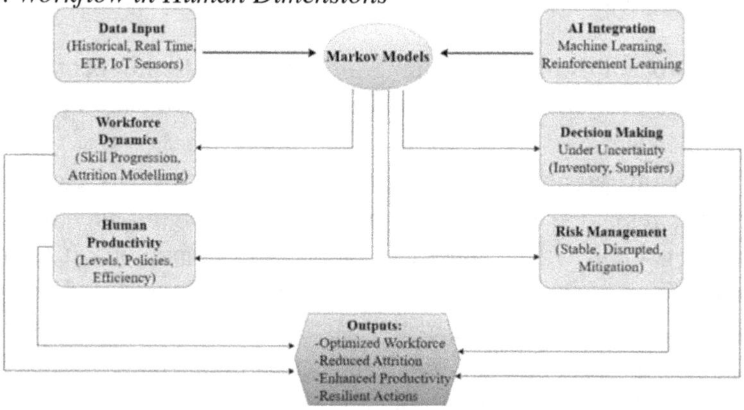

- **Workforce Dynamics:** Markov models are effective analytical tools for examining and forecasting workforce dynamics, including attrition, skill development, and productivity, by leveraging probabilistic transitions between defined states. These models provide valuable insights for workforce management and strategic long-term planning. In workforce attrition modeling, employees are classified into states such as active employment, resignation, retirement, or termination, with transition probabilities quantifying the likelihood of moving between these states, such as an active employee resigning or receiving a promotion. This approach enables the prediction of attrition rates and supports the development of hiring and retention strategies. For skill development and progression, skill levels such as novice, intermediate, expert, or mastery are treated as states, with transition probabilities reflecting the likelihood of advancement through training, mentoring, or experiential learning. These models assist in optimizing training programs to expedite skill growth and facilitate succession planning by identifying skill gaps and development trajectories. Overall, Markov models highlight areas for targeted interventions to enhance retention, skill development, and performance while reducing costs associated with turnover, training inefficiencies, and underperformance.

- **Decision-Making under Uncertainty:** Decision-making under uncertainty involves evaluating and forecasting the decisions of individuals or entities operating in environments characterized by unpredictability, such as supply chain operations. Factors such as demand variability, supplier reliability, and fluctuating market conditions contribute to this complexity. Human decisions ranging from order placements and inventory management to supplier selection are shaped by variables such as past experiences, risk preferences, and the availability of information. Behavioral modeling techniques, including game theory and decision trees, are employed to analyze these choices. Advanced methods, such as integrating Markov models with artificial intelligence (AI) and machine learning (ML) approaches like neural networks and reinforcement learning, enable real-time decision predictions by identifying patterns in historical data. These predictive capabilities help mitigate inefficiencies arising from misaligned decisions or inadequate communication, enhancing operational efficiency and strategic alignment in supply chain processes.

5. ECOLOGICAL DIMENSIONS IN SUPPLY CHAIN MANAGEMENT

Markov models play a crucial role in the ecological dimensions of AI-driven supply chains by predicting and managing changes in environmental factors, such as resource consumption, waste generation, and carbon emissions. They enable the evaluation of the long-term impacts of sustainability initiatives, including recycling efforts and the adoption of renewable energy sources. By integrating AI and real-time data, Markov models provide actionable insights to optimize resource efficiency, reduce environmental impact, and ensure compliance with evolving sustainability standards. In figure 2 we presented a workflow of integrating Markov models in Ecological dimensions and the detailed discussion is provided after.

- **Sustainability Goals:** Sustainability in supply chains focuses on reducing environmental impacts, such as carbon emissions and waste, while enhancing resource utilization. For carbon emissions reduction, operational states—such as low, medium, or high emissions—can represent different levels of environmental impact, with transitions reflecting changes brought about by interventions like adopting renewable energy sources, optimizing transportation routes, or improving fuel efficiency. This approach enables the forecasting of long-term impacts from transitioning to sustainable energy solutions. In waste reduction, states may represent various stages of waste generation (e.g. minimal, moderate, or high waste) or outcomes of waste management (e.g. recycled, reused, or disposed). Transitions model the probabilities of moving between these states, driven by practices such as recycling programs, process redesigns, or material recovery efforts. These models support the optimization of waste management systems by identifying inefficiencies, simulating the effects of sustainability initiatives, and facilitating data-driven decision-making. Additionally, Markov models identify pathways to achieve resource efficiency in material, energy, and water usage. By systematically applying Markov models, organizations can align environmental responsibility with operational efficiency, ensuring that sustainability goals are integrated into long-term strategic planning.

Figure 2. Workflow in Ecological Dimensions

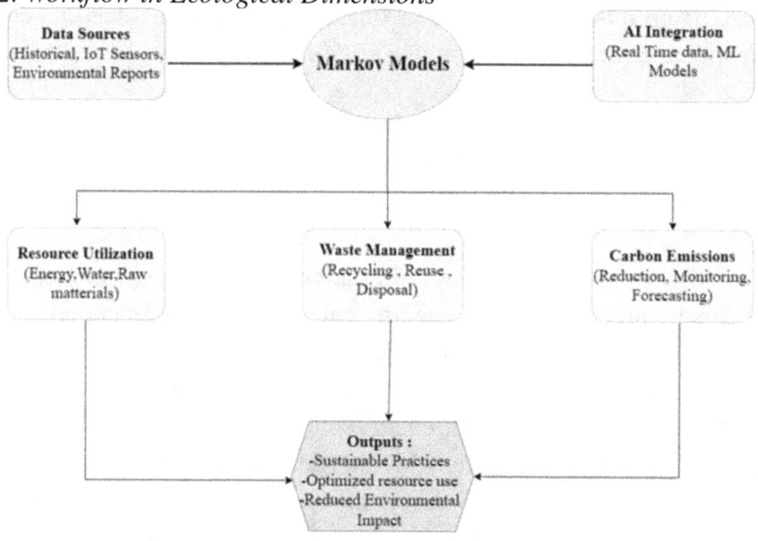

- **Environmental Uncertainty:** Environmental uncertainty in supply chains is driven by factors such as climate unpredictability, resource scarcity, and shifting policy landscapes. Markov models address these uncertainties by simulating probabilistic transitions between states, such as variations in resource availability, emission levels, or regulatory compliance. For example, transitions can represent shifts in resource conditions due to climate change impacts or operational adjustments necessitated by new environmental regulations. By modeling diverse scenarios, Markov models enable organizations to predict long-term outcomes, evaluate mitigation strategies, and adapt to evolving environmental challenges. This approach fosters proactive planning, ensuring that supply chain operations remain resilient and sustainable amid unexpected environmental disruptions.

6. ADVANCED TECHNIQUES AND EXTENSIONS

Markov models excel at handling time-dependent, sequential data, which is prevalent in supply chains. It also provides a clear structure for state transitions, making it easier to integrate with AI techniques like machine learning and reinforcement learning and can handle large-scale, multi-stage supply chain processes when combined with AI-driven computational power. Over time, various methodologies have been developed to integrate Markov Models and Artificial Intelligence (AI) into supply chain management. These methods leverage the probabilistic and sequential decision-making capabilities of Markov models, combined with the predictive and adaptive strengths of AI, to address dynamic challenges in supply chains. Some of the methodologies are mentioned below:

6.1 Markov Decision Processes (MDPs)

A Markov Decision Process (MDP) is a framework for decision-making in stochastic environments, where the outcome depends on the current state and action (Bellman, 1957; Das, 2021). It is defined by states (*S*), actions (*A*), transition probabilities P(s'|s,a), rewards R(s,a) and a discount factor (γ).The goal is to find an optimal policy ($\pi*$) that maximizes the expected cumulative reward:

$$G_t = \sum_{k=0}^{\infty} \gamma^k R\left(s_{t+k}, a_{t+k}\right)$$

as described in (Sutton & Barto, 1998). The Bellman equation is central to solving MDPs. It expresses the value of a state *s* under a policy π in terms of the value of successor states:The optimal value of a state satisfies the Bellman Optimality Equation:

$$V^*(s) = max\left[R(s,a) + \gamma \sum_{s' \in S} P(s'|s,a)\, V^*(s')\right]$$

where $V^*(s)$ is the Value function, representing the expected cumulative reward starting from state s and following policy π^* as introduced in (Bellman, 1957) and further elaborated in (Sutton & Barto, 1998). Solutions to MDPs can be found using techniques like dynamic programming, linear programming, or reinforcement learning (e.g., Q-Learning).

MDPs are mathematical frameworks that enhance AI-driven supply chain systems by enabling optimal sequential decision-making under uncertainty. In supply chain management, where decisions at one stage impact future states and outcomes, MDPs provide a structured way to model and optimize these processes.

The Markov decision process (MDP) provides a framework for addressing decision-making tasks by formulating them as structured problems. By integrating MDPs with methodologies such as dynamic programming, linear programming, and Q-learning, these problems can be effectively analyzed and solved. In 2010 this approach is particularly applied to reverse supply chains, promoting sustainability by ensuring their long-term functionality and enabling the completion of tasks within specified time constraints (Wongthatsanekorn et al., 2010). Later in 2024 a study introduced a framework that integrates sustainability goals and Industry 4.0 technologies into manufacturing systems using Markov Decision Processes. By leveraging real-time data and incorporating sustainability into decision-making, the framework supports strategic optimization and demonstrates its effectiveness through case studies (Sodachi et al., 2024).

6.2 Hidden Markov Models (HMMs)

A Hidden Markov Model (HMM) is a statistical model used to represent systems that have a sequence of observable outputs generated by an underlying sequence of hidden (unobservable) states. It extends the basic Markov chain by introducing observations that are probabilistically dependent on the hidden states (Kadhem, 2024; Team AAC, 2025) is defined by states (S),observations (O),transition probabilities (A),emission probabilities (B) and initial probabilities (π), collectively parameterized as $\lambda = (\pi, A, B)$.

The three core problems in HMMs are:

1. Computing the probability of an observation sequence ($P(O|\lambda)$)
2. Finding the most likely sequence of states (S^*)
3. Learning model parameters (λ^*)

The joint probability of states and observations is given by

$$P(S,O|\lambda) = \pi_{s_1} b_{s_1}(o_1) \prod_{t=1}^{T=1} a_{s_t s_{t+1}} b_{s_{t+1}}(o_{t+1})$$

as described in (Rabiner, 1989). Algorithms like forward (for likelihood), Viterbi (for decoding), and Baum-Welch (for learning) are used to solve HMM-related problems efficiently.

Hidden Markov Models (HMMs) are particularly effective for scenarios where hidden states represent factors like consumer demand or supplier behavior, and observable variables include product sales or system attributes like supply capacity and observation costs. In cases where true demand or supply availability remains unobservable, HMMs provide a robust framework by continuously updating state probabilities as new data becomes available, enabling adaptive decision-making in stochastic environments. Their ability to handle time-dependent processes makes them well-suited for periodic-review inventory models. A 2011 study addressed the unrealistic assumption of perfect information in inventory systems by introducing a partially observed random environment, modeled using an HMM-like framework. The study analyzed two single-item, periodic-review models: one incorporating random supply due to variability in production capacity and transportation availability, and another considering random availability with fixed-ordering costs. By leveraging partially observable states, the research demonstrated the optimality of state-dependent base-stock policies when supply factors are observable, and state-dependent (s, *S*) policies when the availability process is observable (Arifoğlu & Özekici, 2011).

The importance of effective logistics planning and management, particularly for a waste processing company where the size and management of toxic waste deliveries are critical. In 2022 Hidden Markov Models (HMMs) are utilized as a statistical tool in a study to model and predict supply levels, offering a more reliable alternative to classical methods in handling stochastic and partially observed processes. In this approach, the system is represented as a Markov process with hidden states and observable outputs, where the distribution of outputs from hidden states is modeled using a polynomial distribution (Borucka et al., 2022). By incorporating HMMs, the study provides a robust approach to manage inventory systems under imperfect information, ensuring optimal decisions in complex, real-world scenarios.

6.3 Markov Chain Monte Carlo (MCMC)

They use probabilistic sampling methods to simulate uncertainties in supply chains. Markov Chain Monte Carlo (MCMC) methods generate samples from a target distribution $P(x)$, which is often known up to a normalization constant $P(x) = \pi(x) / Z$, where $\pi(x)$ is the unnormalized probability density and $Z = \pi(x)dx$.We can construct a Markov Chain with a transition probability $T(x \rightarrow x')$ that ensures convergence to $P(x)$ as its stationary distribution (Gilks et al., 1995; Kubista et al., 2024). The detailed balance condition,

$$P(x)T(x \rightarrow x') = P(x')T(x' \rightarrow x)$$

guarantees this convergence introduced in (Metropolis et al., 1953) and further developed in (Hastings, 1970). Several algorithms implement MCMC principles, differing in how they define $T(x \rightarrow x')$.The most common are:

1. Metropolis-Hastings Algorithm
2. Gibbs Sampling

Markov Chain Monte Carlo (MCMC) methods play a pivotal role in enhancing sustainable, AI-driven supply chains by addressing uncertainties in renewable energy systems. A novel Markov-chain-based energy storage model optimizes power availability in photovoltaic systems by modeling charge/discharge dynamics and forecasting power output for effective grid integration (Song et al., 2012). Similarly, a simulation framework optimizes renewable energy usage for EV charging by mitigating risks from solar and wind variability, facilitating energy-efficient electromobility in urban settings (Szpytko & Salgado Duarte, 2020). Furthermore, MCMC combined with SARIMA demonstrates exceptional predictability in wind energy systems through time-series analysis, enabling reliable wind speed projections (Al-Duais & Al-Sharpi, 2023). Together, these studies highlight MCMC's versatility in managing variability and improving system reliability. The method excels at handling stochasticity, making it crucial for forecasting and operational planning in renewable energy. Its application spans optimizing storage, balancing energy supply-demand, and improving power output predictability. These capabilities are vital for designing resilient and efficient supply chains powered by renewables. By leveraging MCMC, supply chains can better integrate green energy sources to meet sustainability goals. This makes MCMC a cornerstone of future AI-driven solutions for sustainable energy and mobility.

Thus these are essential tools in AI-driven supply chains due to their ability to model uncertainty, forecast dynamic behaviors, and optimize decision-making processes. MCMC's strength in performing complex probabilistic analyses enables robust forecasting and optimization in uncertain supply chain environments. HMMs are particularly suited for applications such as demand forecasting, inventory management, and customer behavior analysis, as they provide valuable insights into latent state dynamics. MDPs support effective, adaptive, and sustainable supply chain operations by optimizing sequential decision-making processes, including resource allocation, routing, and scheduling, under conditions of uncertainty.

6.4 Machine Learning Integration

Integrating Markov models with Artificial Intelligence (AI) enhances supply chain adaptability to dynamic environments by combining probabilistic modeling with data-driven insights. Machine learning techniques, such as neural networks and reinforcement learning, can be utilized to forecast state transitions based on real-time data inputs, including demand fluctuations and supplier reliability (Chien et al., 2020). Markov models represent the current state of the supply chain and estimate the probabilities of transitioning to other states, such as from optimal inventory levels to stockouts. AI tools can leverage real-time data from sources like IoT sensors, enterprise resource planning (ERP) systems, supply chain management platforms, and external inputs such as weather forecasts or market trends to dynamically refine these transition probabilities (Aljohani, 2023). This integration enables supply chains to respond proactively to evolving conditions, ensuring greater resilience and operational efficiency.

Some of the AI techniques that can be used:

- Reinforcement Learning: Learn optimal policies for supply chain decisions by maximizing long-term rewards (e.g., minimizing costs or delays).
- Deep Learning: Extract complex patterns from large datasets to refine transition probabilities or improve scenario predictions.
- Natural Language Processing (NLP): Analyze unstructured data, such as news reports or policy documents, to assess external factors affecting supply chains.

For example, a retail supply chain might utilize a Markov model to forecast inventory states, such as surplus, optimal, or shortage. AI algorithms analyze real-time sales data alongside external factors, such as weather conditions or market trends, to dynamically update the Markov transition probabilities. This integrated system enables proactive adjustments to inventory replenishment schedules and shipment routes, optimizing stock levels and reducing delays. By incorporating both real-time and historical data, AI enhances the predictive accuracy of Markov models, improving their ability to forecast state transitions. Additionally, real-time updates and scenario analysis allow supply chains to swiftly adapt to disruptions or evolving conditions, while AI-driven Markov models identify strategies to minimize environmental impacts, such as reducing carbon emissions and optimizing resource utilization.

7. CHALLENGES

7.1 Theoretical Challenges

Assumptions: The Markov Property assumes that the future state depends solely on the current state, without considering the sequence of prior states. While this simplifies computations, it may fail to capture the complex dependencies in real-world supply chains, where transitions are often influenced by historical data, emerging trends, or external factors. Similarly, many Markov Models operate under the assumption of stationarity, where transition probabilities remain constant over time. However, in rapidly changing environments, such as disaster scenarios or volatile supply chains, external factors like fluctuating market conditions or weather patterns can significantly alter these probabilities. Additionally, the way states are defined plays a critical role in the model's accuracy. Oversimplifying states to reduce computational demands can lead to the omission of crucial details, whereas overly detailed states can result in excessively large and computationally burdensome models.

Limitations:

Computational Complexity: For large-scale systems, such as global supply chains with numerous nodes and transitions, the state-space can grow exponentially, making the model computationally intensive. This complexity can limit its practicality in real-time applications, particularly during emergencies where rapid decision-making is critical.

Data Dependency: Markov Models rely heavily on accurate transition probabilities derived from historical data. However, in disaster scenarios, reliable and comprehensive data may be limited or unavailable, which can undermine the model's reliability and predictive accuracy.

Limited Decision Scope: While Markov Decision Processes (MDPs) extend the model by incorporating decision-making capabilities, they may still fail to account for external factors or complex, non-linear interactions between different components of the supply chain. This can restrict their effectiveness in addressing more intricate system dynamics.

7.2 Practical Challenges

- *Data Availability and Quality*: Effective modelling depends on high-quality, real-time data, which can be difficult to obtain during disasters due to damaged infrastructure or delays in reporting. Additionally, historical data may not adequately capture unprecedented events, such as pandemics or major natural disasters, limiting the model's applicability in such scenarios.

- *Integration with AI-Driven Systems*: Integrating Markov Models with AI-driven systems, such as machine learning or predictive analytics, often faces challenges due to the need for seamless data integration. Legacy systems, data silos, and the lack of standardized processes within supply chain networks frequently hinder the successful implementation of such integrated solutions.
- *Adaptability to Non-Stationarity*: Real-world supply chains are inherently non-stationary, influenced by dynamic factors such as customer behaviour, market fluctuations, and climate change. While adaptive models exist, integrating these dynamic elements into Markov-based systems remains a significant computational and algorithmic challenge.
- *Scalability Challenges*: For large-scale logistics networks or extensive supply chains, the exponential growth of states and transitions can make real-time analysis impractical without considerable computational resources, posing scalability issues for widespread application.

8. FUTURE SCOPE

- Advancing Markov Models for AI Supply Chains:

Advancing Markov Models for AI-driven supply chains involves addressing limitations and enhancing their capabilities to handle complex, dynamic environments. Incorporating non-Markovian behaviors allows models to account for historical dependencies and long-term effects, improving predictions for time-sensitive events. The integration of deep learning techniques, such as RNNs and transformer models, enables Markov Models to process large datasets and adapt to real-time changes, while hybrid approaches with reinforcement learning optimize decision-making. Additionally, leveraging cloud computing for scalability and fostering interdisciplinary collaboration enhances the robustness and adaptability of these models in global supply chain networks.

- Role in Sustainability and Ethical AI:

Markov Models contribute to sustainability in supply chains by optimizing resource allocation, reducing waste, and aligning supply with demand, minimizing environmental impacts like excess inventory and overproduction. These models support carbon footprint reduction through route and inventory optimization, and facilitate circular economy initiatives by identifying opportunities for recycling and material reuse. Additionally, they ensure equitable resource distribution in underserved areas while promoting transparency and accountability in decision-making, aligning with ethical AI principles. By integrating ethical constraints, Markov Models balance operational efficiency with social responsibility, supporting fairness and sustainability in supply chain operations.

9. CONCLUSION

This chapter explored the pivotal role of Markov models in AI-driven supply chain management, emphasizing their integration within human and ecological dimensions. We began by discussing the foundational concepts of Markov chains, including states, transitions, steady-state analysis, and absorption probabilities, which form the backbone of probabilistic modeling in supply chains. The chapter also delved into the human dimensions of supply chain management, such as workforce dynamics and decision-making under uncertainty, and the ecological dimensions, focusing on sustainability goals like carbon emissions reduction, waste management, and resource optimization.

Furthermore, we examined advanced techniques and extensions of Markov models, including Markov Decision Processes (MDPs) for optimizing sequential decisions, Hidden Markov Models (HMMs) for uncovering latent patterns, and Markov Chain Monte Carlo (MCMC) methods for robust scenario analysis. The integration of these methods with AI and machine learning techniques was highlighted, showcasing their potential to enhance predictive accuracy, adapt to real-time changes, and optimize supply chain operations. Finally, we reflected on the future scope of this field, identifying opportunities for further research and innovation in applying Markov models to address emerging challenges in AI-driven, sustainable, and resilient supply chain management.

By bridging probabilistic modeling with advanced AI capabilities, this chapter underscores the transformative potential of Markov models in building adaptive, efficient, and ecologically responsible supply chains that effectively balance human and environmental priorities.

REFERENCES

Al-Duais, F. S., & Al-Sharpi, R. S. (2023). A unique Markov chain Monte Carlo method for forecasting wind power utilizing a time series model. *Alexandria Engineering Journal*, 74, 51–63. DOI: 10.1016/j.aej.2023.05.019

Aljohani, A. (2023). Predictive analytics and machine learning for real-time supply chain risk mitigation and agility. *Sustainability (Basel)*, 15(20), 15088. DOI: 10.3390/su152015088

Arifoğlu, K., & Özekici, S. (2011). Inventory management with random supply and imperfect information: A hidden Markov model. *International Journal of Production Economics*, 134(1), 123–137. DOI: 10.1016/j.ijpe.2011.04.033

Bellman, R. (1957). *Dynamic programming and Markov processes*. Princeton University Press. Retrieved from https://press.princeton.edu/books/paperback/9780691004864/dynamic-programming

Bellman, R. E. (1957). *Dynamic programming*. Princeton University Press.

Borucka, A., Kozłowski, E., Parczewski, R., Antosz, K., Gil, L., & Pieniak, D. (2022). Supply sequence modelling using hidden Markov models. *Applied Sciences (Basel, Switzerland)*, 13(1), 231. DOI: 10.3390/app13010231

Cauchois, R., Taghipour, A., Kang, D. S., Zoghlami, N., & Abed, M. (2017). Advanced solutions for a supply chain with stochastic information. *Journal of Advanced Management Science*, 5(1), 9–13. Retrieved December 18, 2024, from DOI: 10.18178/joams.5.1.9-13

Chien, C. F., Lin, Y. S., & Lin, S. K. (2020). Deep reinforcement learning for selecting demand forecast models to empower Industry 3.5 and an empirical study for a semiconductor component distributor. *International Journal of Production Research*, 58(9), 2784–2804. DOI: 10.1080/00207543.2020.1733125

Dadsena, K. K., Mathiyazhagan, K., & Taghipour, A. (2021). Analysis of barriers for the build of resilient supply chain networks post-COVID-19. In Managing supply chain risk and disruptions: Post COVID-19 (pp. 79–89). Springer. DOI: 10.1007/978-3-030-72575-4_7

Das, S. (2021). Foundational RL: Solving Markov decision process. Retrieved January 12, 2025, from https://towardsdatascience.com/foundational-rl-solving-markov-decision-process-d90b7e134c0b

Gilks, W. R., Richardson, S., & Spiegelhalter, D. J. (1995). Markov chain Monte Carlo in practice. Chapman & Hall/CRC. Retrieved from https://www.routledge.com/Markov-Chain-Monte-Carlo-in-Practice/Gilks-Richardson-Spiegelhalter/p/book/9780412055515

Glaa, B., Zoghlami, N., & Taghipour, A. (2014). Outsourcing and off-shoring healthcare services: A way to cut the costs and improve quality. In *2014 International Conference on Advanced Logistics and Transport (ICALT)* (pp. 189–194). IEEE. DOI: 10.1109/ICAdLT.2014.6864116

Hastings, W. K. (1970). Monte Carlo sampling methods using Markov chains and their applications. *Biometrika*, 57(1), 97–109. DOI: 10.1093/biomet/57.1.97

Kadhem, S. K. (2024). Mixture models: Parametric, semiparametric, and new directions. Oxford Academic Press. Retrieved from https://academic.oup.com/jrsssa/advance-article-abstract/doi/10.1093/jrsssa/qnae141/7932060

Kubista, N., Braun, D., & Parmigiani, G. (2024). Penetrance estimation in family-based studies with the penetrance R package. https://arxiv.org/pdf/2411.18816

Loivet, W., Taghipour, A., & Kang, D. S. (2020). The rise of green supply chain management: Between complexity and necessity. *Journal of Economics and Business Management*, 8(1), 1–7. DOI: 10.18178/joebm.2020.8.1.604

Lu, X., & Taghipour, A. (2023). A French guide for businesses and governments to integrate the concept of sustainable development into their supply chains. In *Government impact on sustainable and responsible supply chain management* (pp. 1–12). IGI Global. DOI: 10.4018/978-1-6684-9062-4.ch001

Mahfod, J., Canel-Depitre, B., & Taghipour, A. (2019). Quality function deployment-ELECTRE in supplier evaluation. *Journal of Advanced Management Science*, 7(4).

Mbiatem, J., Taghipour, A., & Canel-Depitre, B. (2018). Supplier selection approaches for decision-makers. In *Proceedings of the 8th International Conference on Information Communication and Management* (pp. 108–112).

Merimi, M., & Taghipour, A. (2021). Accelerating the digitalization of the supply chain: An empirical research about the COVID-19 crisis. In *Digitalization of decentralized supply chains during global crises* (pp. 1–24). IGI Global. DOI: 10.4018/978-1-7998-6874-3.ch001

Metropolis, N., Rosenbluth, A. W., Rosenbluth, M. N., Teller, A. H., & Teller, E. (1953). Equation of state calculations by fast computing machines. *The Journal of Chemical Physics*, 21(6), 1087–1092. DOI: 10.1063/1.1699114

Nadia, E. N., Canel-Depitre, B., & Taghipour, A. (2021). The determinants of hotel room rates in Beirut: A hedonic pricing model. International Journal of Trade. *Economics and Finance*, 12(2), 33–42.

Privault, N. (2024). Discrete stochastic processes. Springer. Retrieved from https://link.springer.com/content/pdf/10.1007/978-3-031-65820-4.pdf

Rabiner, L. R. (1989). A tutorial on hidden Markov models and selected applications in speech recognition. *Proceedings of the IEEE*, 77(2), 257–286. DOI: 10.1109/5.18626

Sodachi, M., Pirayesh, A., & Valilai, O. F. (2024). Using Markov decision process model for sustainability assessment in Industry 4.0. *IEEE Access : Practical Innovations, Open Solutions*, 12, 189417–189438. DOI: 10.1109/ACCESS.2024.3514786

Song, J., Krishnamurthy, V., Kwasinski, A., & Sharma, R. (2012). Development of a Markov-chain-based energy storage model for power supply availability assessment of photovoltaic generation plants. *IEEE Transactions on Sustainable Energy*, 4(2), 491–500. DOI: 10.1109/TSTE.2012.2207135

Sutton, R. S., & Barto, A. G. (1998). *Reinforcement learning: An introduction*. MIT Press.

Szpytko, J., & Salgado Duarte, Y. (2020). Markov chain Monte Carlo simulation model for risk assessment of power systems for electromobility use. *Journal of KONBiN*, 50(1), 1–9. DOI: 10.2478/jok-2020-0002

Taghipour, A. (2014). Improving the plan of a manufacturing network with non-integrated business units. *International Journal of Applied Logistics*, 5(2), 1–11. DOI: 10.4018/ijal.2014040101

Taghipour, A. (2020). *Demand forecasting and order planning in supply chains and humanitarian logistics*. IGI Global.

Taghipour, A. (2023). Government impact on sustainable and responsible supply chain management. IGI Global.

Taghipour, A., Abed, M., & Zoghlami, N. (2015). Design for remanufacturing respecting reverse logistics processes: A review. In 2015 4th International Conference on Advanced Logistics and Transport (ICALT) (pp. 299–304). IEEE. DOI: 10.1109/ICAdLT.2015.7136602

Taghipour, A., & Beneteau-Piet, C. (2020). Sustainable supply chain management performance. *International Journal of Innovation, Management and Technology*, 11(6), 165–169. DOI: 10.18178/ijimt.2020.11.6.897

Taghipour, A., Foukolaei, P. Z., Ghaedi, M., & Khazaei, M. (2023). Sustainable multi-objective models for waste-to-energy and waste separation site selection. *Sustainability (Basel)*, 15(22), 15764. DOI: 10.3390/su152215764

Taghipour, A., & Frayret, J.-M. (2011). Supply chain coordination planning: A review. Work Paper, École Polytechnique de Montréal.

Taghipour, A., & Frayret, J.-M. (2013). Coordination of operations planning in supply chains: A review. *International Journal of Business Performance and Supply Chain Modelling*, 5(3), 272–307. DOI: 10.1504/IJBPSCM.2013.055729

Taghipour, A., & Frayret, J.-M. (2013). An algorithm to improve operations planning in decentralized supply chains. In *2013 International Conference on Advanced Logistics and Transport* (pp. 100–103). IEEE. DOI: 10.1109/ICAdLT.2013.6568442

Taghipour, A., Lu, X., & Smahi, M. (2022). Opportunities and challenges for society 5.0 in procurement. In *Blockchain technology in supply chain management for society 5.0* (pp. 91–100). CRC Press. DOI: 10.1201/9781003177432-9

Taghipour, A., & Merimi, M. (2021). Digital transformation of supply chains during crisis: COVID-19. In *11th Annual International Conference on Industrial Engineering and Operations Management* (pp. 7493–7501). IEOM. DOI: 10.46254/AN11.20211298

Taghipour, A., Padash, A., Etemadi, V., Khazaei, M., & Ebrahimi, S. (2024). Sustainable and circular hotels and the water–food–energy nexus: Integration of agrivoltaics, hydropower, solar cells, water reservoirs, and green roofs. *Sustainability (Basel)*, 16(5), 1985. DOI: 10.3390/su16051985

Taghipour, A., Ramezani, M., Khazaei, M., Roohparvar, V., & Hassannayebi, E. (2023). Smart transportation behavior through the COVID-19 pandemic: A ride-hailing system in Iran. *Sustainability (Basel)*, 15(5), 4178. DOI: 10.3390/su15054178

Taghipour, A., Sohrabi, A., Ghaedi, M., & Khazaei, M. (2023). A robust vaccine supply chain model in pandemics: Case of COVID-19 in Iran. *Computers & Industrial Engineering*, 183, 109465. DOI: 10.1016/j.cie.2023.109465

Team, A. A. C. (2025). Hidden Markov model in machine learning. Retrieved January 12, 2025, from https://www.appliedaicourse.com/blog/hidden-markov-model-in-machine-learning/

United Nations AI for Good. (2024). Cracking the code on supply chain sustainability: How AI and data are creating more responsible operations. Retrieved December 18, 2024, from https://medium.com/@UN_AIforGood/cracking-the-code-on-supply-chain-sustainability-how-ai-and-data

Wongthatsanekorn, W., Realff, M. J., & Ammons, J. C. (2010). Multi-time scale Markov decision process approach to strategic network growth of reverse supply chains. *Omega*, 38(1–2), 20–32. DOI: 10.1016/j.omega.2009.02.001

Chapter 17
Predicting E-Consumer Preferences for Collection and Delivery Points Using Deep Learning

Alaa Eddine El Moussaoui
https://orcid.org/0000-0002-5749-7682
Higher School of Technology, Hassan 2 University, Morocco

ABSTRACT

Collection and delivery points (CDPs) are crucial for addressing last-mile logistics challenges, particularly with the rapid growth of e-commerce. This study examines factors influencing e-consumers' preferences for CDPs in Morocco, using deep learning (DL) algorithms to develop a predictive model. A dataset of 3,000 responses, including age, income, and accessibility to CDPs, was used to train the models. Convolutional neural networks (CNN), long short-term memory (LSTM), and FT-transformer were employed to classify preferences. Hyperparameter tuning optimized these models, with the FT-transformer achieving 92% accuracy, outperforming CNN and LSTM. The findings offer actionable insights for logistics providers to better meet consumer needs, enhancing satisfaction and operational efficiency. The chapter also discusses the study's limitations and suggests future research directions in last-mile logistics.

DOI: 10.4018/979-8-3693-7478-8.ch017

1. INTRODUCTION

The rapid expansion of e-commerce has profoundly transformed global supply chains, particularly the segment known as the "last mile," which refers to the final stage of the delivery process to end consumers (El Moussaoui et al, 2022 ; El Moussaoui et al, 2023). This segment is fraught with challenges, including inefficiencies in route planning, delivery delays, and high operational costs. Additionally, urban areas face environmental concerns such as increased emissions and traffic congestion stemming from multiple delivery vehicles navigating densely populated zones (Luong et al, 2024). Addressing these challenges requires innovative solutions that balance efficiency, cost-effectiveness, and sustainability. CDPs have emerged as a viable alternative to traditional home deliveries (Tadić & Veljović, 2021). These intermediary locations, such as supermarkets, service stations, and post offices, allow consumers to collect packages at their convenience, thereby reducing the frequency of failed deliveries and minimizing transportation inefficiency (Amjad, 2022). CDPs also promote sustainable practices by consolidating deliveries and reducing vehicle trips in urban areas. Despite these advantages, the success of CDPs hinges on accurately understanding and predicting consumer preferences for specific CDP types (Vural & Aktepe, 2022).

In the context of Morocco, the rise of online shopping has necessitated improvements in last-mile logistics. Moroccan consumers exhibit diverse preferences influenced by factors such as age, profession, and geographic location (El Moussaoui et al, 2022). However, there is limited research on the behavioral drivers behind CDP choices in the Moroccan market. Existing studies often rely on traditional machine learning (ML) models or small datasets, which fail to capture the complexity of consumer decision-making processes (Taghipour, 2024 ; Moussaoui et al, 2024). This study addresses these gaps by leveraging advanced DL techniques and a robust dataset comprising 3000 responses. By adopting DL models such as CNN, LSTM, and FT-Transformer, we aim to explore the underlying patterns in consumer behavior that influence CDP selection. These models, renowned for their ability to process and interpret large, complex datasets, are ideally suited for this application.

This study addresses critical logistical and consumer behavior questions: Which demographic segments prefer specific CDPs ? How do geographic and economic factors influence CDP choice? By answering these questions, this research bridges theoretical gaps in understanding consumer decision-making and offers a robust framework for predicting e-consumer preferences. Specifically, this paper contributes by providing a detailed statistical analysis of Moroccan e-consumer behavior, leveraging one of the largest datasets of its kind. It implements advanced DL models namely CNN, LSTM, and FT-Transformer to predict CDP preferences with precision. Additionally, the insights generated extend the understanding of consumer behavior and technological applications in logistics, enriching the growing body of knowledge in last-mile logistics.

The remainder of this paper is structured as follows: Section 2 reviews the existing literature on CDPs and the application of DL in logistics. Section 3 outlines the methodology, including dataset creation, processing and models. Section 4 presents experimental results and discusses the findings. Finally, Section 5 concludes with practical implications and future research directions.

2. LITERATURE REVIEW

2.1 CDPs: Innovations in Last Mile Logistics

CDPs (Collection and Delivery Points) have emerged as a pivotal innovation in addressing the pressing challenges associated with last-mile logistics, which represent the final step of the delivery process and often account for a substantial portion of logistics costs (Zenezini et al, 2018). These challenges include inefficiencies in delivery processes, the environmental burden of increased vehicle emissions, and the ever-rising expectations of consumers for swift and flexible delivery options (El Moussaoui et al, 2022). CDPs, classified into unattended points such as parcel lockers and attended points like service counters in supermarkets or post offices, serve as centralized hubs for consumers to retrieve their orders (Hubner et al, 2016). By reducing the dependency on home deliveries, CDPs offer substantial benefits in terms of flexibility and convenience for end-users. From an operational perspective, CDPs streamline delivery logistics by consolidating multiple orders into a single drop-off point. This aggregation significantly lowers the number of delivery trips required, reducing fuel consumption and minimizing greenhouse gas emissions (El Moussaoui, 2024). Furthermore, for logistics providers, the cost-saving potential of CDPs is considerable, as they optimize delivery routes, decrease instances of failed delivery attempts, and reduce operational costs associated with individual deliveries. Consumers also benefit, particularly those with irregular schedules or limited availability to receive deliveries at home, finding in CDPs a solution that aligns with their lifestyles. This added convenience is particularly valuable in urban areas, where the density of deliveries often leads to congestion and delays.

Despite their potential, implementing CDPs comes with a set of challenges that can affect their efficiency and consumer acceptance. High-density urban areas, often ideal for deploying CDPs due to their large customer bases, present logistical difficulties such as limited space for installation and heightened competition among logistics providers. Accessibility, security, and the degree of consumer trust also play pivotal roles in determining the success of CDPs (Kedia et al, 2017 ; Kedia et al, 2019). Poorly managed CDPs can result in operational inefficiencies, leading to longer wait times for consumers and lower satisfaction levels. Moreover, the placement of CDPs requires strategic consideration to maximize convenience for users while minimizing disruptions in urban planning (Silva et al, 2023). Research suggests that accessibility factors such as proximity to residential areas, availability of parking, and integration with public transport networks significantly influence their adoption. The adoption of CDPs varies significantly based on geographic and demographic factors. In Morocco, where e-commerce is rapidly expanding, CDPs have started gaining popularity (El Moussaoui et al, 2024). Common CDP types include supermarkets, service stations, and post offices, each catering to specific consumer needs and preferences. For example, urban residents often prefer supermarkets due to their convenience and the availability of ancillary services like grocery shopping. Meanwhile, commuters tend to favor service stations that align with their daily travel routes. Nevertheless, demographic factors such as income levels, vehicle ownership, and proximity to urban centers also influence the adoption of CDPs. Socioeconomic factors further shape consumer attitudes, as individuals with lower incomes may prioritize cost-effectiveness over proximity, while higher-income groups might value convenience and additional services (El Moussaoui et al, 2023). Despite their growing prominence in Morocco, limited research has been conducted to understand the perspectives of Moroccan consumers toward CDPs. This represents a significant gap that this study seeks to address by exploring consumer preferences, challenges, and opportunities related to CDPs in the Moroccan context. The findings of this study aim to provide actionable insights for logistics providers and policymakers to enhance the design and implementation of CDPs, ensuring they align with local needs and priorities.

2.2 Application of Deep Learning (DL) in Last Mile Logistics

Deep Learning (DL) has revolutionized the logistics sector, offering transformative solutions to complex challenges in last-mile delivery operations (Andrei et al, 2024). Unlike traditional machine learning (ML) algorithms that require manual feature extraction, DL models can autonomously learn hierarchical representations from raw data, enabling more nuanced and accurate analysis (Taghipour, 2024 ; Mienye & Swart, 2024). This capability is particularly relevant in logistics, where data often comes in large, diverse, and complex forms. For instance, Convolutional Neural Networks (CNNs), originally developed for image recognition, have been adapted for logistics tasks such as demand forecasting, route optimization, and geographic trend analysis (Dalal et al, 2024). CNNs excel at capturing spatial relationships within structured datasets, making them valuable for understanding customer preferences across different geographic locations and optimizing warehouse layouts to enhance operational efficiency (El Moussaoui et al, 2022 ; El Moussaoui, 2024). Furthermore, CNNs are capable of identifying patterns that are not immediately apparent, such as demand fluctuations in specific regions or the impact of weather conditions on delivery times, making them a vital tool for predictive analytics in logistics.

Recurrent Neural Networks (RNNs), particularly Long Short-Term Memory (LSTM) networks, are tailored for sequential data and have been employed to predict demand patterns, optimize delivery schedules, and analyze temporal trends in consumer behavior (Bouktif et al, 2020 ; El Moussaoui & Chakir, 2023). For example, RNNs are effective in modeling the dynamics of online shopping habits, providing insights into peak shopping periods that can inform delivery planning and resource allocation. The ability of LSTM networks to capture long-term dependencies within datasets makes them particularly suited for analyzing historical data, enabling logistics providers to anticipate seasonal demand changes and adjust resources accordingly (Alqatawna et al, 2023). Transformers, a recent and highly influential development in DL, have gained prominence in logistics for their ability to handle time-series data and complex datasets. By leveraging attention mechanisms, Transformers efficiently capture both local and global dependencies within data, making them suitable for tasks such as anomaly detection, real-time demand forecasting, and customer segmentation. Their efficiency in processing multifaceted datasets positions Transformers as a cutting-edge tool for analyzing factors influencing CDP preferences, such as demographic and behavioral variables. Transformers also excel in real-time decision-making, enabling logistics providers to respond dynamically to changes in demand or disruptions in delivery networks.

However, the application of DL to CDP-related decision-making remains underexplored in the literature. Most existing studies have relied on traditional statistical methods or basic ML models, which often fall short in capturing the intricate interplay of factors that drive consumer choices. For instance, linear regression models may identify broad trends but lack the sophistication to account for nuanced interactions among demographic, geographic, and behavioral variables. Similarly, basic clustering algorithms may group consumers based on superficial similarities without accounting for deeper behavioral insights. This study seeks to address this gap by employing advanced DL techniques, including CNNs, LSTMs, and FT-Transformers, to develop predictive models that offer high accuracy in analyzing CDP preferences. These models aim to uncover latent patterns within the data, such as correlations between consumer behavior and geographic accessibility or the impact of marketing campaigns on CDP adoption rates. By doing so, it aims to advance both academic research and practical applications in the logistics domain, showcasing the untapped potential of DL in optimizing last-mile delivery operations. Furthermore, the study highlights the need for a more comprehensive approach to integrating DL into logistics systems, emphasizing the transformative impact of these technologies on enhancing operational efficiency, reducing costs, and improving consumer satisfaction. By addressing these critical gaps, the research aims to contribute to a deeper understanding of how DL can be leveraged to address real-world challenges in last-mile logistics, particularly in emerging markets like Morocco where logistical frameworks are still evolving.

3.METHODOLOGY

3.1 Dataset

The dataset used in this study was collected through an extensive online survey distributed across multiple Moroccan social media platforms (Facebook groups, Linkedin,...). The survey was meticulously designed to capture a diverse set of responses from 3000 e-consumers across Morocco, ensuring representation from various age groups, professions, and geographic regions. The questionnaire was bilingual, available in Arabic and French, to accommodate Morocco's linguistic diversity and maximize response rates. The survey gathered information on demographic attributes such as age, gender, academic level, and profession, in addition to preferences for different types of CDPs, including supermarkets, service stations, and post offices. Special attention was given to sampling from both urban and rural areas to capture the nuances in consumer behavior driven by geographic location and lifestyle. This diversity in the dataset ensures a robust foundation for training and evaluating the DL models. The data collection process adhered to strict ethical guidelines, ensuring anonymity and voluntary participation. Respondents were asked detailed questions about their delivery preferences, including the reasons for choosing specific CDPs and their accessibility constraints. These insights were augmented with socioeconomic indicators such as income ranges and vehicle ownership to provide a holistic understanding of consumer behavior. The dataset was partitioned into training and testing subsets using an 80:20 ratio to facilitate model development and performance evaluation. As shown in the Table 1, each CDP type was equally represented, mitigating class imbalance and ensuring the models' ability to generalize across categories:

Table 1. Distribution of answers in each class

CDP Type	Responses	Training set (80%)	Testing set (20%)
Supermarket	1000	800	200
Service station	1000	800	200
Post office	1000	800	200
Total	3000	2400	600

3.2 Data Preprocessing

Data preprocessing is a crucial step in preparing raw data for use in DL models, ensuring consistency and improving dataset quality for training and evaluation. The preprocessing pipeline began with feature selection using Pearson correlation analysis to identify variables most relevant to predicting the target variable (CDP type). Features with a correlation coefficient exceeding a threshold of 0.3 were retained, ensuring the dataset remained compact and informative. For continuous features such as age and income, z-score normalization was applied to standardize their ranges, allowing for a consistent scale across all features. Categorical variables like gender and academic level were transformed into numerical representations using one-hot encoding, generating binary columns for each category, enabling the models to utilize this data effectively.

The dataset was then split into training and testing subsets at an 80:20 ratio, ensuring that the training set was used for model development, while the testing set provided an independent evaluation of model performance. This solid preprocessing pipeline laid the foundation for successful model training.

3.3 Model-Specific Preprocessing

In this study, each DL model required unique preprocessing steps tailored to its architecture. For the CNN model, the tabular data was treated using 1D convolutions, where continuous features such as age, income, and vehicle ownership were normalized using min-max scaling to a range of [0, 1]. This allowed the model to effectively capture relationships across features. The LSTM model required the data to be arranged sequentially to enable the capture of potential feature dependencies, even though the dataset is not inherently sequential. Continuous features were again normalized using min-max scaling to stabilize gradient flow during backpropagation and ensure faster convergence. For the FT-Transformer model, categorical features like gender and academic level were embedded into high-dimensional vectors using an embedding layer, while continuous variables were scaled similarly. The self-attention mechanism of the transformer model was leveraged to capture complex relationships between features, though no positional encoding was applied, as the dataset is not sequential in nature. These preprocessing steps were crucial for preparing the data to fully leverage the strengths of each model, ensuring optimal performance in predicting the CDP Type.

3.4 Model Architecture

This study employed three deep learning architectures including CNN, LSTM, and FT-Transformer. Each designed to capture distinct patterns within the dataset. The CNN architecture utilized three 1D convolutional layers with ReLU activation and max-pooling to extract hierarchical features from the tabular data and reduce dimensionality. These layers helped capture local feature dependencies, while max-pooling downsampled the data, facilitating more efficient training. A fully connected layer then mapped the extracted features to the CDP categories, with dropout regularization used to prevent overfitting.

The LSTM model consisted of two stacked layers with 128 hidden units each, aimed at capturing complex interactions between features. While the dataset is not inherently sequential, the LSTM was employed to model potential dependencies and relationships between features that may influence the choice of CDP. Dropout layers were included to mitigate overfitting, and a dense output layer performed classification. Hyperparameters, including the learning rate and batch size, were fine-tuned to optimize model performance.

The FT-Transformer model leveraged a self-attention mechanism to capture both local and global dependencies between features. It consisted of 12 attention heads and 6 encoder layers. Positional encodings were applied to retain relative feature positions and allow the model to better capture interactions between them. A dense output layer was used for classification, while dropout and layer normalization helped improve stability and prevent overfitting. The Transformer's ability to process features in parallel and its robust feature extraction capabilities made it the highest-performing model across all evaluation metrics.

3.5 Hyperparameter Optimization

Hyperparameter optimization plays a crucial role in improving model performance and generalization. In this study, grid search was employed to systematically explore a predefined set of hyperparameters for each DL model. For the CNN model, the following hyperparameters were tuned: the number of convolutional layers (ranging from two to five), kernel sizes (3x3 and 5x5), and dropout rates (0.2 to 0.5). The learning rate was optimized between 0.001 and 0.01, with batch sizes of 16, 32, and

64 evaluated. These settings were selected to strike a balance between computational efficiency and model accuracy.

For the LSTM model, hyperparameters such as the number of hidden units (ranging from 64 to 256) and the number of stacked layers (from one to three) were tested. Dropout rates were also tuned, as was the learning rate (0.001 to 0.01) and batch size (16, 32, and 64). LSTM models are particularly suited for capturing sequential relationships, and hyperparameter tuning was focused on optimizing these dynamics.

For the FT-Transformer model, hyperparameters including the number of attention heads (8 or 12), encoder layers (4 to 6), and feed-forward network dimensions (512 and 1024) were explored. The learning rate was optimized using the Adam optimizer with warm-up steps, and batch sizes of 16 and 32 were tested.

The grid search process was performed in parallel using a high-performance computing cluster, enabling the simultaneous evaluation of multiple configurations. Early stopping was implemented with a patience of five epochs to prevent overfitting, halting training when validation loss ceased to improve. The best hyperparameters were selected based on performance on the validation set.

3.6 Evaluation Metrics

Model performance was evaluated using accuracy, precision, recall, and F1-score. Since the classes in this classification task were balanced, accuracy provided a solid overall measure of model performance. Precision and recall offered deeper insights into the model's ability to correctly classify each category, while the F1-score, as a harmonic mean of precision and recall, was useful for evaluating the trade-off between these two metrics. For this classification task, the following equations were used to calculate each metric:

$$Accuracy = \frac{TP + TN}{TP + TN + FP + FN} \tag{1}$$

where TP = True Positives, TN = True Negatives, FP = False Positives, FN = False Negatives.

The precision measures the proportion of true positive predictions among all positive predictions. It is calculated using Equation 2:

$$\text{Pr}ecision = \frac{TP}{TP + FP} \tag{2}$$

The recall measures the proportion of true positive predictions among all actual positives. It is calculated using Equation 3:

$$\text{Recall} = \frac{TP}{TP + FN} \tag{3}$$

The last metric is the f1-score which can be calculated using Equation 4:

$$F1 - score = 2 * \frac{\text{Precision} * \text{Recall}}{\text{Precision} + \text{Recall}} \tag{4}$$

4. RESULTS AND DISCUSSION

The results of the experiments studying the classification of CDP preferences are reported in terms of accuracy, precision, recall, and F1-score, as shown in Table 2. The FT-Transformer outperformed the CNN and LSTM models in all evaluation metrics, demonstrating its superior ability to capture both local and global dependencies within the data.

Table 2. Performance of models

Model	Accuracy (%)	Precision (%)	Recall (%)	F1-Score (%)
CNN	89	88	87	88
LSTM	91	90	89	90
FT-Transformer	92	91	91	91

4.1 Model Performance

The FT-Transformer model demonstrated the best performance, achieving an accuracy of 92% and an F1-score of 91%. This remarkable result can be attributed to its self-attention mechanism, which enables the model to identify complex relationships between features and process data in parallel. Unlike CNNs, which specialize in extracting localized spatial features, or LSTMs, which excel at sequential data processing, FT-Transformer combines the strengths of both and extends them further by attending to global dependencies across input features. This capacity made the FT-Transformer particularly effective in capturing the intricate patterns that influence CDP preferences.

The LSTM model, while slightly behind the FT-Transformer, achieved an impressive accuracy of 91%. Its ability to capture sequential dependencies in consumer behavior trends was essential for modeling temporal aspects of CDP preferences. However, LSTM's inherent difficulty in handling long-range dependencies, particularly with complex datasets containing both sequential and non-sequential information, slightly hindered its performance compared to the Transformer. Despite this, LSTM remains a strong contender for tasks where temporal dependencies are significant.

The CNN model, with an accuracy of 89%, demonstrated solid performance in identifying structured patterns within the data. However, it struggled with the complex relationships between demographic and geographic variables. CNNs are more suited to problems where spatial or grid-like data is prevalent, and while they can recognize local patterns, they may struggle to capture the more complex, non-linear interactions present in mixed data types like demographics and geographic features.

4.2 Consumer Preferences for CDPs

The analysis of FT-Transformer model predictions (Figure 1) provided valuable insights into consumer preferences for various types of Collection and Delivery Points (CDPs). These preferences were strongly influenced by demographic and situational factors, highlighting the diverse needs of consumers based on their geographic, lifestyle, and mobility characteristics:

- Supermarkets: Young, urban consumers displayed a pronounced preference for supermarkets as their primary CDP choice. These locations are highly accessible within city environments and cater to the fast-paced lifestyles of this demographic. The integration of CDP services with grocery shopping creates a dual-purpose visit, enhancing convenience and reducing time spent on errands. Additionally, the extended operating hours of supermarkets cater to the schedules of urban consumers, further solidifying their appeal.
- Service Stations: Motorized consumers, particularly those who commute regularly, demonstrated a strong preference for service stations as CDPs. These facilities, strategically situated along major roads and at urban peripheries, provide ease of access during daily commutes. For this demographic, service stations offer a practical solution to integrate package collection with routine travel patterns, minimizing detours and saving time. Their ability to operate 24/7 further adds to their attractiveness, accommodating consumers with varied schedules.

Figure 1. Confusion matrices illustrating the FT-Transformer model's prediction performance across different consumer groups: Young, Urban, Motorized, and Rural Consumers

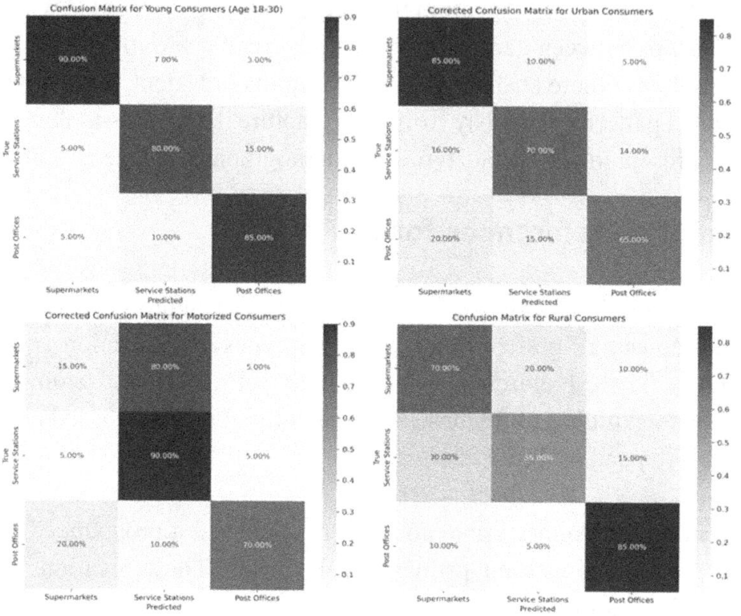

- Post Offices: In rural areas, consumers predominantly selected post offices as their preferred CDP. This trend reflects the limited availability of alternative options in these regions and underscores the longstanding trust placed in post offices as reliable service providers. Post offices often serve as community hubs, which enhances their accessibility and familiarity. Their established presence in rural locales ensures dependable service, meeting the unique logistical challenges faced by consumers in these areas.

5. THEORETICAL, MANAGERIAL, AND PRACTICAL IMPLICATIONS

This study offers profound contributions to the theoretical understanding of logistics and e-commerce, managerial practices, and practical applications in the field of last-mile logistics and consumer behavior prediction.

5.1 Theoretical Implications

The research makes a significant contribution to the theoretical domain by integrating advanced deep learning techniques with the study of consumer behavior in logistics. The application of the FT-Transformer model—a state-of-the-art deep learning architecture—demonstrates its potential to handle complex datasets with higher accuracy and reliability compared to conventional models like CNN and LSTM. This finding enriches the academic discourse on the applicability of transformer-based models in real-world scenarios, particularly within the logistics sector. By exploring demographic and geographic factors influencing consumer preferences for collection and delivery points (CDPs), the study provides a novel perspective on the interplay between technological innovation and behavioral analytics. The detailed examination of urban, motorized, and rural populations' preferences introduces a layered understanding of how geographic and demographic variables shape consumer decision-making in the context of last-mile logistics. This dual emphasis on technology and behavior bridges a critical gap in the literature, offering a foundation for interdisciplinary studies that could explore the integration of additional data types—such as psychological or socio-economic factors—to further refine theoretical models.

5.2 Managerial Implications

From a managerial perspective, the findings of this study have direct and actionable implications for logistics service providers aiming to optimize their operations and improve customer satisfaction. Managers can harness the insights from predictive models to design CDP networks that are not only operationally efficient but also aligned with consumer expectations. For instance, urban consumers, who typically prioritize convenience and accessibility, may benefit from strategically placed CDPs near public transport hubs or high-density residential areas. Conversely, rural consumers, who may face challenges related to infrastructure and accessibility, could be better served by CDPs designed with reliability and coverage in mind. By tailoring services to these distinct consumer segments, logistics providers can not only enhance customer satisfaction but also achieve cost efficiencies by minimizing failed deliveries and optimizing route planning. Additionally, the study highlights the strategic importance of adopting advanced deep learning models like the FT-Transformer. These models allow for better demand forecasting, resource allocation, and customer segmentation, empowering managers to make data-driven decisions that enhance operational agility and competitiveness. The research also underscores the importance of continuous learning and adaptation in managerial practices, as the evolving consumer landscape necessitates dynamic and responsive strategies.

5.3 Practical Implications

On a practical level, this study serves as a compelling demonstration of how cutting-edge deep learning techniques can address pressing logistical challenges. For policymakers and urban planners, the findings provide a data-driven framework for designing logistics infrastructure that meets the needs of diverse populations. For example, the identification of geographic patterns in consumer preferences can inform zoning decisions, ensuring that CDPs are located in areas that maximize accessibility and utility for the target population. This approach not only supports the efficient functioning of logistics systems but also contributes to broader societal goals such as reducing carbon emissions through optimized last-mile delivery routes. Moreover, the study highlights the potential for public-private partnerships in leveraging deep learning technologies to enhance urban logistics. Collaboration between logistics firms and local authorities could facilitate the integration of dynamic data sources—such as traffic and weather information—into operational planning, creating adaptive systems that respond to real-time conditions. Additionally, the research provides a practical roadmap for logistics companies to implement advanced predictive models. By investing in the necessary infrastructure and expertise to deploy models like the FT-Transformer, companies can gain a competitive edge in a rapidly evolving market. These practical applications underscore the transformative potential of integrating deep learning into logistics, offering tangible benefits for businesses, consumers, and society at large.

6. CONCLUSION

This chapter presents a comprehensive exploration of the effectiveness of deep learning models in predicting consumer preferences for collection and delivery points (CDPs) in Morocco. Among the models evaluated, the FT-Transformer emerged as the most effective, demonstrating superior accuracy and F1-scores compared to CNN and LSTM. This finding underscores the transformative potential of transformer-based architectures in addressing complex logistical challenges. By analyzing demographic and geographic factors, the study uncovered significant patterns in consumer preferences, revealing distinct needs among urban, motorized, and rural populations. These insights provide valuable guidance for optimizing last-mile delivery systems, emphasizing the critical role of tailored solutions in enhancing customer satisfaction and operational efficiency. The study's findings highlight the intersection of technological innovation and consumer behavior, offering a robust framework for future research and practical applications in the field of last-mile logistics.

Despite its valuable contributions, the study is not without limitations. One significant limitation is the geographic specificity of the research, as the models were validated within the context of Morocco. This geographic focus raises questions about the generalizability of the findings to other regions with different consumer behaviors and logistical challenges. Another limitation is the exclusion of real-time dynamic factors such as traffic congestion, weather conditions, and urban mobility patterns. These variables are critical for real-world logistics operations and could enhance the accuracy and adaptability of predictive models. Additionally, the study primarily focused on demographic and geographic factors, leaving unexplored the potential influence of psychological, socio-economic, and cultural variables on consumer preferences. These limitations highlight the need for a more holistic approach in future research to capture the multifaceted nature of e-consumer behavior.

Building on the insights and limitations of this study, future research could pursue several promising directions. Expanding the geographic scope of the research to include diverse markets and regions would provide a more comprehensive understanding of consumer preferences and enhance the generalizability of the findings. Integrating real-time dynamic data sources, such as traffic and weather information, into predictive models could significantly improve their accuracy and responsiveness. This approach would enable logistics providers to develop adaptive systems capable of responding to real-world conditions, thereby optimizing last-mile delivery operations. Furthermore, future studies could explore the role of psychological and socio-economic factors in shaping consumer preferences, offering a more holistic view of e-consumer behavior. Investigating the application of other advanced deep learning models or hybrid approaches could also yield valuable insights, further advancing the state of the art in predictive modeling for logistics. By addressing these areas, future research can build on the foundation established by this study, driving innovation and excellence in the field of last-mile logistics.

REFERENCES

Alqatawna, A., Abu-Salih, B., Obeid, N., & Almiani, M. (2023). Incorporating Time-Series Forecasting Techniques to Predict Logistics Companies' Staffing Needs and Order Volume. *Computation (Basel, Switzerland)*, 11(7), 141. DOI: 10.3390/computation11070141

Amjad, S. (2022). Role of Logistical Practices in Quality Service Delivery at Supermarkets: A Case Study from Pakistan: Logistics and Quality Service Delivery. *South Asian Journal of Operations and Logistics*, 1(1), 39–56. DOI: 10.57044/SAJOL.2022.1.1.2204

Andrei, N., Scarlat, C., & Ioanid, A. (2024). Transforming E-commerce logistics: Sustainable practices through autonomous maritime and last-mile transportation solutions. *Logistics (Basel)*, 8(3), 71. DOI: 10.3390/logistics8030071

Bouktif, S., Fiaz, A., Ouni, A., & Serhani, M. A. (2020). Multi-sequence LSTM-RNN deep learning and metaheuristics for electric load forecasting. *Energies*, 13(2), 391. DOI: 10.3390/en13020391

Dalal, S., Lilhore, U. K., Simaiya, S., Radulescu, M., & Belascu, L. (2024). Improving efficiency and sustainability via supply chain optimization through CNNs and BiLSTM. *Technological Forecasting and Social Change*, 209, 123841. DOI: 10.1016/j.techfore.2024.123841

El Moussaoui, A. E. (2024). Factors influencing the entry of logistics service providers (LSPs) into the logistics clusters: a case study of Casablanca city. Management & Sustainability: An Arab Review.

El Moussaoui, A. E. (2024). Impact of logistics pooling on reduction of CO2 emissions in last-mile logistics. Management & Sustainability: An Arab Review.

El Moussaoui, A. E., Benbba, B., & El Andaloussi, Z. (2023). Impact of logistics performance on the store image, consumer satisfaction and loyalty: A quantitative case study. *Arab Gulf Journal of Scientific Research*, 41(3), 226–239. DOI: 10.1108/AGJSR-09-2022-0201

El Moussaoui, A. E., Benbba, B., Jaegler, A., & El Andaloussi, Z. (2022, May). City Logistics: An Evolutionary, Successful & Collaborative City. In 2022 14th International Colloquium of Logistics and Supply Chain Management (LOGISTIQUA) (pp. 1-7). IEEE.

El Moussaoui, A. E., Benbba, B., Jaegler, A., & El Andaloussi, Z. (2022, May). How to improve the urban logistics sector in large cities of developing countries: The case of Casablanca. In 2022 14th International Colloquium of Logistics and Supply Chain Management (LOGISTIQUA) (pp. 1-6). IEEE.

El Moussaoui, A. E., Benbba, B., Jaegler, A., El Andaloussi, Z., & El Amrani, L. (2022, May). Last mile logistics: Challenges & improvement ways. In 2022 14th International Colloquium of Logistics and Supply Chain Management (LOGISTIQUA) (pp. 1-7). IEEE.

El Moussaoui, A. E., Benbba, B., Jaegler, A., El Moussaoui, T., El Andaloussi, Z., & Chakir, L. (2023). Consumer perceptions of online shopping and willingness to use pick-up points: A case study of Morocco. *Sustainability (Basel)*, 15(9), 7405. DOI: 10.3390/su15097405

El Moussaoui, A. E., El Moussaoui, T., Benbba, B., Chakir, L., Jaegler, A., & El Andaloussi, Z. (2024). Sustainable effects of information sharing between distribution logistics actors: A qualitative case study. *Environment, Development and Sustainability*, 1–19. DOI: 10.1007/s10668-023-04387-3

El Moussaoui, A. E., El Moussaoui, T., Benbba, B., Jaegler, A., & El Andaloussi, Z. (2022). Understanding the choice of collection & delivery point by the E-consumer via a machine learning model: Moroccan case study. *Procedia Computer Science*, 210, 204–211. DOI: 10.1016/j.procs.2022.10.138

El Moussaoui, T., & Chakir, L. (2023, September). BERT-CLSTM model for the classification of Moroccan commercial courts verdicts. In 2023 18th Conference on Computer Science and Intelligence Systems (FedCSIS) (pp. 281-284). IEEE. DOI: 10.15439/2023F3561

Hübner, A. H., Kuhn, H., & Wollenburg, J. (2016). Last mile fulfilment and distribution in omni-channel grocery retailing: A strategic planning framework. *International Journal of Retail & Distribution Management*, 44(3). Advance online publication. DOI: 10.1108/IJRDM-11-2014-0154

Kedia, A., Kusumastuti, D., & Nicholson, A. (2017). Acceptability of collection and delivery points from consumers' perspective: A qualitative case study of Christchurch city. *Case Studies on Transport Policy*, 5(4), 587–595. DOI: 10.1016/j.cstp.2017.10.009

Kedia, A., Kusumastuti, D., & Nicholson, A. (2019). Establishing collection and delivery points to encourage the use of active transport: A case study in New Zealand using a consumer-centric approach. *Sustainability (Basel)*, 11(22), 6255. DOI: 10.3390/su11226255

Luong, D. V., Lopes de Sousa Jabbour, A. B., & Nguyen, N. M. (2024). The nexus between collection-and-delivery points and consumers' adoption intention: The role of country culture. *Cogent Business & Management*, 11(1), 2423890. DOI: 10.1080/23311975.2024.2423890

Mienye, I. D., & Swart, T. G. (2024). A Comprehensive Review of Deep Learning: Architectures, Recent Advances, and Applications. *Information (Basel)*, 15(12), 755. DOI: 10.3390/info15120755

Moussaoui, T. E., Loqman, C., & Boumhidi, J. (2024, May). Flat and Nested Named Entity Recognition in Arabic Language. In 2024 4th International Conference on Innovative Research in Applied Science, Engineering and Technology (IRASET) (pp. 1-7). IEEE. DOI: 10.1109/IRASET60544.2024.10548518

Silva, V., Amaral, A., & Fontes, T. (2023). Sustainable urban last-mile logistics: A systematic literature review. *Sustainability (Basel)*, 15(3), 2285. DOI: 10.3390/su15032285

Tadić, S., & Veljović, M. (2021). Home delivery: A framework for structuring. *IJTTE. International Journal for Traffic and Transport Engineering*, 11(1).

Taghipour, A. (2024). Big Data: The New Gold Mine for Supermarkets. In Applications of New Technology in Operations and Supply Chain Management (pp. 98-106). IGI Global.

Taghipour, A. (2024). Supply Chains and Artificial Intelligence Chatbots. In Applications of New Technology in Operations and Supply Chain Management (pp. 260-268). IGI Global. DOI: 10.4018/979-8-3693-1578-1.ch013

Vural, C. A., & Aktepe, Ç. (2022). Why do some sustainable urban logistics innovations fail? The case of collection and delivery points. *Research in Transportation Business & Management*, 45, 100690. DOI: 10.1016/j.rtbm.2021.100690

Zenezini, G., Lagorio, A., Pinto, R., De Marco, A., & Golini, R. (2018). The collection-and-delivery points implementation process from the courier, express and parcel operator's perspective. *IFAC-PapersOnLine*, 51(11), 594–599. DOI: 10.1016/j.ifacol.2018.08.383

Chapter 18
System Experts in AI

Kapil Vhatkar
Dr. D.Y. Patil Institute of Technology, Pimpri, India

Yatri Davda
https://orcid.org/0000-0003-4198-3499
Dr. D.Y. Patil Institute of Technology, Pimpri, India

Aniket Sodhi
Dr. D.Y. Patil Institute of Technology, Pimpri, India

ABSTRACT

An Expert System in AI is software designed to mimic a human expert's decision-making abilities by applying logic to a vast knowledge base. It provides recommendations or decisions similar to those of professionals in fields like medicine, law, or engineering. The system consists of a knowledge base with domain-specific information and an inference engine that applies rules, often through "if-then" logic, to draw conclusions. Its strengths include efficient, continuous analysis and reliable advice, free from human fatigue or error, especially when expertise is scarce, such as in medical diagnoses. The system also allows non-experts to input queries via a user interface, offering real-time support based on predefined knowledge. Some systems include explanation features to clarify decision-making processes.

DOI: 10.4018/979-8-3693-7478-8.ch018

1. INTRODUCTION TO EXPERT SYSTEMS

1.1. Definition of System Experts

System Experts, or Expert Systems are one branch of artificial intelligence, created to emulate the ability of a human expert in a specific area to think and make decisions. These systems are designed with the capacity to solve very complex problems that usually entail human knowledge and reasoning by an expert. An expert system is a hybrid of huge knowledge base, which comprises both rules and facts that are domain-specific, and an inference engine that uses logical reasoning on the input to produce answers. Ultimately, its objective is to emulate how the thought process of human experts would be or to provide accurate advice or recommendation of equal status with that of an expert (Buchanan & Shortliffe, 1984; Taghipour & Frayret, 2013a, and 2013b).

Such systems are used extensively in medicine, engineering, finance, and customer support since their usability depends on application-specific knowledge in these areas. Experts can then base decisions on facts since they input queries that are then replied to in an expert-like manner. The knowledge base is developed from information acquired from human experts. Methods of data-driven reasoning like forward chaining or goal-driven reasoning like backward chaining by the inference engine are then applied to the knowledge base in resolving problems (Nilsson, 1998).

Unlike most modern AI systems, which learn from experience based on data, system experts use preprogrammed rules and logic. That means they only adapt to new situations unless manually updated. Although they have major limitations about flexibility and learning, expert systems excel at providing consistent, accurate, and fast solutions and are invaluable in applications where expert knowledge is limited or in high demand.

1.2. Historical Development of Expert Systems

The expert system traces its historical development back to the 1960s and 1970s by focusing on creating programs that could then simulate a kind of human expertise in a particular domain. Even the early AI had focused on general problem-solving, but within an incredibly short period of time, researchers recognized the importance of specialized knowledge toward solving complex problems in the real world. It led to the development of expert systems-the first integration of a structured knowledge base with logical inference rules to mimic human decision-making (Zoghlami et al., 2016; Cauchois et al., 2017; Jackson, 1998).

The first notable expert system was DENDRAL, developed in the mid-1960s at Stanford University to help chemists analyze molecular structures

However, with the advancement of AI further, the expert system approach came into jeopardy as the expert system was bound to static knowledge bases and did not have the capability to learn from new data. Nevertheless, despite these above constraints, expert systems actually provided the basis for state-of-the-art modern AI techniques, and it continues to play a vital role in areas that mandate rule-based reasoning (Dadsena et al., 2021; Loivet et al., 2020; Taghipour, 2020; Taghipour, 2014).

1.3. Importance of Expert Systems in AI

Marking an important point in the development course of AI, expert systems are likely to play a crucial role in the expansion of human expertise in various fields. It is significant since it allows for the possibility of automating decision-making processes carried out by human experts on matters that otherwise seemed impossible to automate; hence, it increases efficiency, accuracy, and consistency in solving hard problems. Simulating human reasoning, expert systems can be introduced by a structured knowledge base combined with an inference engine. Therefore, even non-experts can have expert-level guiding help in health care, finance, or engineering, among many other areas.

Medical: Expert systems have been applied in the diagnosis process; here, accurate recommendations were made based on patient data and rules of medicine. It applies to finance tasks such as risk and portfolio management based on large volumes of financial data-things that human experts cannot possibly handle quickly or reliably. In these scenarios, the systems offer great value where specialist information either is unavailable or costly to receive, thereby bringing the advice itself to a larger audience.

The other major strength of expert systems is consistency. Expert systems controlled by pre-deliberated rules do not get tired, biased, or inconsistent in reaching their decisions, but human experts do. Expert systems have uniform resultant decisions hence obtaining accuracy and objectivity in their decision-making. The productivity is thus increased by expert systems since they can rapidly evolve and process complex data. Emergency management and technical troubleshooting industries involve such immediate decisions.

Important in the broad domain of artificial intelligence is the kind of expert systems that mimic human intelligence to solve complex problems in a specialized field. Their importance lies in the consistent, reliable, and accurate decision it gives them concerning areas such as health, finance, and engineering. Expert systems improve efficiency, minimize human error, and make expert-level knowledge available to the uninitiated. They allow quick responses to questions, streamline operations, and increase productivity. The expert systems go ahead and bridge the human intelligence gap with the machine capabilities and are what advance innovation and sound decision-making in everything going on in the industries (Merimi, & Taghipour, 2021; Taghipour& Frayret, 2011; Mbiatem et al., 2018; Nadia et al., 2021; Taghipour et al., 2022; Taghipour & Merimi, 2021).

2. CORE COMPONENTS OF EXPERT SYSTEMS

2.1. Knowledge Base

A knowledge base is the core of an artificial intelligence expert system. It is a store or a repository of specific knowledge or expertise and refers to that which the system exploits in order to solve complex problems. Facts, rules, and heuristics, mostly provided by human experts, constitute it and are based on information regarding some particular domain. In this part, facts and rules are structured so the system can come to certain conclusions logically based on the input it receives by a process of reasoning.

The knowledge base may be thought of as the "memory" or "intelligence" of the system because it will contain all the relevant information needed in order to solve problems. It contains two types of knowledge: declarative knowledge information about what is known within a domain-and procedural knowledge, containing the rules to apply the facts in appropriate ways for various situations. These procedures often come in terms of "if-then" rules, where some combination of conditions leads to some kind of action or conclusion.

Among these, proper completeness and accuracy of the information to be entered in a knowledge base pose significant difficulty. The quality of the output from an expert system depends entirely on the quality of the knowledge base on which it is based. In order to feed the knowledge base with detailed and accurate information, cooperation has to be done with human experts. However, a weakness of this is the static knowledge-base; it does not learn or adapt without the intervention of the user while updating the system. This limits the response capability of the system in dynamic application environments.

Despite this weakness, the knowldege base is one vital feature in the consistency, reliability, and efficiency of expert systems to simulate an expert decision-making process in fields including medicine, finance, and engineering.

2.2. Inference Engine

The Inference Engine is the central reasoning component of an AI expert system. It is used to interpret and apply knowledge in a system's knowledge base with which to solve problems or make decisions. In other words, the inference engine is the system's "brain," wherein different conclusions are drawn logically from the information entered by the user and the knowledge base rules.

There are two prime methods deployed by inference engines: forward chaining and backward chaining. In forward chaining, the inference engine starts with the available data input and uses the applicable rules within the knowledge base to draw new conclusion or make any logical derivation. Such a method is indeed data-driven, hence useful for situations where a certain input has been given to the system by the user, and the system uses it in order to find solutions most applicable.

In backward chaining, the system actually begins with a goal or hypothesis, and then it moves backward to investigate whether the available data and rules could support that given goal. This methodology is goal-driven and usually applied when the goal is to confirm a kind of diagnosis by using symptoms and knowledge in the field of medicine, that is, medical expert systems.

The success of such an expert system strongly relies on the inference engine, its ability to apply logical reasoning efficiently and precisely. Combining stored knowledge with techniques of reasoning is offered by inference engines to help systems replicate expert decision-making, reliable solutions in fields across the board.

2.3. User Interface

An expert system has a User Interface (UI), which is very important in facilitating the relationship between the user and the system. It enables a non-expert to input data, questions, or a problem and get expert-level advice or solutions for that particular issue. It acts like a point of communication; otherwise, it will change the inputs from the users into a specific form that the system's inference engine can process based on the knowledge base.

A good UI should ensure easy use, so the system can be accessed by people who are not really aware of deep technical knowledge.

Moreover, most sophisticated expert systems come equipped with an explanation facility as part of the UI. This enables users to know how the system arrived at its output. Such transparency allows for an increased amount of trust in the results that are being provided by the system. In simple words, it is the ultimate goal of the UI: to make expert systems user-friendly so that people outside the scope of the discipline can apply these to the level of expert decision without having the technical knowledge to be extensively purposeful. Effective UI is thus a prerequisite for usability and eventual success of the user.

3. HOW EXPERT SYSTEMS WORK

3.1. Logical Reasoning in Expert Systems

Logical reasoning is one of the critical parts of an expert system within AI. It serves well for a system to behave more like the human mind, reaching decisions and solving problems appropriately. In the broadest form, it is nothing but using formal logic in order to determine conclusions from the data held within the system's knowledge base. The inference engine is basically the tool for doing this sort of reasoning, using rules and facts in order to apply the data input to gain new information.

There are two broad categories of logical reasoning in expert systems: forward chaining and backward chaining. Forward chaining consists of knowing some facts and applying inference rules to deduce new conclusions or actions. This approach is data-driven if the system needs to check several possibilities related to what it has received as input at the first step. For example, using forward chaining in a medical diagnosis system helps narrow down the possible diseases with which the patient is suffering, given his symptoms.

On the other hand, starting from a goal or hypothesis and then looking backward to see whether the facts and available rules can be used to support such a goal is referred to as backward chaining. It is a very useful goal-driven methodology for the diagnosis problem-for example, confirm a disease using some symptom and corresponding medical rules (Taghipour et al., 2023b; Taghipour et al., 2015 ; Taghipour et al., 2023c; Lu & Taghipour, 2023; Taghipour et al., 2023a; Taghipour et al., 2024).

3.2. Forward Chaining

Forward chaining is one of the reasoning techniques applied in AI expert systems to draw conclusions from known facts and rules. It starts from the given information by successively applying relevant inference rules, which systematically generates new facts through this process until a conclusion is deduced or cannot make any more conclusions. The process begins from when the user feeds special data into the system, which are, then processed by the inference engine with rules defined in the knowledge base.

In forward chaining each time the rule is being triggered by the facts at that particular moment, the system generates new facts based on the output of that rule. For example, if a patient has symptoms like fever and cough, in a medical diagnosis expert system, the system can use rules stating "if fever and cough are present, then consider flu," to conclude that the patient might have influenza. The process is repeated till a specific conclusion or all applicable rules have been used up.

One of the greatest strengths of forward chaining is that it follows several paths at once, which makes it suitable for applications where the goal might not be predefined but can be pursued in several paths. This sometimes can also lead to information overload if there are many rules activated, and as such, proper output management becomes necessary. Generally, forward chaining allows expert systems to provide decisions based on the given data, which are comprehensive and real-time, thus enhancing many decisions.

3.3. Backward Chaining

Backward chaining is yet another flavor of reasoning applied in expert systems by artificial intelligence (AI). It starts from a hypothesis or a desired outcome and continues to verify whether known facts and rules will validate that result or not. Where forward chaining begins with known facts and derives conclusions from those facts, backward chaining is more suitable to diagnostic applications where one is searching for the presence or absence of a diagnosed situation on the basis of observed symptoms or data.

In a backward chaining process, the inference engine begins by identifying what it wants to achieve, perhaps the diagnosis of a medical condition. It then returns and goes through the knowledge base looking for applicable rules that are in fact actually a step toward that conclusion. For example, if the objective is to ascertain whether the patient has flu, then the system will search for rules such as "If flu is suspected, then check for symptoms such as fever and cough." If the symptoms exist then, the system's conclusion points toward the fact that the flu could be the diagnosis.

Its strength lies in the fact that it provides only the relevant facts to back up a warranted hypothesis with minimal, if any, redundancy in computation. Its light shines brightest in very complicated decision situations, such as syllogistic legal arguments, diagnosis of medical ailments, or to track down technical bugs.

4. APPLICATIONS OF EXPERT SYSTEMS

4.1. Expert Systems in Healthcare

Healthcare Expert Systems An expert system is a particular type of artificial intelligence program applied to help a health care practitioner with the diagnosis, treatment, and management

One famous early example is MYCIN, which was one of the first expert systems and was utilized for diagnosis and advice for patients with bacterial infections and the appropriate antibiotics to be used. MYCIN confirmed the possibility that expert systems could indeed support practitioners in making clinical decisions that were not only evidence-based, but were relevant to specific instances of individual patients. The modern healthcare expert system usually transcends mere diagnosis and encompasses treatment planning, risk analysis, and patient management.

They have vast knowledge bases of medical literature, clinical guidelines, and history patient data that allow the systems to analyze symptoms, lab results, and history. By using methods of forward and backward chaining, the inference engine draws conclusions from its ideas and generates suggestions for diagnosis, treatment protocols, and alerts against potential complications.

Using expert systems in healthcare improves the efficiency and accuracy levels of diagnosis while greatly reducing error levels and also improving the general quality of care. At the course of time, changes in technology further advance expert systems whose integration with electronic health records and digital tools put them at center stage in modern health care setup.

4.2. Expert Systems in Finance

Expert systems in finance are specialized artificial intelligence applications designed to guide financial professionals and institutions in informed decisions about investments, risk management, and financial analysis. This system simulates human expertise to scrutinize large amounts of financial data, market trends, and economic indicators so as to provide information that would be very insightful.

Examples of such software are Expert Choice and FolioBeyond, which support investment decision making, portfolio construction, and portfolio rebalancing by investors, depending on changing market conditions.

Another important application is that of credit risk assessment. It studies the borrower's credit history and loan behavior by conducting analysis through expert systems to determine the likelihood of default. Such automation improves efficiency in the underwriting process while allowing loans to be approved faster based on strict standards for risk management.

Such expert systems thus provide better accuracy in stating the facts by way of data analysis and, therefore, reduce the chances of human errors, improve decision making, and increase operational efficiency. However, with the increasing complexities of the financial markets, these systems will be crucial for guiding the management of finance concerning facing the risks and opportunities in financial markets.

4.3. Expert Systems in Engineering

Engineering expert systems are advanced applications of artificial intelligence that can support engineers of all disciplines by providing them with the type of knowledge and decision-making powers to match experts in their field. These knowledge bases for an engineering expert system are vast and house principles of engineering, standards of design, along with the best practices in solving complex problems.

One of the most common uses of expert systems in engineering is applied in design optimization. For instance, systems like CAD-based expert systems can facilitate the design structure or even components depending upon material properties, load requirements, and environmental factors. They assist engineers in finding the most appropriate design configurations while meeting various safety standards and regulations.

Application to troubleshooting and diagnostics Another application is in troubleshooting and diagnostics. Expert systems can interpret data from sensors and equipment in the diagnosis of faults in machinery or systems. It provides step-by-step guidance for the maintenance or repair of the machinery or systems, thus saving engineers' time and efforts in troubleshooting efficiently.

These systems enhance productivity while minimizing the errors, hence increasing the quality of engineering projects through logical reasoning and presupposed rules. The ever-developing technology keeps putting these systems with other digital tools and soft wares in line as time goes by and hence become more precious to civil, mechanical, and electrical engineers in coming up with innovations and efficiency in engineered activities (Taghipour, 2023; Taghipour & Beneteau-Piet, 2020 ; Mahfod et al., 2019).

4.4 Hybrid Expert Systems

It deserves attention that the hybrid expert systems, connecting rule-based frameworks with advanced methods of machine learning, prove more flexible and open to study. For example, medical diagnostic systems have used probabilistic reasoning combined with neural nets to increase preciseness. Hybrid discussions will also tackle on the constant change and adaptation of teaching points within expert systems.

5. ADVANTAGES OF EXPERT SYSTEMS

5.1. Consistency and Reliability

These make expert systems invaluable instruments in various fields of activity. By consistency, one typically understands that for the same input, an expert system, no matter external conditions or variations in the user, will give the same output as a result. This is highly important in application areas like health and finance sectors, where even slight variations in making decisions at strategic or policy levels could lead to fateful consequences. Expert systems primarily work based on a set of structured rules and a comprehensive knowledge base to ensure uniformity and, therefore, unbiased decision-making.

Reliability, in contrast, is the ability of a system to produce accurate, reliable, and consistent results over time. Expert systems are developed to minimize human errors because they rely on well-established rules and validated information. Such a system's reliability allows an organization to depend on the recommendations made by the system, thus improving decision-making activities.

Expert systems can run non-stop without feeling fatigued and with not even one drop in performance even during periods of high demand. Such a system is really very useful for real-time analysis and quick decision making hence also very critical in the scenario which requires real-time analysis and quick decisions. Consistency and reliability together give expert systems the onus to contribute to high-quality, productive solutions that build confidence in high-stake domains like medicine, finance, engineering, etc.

The expert system gives high consistency and reliability. Expert systems always provide uniform responses to the same input, thus eliminating human error due to fatigue, bias, or oversight. Since expert systems are based on structured knowledge base and predefined rules, in the same type of scenario, decisions become uniform. This feature is very vital in healthcare, finance, or engineering fields, where decisions are often demanded to be accurate and repetitive. Furthermore, expert systems do not suffer from the disturbances of emotions or cognition; therefore, they become some reliable sources for delivering expertise advice under various conditions.

5.2. Speed and Efficiency

One of the primary advantages that an expert system brings to AI is speed and efficiency, so that the organizations can quickly take well-informed decisions. The importance associated with expert systems is that it gives an added advantage, which human experts cannot provide. When exposed to a large amount of data, experts take much time analyzing such information and then deriving a conclusion. But a well-designed expert system can process massive volumes of information within an instant. This characteristic is extremely invaluable for environments where the speed of decision is essential. For example, in emergency medicine or financial trading, a person must make quick decisions.

The source of strength underlying the effectiveness of an expert system is its well-structured knowledge base and logical inference engine, enabling it to execute a set of predefined rules and algorithms with speed on input data. For instance, a health provider can analyze real-time symptoms and their medical history and lab results through an expert system that will provide immediate recommendations of diagnosis and treatment options. Fast analysis translates into quicker intervention, and possibly saving lives, too.

Expert systems also run continuously, without the characteristic human expert's fatigue, thus providing effective and consistent performance over time. This efficiency contributes to increased productivity and allows organizations better resource allocation, consequently reducing the cost of operations. Generally, the speed and efficiency of expert systems render them indispensible tools in enhancing decision-making processes in different entities as it makes a business respond to challenges and opportunities quickly.

5.3. Accessibility to Non-Experts

These systems are constructed to present expert-level knowledge and decision-making capabilities in a user-friendly manner so that anyone with no specific training or expertise can take advantage of the gateways into valuable insights and recommendations. Democratization in areas like health care, finance, and engineering, where critical insight often goes missing, is of particular significance.

Additionally, most expert systems have explanation facilities that would explain their rationale in arriving at the recommendations, thus further facilitating better understanding and acceptance of the users. This ease of access promotes teamwork between experts and non-experts, where teams can base informed decisions and work much more effectively.

Then expert systems also become an important learning resource as organizations begin to use digital tools more and more extensively, for presenting intricate concepts and thereby enabling the user to learn from it and develop their knowledge in a specific subject matter while simultaneously navigating their own domains more effectively. In other words, access to expert systems significantly enhances decision-making capabilities in all spheres.

This democratization of expertise is particularly useful in critical fields such as health care, finance, and engineering where expert knowledge needs to be available for a decision to be made effectively. For example, a healthcare expert system would help a general practitioner diagnose a condition that they are not trained to treat by coming up with an analysis of symptoms.

Besides that, most expert systems have the explanation facilities that explain why they draw certain conclusions thereby enhancing even the understanding level of the end-users that will ultimately develop confidence in the outcome. More accessible knowledge about experts makes even non-experts make informed decisions. Therefore, this leads to better results as well as more collaborations across different industries (Nilsson, 1998).

6. LIMITATIONS OF EXPERT SYSTEMS

6.1. Static Knowledge Base

One of the essentials of expert systems in AI is a static knowledge base, which is a fixed base of facts, rules, and heuristics derived from human expertise in a specific domain. Since such systems do not learn or adapt to new knowledge, it does not evolve on its own unless it is updated by humans either by knowledge engineers or a domain expert. This has both advantages and limitations.

The advantage of a static knowledge base is that it is reliable and uniform. Since all the information is predefined and well-structured, one would expect the same response to similar questions; hence, the outputs produced by the system are predictable and based on preestablished knowledge.

This is ineffective in rapidly changing fields. Moreover, static knowledge bases get problematic to handle unstructured or vague data such that it constrains the flexibility of the system. Thus, static knowledge bases make a robust foundation for the expert systems, but the system must be updated periodically as new data arrives to maintain continuation and relevance over time.

In a static knowledge base, facts and rules are known before the base is developed. It therefore provides a fixed repository for expert systems where reliable and consistent output is ensured based on established facts and rules. However, this technique can fail without proper updates because it deteriorates as time progresses. Continuous maintenance is usually necessary in expert systems to have the most up-to-date facts and rules.

6.2. Lack of Learning Capabilities

One of the major drawbacks is that it poorly learns. Expert systems, based on a static knowledge base created from predefined rules and facts, are actually very far from the modern AI approach, which goes from machine learning to deep learning.

As the industries move along with their progress and new knowledge is gained, an expert system may not produce relevant information while giving decision-making advice, and thus there is a possibility of mismatch in the decisions provided. For example, in the medical sector, where if a new disease or change in prescription treatment has occurred, it may not be incorporated into the system and thus has the possibility of adversely affecting patient care.

It is also impossible for expert systems to show enhanced performance over time because of the failure of learning from experience. They do not improve their decision making based on past experiences, and thus are less effective than adaptive systems that grow and learn with new data.

Generally, since they lacked the learning capacity, expert systems were much less versatile and adaptable; they needed constant updates and maintenance in order to keep pace with rapid and changing environments. This, therefore reflects the need for the integration of mechanisms for learning in AI applications of the future.

6.3. Difficulties With Unstructured Data

A significant weakness of expert systems is that they cannot handle unstructured data very well. Expert systems are basically created to perform tasks with structured data that essentially consists of information structured in preformatted forms, such as files in databases or spreadsheets; hence, it can be labeled and processed in a neat fashion. Still, most the data available today is unstructured or text, images, audio, and video, particularly in health, finance, and engineering fields.

An expert system cannot break down unstructured data into meaningful insights or usable decisions. Consider the healthcare example: many clinical notes, radiology reports, and patient histories are written in natural language. If an advanced expert system could not interpret natural language processing, it would not make sense of texts, thus potentially failing to identify diagnoses and giving incorrect treatments.

Besides, unstructured data may not be clear-cut and context-dependent with many different things that can cause stiff systems based on predefined rules to suffer when they try to apply their predetermined knowledge, thus limiting the overall effectiveness of the whole system.

One of the significant limitations on AI expert systems is the challenge of handling unstructured data. Expert systems, basically built to manipulate structured data, organized in some predefined format such as databases or spreadsheets, find it extremely difficult to handle the enormous amount of unstructured data found in the real world-from free text to images and audio to video.

That makes a system inapplicable and ineffective in many industries, thereby reducing its overall utility. However, this limitation can be overcome only with the help of advanced technologies, especially the integration of technologies called NLP and computer vision, that allow expert systems to interpret and analyze unstructured information properly.

7. FUTURE OF EXPERT SYSTEMS

7.1. Integrating Machine Learning

With continuous development in the field of machine learning, the future of expert systems is increasingly harnessed to vastly increase their adaptability, learning capacity, and overall effectiveness. Traditional expert systems rely on static knowledge bases and predefined rules, even though they can be pretty effective in certain domains-they still tend to limit experts' flexibility in dealing with new information or some evolving scenarios. The systems can break free from shackles that regard them as simply static and rule-based, thereby becoming dynamic self-improving entities that can know from data through machine learning.

The machine learning is an evolution of this expert system in which it constantly updates its knowledge base from massive datasets to identify patterns and decide whether to change decision-making processes based on feedback and real-world outcomes. For instance, an ML-augmented expert system could perhaps, in health sectors improve its own diagnostic accuracy as it learns from patient outcomes, new medical research, and change in the treatment protocols without human intervention requiring reloading of programs. Such a property for learning and improvement over time makes the system more efficient and reliable in a changing environment.

Ultimately, marrying deep domain expertise in the form of the traditional expert system with adaptability in learning like machine learning will lead to much smarter and self-governing systems. Such more capable systems should help render accurate, timely solutions to improve decision-making in fields such as healthcare and finance, engineering, and others while being relevant to dynamic, data-driven environments.

7.2. Overcoming Current Challenges

Super over the current situation, one would face no problems to get an assured future for expert systems. The key problem is a static knowledge base. A classical expert system relies on pre-defined arguments represented in terms of rules and facts.

These were soon outdated in swiftly changing fields such as medicine or finance. Updating was imperative to keep an expert system up to date and would prove clumsy and resource-intensive. This is attained by including machine learning (ML) as a core solution. Machine learning enables expert systems to automatically learn from new data with the aid of computers adapting and thereby keeping the system current without requiring constant manual intervention (Turban & Aronson, 2001).

Another is the problem of manipulating unstructured data. Most of the data found in industries such as health, finance, and law are free text, images, or audio data that is normally cumbersome to manipulate by legacy expert systems. The better a computer system's ability with natural language processing and computer vision, the better it may be able to process free text, images, or audio - making an expert system even more versatile in a whole range of applications.

Scalability: Scalability is one challenge where high volumes of data have to be processed in real time by the expert system. The inclusion of cloud computing and distributed systems can ensure increased scalability alongside better handling of large volumes and even intricate data sets.

There also is the issue of rigidity in decision-making. The traditional expert systems are rule-bound and hence not able to address uncertain and ambiguous situations. Hybrid systems combining expert systems with probabilistic reasoning or with fuzzy logic may enable more flexible context-sensitive decision-making.

These will ensure that expert systems are robust, relevant, and up to the task of coping with complex, dynamically changing problems in any field.

7.3. Ethical Implications in AI Decision-Making

The fast inductive use of AI in decision making, in particular through expert systems, generates a lot of ethical implications that must be dealt with responsibly. Perhaps the biggest concern is regarding the issue of bias and fairness. Expert systems are based on knowledge bases and rules created by human beings. If the inputs into these are biased or incomplete, the output is likely to be unfair. For instance, the biased recruiting or criminal justice data might lead to making unfair and discriminating selections targeting a given race, gender and socio-economic status. Therefore, periodical audits of AI systems are required to bring in transparency to mitigate such risks.

The moment that the consequence of accepting the recommendation by a system is undesirable, are the developers? Is it the system? Is it the user? How does one trace the accountability lines or explain decisions made by AI when those decisions are wrong?.

Privacy is another major issue. Most AI systems, including expert systems, need a lot of data to be meaningful. Much of that usually involves personal data in a sense, and thus there are problems on security regarding the use of such based on consent. Respect for the rights of people involves data protection and ethical use.

Autonomy and human oversight. Advanced knowledge systems introduce the risk of over-reliance on AI systems, and the risk is that somehow important decisions are delegated to a machine at its discretion. It rests upon the responsibility of ensuring humans remain in control of decision-making in AI processes with human oversight of technology so it serves humanity responsibly and maintains the ethical accountability of doing so.

7.4. Expert Systems and Adaptive Learning

Expert systems have been traditionally static, depending purely on what was coded in the knowledge base and working only according to the predefined rules that somewhat limit their ability to perform dynamically. But with adaptive learning, the ability for flexibility in expert systems could be improved by making a system learn from new data and evolve. The integration would eliminate one of the major limitations of the conventional expert systems-the limitation that its knowledge base cannot change automatically or that processes improve themselves.

Adaptive learning allows expert systems to update their own performance given the patterns discovered in big data, insights gathered from feedback, and thus adjustments to rules. Such dynamic conditions requiring domains-including healthcare, finance, and cybersecurity-intersect uniquely with adaptive learning. For example, an adaptive expert system in healthcare could learn over time given outcome data from patients and new medical research to better diagnose and make recommendations for treatment. This is in contrast with traditional expert systems, which have to be updated manually in order to incorporate new knowledge.

The abilities of machine learning techniques such as supervised or unsupervised learning allow expert systems to process new data and adapt their reasoning accordingly. They are better also at handling previously unseen situations. In addition, adaptive learning further enhances the capability of an expert system to deal with the problem of uncertainty and ambiguity in making its decisions in complex, real-world situations.

The union of expert systems and adaptive learning creates a more intelligent, self-improving AI model. Hybrid methodology will ensure the increase in the length and relevance of the system, cut down the costs for its maintenance and dependence on human updates, making it a much stronger tool for industrial progress - that thrives on continuous innovation and adaptability (Russell & Norvig, 2016).

7.6 Actual Failures in Life

This is the way the notable failures obtained could actually be traced back to its low-grade data or outdated rule-bases. A resolution to the confinements could be the well-set demonstrations for what would look like in the future in applications and practices.

8. CONCLUSION

In summary, expert systems represent one of the most important ideas developed by AI. They represent an advanced model to mimic human knowledge and are applied in solving complex tasks in various industries. Their precise implementation, using a knowledge base and an inference engine, is sure to present decisions consistent, reliable, and at the same level as experts. Consistency plays an important role in industries such as health care, finance, engineering, and legal fields where accuracy has to be high and the results should be repeatable. The main appeal of automation is that it can make decision-making much easier, thereby increasing productivity, diminishing human error, and making specialized knowledge more accessible to non-experts, democratizing expert-level advice.

Despite such strengths, expert systems do possess weaknesses. Because they rely on static knowledge bases, these easily become outdated, and managing unstructured data or learning through experience is not possible, which renders them less adaptive in dynamic environments. Lack of learning ability makes the integration of these in future with machine learning as well as adaptive technologies for evolving and becoming effective even in changing scenarios absolutely necessary.

In the near future, much more could be done with expert systems if they were integrated with machine learning, natural language processing (NLP), and other AI techniques, making them adaptable and resourceful enough to handle unstructured data. Thus, the core strengths of expert systems will be maintained and problems that exist concerning them now will be overcome. Going forward, expert systems will continue to play a crucial role in AI, providing undeniably reliable support for decision-making in increasingly complex and data-intensive environments, expanding their capabilities moving forward.

REFERENCES

Buchanan, B. G., & Shortliffe, E. H. (1984). *Rule-Based Expert Systems: The MYCIN Experiments of the Stanford Heuristic Programming Project*. Addison-Wesley.

Cauchois, R., Taghipour, A., Kang, D., Zoghlami, N., & Abed, M. (2017). Advanced solutions for a supply chain with stochastic information. *Journal of Advanced Management Science*, 5(1), 9–13. DOI: 10.18178/joams.5.1.9-13

Dadsena, K. K., Mathiyazhagan, K., & Taghipour, A. (2021). Analysis of barriers for the build the resilient supply chain networks post-COVID-19. In *Managing Supply Chain Risk and Disruptions: Post COVID-19* (pp. 79–89). Springer International Publishing. DOI: 10.1007/978-3-030-72575-4_7

Glaa, B., Zoghlami, N., & Taghipour, A. (2014, May). Outsourcing and off-shoring healthcare services: A way to cut the costs and improve quality. In 2014 international conference on advanced logistics and transport (ICALT) (pp. 189-194). IEEE.

Jackson, P. (1998). *Introduction to Expert Systems* (2nd ed.). Addison-Wesley.

Loivet, W., Taghipour, A., & Kang, D. S. (2020). The rise of green supply chain management: Between complexity and necessity. *J. Econ. Bus. Manag*, 8(1), 1–7. DOI: 10.18178/joebm.2020.8.1.604

Lu, X., & Taghipour, A. (2023). A French Guide for Businesses and Governments to Integrate the Concept of Sustainable Development Into Their Supply Chains. In *Government Impact on Sustainable and Responsible Supply Chain Management* (pp. 1–12). IGI Global. DOI: 10.4018/978-1-6684-9062-4.ch001

Mahfod, J., Canel-Depitre, B., & Taghipour, A. (2019). Quality function deployment-ELECTRE in supplier evaluation. Journal of Advanced Management Science, 7(4).

Mbiatem, J., Taghipour, A., & Canel-Depitre, B. (2018, August). Supplier selection approaches for decision makers. In *Proceedings of the 8th International Conference on Information Communication and Management* (pp. 108-112).

Merimi, M., & Taghipour, A. (2021). Accelerating the digitalization of the supply chain: An empirical research about COVID-19 crisis. In *Digitalization of decentralized supply chains during global crises* (pp. 1–24). IGI Global. DOI: 10.4018/978-1-7998-6874-3.ch001

Nadia, E. N., Beatrice, C. D., & Atour, T. (2021). The determinants of hotel room rates in Beirut: A hedonic pricing model. International Journal of Trade. *Economics and Finance*, 12(2), 33–42.

Nilsson, N. J. (1998). *Artificial Intelligence: A New Synthesis*. Morgan Kaufmann.

Russell, S., & Norvig, P. (2016). *Artificial Intelligence: A Modern Approach* (3rd ed.). Pearson.

Taghipour, A. (2014). Improving the plan of a manufacturing network with non-integrated business units. *International Journal of Applied Logistics*, 5(2), 1–11. DOI: 10.4018/ijal.2014040101

Taghipour, A. (Ed.). (2020). *Demand forecasting and order planning in supply chains and humanitarian logistics*. IGI Global.

Taghipour, A. (Ed.). (2023). *Government Impact on Sustainable and Responsible Supply Chain Management*. IGI Global. DOI: 10.4018/978-1-6684-9062-4

Taghipour, A., Abed, M., & Zoghlami, N. (2015, May). Design for remanufacturing respecting reverse logistics processes: A review. In 2015 4th International Conference on Advanced Logistics and Transport (ICALT) (pp. 299-304). IEEE. DOI: 10.1109/ICAdLT.2015.7136602

Taghipour, A., & Beneteau-Piet, C. (2020). Sustainable supply chain management performance. *International Journal of Innovation, Management and Technology*, 11(6), 165–169. DOI: 10.18178/ijimt.2020.11.6.897

Taghipour, A., Foukolaei, P. Z., Ghaedi, M., & Khazaei, M. (2023a). Sustainable Multi-Objective Models for Waste-to-Energy and Waste Separation Site Selection. *Sustainability (Basel)*, 15(22), 15764. DOI: 10.3390/su152215764

Taghipour, A., & Frayret, J. M. (2011). Supply chain coordination planning: A review. Work paper, École polytechnique de Montréal.

Taghipour, A., & Frayret, J. M. (2013a, May). An algorithm to improve operations planning in decentralized supply chains. In *2013 International Conference on Advanced Logistics and Transport* (pp. 100-103). IEEE. DOI: 10.1109/ICAdLT.2013.6568442

Taghipour, A., & Frayret, J. M. (2013b). Coordination of operations planning in supply chains: A review. *International Journal of Business Performance and Supply Chain Modelling*, 5(3), 272–307. DOI: 10.1504/IJBPSCM.2013.055729

Taghipour, A., Lu, X., & Smahi, M. (2022). Opportunities and challenges for society 5.0 in procurement. *Blockchain Technology in Supply Chain Management for Society*, 5(0), 91–100. DOI: 10.1201/9781003177432-9

Taghipour, A., & Merimi, M. (2021, March). Digital transformation of supply chains during crisis: covid-19. In 11th Annual International Conference on Industrial Engineering and Operations Management, IEOM (pp. 7493-7501). DOI: 10.46254/AN11.20211298

Taghipour, A., Padash, A., Etemadi, V., Khazaei, M., & Ebrahimi, S. (2024). Sustainable and Circular Hotels and the Water–Food–Energy Nexus: Integration of Agrivoltaics, Hydropower, Solar Cells, Water Reservoirs, and Green Roofs. *Sustainability (Basel)*, 16(5), 1985. DOI: 10.3390/su16051985

Taghipour, A., Ramezani, M., Khazaei, M., Roohparvar, V., & Hassannayebi, E. (2023b). Smart transportation behavior through the COVID-19 pandemic: A ride-hailing system in Iran. *Sustainability (Basel)*, 15(5), 4178. DOI: 10.3390/su15054178

Taghipour, A., Sohrabi, A., Ghaedi, M., & Khazaei, M. (2023c). A robust vaccine supply chain model in pandemics: Case of Covid-19 in Iran. *Computers & Industrial Engineering*, 183, 109465. DOI: 10.1016/j.cie.2023.109465

Turban, E., & Aronson, J. E. (2001). *Decision Support Systems and Intelligent Systems* (6th ed.). Prentice Hall.

Zhang, Y., & Lu, Z. (2015). A survey on the integration of machine learning in expert systems. *Expert Systems with Applications*, 42(21), 7569–7585.

Zoghlami, N., Taghipour, A., Merlo, C., & Abed, M. (2016). Management of divergent production network using decentralised multi-level capacitated lot-sizing models. *International Journal of Shipping and Transport Logistics*, 8(5), 590–604. DOI: 10.1504/IJSTL.2016.078686

Chapter 19
The AI Revolution Implications for Workforce Dynamics and Employment in Supply Chains

Abhijit Chandratreya
https://orcid.org/0000-0001-9403-2094
SCES's Indira Institute of Management, Pune, India

ABSTRACT

This chapter examines the impact of artificial intelligence (AI) on workforce dynamics and employment within supply chains. As AI technologies become increasingly integrated into supply chain operations, they are reshaping job roles, skill requirements, and organizational structures. The chapter explores the dual nature of AI's influence, highlighting both the opportunities for efficiency and innovation and the challenges related to job displacement and workforce adaptation. Through case studies and analysis of current trends, the chapter provides insights into how businesses can strategically manage the transition toward AI-enhanced supply chains. Emphasis is placed on the importance of upskilling and reskilling initiatives to ensure workforce resilience and sustainability in this new era.

DOI: 10.4018/979-8-3693-7478-8.ch019

BACKGROUND

The integration of artificial intelligence (AI) into supply chains has initiated a transformation in workforce dynamics and employment patterns. AI technologies, including machine learning, robotics, and advanced analytics, are being increasingly adopted to enhance efficiency, accuracy, and decision-making processes in supply chain management (Ivanov et al., 2019). This shift towards automation and digitalization has significant implications for the nature of work and the roles of employees within the supply chain sector.

One of the primary impacts of AI on workforce dynamics is the automation of routine and repetitive tasks. AI-powered systems can perform tasks such as inventory management, order processing, and demand forecasting more efficiently than human workers (Chui, Manyika, & Miremadi, 2016). Consequently, there is a reduced demand for low-skilled labor, leading to a potential displacement of workers whose jobs are susceptible to automation. For instance, the use of robotic process automation (RPA) in warehouse operations can significantly reduce the need for manual labor, raising concerns about job losses and the displacement of workers (Frey & Osborne, 2017).

However, the adoption of AI in supply chains also creates opportunities for new types of employment and the evolution of existing roles. As AI systems take over mundane tasks, the demand for high-skilled labor, particularly in areas such as AI development, maintenance, and oversight, increases (Brynjolfsson & McAfee, 2014). Workers are required to possess advanced technical skills and the ability to work alongside AI technologies, leading to a shift in workforce dynamics towards more specialized and knowledge-intensive roles. This shift necessitates a focus on reskilling and upskilling initiatives to prepare the workforce for the changing employment landscape (Manyika et al., 2017).

AI-driven supply chains enable enhanced decision-making and strategic planning capabilities. Advanced analytics and predictive modeling allow for more accurate forecasting and optimization of supply chain operations, which can lead to improved efficiency and reduced costs (Ivanov & Dolgui, 2020). This transformation requires supply chain professionals to develop analytical and strategic skills to interpret AI-generated insights and make informed decisions. As a result, there is a growing demand for professionals with expertise in data analysis, machine learning, and supply chain management.

Despite the potential benefits, the integration of AI in supply chains also poses challenges related to workforce dynamics and employment. The transition to AI-driven processes can exacerbate existing inequalities in the labor market, particularly for workers in regions or industries with limited access to education and training resources (Autor, 2015). Moreover, there are ethical and regulatory considerations regarding the fair implementation of AI technologies and the protection of workers' rights (West, 2018).

The AI revolution is reshaping workforce dynamics and employment in supply chains, presenting both opportunities and challenges. While AI can enhance efficiency and create new high-skilled jobs, it also poses risks of job displacement and increased inequality. Addressing these challenges requires a proactive approach to workforce development, including reskilling and upskilling initiatives, and the establishment of ethical guidelines for AI implementation.

This chapter delves into these themes, providing insights into the multifaceted impact of AI on workforce dynamics and employment in supply chains.

INTRODUCTION

Overview of AI in Supply Chains

Artificial intelligence (AI) is revolutionizing supply chains by enhancing efficiency, accuracy, and agility in operations. AI technologies, such as machine learning, predictive analytics, and automation, are enabling companies to optimize inventory management, streamline logistics, and improve demand forecasting. For instance, AI-powered algorithms can analyze vast amounts of data to predict customer demand patterns, thereby reducing excess inventory and minimizing stockouts (Ivanov et al., 2019). AI-driven automation, including the use of robotics and autonomous vehicles, is transforming warehouse operations and last-mile delivery, leading to faster and more reliable fulfilment processes (Wamba et al., 2020).

The integration of AI in supply chains also facilitates better decision-making and enhances supply chain resilience. By leveraging AI, companies can gain real-time insights into supply chain disruptions and respond proactively to mitigate risks. This predictive capability is particularly crucial in today's dynamic global market, where disruptions such as natural disasters or geopolitical tensions can significantly impact supply chain continuity (Choi, Wallace, & Wang, 2018). AI technologies enable supply chain collaboration by providing a unified platform for stakeholders to share information and coordinate actions, thereby fostering a more interconnected and agile supply chain ecosystem (Dubey et al., 2020).

Importance of Understanding Workforce Implications

AI technologies are reshaping job roles and responsibilities, necessitating a shift in the skill sets required by workers. As automation and machine learning take over routine and repetitive tasks, employees are increasingly expected to perform more complex and cognitive functions (Brynjolfsson & McAfee, 2014). This shift demands significant investments in upskilling and reskilling programs to prepare the workforce for new challenges and opportunities. Organizations that fail to address these changes risk facing a skills gap, which can hinder their ability to fully leverage AI's potential and maintain a competitive edge (Manyika et al., 2017).

The social and economic implications of AI-driven changes in the workforce cannot be overlooked. The displacement of certain job categories due to automation raises concerns about unemployment and income inequality (Acemoglu & Restrepo, 2018). Policymakers, educators, and business leaders need to collaborate in creating strategies that mitigate these negative impacts while promoting inclusive growth. By understanding and addressing workforce implications, stakeholders can develop comprehensive plans that not only enhance operational efficiency but also ensure that the benefits of AI are broadly shared across society (Bessen, 2019). This holistic approach is key to fostering a sustainable and equitable transition to an AI-driven economy.

THE EVOLUTION OF AI IN SUPPLY CHAINS

Historical Context and Technological Milestones

The journey began in the 1950s and 1960s, with the advent of early computing technologies and basic automation. These initial developments laid the groundwork for more advanced AI applications. During the 1980s and 1990s, supply chains began to leverage enterprise resource planning (ERP) systems and advanced planning and scheduling (APS) systems, which integrated various functions such as inventory management, procurement, and production planning (Christopher, 2016). These systems utilized rule-based algorithms and basic data analytics to improve operational efficiency.

The turn of the 21st century marked a significant leap with the introduction of machine learning and predictive analytics in supply chains. Companies began to use these technologies to analyze historical data and predict future trends, leading to more accurate demand forecasting and optimized inventory levels (Chopra & Meindl, 2016). The 2010s saw the rise of big data and the Internet of Things (IoT), which further revolutionized supply chains by enabling real-time tracking and data collection from various sources. This period also witnessed the emergence of blockchain technology, enhancing transparency and security in supply chain transactions (Ivanov, Tsipoulanidis, & Schönberger, 2019). Today, AI-driven supply chains are characterized by the integration of advanced robotics, autonomous vehicles, and sophisticated AI algorithms that continuously learn and adapt to changing conditions, driving unprecedented levels of efficiency and resilience.

Current State of AI Adoption in Supply Chains

Companies across various industries are leveraging AI to optimize their operations, improve decision-making processes, and respond more effectively to market demands and disruptions. Key areas where AI is making a significant impact include demand forecasting, inventory management, and logistics optimization. AI-driven predictive analytics enable businesses to accurately forecast demand, reducing excess inventory and minimizing stockouts, thus improving overall supply chain efficiency (Ivanov et al., 2019).

The use of AI in logistics and transportation is transforming how goods are moved and tracked. Autonomous vehicles, drones, and AI-powered routing algorithms are reducing delivery times and costs while enhancing reliability. Real-time tracking and IoT devices provide granular visibility into supply chain operations, allowing for proactive management of potential issues (Wamba et al., 2020). Additionally, AI is enhancing supplier relationship management by analyzing performance data and predicting risks, leading to more strategic sourcing and procurement decisions.

Despite these advancements, the adoption of AI in supply chains is not without challenges, such as the need for significant investments in technology infrastructure, data privacy concerns, and the requirement for specialized skills to manage and interpret AI systems (Dubey et al., 2020). Nonetheless, the trend towards greater AI integration is expected to continue, driven by the competitive advantages it offers in terms of efficiency, cost savings, and adaptability.

AI TECHNOLOGIES TRANSFORMING SUPPLY CHAINS

Automation and Robotics

These technologies streamline and enhance efficiency across various supply chain functions, from manufacturing and warehousing to transportation and delivery. In manufacturing, automated systems and robotic process automation (RPA) are increasing production speed and consistency while reducing human error and operational costs (Choi, 2020). Warehouses are becoming increasingly sophisticated with the use of autonomous robots that handle tasks such as sorting, packing, and palletizing goods, significantly improving throughput and accuracy (Wang et al., 2021). In transportation, self-driving trucks and drones are beginning to emerge, promising to lower delivery times and costs while enhancing safety. Additionally,

AI-driven robotics are capable of operating 24/7, providing a level of reliability and efficiency that human labor alone cannot match. As these technologies continue to advance, they are expected to further transform supply chains, making them more resilient, responsive, and efficient.

Machine Learning and Predictive Analytics

Machine learning algorithms analyze vast amounts of historical and real-time data to identify patterns and correlations that humans might miss. This capability allows supply chain managers to predict demand fluctuations, optimize inventory levels, and streamline production schedules with unprecedented accuracy (Choi, Wallace, & Wang, 2018).

Predictive analytics further enhances this process by using these insights to forecast future trends and potential disruptions, enabling companies to proactively adjust their strategies. For instance, by predicting which products will be in high demand during specific seasons, businesses can optimize their stock levels to reduce overstock and stockouts, thereby improving operational efficiency and customer satisfaction (Ivanov et al., 2019). The integration of these AI technologies not only enhances efficiency and cost-effectiveness but also builds a more resilient and adaptable supply chain capable of responding swiftly to changes and uncertainties in the market.

Internet of Things (IoT) and Smart Sensors

The Internet of Things (IoT) and smart sensors are the AI technologies transforming supply chains by providing unprecedented levels of visibility, connectivity, and data-driven insights. IoT devices, embedded with smart sensors, enable real-time tracking of goods, monitoring of environmental conditions, and seamless communication across various stages of the supply chain. This real-time data collection allows for proactive management of inventory, ensuring optimal stock levels and reducing the likelihood of overstocking or stockouts (Ben-Daya et al., 2019). Smart sensors in warehouses and vehicles continuously monitor variables such as temperature, humidity, and location, which is crucial for maintaining the quality and integrity of perishable goods and sensitive products (Gubbi et al., 2013). The integration of IoT with AI enhances predictive maintenance of equipment, minimizing downtime and improving operational efficiency. By leveraging the rich data streams generated by IoT and smart sensors, supply chain managers can make informed decisions, anticipate disruptions, and respond swiftly to changing market conditions, thereby driving resilience and agility in supply chain operations (Xu et al., 2014).

Blockchain and AI Integration

Blockchain, with its decentralized and immutable ledger, provides enhanced transparency and security for transactions across the supply chain.

When combined with AI, it enables more robust and reliable data management, facilitating real-time tracking and validation of goods as they move through the supply chain (Kshetri, 2018). AI algorithms can analyze the vast amounts of data recorded on the blockchain to optimize supply chain operations, predict potential disruptions, and improve decision-making processes.

For example, AI can utilize blockchain data to enhance demand forecasting and inventory management by providing a comprehensive and accurate view of supply chain activities (Miller, 2020). This integration not only increases operational efficiency but also fosters greater trust and collaboration among supply chain partners by ensuring data integrity and reducing the risk of fraud. As a result, the synergy between blockchain and AI is driving significant advancements in supply chain transparency, security, and efficiency (Yue et al., 2020).

IMPACT ON WORKFORCE DYNAMICS

Changes in Job Roles and Responsibilities

Traditionally, many supply chain functions, such as inventory management and logistics planning, were carried out manually or with limited automation. However, as AI technologies become more prevalent, these roles are increasingly evolving towards a focus on more complex, value-added tasks (Choi et al., 2018). For instance, routine tasks such as data entry and basic analysis are being automated through AI-driven systems, which frees up employees to engage in strategic planning, problem-solving, and decision-making roles. This shift necessitates a re-evaluation of job descriptions and required skill sets, with an emphasis on competencies related to AI, data analysis, and technology management (Barton & Court, 2019).

AI's ability to process and analyze vast amounts of data has led to the emergence of new roles, such as data scientists and AI specialists, who are responsible for developing, implementing, and managing AI systems (Brynjolfsson & McAfee, 2014). These roles require a blend of technical expertise and business acumen to effectively leverage AI insights for optimizing supply chain performance. As a result, supply chain professionals must adapt to new technologies and continually update their skills to remain competitive in an increasingly automated and data-driven environment (Wamba et al., 2020). This transformation underscores the need for organizations to invest in training and development programs to help employees transition smoothly into their evolving roles and responsibilities.

The Shift from Manual to Cognitive Tasks

Traditionally, many supply chain functions involved manual tasks such as data entry, order processing, and inventory tracking. These tasks were often time-consuming and prone to human error. However, AI technologies, including machine learning algorithms and robotic process automation, are increasingly automating these routine functions, allowing employees to focus on more complex and strategic activities (Brynjolfsson & McAfee, 2014).

As AI takes over manual tasks, there is a corresponding shift towards cognitive tasks that require higher-order thinking, such as data analysis, decision-making, and strategic planning. AI systems can analyze vast amounts of data and provide actionable insights that help optimize inventory levels, forecast demand more accurately, and identify potential supply chain disruptions (Choi et al., 2018). This transition enables supply chain professionals to engage in activities that require creativity, problem-solving skills, and critical thinking—areas where human expertise remains crucial. Consequently, employees are expected to develop new skills related to AI and data analytics to effectively leverage these technologies and drive innovation in supply chain management (Barton & Court, 2019). This shift not only enhances operational efficiency but also transforms job roles, emphasizing the need for ongoing training and adaptation to keep pace with technological advancements.

New Skills and Competencies Required

As artificial intelligence (AI) increasingly integrates into supply chain management, the demand for new skills and competencies among professionals is rising. With the automation of routine tasks, such as data entry and order processing, employees are required to develop new capabilities that align with the evolving technological landscape (Brynjolfsson & McAfee, 2014). Key skills now in demand include data analytics, machine learning, and AI system management. Supply chain professionals must be adept at interpreting data generated by AI systems to make informed decisions and optimize supply chain operations (Choi et al., 2018).

Proficiency in understanding and managing AI technologies is becoming crucial. This includes the ability to work with AI tools and platforms, troubleshoot issues, and implement AI-driven solutions effectively (Barton & Court, 2019). Additionally, soft skills such as problem-solving, strategic thinking, and adaptability are increasingly important as professionals navigate the complexities introduced by AI and automation. The ability to collaborate across departments and with AI systems to enhance decision-making processes is essential for driving innovation and maintaining operational efficiency in a rapidly changing environment (Wamba et al., 2020). Consequently, organizations are investing in training programs and upskilling initiatives to equip their workforce with the necessary competencies to thrive in this new era of supply chain management.

The Role of Human-AI Collaboration

Human-AI collaboration is becoming a cornerstone of modern supply chain management, enhancing efficiency and decision-making while leveraging the strengths of both human expertise and AI technologies. AI systems excel at processing vast amounts of data and identifying patterns, which enables them to perform complex analyses and automate routine tasks with high accuracy (Brynjolfsson & McAfee, 2014). However, human judgment and creativity remain essential for interpreting AI-generated insights and making strategic decisions that align with organizational goals and ethical considerations (Choi et al., 2018).

In practice, human-AI collaboration manifests in several ways. For example, AI can handle data-driven tasks such as demand forecasting and inventory optimization, while human professionals focus on tasks that require strategic thinking, such as negotiating with suppliers or managing customer relationships (Barton & Court, 2019). Additionally, AI can assist in identifying potential issues or inefficiencies in supply chain operations, but humans are needed to validate these findings, address nuanced challenges, and implement effective solutions (Wamba et al., 2020). This collaborative approach not only enhances operational efficiency but also fosters a more adaptive and resilient supply chain, where AI tools augment human capabilities and enable a more agile response to dynamic market conditions and unexpected disruptions.

EMPLOYMENT TRENDS AND CHALLENGES

Job Displacement and Creation

On one hand, AI-driven automation and advanced technologies have led to the displacement of certain job roles, particularly those involving routine and repetitive tasks. For example, roles centered around manual data entry, basic inventory management, and traditional logistics coordination are increasingly being automated, leading to a reduction in demand for these positions (Brynjolfsson & McAfee, 2014). This displacement can contribute to job losses and require workers to transition into new roles, posing significant challenges for those without the requisite skills or training.

On the other hand, AI is also creating new job opportunities by fostering the demand for roles that involve managing, developing, and interpreting AI technologies. Positions such as data scientists, AI specialists, and machine learning engineers are increasingly in demand as organizations seek to leverage AI to enhance their supply chain operations (Barton & Court, 2019).

Roles focused on strategic decision-making, problem-solving, and human-AI collaboration are becoming more prevalent, as companies need professionals who can integrate AI insights with human expertise to drive business outcomes (Choi et al., 2018). These trends underscore the need for comprehensive upskilling and reskilling programs to prepare the workforce for the evolving job market and ensure a smooth transition for displaced workers. Addressing these challenges effectively will be crucial for harnessing the benefits of AI while mitigating its potential adverse effects on employment (Wamba et al., 2020).

Regional and Sectoral Employment Variations

As AI technologies reshape supply chain operations, their impact on employment differs widely depending on geographic location and industry sector.

Regional Variations: In advanced economies, particularly in North America and Western Europe, the adoption of AI has led to significant job displacement in routine, manual tasks, but has also created new opportunities in technology management and data analysis (Choi et al., 2018). These regions have generally invested more in AI technologies and have better access to training and upskilling programs. Conversely, in developing regions, where technology adoption is slower and educational resources are less accessible, the workforce faces greater challenges. These regions often experience higher levels of job displacement without the corresponding opportunities for re-skilling, exacerbating unemployment and economic disparities (Brynjolfsson & McAfee, 2014).

Sectoral Variations: Within different sectors, the impact of AI also varies significantly. In manufacturing and logistics, AI-driven automation has led to substantial reductions in labor-intensive roles, while creating demand for positions related to AI system maintenance and data analytics (Wamba et al., 2020). Conversely, sectors like retail and services are experiencing different dynamics; while AI improves efficiency and customer service, it also requires a shift in skills towards managing AI tools and integrating technology into customer interactions. The challenge in these sectors lies in balancing technology adoption with the human touch that remains crucial for customer satisfaction (Barton & Court, 2019).

The uneven impact of AI on employment underscores the need for targeted policies and initiatives that address regional and sectoral disparities, including tailored education and training programs, and support for affected workers to facilitate a smoother transition into new roles and industries.

Socio-Economic Implications

As AI technologies automate routine and repetitive tasks, there is a notable shift in jobs, with a growing emphasis on roles that require advanced technical skills and cognitive abilities (Brynjolfsson & McAfee, 2014). This transition can lead to job displacement, particularly for workers in roles that are easily automated, such as data entry and basic logistics functions. This displacement can exacerbate income inequality and contribute to economic disparities, as lower-skilled workers may face difficulties transitioning to new roles that demand higher levels of expertise (Acemoglu & Restrepo, 2018).

The benefits of AI-driven improvements in supply chain efficiency and productivity are not always evenly distributed. While organizations that successfully implement AI can achieve significant cost savings and competitive advantages, smaller firms and those in less advanced economies may struggle to keep pace due to limited resources and technological capabilities (Manyika et al., 2017). This disparity can widen the gap between leading and lagging firms, affecting regional economic stability and exacerbating socio-economic inequalities.

To address these challenges, there is a pressing need for targeted policies and programs that support workforce retraining and upskilling, promote inclusive access to technology, and ensure that the economic benefits of AI are broadly shared (Bessen, 2019).

Case Studies of Affected Industries

Case Study 1: Automotive Industry - General Motors

Overview: General Motors (GM) has been at the forefront of integrating AI and automation into its supply chain operations. The company has implemented AI-driven robotics in its manufacturing plants to streamline assembly processes, reduce production time, and enhance product quality.

Impact:

- **Job Displacement**: Automation has led to a reduction in the number of manual assembly line jobs. Workers previously engaged in repetitive tasks are increasingly being replaced by robotic systems.

- **New Roles**: The shift towards automation has created new roles related to the maintenance and management of robotic systems, as well as positions focused on AI system programming and data analysis.
- **Reskilling Initiatives**: GM has invested in reskilling programs for its workforce to help employees transition to new roles within the company, including training in robotics and AI technologies.

Conclusion: While automation has improved operational efficiency and product quality, it has also necessitated a strategic focus on workforce reskilling and the creation of new job opportunities within the company.

Case Study 2: Retail Industry - Walmart

Overview: Walmart, a global leader in retail, has leveraged AI and machine learning to optimize its supply chain operations, including inventory management and demand forecasting. The company uses AI algorithms to analyze sales data, predict demand trends, and manage inventory levels more effectively.

Impact:

- **Operational Efficiency**: AI-driven demand forecasting has significantly reduced stockouts and overstock situations, leading to improved inventory turnover and reduced waste.
- **Job Roles**: The integration of AI has led to changes in job roles, with a greater emphasis on data analysis and strategic planning rather than manual inventory management.
- **Employment Trends**: While some traditional roles in inventory management have been reduced, new positions related to AI system management, data analysis, and technology integration have been created.

Conclusion: Walmart's use of AI has enhanced operational efficiency and customer satisfaction but has also shifted job roles and created a need for new skills in data analysis and AI management.

Case Study 3: Logistics Industry - DHL

Overview: DHL, a leading logistics company, has embraced AI and automation to improve its supply chain operations, including warehouse management and delivery services. The company has implemented AI-powered robots for sorting and packing and uses machine learning algorithms to optimize delivery routes.

Impact:

- **Increased Automation**: AI and robotics have streamlined warehouse operations, reducing the need for manual labor and increasing the speed and accuracy of order fulfillment.
- **Job Displacement and Creation**: The automation of sorting and packing processes has reduced the need for manual labor in warehouses but has created new roles related to robotics maintenance and AI system oversight.
- **Training and Development**: DHL has developed training programs to equip employees with the skills needed to operate and manage advanced AI and robotic systems, ensuring a smooth transition to new roles.

Conclusion: DHL's adoption of AI and automation has enhanced operational efficiency and accuracy but has also highlighted the need for workforce adaptation through training and the creation of new roles in technology management.

STRATEGIES FOR WORKFORCE ADAPTATION

Upskilling and Reskilling Programs

These programs are designed to equip employees with the new skills and competencies required to thrive in evolving job roles and to mitigate the impact of job displacement caused by technological changes.

Upskilling focuses on enhancing the existing skills of employees to keep pace with technological advancements and changes in job requirements. For example, in the context of AI and automation, upskilling might involve training employees in data analysis, AI system management, and advanced software tools. This approach helps employees stay relevant in their current roles and adapt to new technologies without requiring a complete career change (Brynjolfsson & McAfee, 2014).

Reskilling, on the other hand, involves training employees to acquire new skills that enable them to transition to entirely different roles or sectors. As automation and AI reshape job functions, reskilling programs provide employees with the knowledge and capabilities needed to move into emerging fields or roles created by technological advancements. For instance, workers displaced from manual assembly roles due to automation might receive training in robotics maintenance or data science (Barton & Court, 2019).

Organizations and governments are increasingly recognizing the importance of these programs. Companies such as Google and IBM have invested in extensive training programs to upskill their employees and ensure they are prepared for future technological developments (Manyika et al., 2017). Additionally, government initiatives, such as the European Union's Digital Skills and Jobs Coalition, aim to support workforce reskilling and ensure that workers across various industries can transition into roles that align with current market demands (Choi et al., 2018).

Implementing effective upskilling and reskilling programs not only supports workforce adaptation but also fosters a more resilient and innovative workforce. By investing in these strategies, organizations can address skills gaps, enhance employee retention, and drive sustained growth in an increasingly automated and data-driven economy.

Educational and Training Initiatives

As AI transforms job roles and operational processes, there is a pressing need to equip employees with the skills and knowledge required to thrive in this evolving landscape.

1. **Upskilling and Reskilling Programs:** Organizations are increasingly investing in upskilling and reskilling programs to prepare their workforce for new roles created by AI and automation. These programs focus on enhancing existing skills and developing new competencies relevant to AI and data analytics. For example, companies like Google and Microsoft offer specialized training and certification programs in AI, machine learning, and data science, which help employees transition to roles that require these advanced skills (Brynjolfsson & McAfee, 2014). By providing targeted training, organizations can ensure that their employees remain competitive and capable of leveraging new technologies effectively.
2. **Partnerships with Educational Institutions:** Many companies are forming partnerships with universities and technical schools to develop educational programs tailored to the needs of the AI-driven economy. For instance, IBM has collaborated with various educational institutions to create curricula focused on AI and data science, providing students with hands-on experience and industry-relevant knowledge (Barton & Court, 2019). These partnerships help bridge the gap between academic training and practical skills required in the workforce, ensuring that new graduates are well-prepared for careers in technology and data analytics.
3. **Continuous Learning and Development:** Given the rapid pace of technological change, continuous learning and development are essential for workforce adaptation. Many organizations are implementing learning platforms and resources that offer ongoing education opportunities, such as online courses, webinars, and workshops. For example, platforms like Coursera and Udacity offer courses on AI and machine learning that employees can access at their convenience, allowing them to stay updated with the latest advancements and trends (Wamba

et al., 2020). This approach promotes a culture of lifelong learning and helps employees adapt to evolving job requirements and technological innovations.

Public Policy and Government Interventions

As industries across the globe embrace AI technologies, there is a growing need for strategic policies that address the socio-economic impacts on employment and ensure that workers are equipped to thrive in new roles.

1. **Education and Training Programs**: Governments can invest in and promote education and training programs that focus on skills relevant to an AI-driven economy. This includes funding for vocational training, technical education, and higher education programs that emphasize data analytics, AI, robotics, and digital literacy (Bessen, 2019, Taghipour, A. (Ed.). 2023). By aligning educational curricula with industry needs, policymakers can help bridge the skills gap and prepare the workforce for emerging job opportunities.
2. **Reskilling and Upskilling Initiatives**: Public policies can support reskilling and upskilling initiatives through subsidies, grants, or tax incentives for businesses that invest in employee training programs. This can include partnerships between educational institutions and industry to create tailored training programs that address specific skills shortages (Manyika et al., 2017). Additionally, government-funded retraining programs can help displaced workers transition to new roles in growing sectors.
3. **Social Safety Nets**: As the labor market transforms, robust social safety nets are essential to support workers who face job displacement. Governments can enhance unemployment benefits, provide income support, and offer job placement services to help individuals navigate career transitions (Acemoglu & Restrepo, 2018). Policies that support temporary income assistance and career counseling can mitigate the adverse effects of job loss and provide a buffer for workers seeking new employment opportunities.
4. **Encouraging Innovation and Entrepreneurship**: Public policies that foster innovation and entrepreneurship can create new job opportunities and drive economic growth. Governments can support start-ups and small businesses through grants, low-interest loans, and tax incentives, encouraging the development of new technologies and business models that generate employment (Brynjolfsson & McAfee, 2014). By nurturing a dynamic and innovative business environment, policymakers can help diversify the job market and stimulate economic resilience.

5. **Promoting Inclusive Growth**: Ensuring that the benefits of technological advancements are widely shared requires policies that address income inequality and promote inclusive growth. This includes measures to reduce disparities in access to education and technology, as well as efforts to ensure that marginalized and disadvantaged communities benefit from new economic opportunities (Choi et al., 2018). Equitable access to resources and opportunities is critical for creating a balanced and inclusive labor market.

Corporate Responsibility and Ethical Considerations

As businesses integrate advanced technologies into their operations, it is essential to address the ethical implications and ensure that the workforce transition is managed responsibly.

1. **Ethical Considerations in AI Implementation**: Companies must consider the ethical implications of AI technologies, including issues related to privacy, bias, and transparency. Ensuring that AI systems are designed and implemented in a way that respects privacy rights and avoids discriminatory practices is vital for maintaining public trust and regulatory compliance (Dastin, 2018). Ethical AI practices involve establishing clear guidelines for data usage, conducting regular audits to detect and address biases, and being transparent about AI decision-making processes.
2. **Workforce Impact and Reskilling Initiatives**: As AI and automation transform job roles, companies have a responsibility to support their employees through this transition. This includes investing in reskilling and upskilling programs to help workers adapt to new roles created by technological advancements. Providing access to training and development opportunities enables employees to acquire the necessary skills to thrive in a changing job market and reduces the risk of job displacement (Brynjolfsson & McAfee, 2014). Companies should also offer career counseling and support services to assist employees in navigating their career paths during this period of change.
3. **Promoting Inclusivity and Equity**: Ensuring that the benefits of AI and automation are distributed equitably across the workforce is another important aspect of corporate responsibility. This involves addressing potential disparities in access to new technologies and training opportunities and implementing policies that promote diversity and inclusion within the workplace. Companies should strive to create an inclusive environment where all employees can contribute to and benefit from technological advancements (Choi et al., 2018).

4. **Stakeholder Engagement and Transparency**: Engaging with stakeholders, including employees, communities, and policymakers, is essential for addressing the broader socio-economic impacts of AI integration. Companies should be transparent about their AI strategies and the expected impacts on the workforce, actively seeking input from various stakeholders to ensure that their approaches are aligned with societal values and expectations (Barton & Court, 2019).

OUTLOOK

Long-Term Workforce Projections

These projections offer insights into how technological advancements will shape employment patterns, job roles, and skill requirements over the coming decades.

1. **Job Creation and Transformation**: AI and automation are expected to create new job categories and transform existing roles. As routine and repetitive tasks become increasingly automated, there will be a growing demand for jobs that involve managing and maintaining AI systems, data analysis, and strategic decision-making (Brynjolfsson & McAfee, 2014). For instance, roles such as AI system trainers, data scientists, and robotics technicians are projected to see substantial growth. Additionally, industries will see an evolution in traditional roles, with a greater emphasis on skills related to AI oversight and human-AI collaboration (Choi et al., 2018).
2. **Skills and Education Requirements**: The evolving job market will necessitate significant changes in skills and education. Workers will need to acquire competencies in AI technologies, data analytics, and advanced problem-solving to remain competitive. Educational institutions and organizations will play a critical role in providing reskilling and upskilling opportunities to help the workforce adapt to these changes (Manyika et al., 2017). Lifelong learning and continuous professional development will become increasingly important as technological advancements accelerate and job requirements evolve.
3. **Employment Displacement and Economic Shifts**: The automation of certain tasks and roles may lead to job displacement, particularly in sectors where repetitive and manual tasks are prevalent. This shift could result in economic disruptions, particularly for lower-skilled workers who are more susceptible to job loss (Acemoglu & Restrepo, 2018). To mitigate these impacts, it will be essential to develop robust social safety nets, including unemployment benefits and job transition programs, to support displaced workers and facilitate their reentry into the job market.

4. **Long-Term Socio-Economic Implications**: The long-term socio-economic implications of AI and automation include potential changes in income distribution and economic inequality. As high-skilled jobs become more prevalent, there may be a widening gap between those with advanced technical skills and those without (Bessen, 2019). Addressing these disparities will require targeted policies aimed at promoting inclusivity and ensuring that the benefits of technological advancements are broadly shared across society.
5. **Impact on Different Sectors**: Different sectors will experience varying degrees of impact from AI and automation. For example, manufacturing and logistics are likely to see significant automation, while sectors such as healthcare and education may experience more gradual changes, with AI augmenting rather than replacing human roles (Wamba et al., 2020). Understanding sector-specific impacts will be crucial for developing effective strategies to support workforce adaptation and manage transitions.

Emerging Trends and Technologies

Emerging trends and technologies are reshaping the landscape of industries and the workforce, driving significant transformations in how businesses operate and compete. Some of the most impactful trends and technologies are:

1. Artificial Intelligence and Machine Learning

AI and machine learning continue to advance, with applications expanding across various sectors. AI technologies are increasingly used for predictive analytics, automated decision-making, and process optimization. Machine learning algorithms enhance the ability of systems to learn from data and improve over time, enabling more sophisticated analyses and automation of complex tasks (Brynjolfsson & McAfee, 2014). Key applications include natural language processing, computer vision, and autonomous systems.

2. Internet of Things (IoT)

The Internet of Things (IoT) involves the interconnection of devices and systems through the Internet, allowing for real-time data collection and analysis. In supply chains, IoT devices such as sensors and RFID tags provide enhanced visibility and tracking of goods, improving inventory management and reducing operational inefficiencies (Sundararajan, 2019). IoT is also crucial in smart cities, healthcare, and industrial automation, where it facilitates improved monitoring and control.

3. Blockchain Technology

Blockchain technology offers a decentralized and secure way to record transactions and manage data. Its applications in supply chains include enhancing transparency, traceability, and security of transactions. By providing an immutable ledger, blockchain can help prevent fraud, reduce errors, and streamline complex supply chain processes (Kshetri, 2018). The technology is also being explored for use in contract management, identity verification, and financial transactions.

4. Robotic Process Automation (RPA)

Robotic Process Automation (RPA) uses software robots or "bots" to automate repetitive and rule-based tasks. RPA is increasingly employed to improve efficiency and accuracy in processes such as data entry, invoice processing, and customer service. By automating these tasks, organizations can reduce costs and free up human workers to focus on more complex and value-added activities (Lacity et al., 2015).

5. Augmented Reality (AR) and Virtual Reality (VR)

Augmented Reality (AR) and Virtual Reality (VR) technologies are transforming various industries by providing immersive and interactive experiences. In manufacturing and logistics, AR can assist with assembly processes and equipment maintenance by overlaying digital information on physical objects. VR is used for training simulations and design visualization, enabling more effective learning and development (Milgram & Kishino, 1994).

6. 5G Technology

The rollout of 5G technology promises to significantly enhance connectivity, offering faster data transfer speeds, lower latency, and increased network capacity. 5G will enable more robust IoT applications, support the growth of smart cities, and improve communication and collaboration tools. Its impact is expected to be substantial in areas such as autonomous vehicles, telemedicine, and advanced manufacturing (Zhang et al., 2020, Taghipour, A., Lu, X., & Smahi, M. 2022).

7. Quantum Computing

Quantum computing represents a paradigm shift in computational power, leveraging quantum bits (qubits) to perform complex calculations at unprecedented speeds. While still in the experimental phase, quantum computing holds the potential to revolutionize fields such as cryptography, optimization, and drug discovery by solving problems that are currently intractable with classical computers (Arute et al., 2019).

8. Sustainable and Green Technologies

There is a growing emphasis on sustainable and green technologies aimed at reducing environmental impact and promoting energy efficiency. Innovations include renewable energy sources, energy-efficient materials, and technologies for waste reduction and recycling. Businesses are increasingly adopting these technologies to meet regulatory requirements, address consumer preferences, and achieve corporate sustainability goals (Elkington, 1997).

Scenario Analysis: Best and Worst-Case Outcomes

For AI and automation in supply chains, this involves examining both best-case and worst-case scenarios to develop robust strategies and mitigate risks.

Best-Case Scenario: Optimized Efficiency and Growth

1. **Enhanced Operational Efficiency**: In the best-case scenario, the integration of AI and automation in supply chains leads to significant improvements in efficiency and productivity. AI-driven systems optimize inventory management, forecast demand with high accuracy, and automate routine tasks, reducing operational costs and increasing responsiveness (Brynjolfsson & McAfee, 2014).
2. **Job Creation and Skill Development**: Rather than leading to widespread job displacement, technological advancements create new job opportunities in areas such as AI system management, data analytics, and technology integration. Workers benefit from reskilling and upskilling programs, enhancing their capabilities and career prospects (Choi et al., 2018).
3. **Increased Competitiveness and Innovation**: Companies that successfully implement AI and automation gain a competitive edge through enhanced operational capabilities and innovative solutions. This drives economic growth and sets a high standard for industry practices, promoting further technological advancements and investment (Manyika et al., 2017).
4. **Sustainable Practices and Environmental Impact**: AI and automation contribute to more sustainable supply chain practices by optimizing resource usage, reducing waste, and supporting green technologies. This aligns with global sustainability goals and improves corporate social responsibility (Elkington, 1997).

Worst-Case Scenario: Disruption and Inequality

1. **Widespread Job Displacement**: In the worst-case scenario, AI and automation lead to significant job displacement, particularly for low-skilled workers whose roles are easily automated. The resulting unemployment and economic instability disproportionately affect vulnerable populations, exacerbating income inequality and social unrest (Acemoglu & Restrepo, 2018).
2. **Increased Economic Disparities**: The benefits of AI and automation are concentrated among large, technologically advanced firms, while smaller companies and developing regions struggle to keep pace. This deepens economic disparities and limits opportunities for less advantaged businesses and communities (Bessen, 2019).
3. **Ethical and Security Concerns**: The rapid adoption of AI raises ethical and security concerns, including issues related to data privacy, algorithmic bias, and cyber threats. Without proper regulation and oversight, these concerns can lead to misuse of technology, loss of public trust, and potential legal and financial repercussions (Dastin, 2018).
4. **Environmental and Social Challenges**: If not managed responsibly, the deployment of AI and automation could lead to unintended environmental impacts, such as increased electronic waste or energy consumption. Additionally, social challenges may arise from the uneven distribution of technological benefits and the disruption of traditional industries (Wamba et al., 2020).

Strategic Recommendations

1. **Investment in Reskilling and Upskilling**: To mitigate job displacement and enhance workforce adaptability, invest in comprehensive reskilling and upskilling programs. This includes partnerships with educational institutions, on-the-job training, and continuous professional development.
2. **Promote Inclusive Access to Technology**: Ensure equitable access to AI and automation technologies across different regions and business sizes. Support small and medium-sized enterprises (SMEs) with grants, resources, and technological support to level the playing field.
3. **Establish Ethical Guidelines and Regulations**: Develop and enforce ethical guidelines and regulations to address concerns related to AI implementation, including data privacy, algorithmic fairness, and cybersecurity. Engage stakeholders in the creation of policies that balance innovation with social responsibility.

4. **Focus on Sustainable Practices**: Incorporate sustainability into AI and automation strategies by optimizing resource usage, reducing waste, and supporting green technologies. Align corporate practices with global sustainability goals and communicate these efforts to stakeholders.

CONCLUSION

The AI revolution is transforming workforce dynamics and employment in supply chains, ushering in a new era characterized by advanced technological capabilities and shifting job requirements. As artificial intelligence and automation become increasingly integrated into supply chain operations, they drive significant improvements in efficiency, accuracy, and cost-effectiveness. AI technologies, such as predictive analytics and robotics, streamline processes, optimize inventory management, and enhance decision-making, which can lead to substantial operational benefits. However, these advancements also introduce complexities related to workforce adaptation, as traditional roles evolve, and new skill sets become essential. The positive outcomes of this transformation include the creation of innovative job roles and the potential for enhanced productivity, provided that strategic efforts are made to address the associated challenges.

Despite the benefits, the transition to an AI-driven supply chain environment presents considerable challenges, particularly in terms of job displacement and workforce reskilling. The automation of routine tasks and roles may lead to significant disruptions for workers whose positions are susceptible to technological replacement. This necessitates a focused approach to workforce adaptation, including investment in reskilling and upskilling programs to prepare employees for new and evolving roles. Organizations must prioritize ethical considerations, ensuring that the implementation of AI technologies is accompanied by measures to support displaced workers and address socio-economic disparities. Proactive strategies, such as developing comprehensive retraining initiatives and fostering inclusive access to new technologies, are crucial for mitigating the negative impacts on employment.

Looking ahead, the long-term implications of AI in supply chains will largely depend on how effectively organizations and policymakers manage the transition and embrace opportunities for innovation. By balancing technological advancements with a commitment to responsible workforce practices, businesses can leverage AI to drive growth and improve supply chain performance while contributing to a more equitable and sustainable future. Ensuring that the benefits of AI are broadly shared and that ethical considerations guide its implementation, will be key to achieving a positive and resilient workforce dynamic in the evolving landscape of supply chain management. The continued focus on these aspects will determine the overall success and societal impact of the AI revolution.

REFERENCES

Acemoglu, D., & Restrepo, P. (2018). Artificial Intelligence, Automation, and Work. *Economics Working Paper Series*. Retrieved from https://www.nber.org/papers/w24196

Arute, F., Arya, K., Babbush, R., Bacon, D., Bardin, J. C., Barends, R., Biswas, R., Boixo, S., Brandao, F. G. S. L., Buell, D. A., Burkett, B., Chen, Y., Chen, Z., Chiaro, B., Collins, R., Courtney, W., Dunsworth, A., Farhi, E., Foxen, B., & Martinis, J. M. (2019). Quantum supremacy using a programmable superconducting processor. *Nature*, 574(7779), 505–510. DOI: 10.1038/s41586-019-1666-5 PMID: 31645734

Autor, D. H. (2015). Why are there still so many jobs? The history and future of workplace automation. *The Journal of Economic Perspectives*, 29(3), 3–30. DOI: 10.1257/jep.29.3.3

Barton, D., & Court, D. (2019). *Leading in the Age of AI*. McKinsey & Company. Retrieved from https://www.mckinsey.com/business-functions/mckinsey-digital/our-insights/leading-in-the-age-of-ai

Ben-Daya, M., Hassini, E., & Bahroun, Z. (2019). Internet of things and supply chain management: A literature review. *International Journal of Production Research*, 57(15-16), 4719–4742. DOI: 10.1080/00207543.2017.1402140

Bessen, J. (2019). AI and Jobs: The Role of Demand. *National Bureau of Economic Research*. Retrieved from https://www.nber.org/papers/w24235

Brynjolfsson, E., & McAfee, A. (2014). *The Second Machine Age: Work, Progress, and Prosperity in a Time of Brilliant Technologies*. W. W. Norton & Company.

Choi, T. M. (2020). Innovative supply chain management in the age of artificial intelligence. *International Journal of Logistics Research and Applications*, 23(2), 98–101.

Choi, T. M., Wallace, S. W., & Wang, Y. (2018). Big Data Analytics in Operations Management. *Production and Operations Management*, 27(10), 1868–1881. DOI: 10.1111/poms.12838

Chopra, S., & Meindl, P. (2016). *Supply Chain Management: Strategy, Planning, and Operation*. Pearson.

Christopher, M. (2016). *Logistics & Supply Chain Management*. Pearson UK.

Chui, M., Manyika, J., & Miremadi, M. (2016). Where machines could replace humans—And where they can't (yet). *The McKinsey Quarterly*, 3(3), 1–9.

Dastin, J. (2018). Amazon scraps secret AI recruiting tool that showed bias against women. *Reuters*. Retrieved from https://www.reuters.com/article/us-amazon-com-recruitment-insight-idUSKCN1MK08G

Dubey, R., Gunasekaran, A., Childe, S. J., Blome, C., Papadopoulos, T., & Luo, Z. (2020). Big Data and Predictive Analytics and Manufacturing Performance: Integrating Institutional Theory, Resource-Based View and Big Data Culture. *British Journal of Management*, 31(2), 186–210.

Elkington, J. (1997). *Cannibals with Forks: The Triple Bottom Line of 21st Century Business*. Capstone Publishing.

Frey, C. B., & Osborne, M. A. (2017). The future of employment: How susceptible are jobs to computerization? *Technological Forecasting and Social Change*, 114, 254–280. DOI: 10.1016/j.techfore.2016.08.019

Gubbi, J., Buyya, R., Marusic, S., & Palaniswami, M. (2013). Internet of Things (IoT): A vision, architectural elements, and future directions. *Future Generation Computer Systems*, 29(7), 1645–1660. DOI: 10.1016/j.future.2013.01.010

Ivanov, D., & Dolgui, A. (2020). A digital supply chain twin for managing the disruption risks and resilience in the era of Industry 4.0. *Production Planning and Control*, 31(9), 722–738.

Ivanov, D., Dolgui, A., Sokolov, B., Ivanova, M., & Potryasaev, S. (2019). Disruption-driven supply chain (re)-planning and performance impact assessment with consideration of pro-active and recovery policies. *Transportation Research Part E, Logistics and Transportation Review*, 125, 308–330.

Ivanov, D., Tsipoulanidis, A., & Schönberger, J. (2019). *Global Supply Chain and Operations Management: A Decision-Oriented Introduction to the Creation of Value*. Springer.

Kshetri, N. (2018). 1 Blockchain's roles in meeting key supply chain management objectives. *International Journal of Information Management*, 39, 80–89. DOI: 10.1016/j.ijinfomgt.2017.12.005

Lacity, M. C., Willcocks, L. P., & Craig, A. (2015). *Robotic Process Automation: The Next Transformation Lever for Shared Services*. Shared Services & Outsourcing Network.

Manyika, J., Chui, M., Miremadi, M., Bughin, J., George, K., Willmott, P., & Dewhurst, M. (2017). A Future That Works: Automation, Employment, and Productivity. *McKinsey Global Institute*. Retrieved from https://www.mckinsey.com/featured-insights/employment-and-growth/afuture-that-works-automation-employment-and-productivity

Milgram, P., & Kishino, A. (1994). A taxonomy of mixed reality visual displays. *IEICE Transactions on Information and Systems*, E77-D(12), 1321–1329.

Miller, J. (2020). Blockchain and artificial intelligence: Creating synergy for improved supply chain management. *The Journal of Supply Chain Management*, 56(4), 25–32.

Sundararajan, A. (2019). *The Sharing Economy: The End of Employment and the Rise of Crowd-Based Capitalism*. MIT Press.

Taghipour, A. (Ed.). (2023). *Government Impact on Sustainable and Responsible Supply Chain Management*. IGI Global. DOI: 10.4018/978-1-6684-9062-4

Taghipour, A., Lu, X., & Smahi, M. (2022). Opportunities and challenges for Society 5.0 in procurement. *Blockchain Technology in Supply Chain Management for Society*, 5(0), 91–100. DOI: 10.1201/9781003177432-9

Wamba, S. F., Akter, S., Edwards, A., Chopin, G., & Gnanzou, D. (2020). How 'big data' can make big impact: Findings from a systematic review and a longitudinal case study. *International Journal of Production Economics*, 165, 234–246. DOI: 10.1016/j.ijpe.2014.12.031

West, D. M. (2018). *The Future of Work: Robots, AI, and Automation*. Brookings Institution Press.

Xu, L. D., He, W., & Li, S. (2014). Internet of Things in industries: A survey. *IEEE Transactions on Industrial Informatics*, 10(4), 2233–2243. DOI: 10.1109/TII.2014.2300753

Yue, X., Wang, Y., Wang, Y., & Zhang, X. (2020). Blockchain technology for enhancing supply chain resilience: A literature review. *International Journal of Production Economics*, 227, 107678.

Zhang, S., Chen, C., & Liu, H. (2020). 5G Technology and Its Applications: An Overview. *IEEE Access : Practical Innovations, Open Solutions*, 8, 34923–34936.

KEY TERMS AND DEFINITIONS

Artificial Intelligence (AI): AI refers to the simulation of human intelligence in machines designed to perform tasks that typically require human cognition, such as learning, problem-solving, and decision-making. In supply chains, AI is used for predictive analytics, demand forecasting, and optimizing logistics (Brynjolfsson & McAfee, 2014).

Big Data: Big Data refers to large and complex datasets that traditional data processing tools cannot handle efficiently. In supply chains, Big Data analytics enables businesses to derive actionable insights from vast amounts of data, improving decision-making and operational performance (Manyika et al., 2017).

Blockchain Technology: Blockchain is a decentralized digital ledger that records transactions across a network of computers. In supply chains, blockchain enhances transparency and traceability by providing a secure and immutable record of transactions, which helps in reducing fraud and errors (Kshetri, 2018).

Digital Transformation: Digital transformation involves the integration of digital technologies into all aspects of business operations to drive improvements in efficiency, customer experience, and business models. In supply chains, digital transformation includes adopting AI, IoT, and blockchain technologies to enhance operational capabilities (Manyika et al., 2017).

Economic Disparities: Economic disparities refer to the unequal distribution of economic resources and opportunities among different groups. The adoption of AI and automation can exacerbate these disparities if the benefits are not evenly distributed, affecting smaller firms and less advantaged regions (Bessen, 2019).

Ethical AI: Ethical AI involves the development and deployment of AI systems in a manner that is fair, transparent and respects privacy. Ensuring ethical AI practices in supply chains involves addressing biases in algorithms, safeguarding data privacy, and maintaining accountability in AI-driven decisions (Dastin, 2018).

Human-AI Collaboration: Human-AI collaboration involves the integration of human expertise with AI technologies to achieve better outcomes. In supply chains, this collaboration allows humans to leverage AI for data analysis and decision support while focusing on tasks that require human judgment and creativity (Dastin, 2018).

Internet of Things (IoT): IoT refers to the network of interconnected devices and sensors that collect and exchange data. In supply chains, IoT technologies enable real-time tracking of goods, monitor equipment performance, and enhance inventory visibility, leading to improved operational efficiency (Sundararajan, 2019).

Job Displacement: Job displacement refers to the loss of employment due to changes in technology or economic conditions. In supply chains, AI and automation can lead to the displacement of roles that involve repetitive and manual tasks, necessitating strategies for managing workforce transitions (Acemoglu & Restrepo, 2018).

Machine Learning (ML): Machine learning is a subset of AI that involves the development of algorithms that allow systems to learn from and make predictions or decisions based on data. In supply chains, ML algorithms analyze historical data to predict future trends and optimize inventory management (Choi et al., 2018).

Reskilling and Upskilling: Reskilling and upskilling are processes aimed at updating or acquiring new skills to adapt to changing job requirements. In the context of AI and automation, reskilling involves training employees to handle new technologies, while upskilling enhances existing skills to align with advanced technological demands (Choi et al., 2018).

Robotic Process Automation (RPA): RPA involves the use of software robots or "bots" to automate repetitive, rule-based tasks. In supply chains, RPA is employed to streamline processes such as data entry, invoice processing, and order fulfillment, thereby increasing efficiency and reducing errors (Lacity et al., 2015).

Skill Gap: The skill gap refers to the disparity between the skills required by employers and those possessed by the workforce. As AI and automation technologies evolve, there is an increasing need for new skills related to data analysis, AI system management, and technology integration (Brynjolfsson & McAfee, 2014).

Supply Chain Optimization: Supply chain optimization refers to the process of improving the efficiency and effectiveness of supply chain operations through various strategies and technologies. AI and automation play a crucial role in optimizing inventory levels, enhancing demand forecasting, and streamlining logistics (Brynjolfsson & McAfee, 2014).

Workforce Adaptation: Workforce adaptation refers to the ability of employees and organizations to adjust to changes in technology and job requirements. This includes embracing new technologies, developing new skills, and adapting to evolving job roles and responsibilities (Choi et al., 2018).

Chapter 20
The Use of IoV for Vehicle Routing Optimisation: A Systematic Literature Review

Manal El Yadari
https://orcid.org/0009-0000-3827-4494
Sidi Mohammed Ben Abdellah University, Morocco

Fouad Industriels Jawab
https://orcid.org/0000-0002-0508-7278
University Sidi Mohamed Ben Abdellah, Morocco

Imane Moufad
University Sidi Mohamed Ben Abdellah, Morocco

ABSTRACT

Optimising vehicle routing is an important issue that relies on logistics flow management, resource allocation, and transport planning. However, it is influenced by a large number of variables, including freight volumes, distances, traffic conditions, vehicle performance, weather conditions, and infrastructure quality. In this sense, to understand the impact of using IoV on VRP, the authors reviewed the existing literature using the PRISMA method. As a result, they presented IoV and its contribution to vehicle routing optimisation, and more specifically, they deduced the impact of IoV use on economic, energy, environmental, and organisational performance. Finally, they confirmed that IoV aims to optimise vehicle routing performance, manage traffic conditions, and provide communications between vehicles and infrastructure.

DOI: 10.4018/979-8-3693-7478-8.ch020

I. INTRODUCTION

In recent years, the Internet of Vehicles (IoV) has brought significant changes to transportation and logistics (Sahoo et al., 2023). IoV is an extension of the Internet of Things (IoT) that aims to integrate advanced communication technologies with vehicle systems to enable seamless data exchange between vehicles, infrastructure, and other stakeholders (Modica et al., 2023).

This technological advancement is particularly relevant to vehicle routing optimization, which aims to improve the efficiency and effectiveness of vehicle fleets. Vehicle routing problems (VRPs) are complex optimization challenges involving determining the most efficient routes for a fleet of vehicles to deliver goods to various locations (Dantzig & Ramser, 1959).

Traditional methods for solving VRPs have relied heavily on exact and heuristic techniques, which, while effective to some extent, often face limitations in scalability, adaptability, and responsive to real-time data. In this sense, the integration of IoV into vehicle routing optimization presents new opportunities and challenges (Navarro, 2023).

IoV provides real-time data on traffic conditions, vehicle performance, and environmental factors, which aims to develop efficient routing algorithms and decision-making processes.

Coordinating supply chains is essential to optimising the routing process (Gao et al., 2018) and (Taghipour & Frayret, 2013b). This is why several algorithms have been introduced to develop collaborative planning and decision-making.

Basically, the coordination between production organisations also plays an important role in improving the performance of all members of the supply network (Zoghlami et al., 2016). Sharing information is an essential tool for facilitating the coordination of operational planning, that may be used as input for IoV.

However, in many cases, the independent nature of companies in a supply chain makes information sharing challenging, as private data often remains inaccessible to other members. Mechanisms that rely on minimal information sharing introduce added complexity, particularly as the number of partners in the network increases, necessitating innovative approaches to ensure effective coordination (Taghipour, 2014).

Likewise, in the humanitarian logistics context, accurate demand planning supported by standardised consumption data can improve coordination, reduce demand volatility and optimise resource allocation (Taghipour, 2020).

More precisely, (Taghipour & Frayret, 2013a) has used incentive systems to guide the planning decisions of partners and sharing non-strategic data to maintain client privacy. (Mahfod et al., 2019)and (Mbiatem et al., 2018). Moreover, optimum supplier selection also contributes to improve service quality and vehicle delivery efficiency, considering several effective supplier selection criteria.

Similarly, in the healthcare sector, cost-cutting and quality improvement efforts have led to outsourcing and offshoring services, which rely on efficient coordination and decision-making to maximize cost-effectiveness and operational efficiency (Glaa et al., 2014).

Achieving a resilient supply chain network requires addressing barriers and risks effectively, particularly in the face of disruptions like COVID-19. Therefore, adopted the Analytic Hierarchy Process (AHP) to prioritize challenges and guide decision-making, mitigate risks, recover operations, and build more robust networks (Dadsena et al., 2021) and (Taghipour, Sohrabi, et al., 2023).

At the same time, integrating the principles of sustainable development into supply chain processes ensures aligning economic, social, and environmental objectives (Taghipour, 2020) and (Taghipour, 2023). This involves integrating sustainability into production systems, reworking cost structures, and adopting frameworks for corporate social responsibility to meet the present needs without compromising the ability of future generations (Lu & Taghipour, 2023).

Moreover, in this paper (Merimi & Taghipour, 2021), the author studied the supply-demand fluctuations during crises and highlighted the impact of adopting advanced IT systems to enhance transparency and connectivity between buyers and suppliers. As confirmed by, the use of smart transportation system are important for improving service quality(Taghipour, Ramezani, et al., 2023) and (Taghipour, 2021). This transformation aims to create fully digital supply chains, enabling more efficient decision-making and adaptable purchasing contracts in a connected environment.

In this context, the real time data-collection promises to improve route planning by incorporating real-time traffic updates, vehicle-to-vehicle communication, and advanced analytics (Garay-Rondero et al., 2019). Consequently, IoV can lead to more adaptive, efficient, and cost-effective vehicle routing solutions.

Despite the potential benefits, the application of IoV in vehicle routing optimization is still an emerging field. A comprehensive understanding of how IoV technologies influence and improve vehicle routing strategies is essential for advancing both theoretical and practical aspects of transportation logistics (Garay-Rondero et al., 2019). This literature review aims to collect and analyze existing research on the use of IoV for vehicle routing optimization. By examining various methodologies, technological advancements, and case studies, this review seeks to provide a thorough understanding of the current state of the field, identify research gaps, and propose directions for future directions.

The review will cover the evolution of vehicle routing optimization techniques in the context of IoV, evaluate the effectiveness of IoV-based solutions, and highlight the challenges associated with integrating IoV technologies into existing systems.

Through a systematic literature review, this research work aims to provide valuable insights into how IoV can revolutionize vehicle routing optimization and address contemporary logistical challenges.

This paper is structured in four sections. In the first section, we define the research methodology, which consists of a systematic literature review. Then, in the second section, we introduced the vehicle routing problem, classified the different VRP categories and studied the associated VRP solutions. In the third section, we defined the Internet of Vehicles and described the Internet of Vehicles structure. Finally, in the fourth section, we summarise the different areas of use of the IoV on the VRP and highlight its contribution to VRP optimisation.

II. RESEARCH METHODOLOGY

We conducted a systematic literature review that aims to comprehensively identify, evaluate, and synthesize existing research related to the use of Internet of Vehicles for optimizing the routing. In this context, we used the PRISMA method to define the research question, extract data, select relevant research works, reviewed literature and summarized the scientific contribution.

The PRISMA method, a Preferred Reporting Items for Systematic Reviews and Meta-Analyses method used for conducting a systematic literature review. This method incorporates four steps which are:

▪ Identification

We identified relevant research works using the keywords IoV and the vehicle routing problem. We conducted searches using different combinations of these keywords in various databases such as Scopus, Web of Science, and ScienceDirect (TABLE I).

Table 1. Keywords combination

Keywords combination
Internet of Vehicle
Internet of Vehicle and Vehicle routing problem
IoV
IoV and VRP

▪ Screening

In this step, we identified duplicated research works and removed them. Then, we filtered the titles to select only research papers related to IoV and VRP.

▪ Eligibility

We analysed the abstracts of the selected papers and retained only those directly addressing VRP and IoV.

▪ Inclusion

We conducted a full comprehensive review, using the following inclusion criteria: originality, scientific contribution, and quality of the papers, and selected the most relevant papers.

We have identified 2756 research papers and removed 197 duplicated papers. After analyzing the title and abstract, we found 1345 papers related to our research topic. We then conducted a full comprehensive reading and selected 267 papers. Finally, after focusing on scientific contribution, originality, and paper quality, we kept only 100 papers. The Figure 1 summarizes the PRISMA method.

Figure 1. PRISMA Process for a Systematic Literature Review

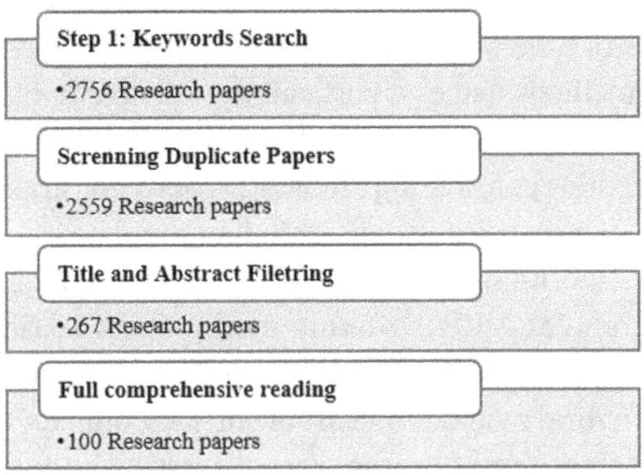

III. VEHICLE ROUTING BACKGROUND

The Vehicle Routing Problem (VRP) is focused on finding the most efficient routes for a fleet of vehicles to deliver goods from suppliers to customers. It aims to ensure optimal service quality, minimize transportation costs, reduce energy consumption, and limit environmental emissions. The VRP was first introduced in 1959 as part of the traveling salesman problem (TSP) (Dantzig & Ramser, 1959), with the goal of finding the most efficient routes for a vehicle fleet while visiting all customers exactly once.

Initially, a linear programming-based approach was suggested to solve the VRP, but later in 1963, other researchers explored the VRP and proposed using exact methods, particularly the branch and bound solution (Clarke & Wright, 1964). While both linear programming and exact methods are powerful tools for solving the VRP, they have limitations related to scalability, flexibility, real-time application, and dealing with complex and dynamic constraints.

To address these limitations, in 1964, a heuristic method, mainly the Clarke-Wright Savings algorithm (Clarke & Wright, 1964), was proposed as a solution for the Vehicle Routing Problem (VRP). Heuristic methods are characterized by their simplicity and ease of implementation, allowing for quick and efficient generation of near-optimal solutions and minimized extensive computational resources.

The algorithm merges iteratively routes based on cost savings, it minimizes the total travel distance or cost. Additionally, the Clarke-Wright Savings Algorithm is adaptable and can be customized to various VRP variants, such as those involving capacity constraints or time windows. Its ability to handle larger problem instances compared to exact methods makes it particularly valuable in planning and logistics applications. Similarly, in 1987, a time windows and capacitated VRP was introduced to propose optimal routing considering time and capacity constraints(Solomon, 1987).

Subsequently, metaheuristic methods including simulated annealing (Kirkpatrick et al., 1983), genetic algorithms (Holland, 1975), and Tabu search have been proposed and aim to solve complex VRPs, offering more powerful, flexible, and scalable solutions. As a continuation of the research carried out, several hybrid methods have been proposed, the hybrid methods consist of solving complex VRPs by combining two or more distinct algorithms from the exact, heuristic, and metaheuristic methods.

The use of hybrid methods is more efficient as it allows benefiting from the solution quality of the exact method, the speed and adaptability of the heuristic, and the capabilities of the metaheuristic method (Kuyu & Vatansever, 2024) and (Frías et al., 2023). However, the choice of using each of these methods among others methods depends on the VRP constraints, metrics, data size, and complexity.

In this sense, we have identified the different VRP category and defined the associated metrics. The TABLE II summarized the literature review findings.

Table 2. Introductuion of VRP Categories

VRP Type	Aim	Metrics	References
Capacitated VRP (CVRP)	Optimize vehicle routing under capacity constraints.	■ Total distance ■ Total cost ■ Number of vehicles	(Dantzig & Ramser, 1959), (Clarke & Wright, 1964),(Xie et al., 2024)
VRP with Time Windows (VRPTW)	Serve customers within specific time windows.	■ Total distance ■ Number of late arrivals ■ Time window	(Solomon, 1987), (Küp et al., 2024),(Ou et al., 2024)
Multi-Depot VRP (MDVRP)	Optimize routes with multiple depots.	■ Total distance ■ Depot utilization ■ Number of vehicles	(Stodola & Kutěj, 2024),(Bouanane et al., 2019),(Osorio-Mora et al., 2021)
VRP with Pickup and Delivery (VRPPD)	Manage both pickup and delivery tasks in routing.	■ Total distance ■ Number of pickups and deliveries ■ Vehicle load	(Martinovic et al., 2008),(Yavary & Sajedi, 2018)
Split Delivery VRP (SDVRP)	Split deliveries across multiple vehicles.	■ Total distance ■ Number of split deliveries ■ Total Cost	(Ma et al., 2013),(Chen & Guo, 2018),(Casazza et al., 2018)
Stochastic VRP (SVRP)	Handle uncertainty in demands and travel times.	■ Total cost ■ Service level	(Li & Jiao, 2022),(Gannouni et al., 2017)
Periodic VRP (PVRP)	Plan routes over a period, considering recurring routing.	■ Total distance ■ Number of visits per customer	(AbdAllah et al., 2017), (Prodhon & Prins, 2008),(Prodhon, 2009)
VRP with Backhauls (VRPB)	Manage deliveries followed by pickups on the return trip.	■ Total distance ■ Number of backhauls ■ Vehicle load	(Wassan et al., 2017),(Tavakkoli-Moghaddam et al., 2005)
Heterogeneous Fleet VRP (HVRP)	Optimize routes for a fleet with different vehicle types.	■ Total distance ■ Fleet occupancy ■ Total cost	(Fadda et al., 2022),(Andani et al., 2023)
Green VRP (GVRP)	Minimize environmental impact, considering fuel consumption or emissions.	■ Total emissions ■ Fuel consumption ■ Total cost	(Kabadurmus & Erdogan, 2022), (HadjTaieb et al., 2022)
Dynamic VRP (DVRP)	Handle real-time changes and the impact oftraffic conditions.	■ Total distance ■ Response time ■ Service level	(Meng et al., 2021),(Daely et al., 2021)
Open VRP (OVRP)	Optimize routes where vehicles without depot return.	■ Total distance ■ Number of open routes ■ Total cost	(Torres et al., 2022)

continued on following page

Table 2. Continued

VRP Type	Aim	Metrics	References
VRP with Access Restrictions (ARVRP)	Plan routes considering access restrictions like road type, time, or vehicle restrictions.	▪ Total distance ▪ Restricted access violations, Total cost	(Ichoua et al., 2003)
VRP with Multiple Trips (MTVRP)	Allow vehicles to make multiple trips during the routing planning.	▪ Total distance ▪ Number of trips ▪ Fleet occupancy	(Rizkiani et al., 2024),(Belver et al., 2017),(Olivera & Viera, 2007)

To identify the most studied VRP category, we have reviewed the literature and measured the occurrence of each VRP category. The figure 2 summarized the conducted occurrence analysis.

Figure 2. Occurrence analysis of VRP categories

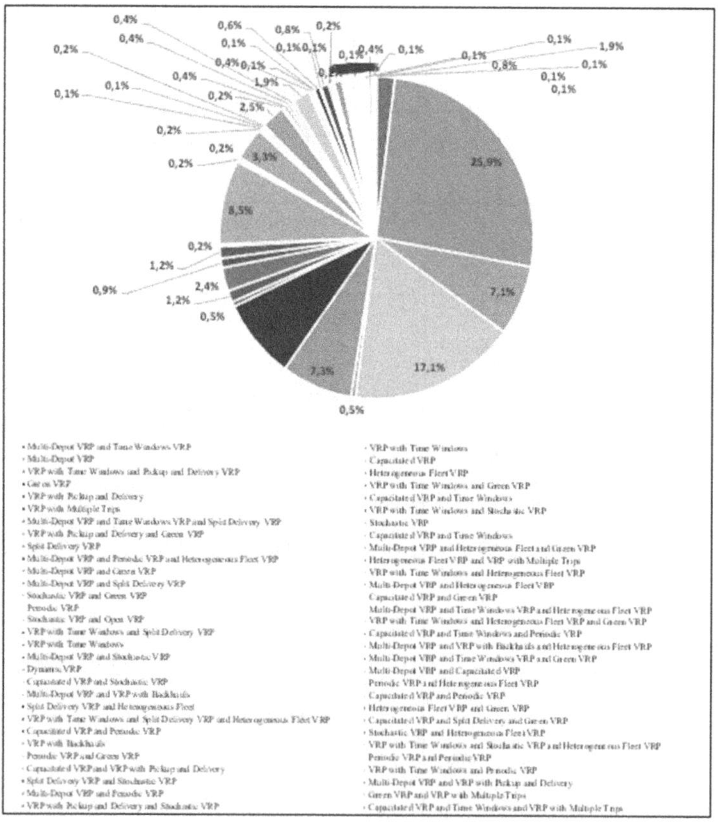

As a result, we observed that the most studied category of VRP is VRP with time windows, mainly 26% of the reviewed research papers mentioned VRP with time window constraints, followed by 17% that studied capacity VRP, 8% introduced stochastic VRP, as well as another 8% introduced green VRP. Finally, the remaining VRPs introduced heterogeneity, split delivery, periodicity, dynamics, multiple trips and other hybrid VRPs that combine these VRPs.

We reviewed the literature and identified the different exiting method that aims to solve the Vehicle Routing Problem (VRP). Our review covered a wide range of approaches, including traditional algorithms and modern techniques. Each approach offers unique solutions addressing the VRP complexity.

As a result, we identified four research solution categories which are: exact algorithms, heuristic approaches, metaheuristics, and hybrid models, these solutions aim to improve route optimisation, minimise costs and improve efficiency. We have summarised our results in TABLE III.

Table 3. Introduction of VRP Solutions

Methods	Category	Algorithms	Metrics	References
Exact Methods	**Mathematical Optimization**			
	Branch and bound	▪ Depth-First ▪ Search, Breadth ▪ First Search, ▪ Branch-and-Bound	Optimality, Computation time	(Navarro, 2023),(Toth & Vigo, 2001),(Almoustafa et al., 2013),(Dastghaibifard et al., 2008)
	Branch and Cut	▪ Cutting Plane ▪ Gomory Cuts, ▪ Branch-and-Cut	Optimality, Computation time, Scalability	(Bard et al., 1998),(Lysgaard et al., 2004)
	Dynamic Programming (DP)	▪ Bellman-Ford Algorithm, ▪ Floyd-Warshall Algorithm, ▪ Held-Karp Algorithm	Optimality, Computation time, Memory usage	(J. Zhao et al., 2024),(Mahya et al., 2023)
	Column Generation	▪ Dantzig-Wolfe Decomposition, Branch-and-Price	Solution quality, Scalability, Computation time	(M. Yu et al., 2021), (Glize et al., 2018),(Zillies et al., 2016)
	Lagrangian Relaxation	▪ Subgradient Optimization, ▪ Lagrangian Decomposition	Solution quality, Computation time, Convergence	(Sato et al., 2010)

continued on following page

Table 3. Continued

Methods	Category	Algorithms	Metrics	References
Heuristics	**Approximation**			
	Savings Algorithms	■ Clarke-Wright Savings Algorithm	Solution quality, Computation time	(Sun et al., 2022),(Shetty et al., 2018),(Wang et al., 2016), (Clarke & Wright, 1964)
	Constructive Heuristics	■ Nearest Neighbor, ■ Sweep Algorithm,	Solution quality, Computation time	(Chbichib et al., 2011)
	Local Search	■ Variable Neighborhood Search (VNS)	Solution quality, Computation time	(Sultana et al., 2024),(Hashimoto et al., 2022)
Metaheuristics	**Approximation Exploration**			
	Genetic Algorithms (GA)	■ Standard GA, ■ Hybrid GA, ■ Memetic Algorithms	Solution diversity, Solution quality, Computation time	(Ochelska-Mierzejewska et al., 2021),(H. Zhao & Yang, 2020),(Volna, 2016),(Kahveci & Ioannou, 2008)
	Simulated Annealing (SA)	■ Standard SA, ■ Adaptive SA, ■ Hybrid SA	Solution quality, Computation time, Robustness	(V. Yu et al., 2023), (Aurachman et al., 2021),(V. Yu et al., 2016),(Seidgar et al., 2016)
	Tabu Search (TS)	■ Standard TS, ■ Adaptive TS, ■ Hybrid TS	Solution quality, Computation time, Convergence	(Ahmed et al., 2022),(Christopher et al., 2021),(Chen et al., 2019)
	Ant Colony Optimization (ACO)	■ Standard ACO, ■ Hybrid ACO, ■ Ant System, ■ AntNet	Solution quality, Computation time, Convergence	(Yousefikhoshbakht et al., 2023),(Bell & McMullen, 2004),(Huang et al., 2022),(Samsuddin et al., 2020)
	Particle Swarm Optimization (PSO)	■ Standard PSO, ■ Hybrid PSO, ■ Multi-Swarm PSO	Solution quality, Computation time, Convergence	(Ramadhani & Garside, 2021),(Marinakis et al., 2018),(Demirtas et al., 2015),(Li & Ma, 2013)
Hybrid Methods		■ Hybrid GA, ■ Hybrid ACO, ■ Memetic Algorithms	Solution quality, Computation time, Flexibility	(Pandey & Jha, 2023),(Kuyu & Vatansever, 2024),(Frías et al., 2023)

We reviewed the literature and identified the most common VRP solutions. We have deduced that the most widely used methods are hybrid methods, which combine different algorithms to solve VRPs. Primarily, 43% of the articles studied the hybrid VRP solution. Next, 37% used the heuristic method to solve VRPs and 13% used metaheuristics. The remaining percentage is distributed between exact methods, combined Heuristic and Metaheuristic methods (Figure 3).

Figure 3. Occurrence analysis

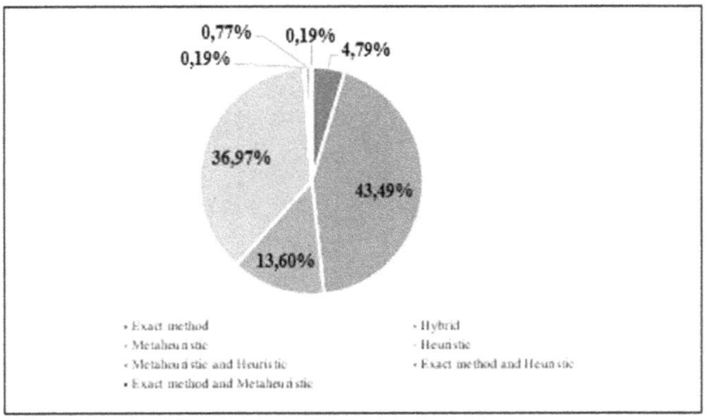

The use of different VRPs solutions depends on the complexity of the studied VRP. However, several other factors are considered when selecting the best solution, such as size, treatment complexity, ease of use, and precision. In this regard, we conducted a literature review to identify the strengths and limitations of each method.

The TABLE IV categorizes the various approaches, highlighting the VRPs solutions strengths, limitations, and typical applications. Through this review, we aim to offer valuable insights into the evolution of VRP solutions and guide future research efforts towards addressing the remaining challenges in this field. Furthermore, we have identified the strength and limitation of each solution algorithm.

Table 4. VRP Solutions strength and limitations

Category	Method	Logic	Strength	Limitation	References
Exact Methods	**Branch and Bound**	**Branching**: divides the problem into subproblems (branches).	Guarantees an optimal and effective solution for Vehicle routing.	Computationally expensive for large-scale problems.	(Navarro, 2023),(Toth & Vigo, 2001),
		Bounding: Uses bounds to eliminate subproblems that cannot improve the current best solution.			
		Pruning: Discards infeasible or non-promising branches.			
Exact Methods	**Branch and Cut**	**Branching**: Similar to branch-and-bound, but enhanced with additional constraints.	Improves efficiency by reducing limits. Suitable for integer programming.	Complex to implement. Still computationally intensive for large problems.	(Lysgaard et al., 2004)
		Cutting Planes: Adds constraints to improve the linear programming relaxation.			
		Bounding and Pruning: Refines bounds and prunes based on the improved relaxation.			

continued on following page

Table 4. Continued

Category	Method	Logic	Strength	Limitation	References
Exact Methods	**Dynamic Programming (DP)**	**Decomposition**: Breaks the problem into overlapping subproblems.	Provides optimal solutions for some variants. Effective for small to medium-sized problems.	Impractical for large-scale problems due to exponential growth of subproblem space.	(Mahya et al., 2023)
		Recursive Solutions: Solves each subproblem recursively and combines solutions.			
		Memoization: Stores solutions of subproblems to avoid redundant computations.			
Exact Methods	**Column Generation**	**Restricted Master Problem**: Solves a simplified version of the problem with a limited set of variables.	Effective for large-scale problems with complex constraints.	Requires solving multiple problems iteratively; can be computationally demanding.	(M. Yu et al., 2021),
		Pricing Problem: Identifies new variables (columns) that improve the solution.			
		Iterative Process: Adds columns to the master problem and resolves.			

continued on following page

Table 4. Continued

Category	Method	Logic	Strength	Limitation	References
Exact Methods	**Lagrangian Relaxation**	**Relaxation**: Incorporates constraints into the objective function with multipliers.	Simplifies complex problems. Useful for large-scale problems with difficult constraints.	Solution quality depends on the choice of multipliers; may not always provide exact solutions.	(Sato et al., 2010)
		Relaxed Problem: Solves the simplified version of the problem.			
		Multiplier Adjustment: Updates multipliers to improve the solution.			
Heuristics	**Savings Algorithms**	**Savings Calculation**: Computes savings from merging routes based on cost reduction.	Simple and easy to implement. Provides good solutions for small to medium-sized problems.	May not always find the optimal solution. Less effective for very large or complex problems.	(Sun et al., 2022), (Shetty et al., 2018)
		Sorting: Orders merges by savings in descending order.			
		Merging: Iteratively merges routes to achieve the highest savings.			
Heuristics	**Constructive Heuristics**	**Incremental Construction**: Builds a solution step-by-step based on criteria such as nearest neighbor or cheapest insertion.	Provides good initial solutions quickly. Simple to implement.	Can get stuck in local optima; may not provide the best overall solution.	(Chbichib et al., 2011)
		Greedy Approach: Makes the best local choice at each step.			

continued on following page

Table 4. Continued

Category	Method	Logic	Strength	Limitation	References
Heuristics	**Local Search**	**Neighbor Generation**: Creates neighboring solutions by making small changes to the current solution.	Can significantly improve solutions; simple to implement.	Local optima can limit effectiveness; may require multiple runs to achieve good results.	(Sultana et al., 2024)
		Evaluation: Assesses the quality of neighbors.			
		Acceptance: Moves to the best neighbor if it improves the current solution.			
Metaheuristics	**Genetic Algorithms (GA)**	**Initialization**: Creates an initial population of solutions.	Good at exploring large solution spaces; can find near-optimal solutions.	Computationally intensive; requires tuning of parameters.	(Ochelska-Mierzejewska et al., 2021),
		Selection: Chooses solutions based on fitness to be parents.			
		Crossover and Mutation: Generates new solutions by combining and altering parent solutions.			
		Iteration: Repeats the process to evolve better solutions.			

continued on following page

Table 4. Continued

Category	Method	Logic	Strength	Limitation	References
Metaheuristics	**Simulated Annealing (SA)**	**Initial Solution**: Starts with an initial solution and temperature. **Neighbor Generation**: Creates neighboring solutions. **Acceptance Probability**: Accepts worse solutions with decreasing probability over time. **Cooling Schedule**: Gradually reduces the temperature.	Effective at escaping local optima; adaptable to various problems.	Can be slow to converge; requires careful parameter tuning.	(V. Yu et al., 2023),
Metaheuristics	**Tabu Search (TS)**	**Memory Structures**: Uses a tabu list to record recent moves and prevent revisiting. **Neighbor Search**: Generates neighboring solutions and evaluates them. **Tabu List Update**: Updates memory structures to avoid cycling. **Iteration**: Continues until a stopping criterion is met.	Effective at escaping local optima; good for complex problems.	Requires careful management of memory structures; sensitive to parameter settings.	(Ahmed et al., 2022)

continued on following page

Table 4. Continued

Category	Method	Logic	Strength	Limitation	References
Metaheuristics	**Ant Colony Optimization (ACO)**	**Pheromone Trails**: Simulates ants laying pheromones to indicate the quality of paths.	Good at finding high-quality solutions for combinatorial problems.	Computationally expensive; performance varies based on problem instance and parameter settings.	(Yousefikhoshbakht et al., 2023),
		Solution Construction: Ants build solutions based on pheromone levels and heuristic information.			
		Pheromone Update: Adjusts pheromone levels based on solution quality.			
Metaheuristics	**Particle Swarm Optimization (PSO)**	**Initialization**: Creates an initial swarm of particles with random positions and velocities.	Simple to implement; effective for continuous optimization problems.	Can struggle with complex or high-dimensional problems; sensitive to parameter settings.	(Ramadhani & Garside, 2021)
		Velocity and Position Update: Adjusts particles' positions based on personal and global best positions.			
		Iteration: Continues until convergence or a stopping criterion is met.			

continued on following page

Table 4. Continued

Category	Method	Logic	Strength	Limitation	References
Hybrid Methods	**Hybrid Methods**	**Combination**: Integrates multiple methods or techniques to leverage their strengths.	Leverages strengths of different methods; can improve solution quality and efficiency.	Complex to design and implement; balancing different techniques can be challenging.	(Kuyu & Vatansever, 2024),(Frías et al., 2023)
		Adaptation: Customizes the hybrid approach to enhance performance based on problem characteristics.			

The use and effectiveness of these algorithms depend on the characteristics of the studied VRP, mainly solving Vehicle Routing Problems (VRP) are highly dependent on the specific characteristics of the problem, including the type of VRP variable, the scale of the problem and the optimisation objectives.

The choice of algorithm is guided by key parameters such as demand satisfaction, distance minimisation, time efficiency, cost reduction, energy consumption and environmental impact.

To achieve effective coordination, companies can collaborate with partners or even competitors, relying on decentralised systems to deal with stochastic information and variable demand (Cauchois et al., 2017).

Multi-objective optimisation contributes to improving competing demands, notably when it considers realistic constraints such as dynamic traffic conditions and regulatory requirements.

Similarly, optimizing the location of waste sorting and processing centres is essential to minimize transportation, and waiting costs. As confirmed by (Taghipour, Foukolaei, et al., 2023), it led to ensuring routing efficiency and pollution reduction, both environmental and operational. Consequently, the optimal routing algorithm must be carefully adapted to meet these diverse and potentially conflicting objectives.

IV. INTRODUCTION TO IOV

Internet of Vehicles is an extension of Internet of Things, but it is oriented towards vehicles. It consists of connecting vehicles with each other, with the infrastructure, Pedestrian, cloud and everything.

• Vehicle to Vehicle: V2V

It refers to the connection between vehicles on the road, it serves to transmit speed, energy, position and acceleration information using different technologies such as Dedicated Short-Range Communication (DSRC), Cellular Vehicles, Ad Hoc networking, GPS and several other sensors to avoid different risks collision and enhance traffic efficiency (Garay-Rondero et al., 2019).

• Vehicle to Infrastructure: V2I

It consists of connecting vehicles to infrastructure, to exchange data and communicate with infrastructure component such as Traffic signals, road signs, weather conditions…using different sensors, systems and connectors (Garay-Rondero et al., 2019).

• Vehicle to Cloud: V2C

The purpose of this technology is to facilitate communication between vehicles on roads and cloud databases, platforms, and services. This connection relies on the use of various technologies such as cellular networks, telematics, onboard diagnostic devices, cloud infrastructures, APIs, and protocols. These technologies work together to collect, store, and process data to improve the efficiency of freight transport (Agnusdei et al., 2022).

• Vehicle to Pedestrian: V2P

It refers to the exchange of data between vehicles and pedestrians on roads to enhance safety and mitigate the risks associated with collisions and accidents. This relies on the use of smartphone applications, sensors, cameras, and intelligent transportation systems (Agnusdei et al., 2022).

• Vehicle to Everything's: V2X

It consists of vehicles communicating with their surrounding environment, such as emergency call centers, charging stations, using different technologies such as cloud and edge computing, sensors and onboard systems, global navigation satellite systems, and Wireless Local Area Network (WLAN) (Lagorio et al., 2023).

The Internet of Vehicles (IoV) has undergone significant advancements, evolving from basic vehicle communication systems into a complex network of interconnected technologies. Initially, vehicle communication focused on systems like On-Board

Diagnostics (OBD) and Automatic Crash Notification (ACN), providing essential data on vehicle performance and safety (Cheng, 2009). In the early 2000s, Vehicle-to-Infrastructure (V2I) and Vehicle-to-Vehicle (V2V) communication technologies emerged, enabling vehicles to exchange information with roadside infrastructure and other vehicles(Zhang et al., 2023).

These technologies aimed to improve traffic management and road safety (Soleimani et al., 2020). Then during the 2010s, IoV integrated with the broader Internet of Things (IoT) ecosystem, utilizing advanced sensors, cloud computing, and big data analytics. The IoV integration allowed for real-time data processing and advanced route optimization strategies (Rana et al., 2020; Rahman et al., 2019). In the current decade, the introduction of 5G connectivity has further advanced IoV, offering enhanced data transfer rates and reduced latency (Anwar et al., 2021).

Basically, the use of IoV supports the development of complex applications such as autonomous vehicles and comprehensive smart city integration (Rout et al., 2020). The IoV developments signify a continuous progression toward a more connected and intelligent transportation system, capable of optimizing vehicle routing and overall mobility in unprecedented ways.

V. IOV CONTRIBUTION ON VRP OPTIMIZATION: AN OVERVIEW

The integration of the Internet of Vehicles (IoV) into vehicle routing optimization represents a significant advancement in the evolution of transportation and logistics management (Soleimani et al., 2020) (TABLE V). IoV comprises technologies that connect the vehicles with each other, infrastructure, and various data sources, creating a highly dynamic and interconnected transportation ecosystem (Adelantado et al., 2022).

The connectivity facilitates real-time data exchange, enhances situational awareness, and enables more precise and adaptive routing decisions. By using IoV technologies such as Vehicle-to-Vehicle (V2V) and Vehicle-to-Infrastructure (V2I) communication, predictive analytics, and advanced sensors, vehicle routing problems (VRP) can be optimized in ways that traditional methods could not achieve.

For example, the use of IoV contributes to controlling real-time traffic and dynamic route optimization, it improves responsiveness to changing traffic conditions, while cloud-based analytics and vehicle-to-cloud (V2C) communication enable sophisticated data processing and decision-making (Zhang et al., 2023).

Furthermore, the integration of Vehicle-to-Pedestrian (V2P) communication and energy-efficient routing contributes to enhanced safety and reduced environmental impact. This overview aims to explore how these IoV technologies address the challenges of VRP optimization, highlighting the transformative potential of IoV in achieving more efficient, safe, and sustainable transportation systems(Soleimani et al., 2020).

Table 5. the use of IoV for VRP optimization

Aspect	Objectif	Contribution	Applications	References
Real-Time Traffic Data	Enhances route planning by providing up-to-date information on traffic conditions.	Allows dynamic route adjustments to avoid traffic congestion and delays.	Adaptive routing systems that alter routes based on current traffic conditions.	(Zheng et al., 2022),(Gialos & Zeimpekis, n.d.),(Soleimani et al., 2020),(Anwar et al., 2021),(Mohamed & Feham, 2018), (Parveen et al., 2023)
Vehicle-to-Vehicle (V2V) Communication	Facilitates communication between vehicles to share information about road conditions, obstacles, and traffic.	Improves coordination among vehicles, reducing collisions and optimizing traffic flow.	Platooning strategies where vehicles travel in synchronized groups to improve efficiency.	(Zhang et al., 2023),(Zorkany et al., 2020),(Alsudani & Ozturk, 2023)
Vehicle-to-Infrastructure (V2I) Communication	Connects vehicles with traffic lights, road signs, and other infrastructure for better route management.	Enables vehicles to receive real-time updates from traffic management systems, improving route optimization.	Systems that adjust vehicle speed based on traffic light signals and road conditions.	(Alsudani & Ozturk, 2023)
Vehicle-to-Pedestrian (V2P) Communication	Enhances safety and routing by enabling communication between vehicles and pedestrians.	Provides alerts to both drivers and pedestrians about potential conflicts, improving safety.	Systems that alert drivers to pedestrians crossing the road and provide warnings to pedestrians about approaching vehicles.	(He et al., 2017),(Sewalkar & Seitz, 2019)
Real-Time Fleet Management	Improves fleet management by tracking vehicle locations, statuses, and performance in real-time.	Enhances decision-making for dispatching, route optimization, and maintenance scheduling.	Fleet management systems that optimize routes based on vehicle performance and availability.	(Adelantado et al., 2022),(Gheni et al., 2024)

continued on following page

Table 5. Continued

Aspect	Objectif	Contribution	Applications	References
Predictive Analytics	Uses historical and real-time data to predict future traffic conditions and vehicle performance.	Allows for proactive route planning and risk management based on predicted data trends.	Predictive models that forecast traffic patterns and adjust routes accordingly.	(Arooj et al., 2021),
Dynamic Route Optimization	Adapts routes in real-time based on changing conditions and operational needs.	Offers flexibility to adjust routes dynamically in response to unexpected events or changes in conditions.	Systems that continuously recalculate routes based on real-time data.	(Chowdhary & Kaur, 2018),(Dhanare et al., 2022)
Energy Efficiency	Optimizes routes to improve fuel consumption.	Incorporates data on vehicle energy consumption and vehicle characteristics to enhance route planning.	Energetic routing systems that prioritize routes based on energy efficiency.	(Shurrab et al., 2022),(Hyeon et al., 2022),(Jegatheesan & Arumugam, 2024),(Shao, 2022)
Environmental efficiency	Reduce environmental emissions.	Incorporates data on environmental emissions to select optimal routing planning.	Eco- routing systems that prioritize routes based on environmental emissions.	(Alsudani & Ozturk, 2023)
Enhanced Safety	Increases safety by providing alerts and warnings based on real-time data and communication between vehicles.	Reduces accident rates and improves overall road safety through advanced warning systems and vehicle coordination.	Safety systems that alert drivers to potential hazards and unsafe conditions.	(Reddy & Muralidhar, 2023)
Integration with Smart Cities	Integrates vehicle routing systems with broader smart city infrastructure for holistic optimization.	Enables coordination between transportation systems, urban planning, and infrastructure management.	City-wide traffic management systems that optimize transportation across various modalities.	(Kanthavel et al., 2021),(Dey et al., 2016)

continued on following page

Table 5. Continued

Aspect	Objectif	Contribution	Applications	References
Vehicular Ad Hoc Networks (VANETs)	Facilitates real-time, decentralized communication between vehicles and infrastructure.	Enhances data exchange and decision-making for traffic management and route optimization.	VANET-based systems for real-time traffic updates and route adjustments.	(Eze et al., 2014),(Soleimani et al., 2020)
Advanced Sensors and IoT Devices	Utilizes advanced sensors and IoT devices for detailed environmental and vehicle data collection.	Provides granular data on vehicle conditions, road conditions, and environmental factors.	IoT-enabled sensors that monitor vehicle health and environmental conditions for route optimization.	(Pendor & Tasgaonkar, 2016),(Rout et al., 2020)
Cloud-Based Analytics	Leverages cloud computing for processing large volumes of vehicle and traffic data.	Provides scalable and powerful data processing and analytics capabilities for enhanced route optimization.	Cloud-based platforms that aggregate and analyze data from multiple sources for route planning.	(Gheni et al., 2024),(Qin et al., 2012)
Vehicle-to-Cloud (V2C) Communication	Connects vehicles to cloud services for enhanced data analysis and storage.	Enables vehicles to send data to cloud servers for advanced analytics and optimization.	Cloud-based analytics platforms that provide insights and optimizations based on vehicle data.	(Teng & Sato, 2024), (Teng & Sato, 2024)

As a result of the conducted literature review, we observed that the integration of the Internet of Vehicles (IoV) significantly improves the optimisation of vehicle routing in various dimensions. The IoV helps to improve the economic, energy and environmental performance of vehicle routing. In economic terms, the IoV enables more efficient route planning and fleet management by exploiting real-time data on traffic conditions, vehicle status and operational requirements.

The integration of IoV reduces operational costs, such as fuel consumption and vehicle maintenance, by optimising routes to avoid congestion and minimise down-time (Rana et al., 2020). From an energy perspective, IoV technologies facilitate eco-friendly routes by integrating data on energy consumption and environmental conditions, enabling the selection of routes that minimise fuel consumption and reduce emissions.

Advanced analytics and real-time adjustments contribute to energy-efficient driving patterns and optimised fuel use (Rahman et al., 2019). Environmentally, the IoV supports sustainable transport practices by integrating data on traffic flow and vehicle emissions, promoting routes that reduce overall environmental impact.

It contributes to adjusting routes based on traffic and weather conditions further enhances environmental benefits by reducing the carbon footprint associated with vehicle operations (Wu et al., 2021). These advances demonstrate that IoV not only improves the efficiency and cost-effectiveness of vehicle routing, but also aligns with sustainability and environmental management objectives (TABLE VI).

Table 6. Review of the Impact of IoV application on VRP

Impact Type	IoV Tool	Description	IoV Impact	References
Economic Performance	**V2I (Vehicle-to-Infrastructure) Communication**	V2I communication enables vehicles to interact with infrastructure (e.g., traffic signals, road signs) to optimize traffic flow, minimize travel time, reduce congestion, and lower accident risks.	▪ Cost Reduction, ▪ Operational Efficiency	(Parveen et al., 2023), (Alsudani & Ozturk, 2023)
	Dynamic Route Optimization Algorithms	Algorithms that utilize data from V2V, V2I, V2C, and VANET to adjust routes in real-time based on current traffic conditions, vehicle status, and historical data.	▪ Cost Reduction, ▪ Increased Revenue	(Gheni et al., 2024)
	Fleet Management Software	Software that leverages V2V, V2I, and VANET data to track vehicle locations, monitor performance, and optimize scheduling and resource management.	▪ Operational Efficiency	(Adelantado et al., 2022)
Energy Performance	**Eco-Driving Assistance Systems**	Systems providing real-time feedback to drivers using V2V, V2I, and V2C data to optimize driving habits for fuel efficiency.	▪ Fuel Efficiency	(Jegatheesan & Arumugam, 2024),(Shao, 2022)
	Energy-Efficient Routing Tools	Tools that utilize V2I, V2V, and V2E data to calculate routes that minimize energy use and fuel consumption.	▪ Energy Consumption	(Shurrab et al., 2022)

continued on following page

Table 6. Continued

Impact Type	IoV Tool	Description	IoV Impact	References
Environmental Performance	**Emission Monitoring Systems**	Systems that monitor vehicle emissions using V2V, V2I, and V2E data to allow real-time adjustments to reduce pollutants.	▪ Emission Reduction	(Alsudani & Ozturk, 2023),
	Green Routing Algorithms	Algorithms using V2I, V2V, and VANET data to select routes that minimize environmental impact by reducing fuel consumption and emissions.	▪ Sustainable Practices	(Shurrab et al., 2022)
Organizational Performance	**Advanced Telematics Systems**	Systems leveraging V2V, V2I, V2C, and VANET data to provide comprehensive insights into vehicle performance, including GPS tracking and diagnostics.	▪ Enhanced Decision-Making	(Eze et al., 2014)
	Customer Relationship Management (CRM) Systems	CRM systems integrating data from V2V, V2I, and V2C to provide accurate delivery times, real-time tracking, and enhanced customer communication.	▪ Improved Customer Satisfaction	(Zhang et al., 2023)
	Adaptive Logistics Platforms	Platforms using data from V2V, V2I, V2C, and VANET to adapt logistics operations based on real-time traffic conditions and demand fluctuations.	▪ Operational Flexibility	(Soleimani et al., 2020)
	Vehicle-to-Energy Stations (V2E)	V2E communication facilitates interactions between vehicles and energy stations for optimal charging and energy supply management.	▪ Energy Supply Management	(Jegatheesan & Arumugam, 2024)

The application of Internet of Vehicles (IoV) technologies, such as Vehicle-to-Infrastructure (V2I), Vehicle-to-Vehicle (V2V), Vehicle-to-Cloud (V2C) and Vehicular Ad-hoc Networks (VANET), has significantly improved vehicle routing problems (VRP) in various performance areas.

In economic terms, V2I communication and dynamic route optimisation algorithms reduce operating costs by improving traffic flow and minimising delays, while fleet management software improves efficiency by providing real-time information on vehicle performance. In terms of energy performance, eco-driving assistance systems and energy-efficient routing tools promote energy efficiency and reduce overall energy consumption by optimising driving habits and selecting the most efficient routes.

Environmentally, emissions monitoring systems and green routing algorithms use V2V, V2I and VANET data to reduce greenhouse gas emissions and support sustainable practices. On an organisational level, advanced telematics systems, customer relationship management systems and adaptive logistics platforms enable better decision-making, improve customer satisfaction and enhance operational flexibility by using real-time data from various IoV communication channels.

In addition, vehicle-to-energy station (V2E) communication supports efficient energy management, particularly for electric vehicles, by facilitating optimal management of charging and energy supply. Overall, IoV technologies play a crucial role in optimising GRP by improving economic, energy, environmental and organisational performance throughout the logistics sector.

On the other side, green supply chain management is a potential and essential factor to consider in the decision-making process. In this sense, integrating GSCM with emerging technologies such as the Internet of Vehicles (IoV) is relevant to optimize freight transport activities. further aligning with green supply chain objectives,the use of IoV contributes to driving sustainable logistics (Loivet et al., 2020).

Additionally, the concept of design for remanufacturing complements these efforts by focusing on innovative approaches to product and process design, enabling the reconstruction and reuse of products(Taghipour et al., 2015).

To further improve the use of new technologies, the Society 5.0 concept contributes to enhancing this dynamic by integrating cutting-edge technologies such as AI, blockchain, IoV, and robotics, enabling real-time visibility and control over supply chain operations (Taghipour et al., 2022).

By merging GSCM with IoV technologies within the Society 5.0 framework, freight transport activities can be optimized to achieve greater sustainability.

CONCLUSION

The conducted literature review on the use of the Internet of Vehicles (IoV) and vehicle routing problem reveals that IoV technologies are revolutionising the logistics and transport sectors. The integration of vehicles in a coherent, intelligent network enables real-time data exchange between vehicles, infrastructure and central systems. The interconnected environment facilitates dynamic and adaptive routing decisions, allowing vehicles to react instantly to changing traffic conditions, road hazards and vehicle performance measurements. As a result, IoV can significantly improve demand satisfaction by guaranteeing on-time deliveries, reducing journey times through optimised route planning, and lowering operational costs by minimising fuel consumption and vehicle downtime. In addition, the in-vehicle computing reduces environmental impact by optimising energy use and minimising emissions, which in turn encourages the development of greener logistics practices.

However, the research paper highlights the critical challenges that need to be addressed to fully exploit the IoV's potential in optimising vehicle routes. One of the main concerns is the need for robust data security protocols to protect the large amounts of sensitive information transmitted over IoV networks. In addition, it is essential to achieve seamless interoperability between the different IoV systems and devices to create a unified network capable of operating effectively across different platforms and technologies. The complexity of managing and processing the immense volumes of real-time data generated by IoV systems is another major obstacle. Advanced data analytics, AI techniques and edge computing play a crucial role in overcoming this challenge, enabling real-time data processing and analysis to support instant decision-making.

The study also highlights the importance of continued innovation in IoV technologies to address these challenges and open up new possibilities in optimising vehicle routing. Future research is expected to explore the integration of machine learning, predictive analytics and distributed computing algorithms into IoV frameworks to improve their ability to predict and adapt to changing conditions. In parallel, the development of standardised protocols and frameworks for IoV implementation will be essential to ensure interoperability across different regions.

REFERENCES

AbdAllah, A. M. F. M., Essam, D. L., & Sarker, R. A. (2017). On Solving Periodic Re-Optimization Dynamic Vehicle Routing Problems. *Applied Soft Computing*, 55, 1–12. Advance online publication. DOI: 10.1016/j.asoc.2017.01.047

Adelantado, F., Ammouriova, M., Herrera, E., Juan, A. A., Shinde, S. S., & Tarchi, D. (2022). Internet of Vehicles and Real-Time Optimization Algorithms: Concepts for Vehicle Networking in Smart Cities. *Vehicles (Basel)*, 4(4), 1223–1245. DOI: 10.3390/vehicles4040065

Agnusdei, G. P., Gnoni, M. G., Sgarbossa, F., & Govindann, K. (2022). Challenges and perspectives of the Industry 4.0 technologies within the last-mile and first-mile reverse logistics: A systematic literature review. *Research in Transportation Business & Management*, 45, 100896. DOI: 10.1016/j.rtbm.2022.100896

Ahmed, A. A. A., Singhal, S., Prakaash, A. S., Dayupay, J., Rahadi, I., Marhoon, H. A., Iswanto, A. H., Abbas, S. F., & Aravindhan, S. (2022). A Mathematical Model for the Vehicles Routing Problem with Multiple Depots, Considering the Possibility of Return Using the Tabu Search Algorithm. *Foundations of Computing and Decision Sciences*, 47(4), 359–370. DOI: 10.2478/fcds-2022-0019

Almoustafa, S., Hanafi, S., & Mladenovic, N. (2013). Multistart Branch and Bound for Large Asymmetric Distance-Constrained Vehicle Routing Problem. In *Springer Proceedings in Mathematics and Statistics* (Vol. 31, pp. 15–38). DOI: 10.1007/978-1-4614-5134-1_2

Alsudani, M., & Ozturk, T. (2023). Wireless Communication Between Vehicles: Exploring the Potential of V2V and V2X Communication for Improved Efficiency, Safety, and Sustainability. *JOM: The Journal of the Minerals. Metals & Materials Society*, 3(1), 5.

Andani, S., Zarlis, M., Mawengkang, H., & Wage, S. (2023). Optimization of heterogeneous VRP model by considering recharging station using mixed integer programming. *Telkomnika*, 21(6), 1267. DOI: 10.12928/telkomnika.v21i6.25313

Anwar, M., Wang, S., Akram, M., Raza Naqvi, S. S., & Mahmood, S. (2021). 5G Enabled MEC: A Distributed Traffic Steering for Seamless Service Migration of Internet of Vehicles. *IEEE Internet of Things Journal, PP*, 1–1. DOI: 10.1109/JIOT.2021.3084912

Arooj, A., Farooq, S., Akram, A., Iqbal, R., Sharma, A., & Dhiman, G. (2021). Big Data Processing and Analysis in Internet of Vehicles: Architecture, Taxonomy, and Open Research Challenges. *Archives of Computational Methods in Engineering*, 29(2), 793–829. Advance online publication. DOI: 10.1007/s11831-021-09590-x

Aurachman, R., Baskara, D., & Habibie, J. (2021). Vehicle routing problem with simulated annealing using python programming. *IOP Conference Series. Materials Science and Engineering*, 1010(1), 012010. DOI: 10.1088/1757-899X/1010/1/012010

Bard, J. F., Huang, L., Dror, M., & Jaillet, P. (1998). A branch and cut algorithm for the VRP with satellite facilities. *IIE Transactions*, 30(9), 821–834. DOI: 10.1080/07408179808966528

Bell, J., & McMullen, P. (2004). An Ant Colony Optimization Approach to the Vehicle Routing Problem. *Advanced Engineering Informatics*, 18(1), 41–48. DOI: 10.1016/j.aei.2004.07.001

Belver, G. A., Lopez, R., & Fuente, A. (2017). *Solving Vehicle Routing Problem with Multiple Trips using Iterative Local Search with Variable Neighborhood Search*. https://www.semanticscholar.org/paper/Solving-Vehicle-Routing-Problem-with-Multiple-Trips-Belver-G%C3%B3mez/532f4eb6829e75ea819cdd250f049f5786b3d476

Bouanane, K., Benadada, Y., & Bencheikh, G. (2019). Multi-Depots Vehicle Routing Problem with Simultaneous Delivery and Pickup and Inventory Restrictions: Formulation and Resolution. *International Journal of Advanced Computer Science and Applications*, 10(2), 110–120. DOI: 10.14569/IJACSA.2019.0100215

Casazza, M., Ceselli, A., & Wolfler Calvo, R. (2018). A branch and price approach for the Split Pickup and Split Delivery VRP. *Electronic Notes in Discrete Mathematics*, 69, 189–196. DOI: 10.1016/j.endm.2018.07.025

Cauchois, R., Taghipour, A., Kang, D., Zoghlami, N., & Abed, M. (2017). Advanced Solutions for a Supply Chain with Stochastic Information. *Journal of Advanced Management Science*, 5(1), 9–13. DOI: 10.18178/joams.5.1.9-13

Chbichib, A., Mellouli, R., & Chabchoub, H. (2011, May 1). *Profitable Vehicle Routing Problem with Multiple Trips: Modeling and constructive heuristics*. 2011 4th International Conference on Logistics, LOGISTIQUA'2011. DOI: 10.1109/LOGISTIQUA.2011.5939450

Chen, G., Gui, P., Ding, T., Na, S., & Zhou, Y. (2019). Optimization of Transportation Routing Problem for Fresh Food by Improved Ant Colony Algorithm Based on Tabu Search. *Sustainability (Basel)*, 11(23), 6584. DOI: 10.3390/su11236584

Chen, R.-M., & Guo, J.-C. (2018).. . *Optimal Routing for Split Delivery Vehicle Routing Problems*, 667–670, 667–670. Advance online publication. DOI: 10.1109/IIAI-AAI.2018.00140

Cheng, L. (2009). *A genetic algorithm for the vehicle routing problem with time windows.* https://www.semanticscholar.org/paper/A-genetic-algorithm-for-the-vehicle-routing-problem-Cheng/daa1ab8d3abf27f78540342ce20c4b0a6437d168

Chowdhary, N., & Kaur, P. (2018). *Dynamic Route Optimization Using Nature-Inspired Algorithms in IoV.*, DOI: 10.1007/978-981-10-5828-8_47

Christopher, Y., Wahyuningsih, S., & Satyananda, D. (2021). Study of variable neighborhood descent and tabu search algorithm in VRPSDP. *Journal of Physics: Conference Series*, 1872(1), 012002. DOI: 10.1088/1742-6596/1872/1/012002

Clarke, G., & Wright, J. W. (1964). Scheduling of Vehicles from a Central Depot to a Number of Delivery Points. *Operations Research*, 12(4), 568–581. DOI: 10.1287/opre.12.4.568

Dadsena, K. K., Mathiyazhagan, K., & Taghipour, A. (2021). Analysis of Barriers for the Build the Resilient Supply Chain Networks Post-COVID-19. In Sakthivel, A. R., Kandasamy, J., & Davim, J. P. (Eds.), *Managing Supply Chain Risk and Disruptions: Post COVID-19* (pp. 79–89). Springer International Publishing., DOI: 10.1007/978-3-030-72575-4_7

Daely, P., Aruan, Y., Lee, J. M., & Kim, D.-S. (2021).. . *Dynamic VRP Optimization Using Discrete PSO in Edge Computing Environment*, 654–656, 654–656. Advance online publication. DOI: 10.1109/ICTC52510.2021.9620744

Dantzig, G. B., & Ramser, J. H. (1959). The Truck Dispatching Problem. *Management Science*, 6(1), 80–91. DOI: 10.1287/mnsc.6.1.80

Dastghaibifard, G. H., Ansari, E., Sheykhalishahi, S. M., Bavandpouri, A., & Ashoor, E. (2008). *A Parallel Branch and Bound Algorithm for Vehicle Routing Problem.*

Demirtas, Y. E., Ozdemir, E., & Demirtas, U. (2015). A particle swarm optimization for the dynamic vehicle routing problem. *2015 6th International Conference on Modeling, Simulation, and Applied Optimization (ICMSAO)*, 1–5. DOI: 10.1109/ICMSAO.2015.7152224

Dey, K. C., Rayamajhi, A., Chowdhury, M., Bhavsar, P., & Martin, J. (2016). Vehicle-to-vehicle (V2V) and vehicle-to-infrastructure (V2I) communication in a heterogeneous wireless network – Performance evaluation. *Transportation Research Part C, Emerging Technologies*, 68, 168–184. DOI: 10.1016/j.trc.2016.03.008

Dhanare, R., Nagwanshi, K., & Varma, S. (2022). A Study to Enhance the Route Optimization Algorithm for the Internet of Vehicle. *Wireless Communications and Mobile Computing*, 2022, 1–20. DOI: 10.1155/2022/1453187

Eze, E. C., Zhang, S., & Liu, E. (2014). Vehicular ad hoc networks (VANETs): Current state, challenges, potentials and way forward. *2014 20th International Conference on Automation and Computing*, 176–181. DOI: 10.1109/IConAC.2014.6935482

Fadda, P., Mancini, S., Serra, P., & Fancello, G. (2022). The Heterogeneous Fleet Vehicle Routing Problem with Draft Limits. *Computers & Operations Research*, 149, 106024. DOI: 10.1016/j.cor.2022.106024

Frías, N., Johnson Parejas, F., & Valle, C. (2023). Hybrid Algorithms for Energy Minimizing Vehicle Routing Problem: Integrating Clusterization and Ant Colony Optimization. *IEEE Access, PP*, 1–1. DOI: 10.1109/ACCESS.2023.3325787

Gannouni, A., Rachid, E., & Talbi, E.-G. (2017). Solving stochastic multiobjective vehicle routing problem using probabilistic metaheuristic. *MATEC Web of Conferences, 105*, 00001. DOI: 10.1051/matecconf/201710500001

Gao, Y., Li, Z., Taghipour, A., Canel, B., & Kang, D. S. (2018). Supply Chain Coordination: A Review. *Journal of Advanced Management Science*, 6(4), 213–217. DOI: 10.18178/joams.6.4.213-217

Garay-Rondero, C. L., Martinez-Flores, J. L., Smith, N. R., Caballero Morales, S. O., & Aldrette-Malacara, A. (2019). Digital supply chain model in Industry 4.0. *Journal of Manufacturing Technology Management*, 31(5), 887–933. DOI: 10.1108/JMTM-08-2018-0280

Gheni, H. M., AbdulRahaim, L. A., & Abdellatif, A. (2024). Real-time driver identification in IoV: A deep learning and cloud integration approach. *Heliyon*, 10(7), e28109. DOI: 10.1016/j.heliyon.2024.e28109 PMID: 38560228

Gialos, A., & Zeimpekis, V. (n.d.). *Vision picking technology: Defining design parameters via a systematic literature review.*

Glaa, B., Zoghlami, N., & Taghipour, A. (2014). Outsourcing and off-shoring healthcare services: A way to cut the costs and improve quality. *2014 International Conference on Advanced Logistics and Transport (ICALT)*, 189–194. DOI: 10.1109/ICAdLT.2014.6864116

Glize, E., Jozefowiez, N., & Ngueveu, S. (2018). *An Exact Column Generation-Based Algorithm for Bi-objective Vehicle Routing Problems: 5th International Symposium, ISCO 2018, Marrakesh, Morocco, April 11–13, 2018, Revised Selected Papers* (pp. 208–218). DOI: 10.1007/978-3-319-96151-4_18

HadjTaieb, S., Hani, Y., Loukil, T. M., & Mhamedi, A. E. (2022). Green VRP applied to Home Health Care problem. *IFAC-PapersOnLine*, 55(10), 3154–3159. DOI: 10.1016/j.ifacol.2022.10.214

Hashimoto, H., Hu, Y., Sugiura, T., Takada, Y., & Yagiura, M. (2022)... *An Iterated Local Search Algorithm for Commuting Bus Routing Problem with Latest Arrival Time Constraint*, 0791–0795, 0791–0795. Advance online publication. DOI: 10.1109/IEEM55944.2022.9989602

He, S., Li, J., & Qiu, T. (2017). Vehicle to Pedestrian Communication Modeling and Collision Avoiding Methodology in Connected Vehicle Environment. *Transportation Research Record: Journal of the Transportation Research Board*, 2621(1), 21–30. DOI: 10.3141/2621-03

Holland, J. H. (1975). *Adaptation in natural and artificial systems: An introductory analysis with applications to biology, control, and artificial intelligence* (pp. viii, 183). U Michigan Press.

Huang, S.-H., Huang, Y.-H., Blazquez, C., & Chen, C.-Y. (2022). Solving the vehicle routing problem with drone for delivery services using an ant colony optimization algorithm. *Advanced Engineering Informatics*, 51, 101536. DOI: 10.1016/j.aei.2022.101536

Hyeon, E., Han, J., Shen, D., Karbowski, D., Kim, N., & Rousseau, A. (2022). Potential Energy Saving of V2V-Connected Vehicles in Large-Scale Traffic. *IFAC-PapersOnLine*, 55(24), 78–83. DOI: 10.1016/j.ifacol.2022.10.265

Ichoua, S., Gendreau, M., & Potvin, J.-Y. (2003). Vehicle Dispatching With Time—Dependent Travel Times. *European Journal of Operational Research*, 144(2), 379–396. DOI: 10.1016/S0377-2217(02)00147-9

Jegatheesan, D., & Arumugam, C. (2024). SIoV-FTFSA-CAOA: A fuzzy trust-based approach for enhancing security and energy efficiency in social internet of vehicles. *Wireless Networks*, 30(4), 1–20. DOI: 10.1007/s11276-023-03626-9

Kabadurmus, O., & Erdogan, M. (2022). Bi-Objective Green Vehicle Routing Problem Minimizing Carbon Emissions and Maximizing Service Level. *Gazi Üniversitesi Mühendislik-Mimarlık Fakültesi Dergisi*, 38(1), 103–112. Advance online publication. DOI: 10.17341/gazimmfd.633583

Kahveci, N., & Ioannou, P. (2008, August 18). *Genetic Algorithms for Shortest Path Routing of Autonomous Gliders*. DOI: 10.2514/6.2008-6646

Kanthavel, D., Sangeetha, S. K. B., & Keerthana, K. P. (2021). An empirical study of vehicle to infrastructure communications—An intense learning of smart infrastructure for safety and mobility. *International Journal of Intelligent Networks*, 2, 77–82. DOI: 10.1016/j.ijin.2021.06.003

Kirkpatrick, S., Gelatt, C. D.Jr, & Vecchi, M. P. (1983). Optimization by Simulated Annealing. *Science*, 220(4598), 671–680. DOI: 10.1126/science.220.4598.671 PMID: 17813860

Küp, E. T., Cebeci, S., Bayram, B., Aydın, G., Bozkaya, B., & Akhavan-Tabatabaei, R. (2024). An Integrated Framework for Dynamic Vehicle Routing Problems with Pick-up and Delivery Time Windows and Shared Fleet Capacity Planning. *Symmetry*, 16(4), 4. Advance online publication. DOI: 10.3390/sym16040505

Kuyu, Y. Ç., & Vatansever, F. (2024). A hybrid approach of ALNS with alternative initialization and acceptance mechanisms for capacitated vehicle routing problems. *Cluster Computing*, 27(10), 13583–13606. Advance online publication. DOI: 10.1007/s10586-024-04643-9

Lagorio, A., Cimini, C., Pinto, R., & Cavalieri, S. (2023). 5G in Logistics 4.0: Potential applications and challenges. *Procedia Computer Science*, 217, 650–659. DOI: 10.1016/j.procs.2022.12.261

Li, Z., & Jiao, P. (2022). Two-stage stochastic programming for the inventory routing problem with stochastic demands in fuel delivery. *International Journal of Industrial Engineering Computations*, 13(4), 507–522. DOI: 10.5267/j.ijiec.2022.7.004

Li, Z., & Ma, Y. (2013).. . *Particle Swarm Optimization Based on the Average Optimal Information for Vehicle Routing Problem*, 51–54. Advance online publication. DOI: 10.1109/ISCID.2013.20

Loivet, W., Taghipour, A., & Kang, D. S. (2020). The Rise of Green Supply Chain Management: Between Complexity and Necessity. *Journal of Economics. Business and Management*, 8(1), 1–7. DOI: 10.18178/joebm.2020.8.1.604

Lu, X., & Taghipour, A. (2023). A French Guide for Businesses and Governments to Integrate the Concept of Sustainable Development Into Their Supply Chains. In *Government Impact on Sustainable and Responsible Supply Chain Management* (pp. 1–12). IGI Global Scientific Publishing., DOI: 10.4018/978-1-6684-9062-4.ch001

Lysgaard, J., Letchford, A. N., & Eglese, R. W. (2004). A new branch-and-cut algorithm for the capacitated vehicle routing problem. *Mathematical Programming*, 100(2), 423–445. DOI: 10.1007/s10107-003-0481-8

Ma, Y., Tang, J., & Guan, J. (2013). Split Delivery Weighted Vehicle Routing Problem: Comparison perspective. *Chinese Control Conference, CCC*, 8436–8440.

Mahfod, J., Canel-Depitre, B., & Taghipour, A. (2019). Quality Function Deployment-ELECTRE in Supplier Evaluation. *Journal of Advanced Management Science*, 131–135, 131–135. Advance online publication. DOI: 10.18178/joams.7.4.131-135

Mahya, A., Khabibah, S., & Surarso, B. (2023). *Optimization of vehicle routing problem with time windows using nearest neighbor algorithm and dynamic programming (case study: Pekalongan city Environmental Agency)*. 020007. DOI: 10.1063/5.0140518

Marinakis, Y., Marinaki, M., & Migdalas, A. (2018). Particle swarm optimization for the vehicle routing problem: A survey and a comparative analysis. In *Handbook of Heuristics* (pp. 1163–1196). DOI: 10.1007/978-3-319-07124-4_42

Martinovic, G., Aleksi, I., & Baumgartner, A. (2008). Single-Commodity Vehicle Routing Problem with Pickup and Delivery Service. *Mathematical Problems in Engineering*, 2008(1), 697981. Advance online publication. DOI: 10.1155/2008/697981

Mbiatem, J., Taghipour, A., & Canel-Depitre, B. (2018). Supplier Selection Approaches for Decision Makers. *Proceedings of the 8th International Conference on Information Communication and Management*, 108–112. DOI: 10.1145/3268891.3268895

Meng, X., Li, Z., & Tang, J. (2021). *A Dynamic VRP with Varying Transportation Costs and Its Solution Strategy*. 1–6. DOI: 10.1109/ICNSC52481.2021.9702224

Merimi, M., & Taghipour, A. (2021). Accelerating the Digitalization of the Supply Chain: An Empirical Research About COVID-19 Crisis. In *Digitalization of Decentralized Supply Chains During Global Crises* (pp. 1–24). IGI Global Scientific Publishing. DOI: 10.4018/978-1-7998-6874-3.ch001

Modica, T., Colicchia, C., Tappia, E., & Melacini, M. (2023). Empowering freight transportation through Logistics 4.0: A maturity model for value creation. *Production Planning and Control*, 34(12), 1149–1164. DOI: 10.1080/09537287.2021.1988176

Mohamed, B., & Feham, M. (2018). *Greedy Traffic Aware Routing Protocol Implementation in the Context of Internet of Vehicles*. 1–6. DOI: 10.1145/3284557.3284698

Navarro, M. (2023, January 1). *Comparison Between the Branch and Bound and Evolutionary Algorithms to Solve the Classic Vehicle Routing Problem Known as VRP*. DOI: 10.18687/LACCEI2023.1.1.1499

Ochelska-Mierzejewska, J., Poniszewska-Maranda, A., & Marańda, W. (2021). Selected Genetic Algorithms for Vehicle Routing Problem Solving. *Electronics (Basel)*, 10(24), 3147. DOI: 10.3390/electronics10243147

Olivera, A., & Viera, O. (2007). Solving the vehicle routing problem with multiple trips by adaptive memory programming. *International Journal of Logistics Systems and Management - Int J Logist Syst Manag, 3*. DOI: 10.1504/IJLSM.2007.013213

Osorio-Mora, A., Soto-Bustos, M., Gatica, G., Palominos, P., & Linfati, R. (2021). The Multi-Depot Cumulative Vehicle Routing Problem With Mandatory Visit Times and Minimum Delayed Latency. *IEEE Access : Practical Innovations, Open Solutions*, 9, 27210–27225. DOI: 10.1109/ACCESS.2021.3058242

Ou, J., Liu, X., Xing, L., Lv, J., Hu, Y., Zheng, J., Zou, J., & Li, M. (2024). Solving many-objective delivery and pickup vehicle routing problem with time windows with a constrained evolutionary optimization algorithm. *Expert Systems with Applications*, 255, 124712. DOI: 10.1016/j.eswa.2024.124712

Pandey, K., & Jha, C. K. (2023). A hybrid optimal solution for the combinatorial problem of vehicle routing. *International Journal of Intelligent Engineering Informatics*, 11(4), 337–352. DOI: 10.1504/IJIEI.2023.136104

Parveen, K., Kumar, S., Singh, R. P., Kumar, A., Yaduwanshi, R., & Dora, D. P. (2023). TS-CAGR:Traffic sensitive connectivity-aware geocast routing protocol in internet of vehicles. *Ad Hoc Networks*, 147, 103210. DOI: 10.1016/j.adhoc.2023.103210

Pendor, R. B., & Tasgaonkar, P. P. (2016). An IoT framework for intelligent vehicle monitoring system. *2016 International Conference on Communication and Signal Processing (ICCSP)*, 1694–1696. DOI: 10.1109/ICCSP.2016.7754454

Prodhon, C. (2009). *An ELSxPath Relinking Hybrid for the Periodic Location-Routing Problem*. 15–29. DOI: 10.1007/978-3-642-04918-7_2

Prodhon, C., & Prins, C. (2008). *A Memetic Algorithm with Population Management (MA|PM) for the Periodic Location-Routing Problem. 5296*, 43–57. DOI: 10.1007/978-3-540-88439-2_4

Qin, Y., Huang, D., & Zhang, X. (2012). VehiCloud: Cloud Computing Facilitating Routing in Vehicular Networks. *2012 IEEE 11th International Conference on Trust, Security and Privacy in Computing and Communications*, 1438–1445. DOI: 10.1109/TrustCom.2012.16

Ramadhani, B., & Garside, A. (2021). Particle Swarm Optimization Algorithm to Solve Vehicle Routing Problem with Fuel Consumption Minimization. *Jurnal Optimasi Sistem Industri*, 20(1), 1–10. DOI: 10.25077/josi.v20.n1.p1-10.2021

Reddy, K., & Muralidhar, A. (2023). Machine Learning-Based Road Safety Prediction Strategies for Internet of Vehicles (IoV) Enabled Vehicles: A Systematic Literature Review. *IEEE Access, PP*, 1–1. DOI: 10.1109/ACCESS.2023.3315852

Rizkiani, F., Geraudy, I., & Imran, A. (2024). Solving the Vehicle Routing Problem with Multiple Trips and Simultaneous Delivery-Pickup for Drinking Water Distribution Company. *E3S Web of Conferences, 484*. DOI: 10.1051/e3sconf/202448401003

Rout, R. R., Vemireddy, S., Raul, S. K., & Somayajulu, D. V. L. N. (2020). Fuzzy logic-based emergency vehicle routing: An IoT system development for smart city applications. *Computers & Electrical Engineering*, 88, 106839. DOI: 10.1016/j.compeleceng.2020.106839

Sahoo, S., Kumar, S., Abedin, M. Z., Lim, W. M., & Jakhar, S. K. (2023). Deep learning applications in manufacturing operations: A review of trends and ways forward. *Journal of Enterprise Information Management*, 36(1), 221–251. DOI: 10.1108/JEIM-01-2022-0025

Samsuddin, S., Othman, M., & Mi Yusuf, L. (2020). Utilizing Ant Colony Optimization and Intelligent Water Drop for Solving Multi Depot Vehicle Routing Problem. *IOP Conference Series. Materials Science and Engineering*, 864(1), 012095. DOI: 10.1088/1757-899X/864/1/012095

Sato, T., Morita, T., & Murata, T. (2010). Lagrangian relaxation method for network flow modeled crew and vehicle rescheduling. *2010 2nd International Conference on Advanced Computer Control*, 403–408. 2010 2nd International Conference on Advanced Computer Control. DOI: 10.1109/ICACC.2010.5486971

Seidgar, H., Abedi, M., Rad, S., & Rezaeian, J. (2016). An efficient hybrid of genetic and simulated annealing algorithms for multi server vehicle routing problem with multi entry. *International Journal of Industrial and Systems Engineering*, 24(3), 333. DOI: 10.1504/IJISE.2016.079823

Sewalkar, P., & Seitz, J. (2019). Vehicle-to-Pedestrian Communication for Vulnerable Road Users: Survey, Design Considerations, and Challenges. *Sensors (Basel)*, 19(2), 358. Advance online publication. DOI: 10.3390/s19020358 PMID: 30658392

Shao, X. (2022). Improved energy-efficient routing architecture for traffic management system using a hybrid meta-heuristic algorithm in Internet of vehicles. *Journal of High Speed Networks*, 28(4), 1–10. DOI: 10.3233/JHS-222003

Shetty, A., Puthusseri, K., Shetty, A., & Mani, R. (2018). *A Savings Algorithm Approach to the Truck Routing and Loading Problem*. 284–290. DOI: 10.1109/ICRCICN.2018.8718688

Shurrab, M., Singh, S., Otrok, H., Mizouni, R., Khadkikar, V., & Zeineldin, H. (2022). An Efficient Vehicle-to-Vehicle (V2V) Energy Sharing Framework. *IEEE Internet of Things Journal, 9*(7), 5315–5328. *IEEE Internet of Things Journal*. Advance online publication. DOI: 10.1109/JIOT.2021.3109010

Soleimani, P., Marvasti, M. R. B., & Ghorbanzadeh, P. (2020). A Hybrid Traffic Management Method Based on Combination of IOV and VANET Network in Urban Routing for Emergency Vehicles. *2020 4th International Conference on Smart City, Internet of Things and Applications (SCIOT)*, 58–65. DOI: 10.1109/SCIOT50840.2020.9250198

Solomon, M. M. (1987). Algorithms for the Vehicle Routing and Scheduling Problems with Time Window Constraints. *Operations Research*, 35(2), 254–265. DOI: 10.1287/opre.35.2.254

Stodola, P., & Kutěj, L. (2024). Multi-Depot Vehicle Routing Problem with Drones: Mathematical formulation, solution algorithm and experiments. *Expert Systems with Applications*, 241, 122483. DOI: 10.1016/j.eswa.2023.122483

Sultana, N., Chan, J., Abbasi, B., Sarwar, T., & Qin, A. K. (2024). Learning to Guide Local Search Optimisation for Routing Problems. *Operations Research Letters*, 55, 107136. DOI: 10.1016/j.orl.2024.107136

Sun, Q., Tyagu, T., & Noche, B. (2022). Savings matrix algorithm to address vehicle routing problem. In Ye, X., Falcone, F., & Cui, H. (Eds.), *2nd International Conference on Internet of Things and Smart City (IoTSC 2022)* (p. 163). SPIE. DOI: 10.1117/12.2637152

Taghipour, A. (2014). Improving the Plan of a Manufacturing Network with Non-Integrated Business Units. *International Journal of Applied Logistics*, 5(2), 1–11. DOI: 10.4018/ijal.2014040101

Taghipour, A. (Ed.). (2020). *Demand forecasting and order planning in supply chains and humanitarian logistics*. IGI Global.

Taghipour, A. (2020). Sustainable Supply Chain Management Performance. *International Journal of Innovation, Management and Technology*, 11(6), 165–169. DOI: 10.18178/ijimt.2020.11.6.897

Taghipour, A. (2021). Digital transformation of supply chains during crisis: Covid19. *11th Annual International Conference on Industrial Engineering and Operations Management, IEOM 2021*. https://www.academia.edu/79144632/Digital_transformation_of_supply_chains_during_crisis_Covid19

Taghipour, A. (Ed.). (2023). *Government Impact on Sustainable and Responsible Supply Chain Management*. IGI Global. DOI: 10.4018/978-1-6684-9062-4

Taghipour, A., Abed, M., & Zoghlami, N. (2015). Design for remanufacturing respecting reverse logistics processes: A review. *2015 4th International Conference on Advanced Logistics and Transport (ICALT)*, 299–304. DOI: 10.1109/ICAdLT.2015.7136602

Taghipour, A., Foukolaei, P. Z., Ghaedi, M., & Khazaei, M. (2023). Sustainable Multi-Objective Models for Waste-to-Energy and Waste Separation Site Selection. *Sustainability (Basel)*, 15(22), 15764. DOI: 10.3390/su152215764

Taghipour, A., & Frayret, J. M. (2013a). An algorithm to improve operations planning in decentralized supply chains. *2013 International Conference on Advanced Logistics and Transport*, 100–103. DOI: 10.1109/ICAdLT.2013.6568442

Taghipour, A., & Frayret, J. M. (2013b). Coordination of operations planning in supply chains: A review. *International Journal of Business Performance and Supply Chain Modelling*, 5(3), 272. DOI: 10.1504/IJBPSCM.2013.055729

Taghipour, A., Lu, X., & Smahi, M. (2022). Opportunities and Challenges for Society 5.0 in Procurement. In *Blockchain Technology in Supply Chain Management for Society 5.0*. CRC Press. DOI: 10.1201/9781003177432-9

Taghipour, A., Ramezani, M., Khazaei, M., Roohparvar, V., & Hassannayebi, E. (2023). Smart Transportation Behavior through the COVID-19 Pandemic: A Ride-Hailing System in Iran. *Sustainability (Basel)*, 15(5), 5. Advance online publication. DOI: 10.3390/su15054178

Taghipour, A., Sohrabi, A., Ghaedi, M., & Khazaei, M. (2023). A robust vaccine supply chain model in pandemics: Case of Covid-19 in Iran. *Computers & Industrial Engineering*, 183, 109465. DOI: 10.1016/j.cie.2023.109465

Tavakkoli-Moghaddam, R., Rabbani, M., Saremi, A., & Safaei, N. (2005). *Solving the backhaul vehicle routing problem by genetic algorithms.*

Teng, R., & Sato, K. (2024). A Fundamental Study of Reliable Vehicle-to-Cloud Communication Using Multiple Paths with Redundancy Mitigation. *Applied Sciences (Basel, Switzerland)*, 14(7), 2841. DOI: 10.3390/app14072841

Torres, F., Gendreau, M., & Rei, W. (2022). Crowdshipping: An open VRP variant with stochastic destinations. *Transportation Research Part C, Emerging Technologies*, 140, 103677. DOI: 10.1016/j.trc.2022.103677

Toth, P., & Vigo, D. (2001). An overview of vehicle routing problems. In *The vehicle routing problem* (pp. 1–26). Society for Industrial and Applied Mathematics.

Volna, E. (2016). *Genetic algorithms for the vehicle routing problem. 1738*, 120002. DOI: 10.1063/1.4951885

Wang, X., Shu-Zhi, Z., Hao, C., & Yan, L. (2016). *An improved savings method for vehicle routing problem*. 1–4. DOI: 10.1109/CCSSE.2016.7784340

Wassan, N., Wassan, N., Nagy, G., & Salhi, S. (2017). The Multiple Trip Vehicle Routing Problem with Backhauls. *Computers & Operations Research*, 78(C), 454–467. DOI: 10.1016/j.cor.2015.12.017

Xie, N., Lee, X., Cai, D., Saito, Y., Asai, N., & Lau, H. C. (2024). *A Feasibility-Preserved Quantum Approximate Solver for the Capacitated Vehicle Routing Problem*. /arXiv.2308.08785DOI: 10.21203/rs.3.rs-4274870/v1

Yavary, A., & Sajedi, H. (2018). *Solving Dynamic Vehicle Routing Problem with Pickup and Delivery by CLARITY Method*. 207–212. DOI: 10.1109/INES.2018.8523908

Yousefikhoshbakht, M., Didehvar, F., Rahmati, F., & Ahmed, Z. (2023). *Fixed fleet open vehicle routing problem: Mathematical model and a modified ant colony optimization*.

Yu, M., Nagarajan, V., & Shen, S. (2021). Improving Column Generation for Vehicle Routing Problems via Random Coloring and Parallelization. *INFORMS Journal on Computing*, 34(2), 953–973. Advance online publication. DOI: 10.1287/ijoc.2021.1105

Yu, V., Lin, S.-W., Zhou, L., & Baldacci, R. (2023). A fast simulated annealing heuristic for the multi-depot two-echelon vehicle routing problem with delivery options. *Transportation Letters*, 1–12. DOI: 10.1080/19427867.2023.2257923

Yu, V., Redi, A. A. N. P., Agustina, Y., & Wibowo, O. (2016). A Simulated Annealing Heuristic for the Hybrid Vehicle Routing Problem. *Applied Soft Computing*, 53, 119–132. Advance online publication. DOI: 10.1016/j.asoc.2016.12.027

Zhang, Z., Zhou, R., Wang, M., & Liu, L. (2023). V2V Composite Routing Algorithm for Internet of Vehicles Based on Improved Q-Learning. *2023 China Automation Congress (CAC)*, 446–450. DOI: 10.1109/CAC59555.2023.10451920

Zhao, H., & Yang, Y. (2020). The research on two phase pickup vehicle routing based on the K-means++ and genetic algorithms. *International Journal of Web Engineering and Technology*, 15(1), 32. DOI: 10.1504/IJWET.2020.107685

Zhao, J., Poon, M., Tan, V. Y. F., & Zhang, Z. (2024). A hybrid genetic search and dynamic programming-based split algorithm for the multi-trip time-dependent vehicle routing problem. *European Journal of Operational Research*, 317(3), 921–935. DOI: 10.1016/j.ejor.2024.04.011

Zheng, T., Glock, C. H., & Grosse, E. H. (2022). Opportunities for using eye tracking technology in manufacturing and logistics: Systematic literature review and research agenda. *Computers & Industrial Engineering*, 171, 108444. DOI: 10.1016/j.cie.2022.108444

Zillies, J., Westphal, S., Thakur, D., Kumar, V., Pappas, G., & Scheidt, D. (2016, October 1). *A Column Generation Approach for Optimized Routing and Coordination of a UAV Fleet*. DOI: 10.1109/SSRR.2016.7784326

Zoghlami, N., Taghipour, A., Merlo, C., & Abed, M. (2016). Management of divergent production network using decentralised multi-level capacitated lot-sizing models. *International Journal of Shipping and Transport Logistics*, 8(5), 590. DOI: 10.1504/IJSTL.2016.078686

Zorkany, M., Yasser, A., & Galal, A. I. A. (2020). Vehicle To Vehicle "V2V" Communication: Scope, Importance, Challenges, Research Directions and Future. *The Open Transplantation Journal*, 14(1), 86–98. DOI: 10.2174/1874447802014010086

Compilation of References

Aarikka-Stenroos, L., Chiaroni, D., Kaipainen, J., & Urbinati, A. (2022). Companies' circular business models enabled by supply chain collaborations: An empirical-based framework, synthesis, and research agenda. *Industrial Marketing Management*, 105, 322–339. DOI: 10.1016/j.indmarman.2022.06.015

Abakouy, K., & Berrada, I. (2018). L'analyse prédictive des campagnes email marketing avec le Big Data Mining. *Actes du Colloque International VSST*, Toulouse, France.

Abbasi, S., Haque Jami, R., & Jiisun, M. S. H. (2023). White Paper-Framework for Integrated Multisector Smart Logistics. *Framework for Integrated Multisector Smart Logistics*, 1-34.

AbdAllah, A. M. F. M., Essam, D. L., & Sarker, R. A. (2017). On Solving Periodic Re-Optimization Dynamic Vehicle Routing Problems. *Applied Soft Computing*, 55, 1–12. Advance online publication. DOI: 10.1016/j.asoc.2017.01.047

Abdul, C., Wang, W., & Li, Y. (2020). The impact of technology on recruitment process. *Issues in Information Systems*, 21(4).

Abdullah, R., Mohamad, M. N., & Thurasamy, R. (2019). Towards sustainable performance: Promoting eco-design in green supply chain management practices. *International Journal of Supply Chain Management*, 8(3), 609–616.

Abdussamad, Z., Agyei, I. T., Döngül, E. S., Abdussamad, J., Raj, R., & Effendy, F. (2022). Impact of internet of things (IOT) on human resource management: A review. *Materials Today: Proceedings*, 56, 3534–3543. DOI: 10.1016/j.matpr.2021.11.247

Acemoglu, D., & Restrepo, P. (2018). Artificial Intelligence, Automation, and Work. *Economics Working Paper Series*. Retrieved from https://www.nber.org/papers/w24196

Acemoglu, D., & Restrepo, P. (2018). Artificial Intelligence, Automation, and Work. NBER Working Paper No. 24196.

Addy, A. (2023). Artificial intelligence in the supply chain management for vaccine distribution in the West African healthcare sector with a focus on Ghana. *International Journal of Legal Science Innovation, 10.*

Adelantado, F., Ammouriova, M., Herrera, E., Juan, A. A., Shinde, S. S., & Tarchi, D. (2022). Internet of Vehicles and Real-Time Optimization Algorithms: Concepts for Vehicle Networking in Smart Cities. *Vehicles (Basel)*, 4(4), 1223–1245. DOI: 10.3390/vehicles4040065

Adenekan, O. A., Solomon, N. O., Simpa, P., & Obasi, S. C. (2024). Enhancing manufacturing productivity: A review of AI-Driven supply chain management optimization and ERP systems integration. *International Journal of Management & Entrepreneurship Research*, 6(5), 1607–1624. DOI: 10.51594/ijmer.v6i5.1126

Adoue, C., R Beulque, L. Carré, et J. Couteau. 2014. Quelles Stratégies D'entreprise Pour Une éConomie Circulaire Moteur De Croissance? Paris: Institut de l'économie circulaire.

Agarwal, V., Mathiyazhagan, K., Malhotra, S., & Saikouk, T. (2022). Analysis of challenges in sustainable human resource management due to disruptions by Industry 4.0: An emerging economy perspective. *International Journal of Manpower*, 43(2), 513–541. DOI: 10.1108/IJM-03-2021-0192

Agnusdei, G. P., Gnoni, M. G., Sgarbossa, F., & Govindann, K. (2022). Challenges and perspectives of the Industry 4.0 technologies within the last-mile and first-mile reverse logistics: A systematic literature review. *Research in Transportation Business & Management*, 45, 100896. DOI: 10.1016/j.rtbm.2022.100896

Ahmad, H., Asghar, M. Z., Khan, A. S., & Habib, A. (2020). A systematic literature review of personality trait classification from textual content. *Open Computer Science*, 10(1), 175–193. DOI: 10.1515/comp-2020-0188

Ahmed, H., Al Bashar, M., Taher, M. A., & Rahman, M. A. (2024). Innovative Approaches To Sustainable Supply Chain Management In The Manufacturing Industry: A Systematic Literature Review. Global Mainstream Journal of Innovation, Engineering & Emerging Technology, 3(2), 1-13.

Ahmed, A. A. A., Singhal, S., Prakaash, A. S., Dayupay, J., Rahadi, I., Marhoon, H. A., Iswanto, A. H., Abbas, S. F., & Aravindhan, S. (2022). A Mathematical Model for the Vehicles Routing Problem with Multiple Depots, Considering the Possibility of Return Using the Tabu Search Algorithm. *Foundations of Computing and Decision Sciences*, 47(4), 359–370. DOI: 10.2478/fcds-2022-0019

Ajunwa, I., & Greene, D. (2019). Platforms at work: Automated hiring platforms and other new intermediaries in the organization of work. In *Work and labor in the digital age*. Emerald Publishing Limited. DOI: 10.1108/S0277-283320190000033005

Akbari, M., Kok, S. K., Hopkins, J., Frederico, G. F., Nguyen, H., & Alonso, A. D. (2024). The changing landscape of digital transformation in supply chains: Impacts of industry 4.0 in Vietnam. *International Journal of Logistics Management*, 35(4), 1040–1072. DOI: 10.1108/IJLM-11-2022-0442

Akimushkin, C., Amancio, D. R., & Oliveira, O. N.Jr. (2018). On the role of words in the network structure of texts: Application to authorship attribution. *Physica A*, 495, 49–58. DOI: 10.1016/j.physa.2017.12.054

Akono, D., & Fernandes, V. (2009). Impacts du développement durable sur les organisations logistiques. *Management et Avenir*, 6(26), 241–255. DOI: 10.3917/mav.026.0241

Akram, H. W., Abbas, H., & Ciddikie, D. (2024a). Green Manufacturing: A Comprehensive Exploration of Sustainable Practices and Industry 4.0 Integration. In Green Manufacturing for Industry 4.0 (pp. 159-178). Productivity Press.

Akram, H. W., Abbas, H., & Sulaiman, M. A. B. A. (2024b). Logistics & supply chain resilience in context of Oman. In *Reference module in social sciences*. Elsevier. DOI: 10.1016/B978-0-443-13701-3.00055-4

Albert, M. N., & Avenier, M. J. (2011). Légitimation de savoirs élaborés dans une épistémologie constructiviste à partir de l'expérience de praticiens. *Recherches Qualitatives*, 30(2), 22–47. DOI: 10.7202/1084829ar

Al-Duais, F. S., & Al-Sharpi, R. S. (2023). A unique Markov chain Monte Carlo method for forecasting wind power utilizing a time series model. *Alexandria Engineering Journal*, 74, 51–63. DOI: 10.1016/j.aej.2023.05.019

Aliahmadi, A., Nozari, H., & Ghahremani-Nahr, J. (2022). Aiot-based sustainable smart supply chain framework. *International Journal of Innovation in Management. Economics and Social Sciences*, 2(2), 28–38.

Ali, M. A. M., Al-Zahrani, A., & Al-Assaf, A. K. T. (2021). Machine Learning Techniques for Disaster Prediction and Risk Management. *IEEE Access : Practical Innovations, Open Solutions*, 9, 129564–129578.

Aljohani, A. (2023). Predictive analytics and machine learning for real-time supply chain risk mitigation and agility. *Sustainability (Basel)*, 15(20), 15088. DOI: 10.3390/su152015088

Allal-Chérif, O., Aránega, A. Y., & Sánchez, R. C. (2021). Intelligent recruitment: How to identify, select, and retain talents from around the world using artificial intelligence. *Technological Forecasting and Social Change*, 169, 120822. DOI: 10.1016/j.techfore.2021.120822

Almoustafa, S., Hanafi, S., & Mladenovic, N. (2013). Multistart Branch and Bound for Large Asymmetric Distance-Constrained Vehicle Routing Problem. In *Springer Proceedings in Mathematics and Statistics* (Vol. 31, pp. 15–38). DOI: 10.1007/978-1-4614-5134-1_2

Alqatawna, A., Abu-Salih, B., Obeid, N., & Almiani, M. (2023). Incorporating Time-Series Forecasting Techniques to Predict Logistics Companies' Staffing Needs and Order Volume. *Computation (Basel, Switzerland)*, 11(7), 141. DOI: 10.3390/computation11070141

Alsheyadi, A., Baawain, A., & Shaukat, M. R. (2024). E-supply chain coordination and performance impacts: An empirical investigation. *Production & Manufacturing Research*, 12(1), 2379942. DOI: 10.1080/21693277.2024.2379942

Alsudani, M., & Ozturk, T. (2023). Wireless Communication Between Vehicles: Exploring the Potential of V2V and V2X Communication for Improved Efficiency, Safety, and Sustainability. *JOM: The Journal of the Minerals. Metals & Materials Society*, 3(1), 5.

Amjad, S. (2022). Role of Logistical Practices in Quality Service Delivery at Supermarkets: A Case Study from Pakistan: Logistics and Quality Service Delivery. *South Asian Journal of Operations and Logistics*, 1(1), 39–56. DOI: 10.57044/SAJOL.2022.1.1.2204

Amor, R., & Dimyadi, J. (2021). The promise of automated compliance checking. *Developments in the built environment, 5*, 100039.

Ananny, M., & Crawford, K. (2018). Seeing without knowing: Limitations of the transparency ideal and its application to algorithmic accountability. *New Media & Society, 20*(3), 973-989.

Andani, S., Zarlis, M., Mawengkang, H., & Wage, S. (2023). Optimization of heterogeneous VRP model by considering recharging station using mixed integer programming. *Telkomnika*, 21(6), 1267. DOI: 10.12928/telkomnika.v21i6.25313

Andel, T. R., & McDonald, J. T. (2013, October). A systems approach to cyber assurance education. In Proceedings of the 2013 on InfoSecCD'13: Information Security Curriculum Development Conference (pp. 13-19). DOI: 10.1145/2528908.2528920

Andrei, N., Scarlat, C., & Ioanid, A. (2024). Transforming E-commerce logistics: Sustainable practices through autonomous maritime and last-mile transportation solutions. *Logistics (Basel)*, 8(3), 71. DOI: 10.3390/logistics8030071

Annoni, A., Nativi, S., Çöltekin, A., Desha, C., Eremchenko, E., Gevaert, C. M., Giuliani, G., Chen, M., Perez-Mora, L., Strobl, J., & Tumampos, S. (2023). Digital earth: Yesterday, today, and tomorrow. *International Journal of Digital Earth*, 16(1), 1022–1072. DOI: 10.1080/17538947.2023.2187467

Anuradha, J. (2015). A brief introduction on Big Data 5Vs characteristics and Hadoop technology. *Procedia Computer Science*, 48, 319–324. DOI: 10.1016/j.procs.2015.04.188

Anwar, M., Wang, S., Akram, M., Raza Naqvi, S. S., & Mahmood, S. (2021). 5G Enabled MEC: A Distributed Traffic Steering for Seamless Service Migration of Internet of Vehicles. *IEEE Internet of Things Journal, PP*, 1–1. DOI: 10.1109/JIOT.2021.3084912

Archana, T., & Stephen, R. K. (2024). The Future of Artificial Intelligence in Manufacturing Industries. In Industry Applications of Thrust Manufacturing: Convergence with Real-Time Data and AI (pp. 98-117). IGI Global. DOI: 10.4018/979-8-3693-4276-3.ch004

Archana, T., Banu, J. F., Prasad, S., & Shrivastava, P. R. (2023). A Flexible and Investigation Approach for Encrypted Features Space Using Neural Network. *Advances in Science and Technology (Owerri, Nigeria)*, 124, 754–761. DOI: 10.4028/p-nd03y2

Arifoğlu, K., & Özekici, S. (2011). Inventory management with random supply and imperfect information: A hidden Markov model. *International Journal of Production Economics*, 134(1), 123–137. DOI: 10.1016/j.ijpe.2011.04.033

Armengaud, E., Fruhwirth, M., Rothbart, M., Weinzerl, M., & Zembacher, G. (2020). Digitalization as an opportunity to remove silo-thinking and enable holistic value creation. Systems Engineering for Automotive Powertrain Development, 1-28.

Arooj, A., Farooq, S., Akram, A., Iqbal, R., Sharma, A., & Dhiman, G. (2021). Big Data Processing and Analysis in Internet of Vehicles: Architecture, Taxonomy, and Open Research Challenges. *Archives of Computational Methods in Engineering*, 29(2), 793–829. Advance online publication. DOI: 10.1007/s11831-021-09590-x

Arruda, E. H., Melatto, R. A. P. B., Levy, W., & de Melo Conti, D. (2021). Circular economy: A brief literature review (2015–2020). *Sustainable Operations and Computers*, 2, 79–86. DOI: 10.1016/j.susoc.2021.05.001

Arute, F., Arya, K., Babbush, R., Bacon, D., Bardin, J. C., Barends, R., Biswas, R., Boixo, S., Brandao, F. G. S. L., Buell, D. A., Burkett, B., Chen, Y., Chen, Z., Chiaro, B., Collins, R., Courtney, W., Dunsworth, A., Farhi, E., Foxen, B., & Martinis, J. M. (2019). Quantum supremacy using a programmable superconducting processor. *Nature*, 574(7779), 505–510. DOI: 10.1038/s41586-019-1666-5 PMID: 31645734

Ashraf, M. H., Yalcin, M. G., & Hos, R. (2024). Reconceptualizing supply chain flows (SCFLOWS). *International Journal of Logistics Management*. Advance online publication. DOI: 10.1108/IJLM-03-2024-0180

Ashurst, C., Hine, E., Sedille, P., & Carlier, A. (2022, June). Ai ethics statements: analysis and lessons learnt from neurips broader impact statements. In *Proceedings of the 2022 ACM conference on fairness, accountability, and transparency* (pp. 2047-2056). DOI: 10.1145/3531146.3533780

Atilgan, C., & McCullen, P. (2011). Improving supply chain performance through auditing: A change management perspective. *Supply Chain Management*, 16(1), 11–19. DOI: 10.1108/13598541111103467

Attaran, M. (2020, July). Digital technology enablers and their implications for supply chain management. In Supply Chain Forum *International Journal (Toronto, Ont.)*, 21(3), 158–172.

Aurachman, R., Baskara, D., & Habibie, J. (2021). Vehicle routing problem with simulated annealing using python programming. *IOP Conference Series. Materials Science and Engineering*, 1010(1), 012010. DOI: 10.1088/1757-899X/1010/1/012010

Aurez, V., & Georgeault, L. (2016). *Economie Circulaire système économique et finitudes des ressources, 360*. De Boeck Supérieur.

Autor, D. H. (2015). Why are there still so many jobs? The history and future of workplace automation. *The Journal of Economic Perspectives*, 29(3), 3–30. DOI: 10.1257/jep.29.3.3

Aversa, P., Haefliger, S., Hueller, F., & Reza, D. G. (2021). Customer complementarity in the digital space: Exploring Amazon's business model diversification. *Long Range Planning*, 54(5), 101985. DOI: 10.1016/j.lrp.2020.101985

Avlijas, G., Vukanovic Dumanovic, V., & Radunovic, M. (2021). Measuring the effects of automatic replenishment on product availability in retail stores. *Sustainability (Basel)*, 13(3), 1391. DOI: 10.3390/su13031391

Awwad, A. S., Ababneh, O. M. A., & Karasneh, M. (2022). The mediating impact of IT capabilities on the association between dynamic capabilities and organizational agility: The case of the Jordanian IT sector. *Global Journal of Flexible Systems Managment*, 23(3), 315–330. DOI: 10.1007/s40171-022-00303-2 PMID: 37521253

Bai, C., Zhu, Q., & Sarkis, J. (2024). Circular economy and circularity supplier selection: A fuzzy group decision approach. *International Journal of Production Research*, 62(7), 2307–2330. DOI: 10.1080/00207543.2022.2037779

Baisya, R. (2024). Supply Chain and Logistics Management: An Integrated Approach. Taylor & Francis.

Bai, Z., & Bai, X. (2021). Sports big data: Management, analysis, applications, and challenges. *Complexity*, 2021(1), 6676297. DOI: 10.1155/2021/6676297

Balasubramaniam, N., Kauppinen, M., Kujala, S., & Hiekkanen, K. (2020). Ethical guidelines for solving ethical issues and developing AI systems. In *Product-Focused Software Process Improvement: 21st International Conference, PROFES 2020, Turin, Italy, November 25–27, 2020Proceedings*, 21, 331–346.

Banu, J. F., Stephen, R. K., Aditya, N., & Raaghav, L. D. (2024, February). Alzheimer's Disease Detection using Deep Learning Algorithm. In *2024 11th International Conference on Computing for Sustainable Global Development (INDIACom)* (pp. 1531-1536). IEEE. DOI: 10.23919/INDIACom61295.2024.10498963

Bard, J. F., Huang, L., Dror, M., & Jaillet, P. (1998). A branch and cut algorithm for the VRP with satellite facilities. *IIE Transactions*, 30(9), 821–834. DOI: 10.1080/07408179808966528

Barhate, D., Pathak, S., & Dubey, A. K. (2023). Hyperparameter-tuned batch-updated stochastic gradient descent: Plant species identification by using hybrid deep learning. *Ecological Informatics*, 75, 102094. DOI: 10.1016/j.ecoinf.2023.102094

Barton, D., & Court, D. (2019). *Leading in the Age of AI*. McKinsey & Company. Retrieved from https://www.mckinsey.com/business-functions/mckinsey-digital/our-insights/leading-in-the-age-of-ai

Basana, S., Malelak, M., Suprapto, W., Siagian, H., & Tarigan, Z. (2024). The impact of SCM integration on business performance through information sharing, quality integration and innovation system. *Uncertain Supply Chain Management*, 12(1), 435–448. DOI: 10.5267/j.uscm.2023.9.008

Basti, R., Jamoussi, S., Charfi, A., & Hamadou, A. B. (2019, September). Arabic Twitter User Profiling: Application to Cyber-security. In *WEBIST* (pp. 110-117).

Beamon, B. M. (1999). Designing the green supply chain. *Logistics Information Management*, 12(4), 332–342. DOI: 10.1108/09576059910284159

Belhadi, A., Mani, V., Kamble, S. S., Khan, S. A. R., & Verma, S. (2024). Artificial intelligence-driven innovation for enhancing supply chain resilience and performance under the effect of supply chain dynamism: An empirical investigation. *Annals of Operations Research*, 333(2), 627–652. DOI: 10.1007/s10479-021-03956-x PMID: 33551534

Bellesia, F., Mattarelli, E., Bertolotti, F., & Sobrero, M. (2019). Platforms as entrepreneurial incubators? How online labor markets shape work identity. *Journal of Managerial Psychology*, 34(4), 246–268. DOI: 10.1108/JMP-06-2018-0269

Bell, J., & McMullen, P. (2004). An Ant Colony Optimization Approach to the Vehicle Routing Problem. *Advanced Engineering Informatics*, 18(1), 41–48. DOI: 10.1016/j.aei.2004.07.001

Bellman, R. (1957). *Dynamic programming and Markov processes*. Princeton University Press. Retrieved from https://press.princeton.edu/books/paperback/9780691004864/dynamic-programming

Bellman, R. E. (1957). *Dynamic programming*. Princeton University Press.

Belver, G. A., Lopez, R., & Fuente, A. (2017). *Solving Vehicle Routing Problem with Multiple Trips using Iterative Local Search with Variable Neighborhood Search*. https://www.semanticscholar.org/paper/Solving-Vehicle-Routing-Problem-with-Multiple-Trips-Belver-G%C3%B3mez/532f4eb6829e75ea819cdd250f049f5786b3d476

Ben-Daya, M., Hassini, E., & Bahroun, Z. (2019). Internet of things and supply chain management: A literature review. *International Journal of Production Research*, 57(15-16), 4719–4742. DOI: 10.1080/00207543.2017.1402140

Benjamins, R. (2021). A choices framework for the responsible use of AI. *AI and Ethics*, 1(1), 49–53. DOI: 10.1007/s43681-020-00012-5

Bentalha, B. (2024). Employees and Robots in Amazon's Robotic Mobile Fulfillment Systems: A Netnographic Analysis of a Supply Chain Transformation. In Blockchain Applications for Smart Contract Technologies (pp. 188-207). IGI Global. DOI: 10.4018/979-8-3693-1511-8.ch009

Bessen, J. (2019). AI and Jobs: The Role of Demand. *National Bureau of Economic Research*. Retrieved from https://www.nber.org/papers/w24235

Bharadiya, J. P. (2023). The role of machine learning in transforming business intelligence. *International Journal of Computing and Artificial Intelligence*, 4(1), 16–24. DOI: 10.33545/27076571.2023.v4.i1a.60

Binnendijk, H., & Libicki, M. C. (2019). *Artificial Intelligence: Implications for the Future of Defense*. RAND Corporation.

Birhane, A., Ruane, E., Laurent, T., & Brown, S. (2022, June). The forgotten margins of AI ethics. In *Proceedings of the 2022 ACM Conference on Fairness, Accountability, and Transparency* (pp. 948-958). DOI: 10.1145/3531146.3533157

Bishop, J. (2021). Artificial Intelligence in Warfare: Opportunities and Challenges. *Journal of Defense Studies and Resource Management*, 9(1), 1–11.

Bititci, U. S., Martinez, V., Albores, P., & Parung, J. (2004). Creating and managing value in collaborative networks. *International Journal of Physical Distribution & Logistics Management*, 34(3/4), 251–268. DOI: 10.1108/09600030410533574

Blanka, C., Krumay, B., & Rueckel, D. (2022). The interplay of digital transformation and employee competency: A design science approach. *Technological Forecasting and Social Change*, 178, 121575. DOI: 10.1016/j.techfore.2022.121575

Blasch, E., Shen, D., Pham, K. D., & Chen, G. (2015, May). Review of game theory applications for situation awareness. In *Sensors and Systems for Space Applications VIII* (Vol. 9469, pp. 141–150). SPIE.

Bleier, A., & Eisenbeiss, M. (2015). The Impact of Relevance on the Effectiveness of Online Advertising. *Journal of Advertising*, 44(3), 320–328.

Borucka, A., Kozłowski, E., Parczewski, R., Antosz, K., Gil, L., & Pieniak, D. (2022). Supply sequence modelling using hidden Markov models. *Applied Sciences (Basel, Switzerland)*, 13(1), 231. DOI: 10.3390/app13010231

Bouanane, K., Benadada, Y., & Bencheikh, G. (2019). Multi-Depots Vehicle Routing Problem with Simultaneous Delivery and Pickup and Inventory Restrictions: Formulation and Resolution. *International Journal of Advanced Computer Science and Applications*, 10(2), 110–120. DOI: 10.14569/IJACSA.2019.0100215

Bouillot, P. (2017). *Gestion de la relation client: Total relationship management, Big data et Marketing mobile*. HAL Archives Ouvertes.

Bouktif, S., Fiaz, A., Ouni, A., & Serhani, M. A. (2020). Multi-sequence LSTM-RNN deep learning and metaheuristics for electric load forecasting. *Energies*, 13(2), 391. DOI: 10.3390/en13020391

Brun, A., Karaosman, H., & Barresi, T. (2020). Supply chain collaboration for transparency. *Sustainability (Basel)*, 12(11), 4429. DOI: 10.3390/su12114429

Brynjolfsson, E., & McAfee, A. (2014). *The Second Machine Age: Work, Progress, and Prosperity in a Time of Brilliant Technologies*. W. W. Norton & Company.

Buchanan, B. G., & Shortliffe, E. H. (1984). *Rule-Based Expert Systems: The MYCIN Experiments of the Stanford Heuristic Programming Project*. Addison-Wesley.

Buil, I., Catalán, S., & Martínez, E. (2020). Understanding applicants' reactions to gamified recruitment. *Journal of Business Research*, 110, 41–50. DOI: 10.1016/j.jbusres.2019.12.041

Cacciolatti, L., Lee, S. H., & Molinero, C. M. (2017). Clashing institutional interests in skills between government and industry: An analysis of demand for technical and soft skills of graduates in the UK. *Technological Forecasting and Social Change*, 119, 139–153. DOI: 10.1016/j.techfore.2017.03.024

Camarinha-Matos, L. M., Fornasiero, R., Ramezani, J., & Ferrada, F. (2019). Collaborative networks: A pillar of digital transformation. *Applied Sciences (Basel, Switzerland)*, 9(24), 5431. DOI: 10.3390/app9245431

Cappuccio, M. L., Galliott, J. C., Eyssel, F., & Lanteri, A. (2024). Autonomous systems and technology resistance: New tools for monitoring acceptance, trust, and tolerance. *International Journal of Social Robotics*, 16(6), 1–25. DOI: 10.1007/s12369-023-01065-2

Casazza, M., Ceselli, A., & Wolfler Calvo, R. (2018). A branch and price approach for the Split Pickup and Split Delivery VRP. *Electronic Notes in Discrete Mathematics*, 69, 189–196. DOI: 10.1016/j.endm.2018.07.025

Cauchois, R., Taghipour, A., Kang, D., Zoghlami, N., & Abed, M. (2017). Advanced solutions for a supply chain with stochastic information. *Journal of Advanced Management Science*, 5(1), 9–13. DOI: 10.18178/joams.5.1.9-13

Cell, D. I. A. I. (2023). Ethical guidelines for application of Artificial Intelligence in Biomedical Research and Healthcare.

Centobelli, P., Cerchione, R., & Ertz, M. (2020). Agile supply chain management: Where did it come from and where will it go in the era of digital transformation? *Industrial Marketing Management*, 90, 324–345. DOI: 10.1016/j.indmarman.2020.07.011

Chbichib, A., Mellouli, R., & Chabchoub, H. (2011, May 1). *Profitable Vehicle Routing Problem with Multiple Trips: Modeling and constructive heuristics.* 2011 4th International Conference on Logistics, LOGISTIQUA'2011. DOI: 10.1109/LOGISTIQUA.2011.5939450

Cheng, L. (2009). *A genetic algorithm for the vehicle routing problem with time windows.* https://www.semanticscholar.org/paper/A-genetic-algorithm-for-the-vehicle-routing-problem-Cheng/daa1ab8d3abf27f78540342ce20c4b0a6437d168

Chen, G., Gui, P., Ding, T., Na, S., & Zhou, Y. (2019). Optimization of Transportation Routing Problem for Fresh Food by Improved Ant Colony Algorithm Based on Tabu Search. *Sustainability (Basel)*, 11(23), 6584. DOI: 10.3390/su11236584

Chen, R.-M., & Guo, J.-C. (2018).. . *Optimal Routing for Split Delivery Vehicle Routing Problems*, 667–670, 667–670. Advance online publication. DOI: 10.1109/IIAI-AAI.2018.00140

Chen, Y., Zhu, Q., & Sarkis, J. (2023). Green supply chain management practice adoption sequence: A cumulative capability perspective. *International Journal of Production Research*, 61(17), 5918–5933. DOI: 10.1080/00207543.2022.2118891

Chesney, R., & Citron, D. K. (2019). Deepfakes and the New Disinformation War: The Coming Age of Post-Truth Geopolitics. *Foreign Affairs.*

Cheung, K. F., Bell, M. G., & Bhattacharjya, J. (2021). Cybersecurity in logistics and supply chain management: An overview and future research directions. *Transportation Research Part E, Logistics and Transportation Review*, 146, 102217. DOI: 10.1016/j.tre.2020.102217

Chiapello, E., & Godefroy, C. (2017). Big Data, algorithmes et marketing: Rendre des comptes. *Statistique et Société*, 5(3), 17–28.

Chiarello, F., Fantoni, G., Hogarth, T., Giordano, V., Baltina, L., & Spada, I. (2021). Towards ESCO 4.0–Is the European classification of skills in line with Industry 4.0? A text mining approach. *Technological Forecasting and Social Change*, 173, 121177. DOI: 10.1016/j.techfore.2021.121177

Chien, C. F., Lin, Y. S., & Lin, S. K. (2020). Deep reinforcement learning for selecting demand forecast models to empower Industry 3.5 and an empirical study for a semiconductor component distributor. *International Journal of Production Research*, 58(9), 2784–2804. DOI: 10.1080/00207543.2020.1733125

Choi, T. M. (2020). Innovative supply chain management in the age of artificial intelligence. *International Journal of Logistics Research and Applications*, 23(2), 98–101.

Choi, T. M., Wallace, S. W., & Wang, Y. (2018). Big Data Analytics in Operations Management. *Production and Operations Management*, 27(10), 1868–1881. DOI: 10.1111/poms.12838

Chopra, S., & Meindl, P. (2016). *Supply Chain Management: Strategy, Planning, and Operation*. Pearson.

Chowdhary, N., & Kaur, P. (2018). *Dynamic Route Optimization Using Nature-Inspired Algorithms in IoV*., DOI: 10.1007/978-981-10-5828-8_47

Christopher, M. (2016). *Logistics & Supply Chain Management*. Pearson UK.

Christopher, Y., Wahyuningsih, S., & Satyananda, D. (2021). Study of variable neighborhood descent and tabu search algorithm in VRPSDP. *Journal of Physics: Conference Series*, 1872(1), 012002. DOI: 10.1088/1742-6596/1872/1/012002

Chuah, Y. R., Goh, A. V., & Ooi, H. T. (2023). The adoption of automation in Schenker Penang Logistics Centre 5 (plc5) warehouse: A case study (Doctoral dissertation, UTAR).

Chui, M., Manyika, J., & Miremadi, M. (2016). Where machines could replace humans—And where they can't (yet). *The McKinsey Quarterly*, 3(3), 1–9.

Clarke, G., & Wright, J. W. (1964). Scheduling of Vehicles from a Central Depot to a Number of Delivery Points. *Operations Research*, 12(4), 568–581. DOI: 10.1287/opre.12.4.568

Clauss, T., Harengel, P., & Hock, M. (2019). The perception of value of platform-based business models in the sharing economy: Determining the drivers of user loyalty. *Review of Managerial Science*, 13(3), 605–634. DOI: 10.1007/s11846-018-0313-0

Cresci, S. (2020). A Decade of Social Bot Detection. *Communications of the ACM*, 63(10), 72–83. DOI: 10.1145/3409116

Cukier, W. (2019). Disruptive processes and skills mismatches in the new economy: Theorizing social inclusion and innovation as solutions. *Journal of Global Responsibility*, 10(3), 211–225. DOI: 10.1108/JGR-11-2018-0079

Cummings, M. L. (2017). Artificial Intelligence and the Future of Warfare. Center for a New American Security.

Dablanc, L. (2023). Policy and planning for urban logistics. In *The Routledge handbook of urban logistics* (pp. 251–266). Routledge. DOI: 10.4324/9781003241478-23

Dadsena, K. K., Mathiyazhagan, K., & Taghipour, A. (2021). Analysis of barriers for the build the resilient supply chain networks post-COVID-19. In *Managing Supply Chain Risk and Disruptions: Post COVID-19* (pp. 79–89). Springer International Publishing. DOI: 10.1007/978-3-030-72575-4_7

Daely, P., Aruan, Y., Lee, J. M., & Kim, D.-S. (2021)... *Dynamic VRP Optimization Using Discrete PSO in Edge Computing Environment*, 654–656, 654–656. Advance online publication. DOI: 10.1109/ICTC52510.2021.9620744

Dalal, S., Lilhore, U. K., Simaiya, S., Radulescu, M., & Belascu, L. (2024). Improving efficiency and sustainability via supply chain optimization through CNNs and BiLSTM. *Technological Forecasting and Social Change*, 209, 123841. DOI: 10.1016/j.techfore.2024.123841

Damoah, I. S., Ayakwah, A., & Tingbani, I. (2021). Artificial intelligence (ai)-enhanced medical drones in the healthcare supply chain (HSC) for sustainability development: A case study. *Journal of Cleaner Production*, 328, 129598. DOI: 10.1016/j.jclepro.2021.129598

Daneshvar, S., & Inkpen, D. (2018, September). Gender identification in twitter using n-grams and lsa. In *Proceedings of the ninth international conference of the CLEF association (CLEF 2018)*. CEUR-WS.

Dantzig, G. B., & Ramser, J. H. (1959). The Truck Dispatching Problem. *Management Science*, 6(1), 80–91. DOI: 10.1287/mnsc.6.1.80

Das, S. (2021). Foundational RL: Solving Markov decision process. Retrieved January 12, 2025, from https://towardsdatascience.com/foundational-rl-solving-markov-decision-process-d90b7e134c0b

Dastghaibifard, G. H., Ansari, E., Sheykhalishahi, S. M., Bavandpouri, A., & Ashoor, E. (2008). *A Parallel Branch and Bound Algorithm for Vehicle Routing Problem*.

Dastin, J. (2018). Amazon scraps secret AI recruiting tool that showed bias against women. *Reuters*. Retrieved from https://www.reuters.com/article/us-amazon-com-recruitment-insight-idUSKCN1MK08G

Dattner, B., Chamorro-Premuzic, T., Buchband, R., & Schettler, L. (2019). The legal and ethical implications of using AI in hiring. *Harvard Business Review*, 25, 1–7.

Dave, T., Athaluri, S. A., & Singh, S. (2023). ChatGPT in medicine: An overview of its applications, advantages, limitations, future prospects, and ethical considerations. *Frontiers in Artificial Intelligence*, 6, 1169595. DOI: 10.3389/frai.2023.1169595 PMID: 37215063

Demirtas, Y. E., Ozdemir, E., & Demirtas, U. (2015). A particle swarm optimization for the dynamic vehicle routing problem. *2015 6th International Conference on Modeling, Simulation, and Applied Optimization (ICMSAO)*, 1–5. DOI: 10.1109/ICMSAO.2015.7152224

Denyer, D., & Tranfield, D. (2009). Producing a systematic review. In Buchanan, D. A., & Bryman, A. (Eds.), *The Sage handbook of organizational research methods* (pp. 671–689). Sage Publications Ltd.

Derwik, P., Hellström, D., & Karlsson, S. (2016). Manager competences in logistics and supply chain practice. *Journal of Business Research*, 69(11), 4820–4825. DOI: 10.1016/j.jbusres.2016.04.037

Deveshwar, A., & Rathee, R. (2010). Challenges for supply chain management in today's global competitive environment. *International Review of Business Research Papers*, 6(2), 194–203.

Dey, K. C., Rayamajhi, A., Chowdhury, M., Bhavsar, P., & Martin, J. (2016). Vehicle-to-vehicle (V2V) and vehicle-to-infrastructure (V2I) communication in a heterogeneous wireless network – Performance evaluation. *Transportation Research Part C, Emerging Technologies*, 68, 168–184. DOI: 10.1016/j.trc.2016.03.008

Dhanare, R., Nagwanshi, K., & Varma, S. (2022). A Study to Enhance the Route Optimization Algorithm for the Internet of Vehicle. *Wireless Communications and Mobile Computing*, 2022, 1–20. DOI: 10.1155/2022/1453187

Dhiman, N., Singh, H., & Thakur, A. (2023, August). An Efficient Approach for Image Forgery Detection Using Deep Convolutional Neural Network. In *2023 International Conference on Electrical, Electronics, Communication and Computers (ELEXCOM)* (pp. 1-5). IEEE. DOI: 10.1109/ELEXCOM58812.2023.10370304

Dibi, A., El Moussaoui, T., El Moussaoui, A. E., & Loqman, C. (2024). Qualimatric intervention research: interest for management sciences researchers. In *Applying qualitative research methods to management science* (pp. 92–127). IGI Global. DOI: 10.4018/979-8-3693-5543-5.ch005

Diemer, A. (2016). Les symbioses industrielles: Un nouveau champ d'analyse pour l'économie industrielle. *Innovations (Paris)*, 2(50), 65–94. DOI: 10.3917/inno.050.0065

Donada, C., & Fournier, G. (2014). Stratégie industrielle pour un écosystème en émergence: Le cas de la mobilité 2.0, décarbonée, intermodale et collaborative. *Revue d'Economie Industrielle*, (148), 317–348. DOI: 10.4000/rei.6016

Drabiak, K., Kyzer, S., Nemov, V., & El Naqa, I. (2023). AI and machine learning ethics, law, diversity, and global impact. *The British Journal of Radiology*, 96(1150), 20220934. DOI: 10.1259/bjr.20220934 PMID: 37191072

Draissi, Z., Zhanyong, Q., & Raguindin, P. Z. J. (2021). Knowledge mapping of skills mismatch phenomenon: A scientometric analysis. *Higher Education. Skills and Work-Based Learning*, 12(2), 271–293. DOI: 10.1108/HESWBL-02-2021-0019

Dubey, R., Gunasekaran, A., Childe, S. J., Blome, C., Papadopoulos, T., & Luo, Z. (2020). Big Data and Predictive Analytics and Manufacturing Performance: Integrating Institutional Theory, Resource-Based View and Big Data Culture. *British Journal of Management*, 31(2), 186–210.

Dubey, R., Gunasekaran, A., Papadopoulos, T., Childe, S. J., Shibin, K. T., & Wamba, S. F. (2017). Sustainable supply chain management: Framework and further research directions. *Journal of Cleaner Production*, 142, 1119–1130. DOI: 10.1016/j.jclepro.2016.03.117

Dumez, H. (2013). *Méthodologie de la recherche qualitative, 227.* Vuibert.

Durlik, I., Miller, T., Cembrowska-Lech, D., Krzemińska, A., Złoczowska, E., & Nowak, A. (2023). Navigating the sea of data: A comprehensive review on data analysis in maritime IoT applications. *Applied Sciences (Basel, Switzerland)*, 13(17), 9742. DOI: 10.3390/app13179742

Dussart, C. (2017). *Le développement du Big Data signe-t-il la fin des études marketing? ProQuest Dissertations Publishing The Coca-Cola Company. (2022). Bottling Investments Group (BIG).* Coca-Cola Investor Relations.

Eitel-Porter, R. (2021). Beyond the promise: Implementing ethical AI. *AI and Ethics*, 1(1), 73–80. DOI: 10.1007/s43681-020-00011-6

El Moussaoui, A. E. (2024). Factors influencing the entry of logistics service providers (LSPs) into the logistics clusters: a case study of Casablanca city. *Management & Sustainability: An Arab Review.*

El Moussaoui, A. E. (2024). Impact of logistics pooling on reduction of CO2 emissions in last-mile logistics. *Management & Sustainability: An Arab Review.*

El Moussaoui, A. E., Benbba, B., Jaegler, A., & El Andaloussi, Z. (2022, May). City Logistics: An Evolutionary, Successful & Collaborative City. In *2022 14th International Colloquium of Logistics and Supply Chain Management (LOGISTIQUA)* (pp. 1-7). IEEE.

El Moussaoui, A. E., Benbba, B., Jaegler, A., & El Andaloussi, Z. (2022, May). How to improve the urban logistics sector in large cities of developing countries: The case of Casablanca. In *2022 14th International Colloquium of Logistics and Supply Chain Management (LOGISTIQUA)* (pp. 1-6). IEEE.

El Moussaoui, A. E., Benbba, B., Jaegler, A., El Andaloussi, Z., & El Amrani, L. (2022, May). Last mile logistics: Challenges & improvement ways. In *2022 14th International Colloquium of Logistics and Supply Chain Management (LOGISTIQUA)* (pp. 1-7). IEEE.

El Moussaoui, T., & Chakir, L. (2023, September). BERT-CLSTM model for the classification of Moroccan commercial courts verdicts. In 2023 18th Conference on Computer Science and Intelligence Systems (FedCSIS) (pp. 281-284). IEEE. DOI: 10.15439/2023F3561

El Moussaoui, A. E. (2024). Unlocking the drivers behind logistics service providers' participation in Morocco's logistics clusters. *Arab Gulf Journal of Scientific Research*, 42(4), 2086–2102. DOI: 10.1108/AGJSR-07-2023-0343

El Moussaoui, A. E., Benbba, B., & El Andaloussi, Z. (2023). Impact of logistics performance on the store image, consumer satisfaction and loyalty: A quantitative case study. *Arab Gulf Journal of Scientific Research*, 41(3), 226–239. DOI: 10.1108/AGJSR-09-2022-0201

El Moussaoui, A. E., Benbba, B., El Andaloussi, Z., & Jaegler, A. (2022). Logistique urbaine au Maroc: Concepts, état des lieux et pistes d'amélioration. *Revue Française de Gestion Industrielle*, 36(2), 71–88. DOI: 10.53102/2022.36.02.910

El Moussaoui, A. E., Benbba, B., Jaegler, A., El Moussaoui, T., El Andaloussi, Z., & Chakir, L. (2023). Consumer perceptions of online shopping and willingness to use pick-up points: A case study of Morocco. *Sustainability (Basel)*, 15(9), 7405. DOI: 10.3390/su15097405

El Moussaoui, A. E., El Moussaoui, T., Benbba, B., Chakir, L., Jaegler, A., & El Andaloussi, Z. (2024). Sustainable effects of information sharing between distribution logistics actors: A qualitative case study. *Environment, Development and Sustainability*, •••, 1–19. DOI: 10.1007/s10668-023-04387-3

El Moussaoui, A. E., El Moussaoui, T., Benbba, B., Jaegler, A., & El Andaloussi, Z. (2022). Understanding the choice of collection & delivery point by the E-consumer via a machine learning model: Moroccan case study. *Procedia Computer Science*, 210, 204–211. DOI: 10.1016/j.procs.2022.10.138

El Moussaoui, T., Chakir, L., & Boumhidi, J. (2023). Preserving privacy in Arabic judgments: Ai-powered anonymization for enhanced legal data privacy. *IEEE Access : Practical Innovations, Open Solutions*, 11, 117851–117864. DOI: 10.1109/ACCESS.2023.3324288

El Moussaoui, T., & Loqman, C. (2024). Advancements in Arabic Named Entity Recognition: A Comprehensive Review. *IEEE Access : Practical Innovations, Open Solutions*, 12, 180238–180266. DOI: 10.1109/ACCESS.2024.3491897

Elkington, J. (1997). *Cannibals with Forks: The Triple Bottom Line of 21st Century Business*. Capstone Publishing.

Ellahi, R. M., Wood, L. C., & Bekhit, A. E. D. A. (2023). Blockchain-based frameworks for food traceability: A systematic review. *Foods*, 12(16), 3026. DOI: 10.3390/foods12163026 PMID: 37628025

Enăchescu, M. I. (2016). A prototype for an e-recruitment platform using semantic web technologies. *Journal: Informatica Economică*, 20(4), 62–75. DOI: 10.12948/issn14531305/20.4.2016.06

Epiphaniou, G., Bottarelli, M., Al-Khateeb, H., Ersotelos, N. T., Kanyaru, J., & Nahar, V. (2020). Smart distributed ledger technologies in Industry 4.0: Challenges and opportunities in supply chain management. Cyber Defence in the Age of AI, Smart Societies and Augmented Humanity, 319-345.

European Commission (2020). The Impact of Artificial Intelligence on Media Integrity. Regulatory approaches in the EU-the general Data Protection Regulation, Digital Services Act-and their implications for deepfake regulation.

Eze, E. C., Zhang, S., & Liu, E. (2014). Vehicular ad hoc networks (VANETs): Current state, challenges, potentials and way forward. *2014 20th International Conference on Automation and Computing*, 176–181. DOI: 10.1109/IConAC.2014.6935482

Fadda, P., Mancini, S., Serra, P., & Fancello, G. (2022). The Heterogeneous Fleet Vehicle Routing Problem with Draft Limits. *Computers & Operations Research*, 149, 106024. DOI: 10.1016/j.cor.2022.106024

Faliagka, E., Iliadis, L., Karydis, I., Rigou, M., Sioutas, S., Tsakalidis, A., & Tzimas, G. (2014). On-line consistent ranking on e-recruitment: Seeking the truth behind a well-formed CV. *Artificial Intelligence Review*, 42(3), 515–528. DOI: 10.1007/s10462-013-9414-y

Fatima, M., Hasan, K., Anwar, S., & Nawab, R. M. A. (2017). Multilingual author profiling on Facebook. *Information Processing & Management*, 53(4), 886–904. DOI: 10.1016/j.ipm.2017.03.005

Feiferytė-Skirienė, A., & Stasiškienė, Ž. (2021). Seeking circularity: Circular urban metabolism in the context of industrial symbiosis. *Sustainability (Basel)*, 13(16), 9094. DOI: 10.3390/su13169094

Ferrer, X., Van Nuenen, T., Such, J. M., Coté, M., & Criado, N. (2021). Bias and discrimination in AI: A cross-disciplinary perspective. *IEEE Technology and Society Magazine*, 40(2), 72–80. DOI: 10.1109/MTS.2021.3056293

Fiske, A., Henningsen, P., & Buyx, A. (2019). Your robot therapist will see you now: Ethical implications of embodied artificial intelligence in psychiatry, psychology, and psychotherapy. *Journal of Medical Internet Research*, 21(5), e13216. DOI: 10.2196/13216 PMID: 31094356

Frederico, G. F., Garza-Reyes, J. A., Anosike, A., & Kumar, V. (2020). Supply Chain 4.0: Concepts, maturity and research agenda. *Supply Chain Management*, 25(2), 262–282. DOI: 10.1108/SCM-09-2018-0339

Freeney, C. W. (2024). *Case Study on Organizational Adoption of RFID and GPS Technology to Improve Its Supply Chain Process* (Doctoral dissertation, Trident University International).

Frey, C. B., & Osborne, M. A. (2017). The future of employment: How susceptible are jobs to computerization? *Technological Forecasting and Social Change*, 114, 254–280. DOI: 10.1016/j.techfore.2016.08.019

Frías, N., Johnson Parejas, F., & Valle, C. (2023). Hybrid Algorithms for Energy Minimizing Vehicle Routing Problem: Integrating Clusterization and Ant Colony Optimization. *IEEE Access, PP*, 1–1. DOI: 10.1109/ACCESS.2023.3325787

Fritts, M., & Cabrera, F. (2021). AI recruitment algorithms and the dehumanization problem. *Ethics and Information Technology*, 23(4), 791–801. DOI: 10.1007/s10676-021-09615-w

Fulconis, F., Paché, G., & Reynaud, E. (2016). Vers une nouvelle forme de croissance économique: les apports des recherches en logistique et supply chain management. *Revue française de gestion*, (8), 127-149.

Gannouni, A., Rachid, E., & Talbi, E.-G. (2017). Solving stochastic multiobjective vehicle routing problem using probabilistic metaheuristic. *MATEC Web of Conferences, 105*, 00001. DOI: 10.1051/matecconf/201710500001

Gao, Y., Li, Z., Taghipour, A., Canel, B., & Kang, D. S. (2018). Supply Chain Coordination: A Review. *Journal of Advanced Management Science*, 6(4), 213–217. DOI: 10.18178/joams.6.4.213-217

Garay-Rondero, C. L., Martinez-Flores, J. L., Smith, N. R., Caballero Morales, S. O., & Aldrette-Malacara, A. (2019). Digital supply chain model in Industry 4.0. *Journal of Manufacturing Technology Management*, 31(5), 887–933. DOI: 10.1108/JMTM-08-2018-0280

George, A. S., George, A. H., & Baskar, T. (2023). Edge Computing and the Future of Cloud Computing: A Survey of Industry Perspectives and Predictions. *Partners Universal International Research Journal*, 2(2), 19–44.

Gerke, S., Minssen, T., & Cohen, G. (2020). Ethical and legal challenges of artificial intelligence-driven healthcare. In *Artificial intelligence in healthcare* (pp. 295–336). Academic Press. DOI: 10.1016/B978-0-12-818438-7.00012-5

Gheni, H. M., AbdulRahaim, L. A., & Abdellatif, A. (2024). Real-time driver identification in IoV: A deep learning and cloud integration approach. *Heliyon*, 10(7), e28109. DOI: 10.1016/j.heliyon.2024.e28109 PMID: 38560228

Ghosh, R. K., Thakur, R. C., & Ghosh, A. B. (2020). Role of IoT in Disaster Management: A Survey. *IEEE Internet of Things Journal*, 7(8), 6985–6994.

Giachanou, A., Ríssola, E. A., Ghanem, B., Crestani, F., & Rosso, P. (2020). The role of personality and linguistic patterns in discriminating between fake news spreaders and fact checkers. In *Natural Language Processing and Information Systems: 25th International Conference on Applications of Natural Language to Information Systems, NLDB 2020, Saarbrücken, Germany, June 24–26, 2020Proceedings*, 25, 181–192.

Gialos, A., & Zeimpekis, V. (n.d.). *Vision picking technology: Defining design parameters via a systematic literature review.*

Gilks, W. R., Richardson, S., & Spiegelhalter, D. J. (1995). Markov chain Monte Carlo in practice. Chapman & Hall/CRC. Retrieved from https://www.routledge.com/Markov-Chain-Monte-Carlo-in-Practice/Gilks-Richardson-Spiegelhalter/p/book/9780412055515

Glaa, B., Zoghlami, N., & Taghipour, A. (2014, May). Outsourcing and off-shoring healthcare services: A way to cut the costs and improve quality. In 2014 international conference on advanced logistics and transport (ICALT) (pp. 189-194). IEEE.

Glaa, B., Zoghlami, N., & Taghipour, A. (2014, May). Outsourcing and off-shoring healthcare services: A way to cut the costs and improve quality. In *2014 international conference on advanced logistics and transport (ICALT)* (pp. 189-194). IEEE. DOI: 10.1109/ICAdLT.2014.6864116

Glize, E., Jozefowiez, N., & Ngueveu, S. (2018). *An Exact Column Generation-Based Algorithm for Bi-objective Vehicle Routing Problems: 5th International Symposium, ISCO 2018, Marrakesh, Morocco, April 11–13, 2018, Revised Selected Papers* (pp. 208–218). DOI: 10.1007/978-3-319-96151-4_18

Goepp, V., Caillaud, E., & Rose, B. (2013). A Framework for the Design of Knowledge Management Systems in Eco-Design. *International Journal of Production Research*, 51(19), 5803–5823. DOI: 10.1080/00207543.2013.795252

Gokulakrishnan, S. B., & Prakash, A. M. R. (2019). Web-Based Disaster Management System Using GIS and IoT. *IEEE Internet of Things Journal*, 6(2), 2077–2085.

Goldfarb, A., & Tucker, C. (2011). Online Display Advertising: Targeting and Obtrusiveness. *Marketing Science*, 30(3), 389–404. DOI: 10.1287/mksc.1100.0583

Gol, E. S., Stein, M. K., & Avital, M. (2019). Crowdwork platform governance toward organizational value creation. *The Journal of Strategic Information Systems*, 28(2), 175–195. DOI: 10.1016/j.jsis.2019.01.001

Goodfellow, I. (2014). Generative Adversarial Nets. Advances in Neural Information Processing Systems (NeurIPS).

Gregory, R. W., Kaganer, E., Henfridsson, O., & Ruch, T. J. (2018). IT consumerization and the transformation of IT governance. *Management Information Systems Quarterly*, 42(4), 1225–1229. DOI: 10.25300/MISQ/2018/13703

Growthsetting. (2023). How Netflix Enhances User Experience with AI Recommendations.

Gubbi, J., Buyya, R., Marusic, S., & Palaniswami, M. (2013). Internet of Things (IoT): A vision, architectural elements, and future directions. *Future Generation Computer Systems*, 29(7), 1645–1660. DOI: 10.1016/j.future.2013.01.010

Guo, S., Alamudun, F., & Hammond, T. (2016). RésuMatcher: A personalized résumé-job matching system. *Expert Systems with Applications*, 60, 169–182. DOI: 10.1016/j.eswa.2016.04.013

Gupta, A., Singh, R. K., & Gupta, S. (2022). Developing human resource for the digitization of logistics operations: Readiness index framework. *International Journal of Manpower*, 43(2), 355–379. DOI: 10.1108/IJM-03-2021-0175

Gupta, S., Modgil, S., Gunasekaran, A., & Bag, S. (2020, July). Dynamic capabilities and institutional theories for Industry 4.0 and digital supply chain. In Supply Chain Forum *International Journal (Toronto, Ont.)*, 21(3), 139–157.

Habib, M., Bao, Y., & Ilmudeen, A. (2020). The impact of green entrepreneurial orientation, market orientation and green supply chain management practices on sustainable firm performance. *Cogent Business & Management*, 7(1), 1743616. DOI: 10.1080/23311975.2020.1743616

HadjTaieb, S., Hani, Y., Loukil, T. M., & Mhamedi, A. E. (2022). Green VRP applied to Home Health Care problem. *IFAC-PapersOnLine*, 55(10), 3154–3159. DOI: 10.1016/j.ifacol.2022.10.214

Hagendorff, T. (2020). The ethics of AI ethics: An evaluation of guidelines. *Minds and Machines*, 30(1), 99–120. DOI: 10.1007/s11023-020-09517-8

Halim, Z., & Rehan, M. (2020). On identification of driving-induced stress using electroencephalogram signals: A framework based on wearable safety-critical scheme and machine learning. *Information Fusion*, 53, 66–79. DOI: 10.1016/j.inffus.2019.06.006

Halim, Z., & Zouq, A. (2021). On identification of big-five personality traits through choice of images in a real-world setting. *Multimedia Tools and Applications*, 80(24), 33377–33408. DOI: 10.1007/s11042-021-11419-5

Handfield, R., & Linton, T. (2017). The LIVING supply chain: The evolving imperative of operating in real time. John Wiley & Sons.

Hashimoto, H., Hu, Y., Sugiura, T., Takada, Y., & Yagiura, M. (2022)... *An Iterated Local Search Algorithm for Commuting Bus Routing Problem with Latest Arrival Time Constraint*, 0791–0795, 0791–0795. Advance online publication. DOI: 10.1109/IEEM55944.2022.9989602

Hastig, G. M., & Sodhi, M. S. (2020). Blockchain for supply chain traceability: Business requirements and critical success factors. *Production and Operations Management*, 29(4), 935–954. DOI: 10.1111/poms.13147

Hastings, W. K. (1970). Monte Carlo sampling methods using Markov chains and their applications. *Biometrika*, 57(1), 97–109. DOI: 10.1093/biomet/57.1.97

Hastuti, R., & Syafruddin, . (2023). Ethical considerations in the age of artificial intelligence: Balancing innovation and social values. *West Science Social and Humanities Studies*, 1(02), 76–87. DOI: 10.58812/wsshs.v1i02.191

Hazen, B. T., Russo, I., Confente, I., & Pellathy, D. (2021). Supply chain management for circular economy: Conceptual framework and research agenda. *International Journal of Logistics Management*, 32(2), 510–537. DOI: 10.1108/IJLM-12-2019-0332

Heath, T., & Roper, A. (2021). AI in Defense: A Review of the Ethics, Governance, and Policy Issues. *AI & Society*.

Helo, P., & Thai, V. V. (2024). Logistics 4.0–digital transformation with smart connected tracking and tracing devices. *International Journal of Production Economics*, 275, 109336. DOI: 10.1016/j.ijpe.2024.109336

Hernández, E. G. (2024). Towards an ethical and inclusive implementation of artificial intelligence in organizations: a multidimensional framework. arXiv preprint arXiv:2405.01697.

Hernandez-de-Menendez, M., Morales-Menendez, R., Escobar, C. A., & McGovern, M. (2020). Competencies for industry 4.0. *International Journal on Interactive Design and Manufacturing*, 14(4), 1511–1524. DOI: 10.1007/s12008-020-00716-2

Hernández, J. E., & García, J. R. (2021). Data Warehouse with Big Data Technology for Higher Education. *Procedia Computer Science*, 181, 123–130.

He, S., Li, J., & Qiu, T. (2017). Vehicle to Pedestrian Communication Modeling and Collision Avoiding Methodology in Connected Vehicle Environment. *Transportation Research Record: Journal of the Transportation Research Board*, 2621(1), 21–30. DOI: 10.3141/2621-03

Hinings, B., Gegenhuber, T., & Greenwood, R. (2018). Digital innovation and transformation: An institutional perspective. *Information and Organization*, 28(1), 52–61. DOI: 10.1016/j.infoandorg.2018.02.004

Hlady-Rispal, M. (2015). Une stratégie de recherche en gestion: L'étude de cas. *Revue Française de Gestion*, 41(127), 251–266. DOI: 10.3166/RFG.253.251-266

Hmioui, A., Bentalha, B., & Beniaich, R. (2024). Green Supply Chain Management et développement territorial: Une revue de littérature. *Alternatives Managériales Economiques*, 6(2), 52–74.

Hofmann, E., & Rüsch, M. (2017). Industry 4.0 and the current status as well as future prospects on logistics. *Computers in Industry*, 89, 23–34. DOI: 10.1016/j.compind.2017.04.002

Hohenstein, J., Kizilcec, R. F., DiFranzo, D., Aghajari, Z., Mieczkowski, H., Levy, K., Naaman, M., Hancock, J., & Jung, M. F. (2023). Artificial intelligence in communication impacts language and social relationships. *Scientific Reports*, 13(1), 5487. DOI: 10.1038/s41598-023-30938-9 PMID: 37015964

Holland, J. H. (1975). *Adaptation in natural and artificial systems: An introductory analysis with applications to biology, control, and artificial intelligence* (pp. viii, 183). U Michigan Press.

Huang, C., Zhang, Z., Mao, B., & Yao, X. (2022). An overview of artificial intelligence ethics. *IEEE Transactions on Artificial Intelligence*, 4(4), 799–819. DOI: 10.1109/TAI.2022.3194503

Huang, K., Wang, K., Lee, P. K., & Yeung, A. C. (2023). The impact of industry 4.0 on supply chain capability and supply chain resilience: A dynamic resource-based view. *International Journal of Production Economics*, 262, 108913. DOI: 10.1016/j.ijpe.2023.108913

Huang, S.-H., Huang, Y.-H., Blazquez, C., & Chen, C.-Y. (2022). Solving the vehicle routing problem with drone for delivery services using an ant colony optimization algorithm. *Advanced Engineering Informatics*, 51, 101536. DOI: 10.1016/j.aei.2022.101536

Hübner, A. H., Kuhn, H., & Wollenburg, J. (2016). Last mile fulfilment and distribution in omni-channel grocery retailing: A strategic planning framework. *International Journal of Retail & Distribution Management*, 44(3). Advance online publication. DOI: 10.1108/IJRDM-11-2014-0154

Hughes, C., Robert, L., Frady, K., & Arroyos, A. (2019). Middle-skill-level Employees and Technological Environments. In *Managing Technology and Middle-and Low-skilled Employees*. Emerald Publishing Limited. DOI: 10.1108/978-1-78973-077-720191002

Huhns, M. N., & Stephens, L. M. (2001). Automating supply chains. *IEEE Internet Computing*, 5(4), 90–93. DOI: 10.1109/4236.939455

Hunkenschroer, A. L., & Kriebitz, A. (2023). Is AI recruiting (un) ethical? A human rights perspective on the use of AI for hiring. *AI and Ethics*, 3(1), 199–213. DOI: 10.1007/s43681-022-00166-4 PMID: 35909984

Hu, Y., Zeng, J., Chen, Y., & Li, W. (2021). Disaster Risk Assessment Based on Multi- Source Data and Machine Learning. *IEEE Transactions on Geoscience and Remote Sensing*, 59(9), 7785–7796.

Hyeon, E., Han, J., Shen, D., Karbowski, D., Kim, N., & Rousseau, A. (2022). Potential Energy Saving of V2V-Connected Vehicles in Large-Scale Traffic. *IFAC-PapersOnLine*, 55(24), 78–83. DOI: 10.1016/j.ifacol.2022.10.265

Ichoua, S., Gendreau, M., & Potvin, J.-Y. (2003). Vehicle Dispatching With Time—Dependent Travel Times. *European Journal of Operational Research*, 144(2), 379–396. DOI: 10.1016/S0377-2217(02)00147-9

Idowu, A., & Elbanna, A. (2021). Crowdworkers, social affirmation and work identity: Rethinking dominant assumptions of crowdwork. *Information and Organization*, 31(4), 100335. DOI: 10.1016/j.infoandorg.2021.100335

Ishfaq, R., Davis-Sramek, B., & Gibson, B. (2022). Digital supply chains in omnichannel retail: A conceptual framework. *Journal of Business Logistics*, 43(2), 169–188. DOI: 10.1111/jbl.12277

Islam, M. A. (2022). Industry 4.0: Skill set for employability. *Social Sciences & Humanities Open*, 6(1), 100280. DOI: 10.1016/j.ssaho.2022.100280

Ivanov, D., Tsipoulanidis, A., & Schönberger, J. (2019). *Global Supply Chain and Operations Management: A Decision-Oriented Introduction to the Creation of Value*. Springer.

Ivanov, D., & Dolgui, A. (2020). A digital supply chain twin for managing the disruption risks and resilience in the era of Industry 4.0. *Production Planning and Control*, 31(9), 722–738.

Ivanov, D., Dolgui, A., Sokolov, B., Ivanova, M., & Potryasaev, S. (2019). Disruption-driven supply chain (re)-planning and performance impact assessment with consideration of pro-active and recovery policies. *Transportation Research Part E, Logistics and Transportation Review*, 125, 308–330.

Jackson, P. (1998). *Introduction to Expert Systems* (2nd ed.). Addison-Wesley.

Jarrahi, M. H. (2018). Artificial intelligence and the future of work: Human-AI symbiosis in organizational decision making. *Business Horizons*, 61(4), 577–586. DOI: 10.1016/j.bushor.2018.03.007

Javed, A., Basit, A., Ejaz, F., Hameed, A., Fodor, Z. J., & Hossain, M. B. (2024). The role of advanced technologies and supply chain collaboration: During COVID-19 on sustainable supply chain performance. *Discover Sustainability*, 5(1), 46. DOI: 10.1007/s43621-024-00228-z

Jegatheesan, D., & Arumugam, C. (2024). SIoV-FTFSA-CAOA: A fuzzy trust-based approach for enhancing security and energy efficiency in social internet of vehicles. *Wireless Networks*, 30(4), 1–20. DOI: 10.1007/s11276-023-03626-9

Jeong, W. Y., & Choi, M. (2019). Design of recruitment management platform using digital certificate on blockchain. *Journal of Information Processing Systems*, 15(3), 707–716.

Jha, P. K., & Gupta, A. K. (2021). Artificial Intelligence Techniques for Disaster Risk Reduction: A Review. *IEEE Transactions on Engineering Management*, 68(1), 186–200.

Jiang, Y., Tran, T. H., & Williams, L. (2023). Machine learning and mixed reality for smart aviation: Applications and challenges. *Journal of Air Transport Management*, 111, 102437. DOI: 10.1016/j.jairtraman.2023.102437

Johnson, R. D., Stone, D. L., & Lukaszewski, K. M. (2020). The benefits of eHRM and AI for talent acquisition. *Journal of Tourism Futures*, 7(1), 40–52. DOI: 10.1108/JTF-02-2020-0013

Jones, E. C. (2019). The Internet of Things (IoT) Technologies and the Tracking of Supply Chain Assets. In *Supply Chain Engineering and Logistics Handbook* (pp. 419–441). CRC Press. DOI: 10.1201/9781315159096-13

Joo, Y., Hwang, I., Cappellato, L., Ferro, N., Losada, D., & Müller, H. (2019). Author profiling on social media: An ensemble learning model using various features. *Notebook for PAN at CLEF, 2380.*

Ju, K. J., Park, B., & Kim, T. (2016). Causal relationship between supply chain dynamic capabilities, technological innovation, and operational performance. Management and Production Engineering Review, 7.

JUMP. (2024). *The Netflix Effect: Shaping the future of personalization in streaming*Recent Advances and Case Studies.

Jyothy, S. T., & Sarvagya, M. (2022, November). Supply Chain Management in Pharmaceutical Industry Using IOT. In *2022 IEEE North Karnataka Subsection Flagship International Conference (NKCon)* (pp. 1-7). IEEE.

Kabadurmus, O., & Erdogan, M. (2022). Bi-Objective Green Vehicle Routing Problem Minimizing Carbon Emissions and Maximizing Service Level. *Gazi Üniversitesi Mühendislik-Mimarlık Fakültesi Dergisi*, 38(1), 103–112. Advance online publication. DOI: 10.17341/gazimmfd.633583

Kadhem, S. K. (2024). Mixture models: Parametric, semiparametric, and new directions. Oxford Academic Press. Retrieved from https://academic.oup.com/jrsssa/advance-article-abstract/doi/10.1093/jrsssa/qnae141/7932060

Kahveci, N., & Ioannou, P. (2008, August 18). *Genetic Algorithms for Shortest Path Routing of Autonomous Gliders*. DOI: 10.2514/6.2008-6646

Kain, R., & Verma, A. (2018). Logistics management in supply chain–an overview. *Materials Today: Proceedings*, 5(2), 3811–3816. DOI: 10.1016/j.matpr.2017.11.634

Kanthavel, D., Sangeetha, S. K. B., & Keerthana, K. P. (2021). An empirical study of vehicle to infrastructure communications—An intense learning of smart infrastructure for safety and mobility. *International Journal of Intelligent Networks*, 2, 77–82. DOI: 10.1016/j.ijin.2021.06.003

Karakatsanis, I., AlKhader, W., MacCrory, F., Alibasic, A., Omar, M. A., Aung, Z., & Woon, W. L. (2017). Data mining approach to monitoring the requirements of the job market: A case study. *Information Systems*, 65, 1–6. DOI: 10.1016/j.is.2016.10.009

Karakolis, E., Kapsalis, P., Skalidakis, S., Kontzinos, C., Kokkinakos, P., Markaki, O., & Askounis, D. (2022). Bridging the gap between technological education and job market requirements through data analytics and decision support services. *Applied Sciences (Basel, Switzerland)*, 12(14), 7139. DOI: 10.3390/app12147139

Karaoglu, G., Hargittai, E., & Nguyen, M. H. (2022). Inequality in online job searching in the age of social media. *Information Communication and Society*, 25(12), 1826–1844. DOI: 10.1080/1369118X.2021.1897150

Kataria, D., Walid, A., Daneshmand, M., Dutta, A., Enright, M. A., Gu, R., . . . Polk, C. (2023, November). Artificial Intelligence And Machine Learning. In 2023 IEEE Future Networks World Forum (FNWF) (pp. 1-69). IEEE.

Kathuria, Y., Chaki, N. R., Kaur, M., Kumar, S., & Medhavi, R. (2024, October). Responsible AI impact assessment mechanism for India: A robust strategy for effective governance of AI systems in the country. In *AIP Conference Proceedings* (Vol. 3220, No. 1). AIP Publishing. DOI: 10.1063/5.0234670

Kedia, A., Kusumastuti, D., & Nicholson, A. (2017). Acceptability of collection and delivery points from consumers' perspective: A qualitative case study of Christchurch city. *Case Studies on Transport Policy*, 5(4), 587–595. DOI: 10.1016/j.cstp.2017.10.009

Kedia, A., Kusumastuti, D., & Nicholson, A. (2019). Establishing collection and delivery points to encourage the use of active transport: A case study in New Zealand using a consumer-centric approach. *Sustainability (Basel)*, 11(22), 6255. DOI: 10.3390/su11226255

Kembro, J., & Norrman, A. (2022). The transformation from manual to smart warehousing: An exploratory study with Swedish retailers. *International Journal of Logistics Management*, 33(5), 107–135. DOI: 10.1108/IJLM-11-2021-0525

Kern, J. (2021). The digital transformation of logistics: A review about technologies and their implementation status. The digital transformation of logistics: Demystifying impacts of the fourth industrial revolution, 361-403.

Khadem, M., Khadem, A., & Khadem, S. (2023). Application of artificial intelligence in supply chain revolutionizing efficiency and optimization. *International Journal of Industrial Engineering and Operational Research*, 5(1), 29–38.

Khalid, M. I., Ehsan, I., Al-Ani, A. K., Iqbal, J., Hussain, S., Ullah, S. S., & Nayab, . (2023). A comprehensive survey on blockchain-based decentralized storage networks. *IEEE Access : Practical Innovations, Open Solutions*, 11, 10995–11015. DOI: 10.1109/ACCESS.2023.3240237

Khanna, A. H. D. R. P. K. R. V. A. J. (2022). Disaster Management: Applications of Machine Learning and Deep Learning Techniques. *IEEE Access : Practical Innovations, Open Solutions*, 10, 6756–6770.

Kim, J. Y., & Heo, W. (2022). Artificial intelligence video interviewing for employment: Perspectives from applicants, companies, developer and academicians. *Information Technology & People*, 35(3), 861–878. DOI: 10.1108/ITP-04-2019-0173

Kingma, D. P., & Welling, M. (2013). Auto-Encoding Variational Bayes. *International Conference on Learning Representations (ICLR).*

Kingsly Stephen, R., Chandra Sekar, A., & Dinakaran, K. (2019). Sectional Transmission analysis approach for improved reliable transmission and secure routing in wireless sensor networks. *Cluster Computing*, 22(S2, Suppl 2), 3759–3770. DOI: 10.1007/s10586-018-2276-y

Kingsly, S. R., & Archana, T. (2025). Understanding AI and Machine Learning Concepts. In *AI and Machine Learning Applications in Supply Chains and Marketing* (pp. 371–392). IGI Global.

Kirkpatrick, S., Gelatt, C. D.Jr, & Vecchi, M. P. (1983). Optimization by Simulated Annealing. *Science*, 220(4598), 671–680. DOI: 10.1126/science.220.4598.671 PMID: 17813860

Kişi, N. (2022). Exploratory research on the use of blockchain technology in recruitment. *Sustainability (Basel)*, 14(16), 10098. DOI: 10.3390/su141610098

Klumpp, M., & Zijm, H. (2019). Logistics innovation and social sustainability: How to prevent an artificial divide in human–computer interaction. *Journal of Business Logistics*, 40(3), 265–278. DOI: 10.1111/jbl.12198

Kriechel, B., & Vetter, T. (2019). Skills Mismatch Measurement in ETF Partner Countries. *European Training Foundation*. Retrieved in: https://www.etf.europa.eu/sites/default/files/2019-05/Skills%20mismatch%20measurement_ETF%20partner%20countries.pdf

Kshetri, N. (2018). 1 Blockchain's roles in meeting key supply chain management objectives. *International Journal of Information Management*, 39, 80–89. DOI: 10.1016/j.ijinfomgt.2017.12.005

Kubista, N., Braun, D., & Parmigiani, G. (2024). Penetrance estimation in family-based studies with the penetrance R package. https://arxiv.org/pdf/2411.18816

Kuhn, K. M., Meijerink, J., & Keegan, A. (2021). Human Resource Management and the Gig Economy: Challenges and Opportunities at the Intersection between Organizational HR Decision-Makers and Digital Labor Platforms. *Research in Personnel and Human Resources Management*, 39, 1-46.

Kukreja, V., & Sakshi, . (2022). Machine learning models for mathematical symbol recognition: A stem to stern literature analysis. *Multimedia Tools and Applications*, 81(20), 28651–28687. DOI: 10.1007/s11042-022-12644-2

Kumar, A., Kumar, S. A., Dutt, V., Dubey, A. K., & Narang, S. (2023). A hybrid secure cloud platform maintenance based on improved attribute-based encryption strategies. *International Journal of Interactive Multimedia and Artificial Intelligence*, 8(2), 150–157. DOI: 10.9781/ijimai.2021.11.004

Kumar, S., Tiwari, P., & Zymbler, M. (2019). Internet of Things is a revolutionary approach for future technology enhancement: A review. *Journal of Big Data*, 6(1), 1–21. DOI: 10.1186/s40537-019-0268-2

Küp, E. T., Cebeci, S., Bayram, B., Aydın, G., Bozkaya, B., & Akhavan-Tabatabaei, R. (2024). An Integrated Framework for Dynamic Vehicle Routing Problems with Pick-up and Delivery Time Windows and Shared Fleet Capacity Planning. *Symmetry*, 16(4), 4. Advance online publication. DOI: 10.3390/sym16040505

Kuyu, Y. Ç., & Vatansever, F. (2024). A hybrid approach of ALNS with alternative initialization and acceptance mechanisms for capacitated vehicle routing problems. *Cluster Computing*, 27(10), 13583–13606. Advance online publication. DOI: 10.1007/s10586-024-04643-9

Lacity, M. C., Willcocks, L. P., & Craig, A. (2015). *Robotic Process Automation: The Next Transformation Lever for Shared Services*. Shared Services & Outsourcing Network.

Lagorio, A., Cimini, C., Pinto, R., & Cavalieri, S. (2023). 5G in Logistics 4.0: Potential applications and challenges. *Procedia Computer Science*, 217, 650–659. DOI: 10.1016/j.procs.2022.12.261

Lagorio, A., Zenezini, G., Mangano, G., & Pinto, R. (2022). A systematic literature review of innovative technologies adopted in logistics management. *International Journal of Logistics*, 25(7), 1043–1066. DOI: 10.1080/13675567.2020.1850661

Lambrecht, A., & Tucker, C. (2013). When Does Retargeting Work? Information Specificity in Online Advertising. *JMR, Journal of Marketing Research*, 50(5), 561–576. DOI: 10.1509/jmr.11.0503

LaPrade, A., Mertens, J., Moore, T., & Wright, A. (2019). The enterprise guide to closing the skills gap: Strategies for building and maintaining a skilled workforce. *IBM Institute for Business Value*. Retrieved in: https://www.ibm.com/downloads/cas/EPYMNBJA

Le Moigne, R. (2014). L'économie circulaire, comment la mettre en œuvre dans l'entreprise grâce à la reverse supply chain? Paris: Edition DUNOD.

Leimeister, J. M., Zogaj, S., & Durward, D. (2015). New Forms of Employment and IT—Crowdsourcing. In *4th conference of the regulating for decent work network*.

Le, T. T., Tran, P. Q., Lam, N. P., Tra, M. N. L., & Uyen, P. H. P. (2024). Corporate social responsibility, green innovation, environment strategy and corporate sustainable development. *Operations Management Research : Advancing Practice Through Research*, 17(1), 114–134. DOI: 10.1007/s12063-023-00411-x

Li, J., & Sarkis, J. (2022). Product eco-design practice in green supply chain management: A China-global examination of research. *Nankai Business Review International*, 13(1), 124–153. DOI: 10.1108/NBRI-02-2021-0006

Li, L. (2022). Reskilling and upskilling the future-ready workforce for industry 4.0 and beyond. *Information Systems Frontiers*, 1–16. PMID: 35855776

Liu, H., & Pan, W. (2024). A comprehensive literature review of the digital business ecosystem and digital business model with comparison in China, Japan and South Korea. *Asia Pacific Business Review*, 1–25. DOI: 10.1080/13602381.2024.2321198

Liu, L., Guo, F., Zou, Z., & Duffy, V. G. (2024). Application, development and future opportunities of collaborative robots (cobots) in manufacturing: A literature review. *International Journal of Human-Computer Interaction*, 40(4), 915–932. DOI: 10.1080/10447318.2022.2041907

Liu, W., Liu, T., Tang, O., Lee, P. T. W., & Chen, Z. (2024). The impact of digital supply chain announcements disclosing corporate social responsibility information on stock market value. *Industrial Management & Data Systems*, 124(2), 724–760. DOI: 10.1108/IMDS-03-2023-0189

Li, Z., & Jiao, P. (2022). Two-stage stochastic programming for the inventory routing problem with stochastic demands in fuel delivery. *International Journal of Industrial Engineering Computations*, 13(4), 507–522. DOI: 10.5267/j.ijiec.2022.7.004

Li, Z., & Ma, Y. (2013).. . *Particle Swarm Optimization Based on the Average Optimal Information for Vehicle Routing Problem*, 51–54. Advance online publication. DOI: 10.1109/ISCID.2013.20

Loivet, W., Taghipour, A., & Kang, D. S. (2020). The rise of green supply chain management: Between complexity and necessity. *J. Econ. Bus. Manag*, 8(1), 1–7. DOI: 10.18178/joebm.2020.8.1.604

Luong, D. V., Lopes de Sousa Jabbour, A. B., & Nguyen, N. M. (2024). The nexus between collection-and-delivery points and consumers' adoption intention: The role of country culture. *Cogent Business & Management*, 11(1), 2423890. DOI: 10.1080/23311975.2024.2423890

Lu, X., & Taghipour, A. (2023). A French Guide for Businesses and Governments to Integrate the Concept of Sustainable Development Into Their Supply Chains. In *Government Impact on Sustainable and Responsible Supply Chain Management* (pp. 1–12). IGI Global. DOI: 10.4018/978-1-6684-9062-4.ch001

Lysgaard, J., Letchford, A. N., & Eglese, R. W. (2004). A new branch-and-cut algorithm for the capacitated vehicle routing problem. *Mathematical Programming*, 100(2), 423–445. DOI: 10.1007/s10107-003-0481-8

Ma, Y., Tang, J., & Guan, J. (2013). Split Delivery Weighted Vehicle Routing Problem: Comparison perspective. *Chinese Control Conference, CCC*, 8436–8440.

Mahfod, J., Canel-Depitre, B., & Taghipour, A. (2019). Quality function deployment-ELECTRE in supplier evaluation. Journal of advanced management science, 7(4).

Mahfod, J., Canel-Depitre, B., & Taghipour, A. (2019). Quality function deployment-ELECTRE in supplier evaluation. Journal of Advanced Management Science, 7(4).

Mahfod, J., Canel-Depitre, B., & Taghipour, A. (2019). Quality Function Deployment-ELECTRE in Supplier Evaluation. *Journal of Advanced Management Science*, 131–135, 131–135. Advance online publication. DOI: 10.18178/joams.7.4.131-135

Mahfod, J., Canel-Depitre, B., & Taghipour, A. (2019). Quality function deployment-ELECTRE in supplier evaluation. *Journal of Advanced Management Science*, 7(4).

Mahya, A., Khabibah, S., & Surarso, B. (2023). *Optimization of vehicle routing problem with time windows using nearest neighbor algorithm and dynamic programming (case study: Pekalongan city Environmental Agency)*. 020007. DOI: 10.1063/5.0140518

Manifavas, H., & Karamitsos, I. (2022). Blockchain in supply chain management. In *Heterogeneous Cyber Physical Systems of Systems* (pp. 61–94). River Publishers. DOI: 10.1201/9781003338390-4

Manyika, J., Chui, M., Miremadi, M., Bughin, J., George, K., Willmott, P., & Dewhurst, M. (2017). A Future That Works: Automation, Employment, and Productivity. *McKinsey Global Institute*. Retrieved from https://www.mckinsey.com/featured-insights/employment-and-growth/afuture-that-works-automation-employment-and-productivity

Manyika, J., Chui, M., Miremadi, M., Bughin, J., George, K., Willmott, P., & Dewhurst, M. (2017). *A Future That Works: Automation, Employment, and Productivity*. McKinsey Global Institute.

Marchet, G., Melacini, M., Perotti, S., Rasini, M., & Tappia, E. (2018). Business logistics models in omni-channel: A classification framework and empirical analysis. *International Journal of Physical Distribution & Logistics Management*, 48(4), 439–464. DOI: 10.1108/IJPDLM-09-2016-0273

Marinakis, Y., Marinaki, M., & Migdalas, A. (2018). Particle swarm optimization for the vehicle routing problem: A survey and a comparative analysis. In *Handbook of Heuristics* (pp. 1163–1196). DOI: 10.1007/978-3-319-07124-4_42

Marshall, D., McCarthy, L., Heavey, C., & McGrath, P. (2015). Environmental and social supply chain management sustainability practices: Construct development and measurement. *Production Planning and Control*, 26(8), 673–690. DOI: 10.1080/09537287.2014.963726

Martinovic, G., Aleksi, I., & Baumgartner, A. (2008). Single-Commodity Vehicle Routing Problem with Pickup and Delivery Service. *Mathematical Problems in Engineering*, 2008(1), 697981. Advance online publication. DOI: 10.1155/2008/697981

Martins, N., Dominique-Ferreira, S., & Lopes, C. (2022). Design and development of a digital platform for seasonal jobs: Improving the hiring process. *Journal of Global Scholars of Marketing Science*, 32(3), 452–469. DOI: 10.1080/21639159.2020.1808851

Masudin, I., Wastono, T., & Zulfikarijah, F. (2018). The effect of managerial intention and initiative on green supply chain management adoption in Indonesian manufacturing performance. *Cogent Business & Management*, 5(1), 1485212. DOI: 10.1080/23311975.2018.1485212

Maulani, I. E., & Johansyah, C. A. (2023). The development of 5G technology and its implications for the industry. *Devotion: Journal of Research and Community Service*, 4(2), 631–635. DOI: 10.36418/devotion.v4i2.416

Mazhar, T., Irfan, H. M., Khan, S., Haq, I., Ullah, I., Iqbal, M., & Hamam, H. (2023). Analysis of cyber security attacks and its solutions for the smart grid using machine learning and blockchain methods. *Future Internet*, 15(2), 83. DOI: 10.3390/fi15020083

Mbiatem, J., Taghipour, A., & Canel-Depitre, B. (2018). Supplier Selection Approaches for Decision Makers. *Proceedings of the 8th International Conference on Information Communication and Management*, 108–112. DOI: 10.1145/3268891.3268895

Mbiatem, J., Taghipour, A., & Canel-Depitre, B. (2018). Supplier selection approaches for decision-makers. In *Proceedings of the 8th International Conference on Information Communication and Management* (pp. 108–112).

Mbiatem, J., Taghipour, A., & Canel-Depitre, B. (2018, August). Supplier selection approaches for decision makers. In *Proceedings of the 8th International Conference on Information Communication and Management* (pp. 108-112).

McGuinness, S., & Ortiz, L. (2016). Skill gaps in the workplace: Measurement, determinants and impacts. *Industrial Relations Journal*, 47(3), 253–278. DOI: 10.1111/irj.12136

Mendoza-Ramírez, C. E., Tudon-Martinez, J. C., Félix-Herrán, L. C., Lozoya-Santos, J. D. J., & Vargas-Martínez, A. (2023). Augmented reality: Survey. *Applied Sciences (Basel, Switzerland)*, 13(18), 10491. DOI: 10.3390/app131810491

Meng, X., Li, Z., & Tang, J. (2021). *A Dynamic VRP with Varying Transportation Costs and Its Solution Strategy*. 1–6. DOI: 10.1109/ICNSC52481.2021.9702224

Menon, N. G. R., Paul, A. J., & Ghosh, S. K. (2021). Real-Time Disaster Monitoring and Management Using IoT and Machine Learning. *IEEE Access : Practical Innovations, Open Solutions*, 9, 125763–125779.

Mentzer, J. T., DeWitt, W., Keebler, J. S., Min, S., Nix, N. W., Smith, C. D., & Zacharia, Z. G. (2001). Defining supply chain management. *Journal of Business Logistics*, 22(2), 1–25. DOI: 10.1002/j.2158-1592.2001.tb00001.x

Merimi, M., & Taghipour, A. (2021). Accelerating the digitalization of the supply chain: An empirical research about COVID-19 crisis. In *Digitalization of decentralized supply chains during global crises* (pp. 1–24). IGI Global. DOI: 10.4018/978-1-7998-6874-3.ch001

Metropolis, N., Rosenbluth, A. W., Rosenbluth, M. N., Teller, A. H., & Teller, E. (1953). Equation of state calculations by fast computing machines. *The Journal of Chemical Physics*, 21(6), 1087–1092. DOI: 10.1063/1.1699114

Meurisch, C., & Mühlhäuser, M. (2021). Data protection in AI services: A survey. *ACM Computing Surveys*, 54(2), 1–38. DOI: 10.1145/3440754

Mienye, I. D., & Swart, T. G. (2024). A Comprehensive Review of Deep Learning: Architectures, Recent Advances, and Applications. *Information (Basel)*, 15(12), 755. DOI: 10.3390/info15120755

Mikal, J. P., & Hurst, S. (2022). Ethical guidance for conducting health research with online communities: A scoping review of existing guidance. *PLoS One*, 17(1), e0302924. PMID: 38758778

Milgram, P., & Kishino, A. (1994). A taxonomy of mixed reality visual displays. *IEICE Transactions on Information and Systems*, E77-D(12), 1321–1329.

Miller, J. (2020). Blockchain and artificial intelligence: Creating synergy for improved supply chain management. *The Journal of Supply Chain Management*, 56(4), 25–32.

Minbaeva, D. (2021). Disrupted HR? *Human Resource Management Review*, 31(4), 100820. DOI: 10.1016/j.hrmr.2020.100820

Mishra, K., Siwal, S. S., & Thakur, V. K. (2024). E-waste Recycling and Utilization: A Review of Current Technologies and Future Perspectives. *Current Opinion in Green and Sustainable Chemistry*, 47, 100900. DOI: 10.1016/j.cogsc.2024.100900

Mittal, S., Koushik, P., Batra, I., & Whig, P. (2024). AI-Driven Inventory Management for Optimizing Operations With Quantum Computing. In Quantum Computing and Supply Chain Management: A New Era of Optimization (pp. 125-140). IGI Global.

Modica, T., Colicchia, C., Tappia, E., & Melacini, M. (2023). Empowering freight transportation through Logistics 4.0: A maturity model for value creation. *Production Planning and Control*, 34(12), 1149–1164. DOI: 10.1080/09537287.2021.1988176

Mohamed, B., & Feham, M. (2018). *Greedy Traffic Aware Routing Protocol Implementation in the Context of Internet of Vehicles*. 1–6. DOI: 10.1145/3284557.3284698

Mohammed, A. R., Alhajri, H. M., & Aldhahrani, N. A. (2021). Data-Driven Disaster Risk Assessment: A Review of Methodologies and Approaches. *IEEE Transactions on Emerging Topics in Computing*, 9(2), 640–654.

Mood, M. M., Mohaghar, A., & Nesari, Y. (2024). Designing a method to model the socio-technical systems. *International Journal of Information and Decision Sciences*, 16(3), 284–318. DOI: 10.1504/IJIDS.2024.140185

Morgan, D. L., & Nica, A. (2020). Iterative thematic inquiry: A new method for analyzing qualitative data. *International Journal of Qualitative Methods*, 19, 1609406920955118. DOI: 10.1177/1609406920955118

Morgan, J., Halton, M., Qiao, Y., & Breslin, J. G. (2021). Industry 4.0 smart reconfigurable manufacturing machines. *Journal of Manufacturing Systems*, 59, 481–506. DOI: 10.1016/j.jmsy.2021.03.001

Moshood, T. D., Nawanir, G., Sorooshian, S., & Okfalisa, O. (2021). Digital twins driven supply chain visibility within logistics: A new paradigm for future logistics. *Applied System Innovation*, 4(2), 29. DOI: 10.3390/asi4020029

Moumen, I., Rafalia, N., Abouchabaka, J., & Aoufi, M. (2023). Real-time GPS Tracking System for IoT-Enabled Connected Vehicles. In *E3S Web of Conferences* (Vol. 412, p. 01095). EDP Sciences. DOI: 10.1051/e3sconf/202341201095

Moussaoui, T. E., Loqman, C., & Boumhidi, J. (2024, May). Flat and Nested Named Entity Recognition in Arabic Language. In 2024 4th International Conference on Innovative Research in Applied Science, Engineering and Technology (IRASET) (pp. 1-7). IEEE. DOI: 10.1109/IRASET60544.2024.10548518

Mubarik, M. S., & Khan, S. A. (2024). Digital Supply Chain and Sustainability Challenges. In The Theory, Methods and Application of Managing Digital Supply Chains (pp. 133-143). Emerald Publishing Limited. DOI: 10.1108/978-1-80455-968-020241009

Muduli, K., Luthra, S., Kumar Mangla, S., Jabbour, C. J. C., Aich, S., & de Guimarães, J. C. F. (2020). Environmental management and the "soft side" of organisations : Discovering the most relevant behavioural factors in green supply chains. *Business Strategy and the Environment*, 29(4), 1647–1665. DOI: 10.1002/bse.2459

Murray, A., Skene, K., & Haynes, K. (2017). The Circular Economy: An Interdisciplinary Exploration of the Concept and Application in a Global Context. *Journal of Business Ethics*, 140(140), 369–380. DOI: 10.1007/s10551-015-2693-2

Nadia, E. N., Beatrice, C. D., & Atour, T. (2021). The determinants of hotel room rates in Beirut: A hedonic pricing model. International Journal of Trade. *Economics and Finance*, 12(2), 33–42.

Navarro, M. (2023, January 1). *Comparison Between the Branch and Bound and Evolutionary Algorithms to Solve the Classic Vehicle Routing Problem Known as VRP*. DOI: 10.18687/LACCEI2023.1.1.1499

Nayal, K., Kumar, S., Raut, R. D., Queiroz, M. M., Priyadarshinee, P., & Narkhede, B. E. (2022). Supply chain firm performance in circular economy and digital era to achieve sustainable development goals. *Business Strategy and the Environment*, 31(3), 1058–1073. DOI: 10.1002/bse.2935

Negi, S. (2024). Global supply chain competitiveness: The synergistic role of integrated logistics and global sourcing. *Global Business and Organizational Excellence*, 43(4), 111–130. DOI: 10.1002/joe.22247

Netflix Research. (2024). Discovering Creative Insights in Promotional Artwork.

Netflix TechBlog. (2024). Innovating Faster on Personalization Algorithms at Netflix Using Interleaving.

Nguyen, S., Leman, A., Xiao, Z., Fu, X., Zhang, X., Wei, X., Zhang, W., Li, N., Zhang, W., & Qin, Z. (2023). Blockchain-Powered Incentive System for JIT Arrival Operations and Decarbonization in Maritime Shipping. *Sustainability (Basel)*, 15(22), 15686. DOI: 10.3390/su152215686

Nikolaou, I. (2021). What is the Role of Technology in Recruitment and Selection? *The Spanish Journal of Psychology*, 24(e2), 1–6. DOI: 10.1017/SJP.2021.6 PMID: 33536110

Nilsson, N. J. (1998). *Artificial Intelligence: A New Synthesis*. Morgan Kaufmann.

Nissen, V. (2018). Digital transformation of the consulting industry—introduction and overview. *Digital Transformation of the Consulting Industry: Extending the Traditional Delivery Model*, 1-58.

Nomura, S., Imaizumi, S., Areias, A. C., & Yamauchi, F. (2017). Toward labor market policy 2.0: the potential for using online job-portal big data to inform labor market policies in India. *World Bank Policy Research Working Paper*, (7966). Retrieved in: https://papers.ssrn.com/sol3/papers.cfm?abstract_id=2923517

Ochelska-Mierzejewska, J., Poniszewska-Maranda, A., & Marańda, W. (2021). Selected Genetic Algorithms for Vehicle Routing Problem Solving. *Electronics (Basel)*, 10(24), 3147. DOI: 10.3390/electronics10243147

Oliveira-Dias, D., Maqueira-Marín, J. M., Moyano-Fuentes, J., & Tortorella, G. (2024). How does the technology environment impact supply chain strategy and operational performance? *Journal of Manufacturing Technology Management.*

Olivera, A., & Viera, O. (2007). Solving the vehicle routing problem with multiple trips by adaptive memory programming. *International Journal of Logistics Systems and Management - Int J Logist Syst Manag, 3.* DOI: 10.1504/IJLSM.2007.013213

Onik, M. M. H., Miraz, M. H., & Kim, C. S. (2018, April). A recruitment and human resource management technique using blockchain technology for industry 4.0. In *Smart Cities Symposium 2018* (pp. 1-6). IET. DOI: 10.1049/cp.2018.1371

Osorio-Mora, A., Soto-Bustos, M., Gatica, G., Palominos, P., & Linfati, R. (2021). The Multi-Depot Cumulative Vehicle Routing Problem With Mandatory Visit Times and Minimum Delayed Latency. *IEEE Access : Practical Innovations, Open Solutions*, 9, 27210–27225. DOI: 10.1109/ACCESS.2021.3058242

Oswal, N., Ateeq, K., & Mathew, S. (2021, April). Trends in Recruitment Information and Communication System using Artificial Intelligence in Industry 4.0. In *FEMIB* (pp. 111-118).

Ou, J., Liu, X., Xing, L., Lv, J., Hu, Y., Zheng, J., Zou, J., & Li, M. (2024). Solving many-objective delivery and pickup vehicle routing problem with time windows with a constrained evolutionary optimization algorithm. *Expert Systems with Applications*, 255, 124712. DOI: 10.1016/j.eswa.2024.124712

Ozdemir, P., Sevim, A., & Albayrak, T. (2023). Closing the gap between present and future through education: MINE-EMI project. *Case Studies on Transport Policy*, 11, 100936. DOI: 10.1016/j.cstp.2022.100936

Pajic, V., Andrejic, M., & Chatterjee, P. (2024). Enhancing cold chain logistics: A framework for advanced temperature monitoring in transportation and storage. *Mechatron. Intell Transp. Syst*, 3(1), 16–30.

Pandey, K., & Jha, C. K. (2023). A hybrid optimal solution for the combinatorial problem of vehicle routing. *International Journal of Intelligent Engineering Informatics*, 11(4), 337–352. DOI: 10.1504/IJIEI.2023.136104

Pant, P., & Taghipour, A. (2023). Machine learning and blockchain for 5G-enabled IIoT. In *Blockchain Applications in Cryptocurrency for Technological Evolution* (pp. 196–212). IGI Global.

Parveen, K., Kumar, S., Singh, R. P., Kumar, A., Yaduwanshi, R., & Dora, D. P. (2023). TS-CAGR:Traffic sensitive connectivity-aware geocast routing protocol in internet of vehicles. *Ad Hoc Networks*, 147, 103210. DOI: 10.1016/j.adhoc.2023.103210

Patil, K., Garg, V., Gabaldon, J., Patil, H., Niranjan, S., & Hawkins, T. (2024). Firm performance in digitally integrated supply chains: A combined perspective of transaction cost economics and relational exchange theory. *Journal of Enterprise Information Management*, 37(2), 381–413. DOI: 10.1108/JEIM-09-2022-0335

Pazzanese, C. (2020). *Ethical concerns mount as AI takes bigger decision-making role*. Harvard Gazette.

Pejic-Bach, M., Bertoncel, T., Meško, M., & Krstić, Ž. (2020). Text mining of industry 4.0 job advertisements. *International Journal of Information Management*, 50, 416–431. DOI: 10.1016/j.ijinfomgt.2019.07.014

Pendor, R. B., & Tasgaonkar, P. P. (2016). An IoT framework for intelligent vehicle monitoring system. *2016 International Conference on Communication and Signal Processing (ICCSP)*, 1694–1696. DOI: 10.1109/ICCSP.2016.7754454

Pillai, R., & Sivathanu, B. (2020). Adoption of artificial intelligence (AI) for talent acquisition in IT/ITeS organizations. *Benchmarking*, 27(9), 2599–2629. DOI: 10.1108/BIJ-04-2020-0186

Porter, M. E., & Heppelmann, J. E. (2017). Why every organization needs an augmented reality strategy. HBR'S 10 MUST, 85, 2-18.

Pournader, M., Ghaderi, H., Hassanzadegan, A., & Fahimnia, B. (2021). Artificial intelligence applications in supply chain management. *International Journal of Production Economics*, 241, 108250. DOI: 10.1016/j.ijpe.2021.108250

Prakash, N. R., & Kumari, S. S. S. (2020). Predictive Analytics for Disaster Management: A Systematic Review. *IEEE Transactions on Systems, Man, and Cybernetics. Systems*, 50(3), 967–982.

Prasad, A. S. R., Mohanty, D. K., & Sahu, A. K. (2020). A Survey on Disaster Management Using IoT and Big Data. *IEEE Internet of Things Journal*, 7(6), 4714–4728.

Prashar, G., Vasudev, H., & Bhuddhi, D. (2023). Additive manufacturing: Expanding 3D printing horizon in industry 4.0. *International Journal on Interactive Design and Manufacturing*, 17(5), 2221–2235. DOI: 10.1007/s12008-022-00956-4

Prikshat, V., Malik, A., & Budhwar, P. (2021). AI-augmented HRM: Antecedents, assimilation and multilevel consequences. *Human Resource Management Review*, 1–18.

Privault, N. (2024). Discrete stochastic processes. Springer. Retrieved from https://link.springer.com/content/pdf/10.1007/978-3-031-65820-4.pdf

Prodhon, C. (2009). *An ELSxPath Relinking Hybrid for the Periodic Location-Routing Problem*. 15–29. DOI: 10.1007/978-3-642-04918-7_2

Prodhon, C., & Prins, C. (2008). *A Memetic Algorithm with Population Management (MA|PM) for the Periodic Location-Routing Problem. 5296*, 43–57. DOI: 10.1007/978-3-540-88439-2_4

Puckett, J., Boutenko, V., Hoteit, L., Polunin, K., Perapechka, S., Stepanenko, A., & Bikkulova, G. (2020). Fixing the global skills mismatch. *BCG*. Retrieved in: https://www.bcg.com/publications/2020/fixing-global-skills-mismatch

Puhovichova, D., & Jankelova, N. (2020). Changes of human resource management in the context of impact of the fourth industrial revolution. *Industry 4.0, 5*(3), 138-141.

Puskar, J. (2019). Counting on the body: Techniques of embodied digitality. *New Media & Society*, 21(10), 2242–2260. DOI: 10.1177/1461444819843300

Qin, Y., Huang, D., & Zhang, X. (2012). VehiCloud: Cloud Computing Facilitating Routing in Vehicular Networks. *2012 IEEE 11th International Conference on Trust, Security and Privacy in Computing and Communications*, 1438–1445. DOI: 10.1109/TrustCom.2012.16

Rabiner, L. R. (1989). A tutorial on hidden Markov models and selected applications in speech recognition. *Proceedings of the IEEE*, 77(2), 257–286. DOI: 10.1109/5.18626

Rahman, A. U., & Halim, Z. (2022). Predicting the big five personality traits from hand-written text features through semi-supervised learning. *Multimedia Tools and Applications*, 81(23), 33671–33687. DOI: 10.1007/s11042-022-13114-5

Rahman, M. M., Alharbi, A. A., & Khatun, I. M. I. A. (2020). Smart Disaster Management System Using IoT and Big Data Analytics. *IEEE Access : Practical Innovations, Open Solutions*, 8, 26320–26330.

Rahman, M., Aydin, E., Haffar, M., & Nwagbara, U. (2022). The role of social media in e-recruitment process: Empirical evidence from developing countries in social network theory. *Journal of Enterprise Information Management*, 35(6), 1697–1718. DOI: 10.1108/JEIM-12-2019-0382

Rama Murthy, S., Huang, J., & Velu, C. (2023). Business Model Innovation in Additive Manufacturing Equipment Sector. In *Springer Handbook of Additive Manufacturing* (pp. 43–54). Springer International Publishing. DOI: 10.1007/978-3-031-20752-5_3

Ramadhani, B., & Garside, A. (2021). Particle Swarm Optimization Algorithm to Solve Vehicle Routing Problem with Fuel Consumption Minimization. *Jurnal Optimasi Sistem Industri*, 20(1), 1–10. DOI: 10.25077/josi.v20.n1.p1-10.2021

Rana, G., & Sharma, R. (2019). Emerging human resource management practices in Industry 4.0. *Strategic HR Review*, 18(4), 176–181. DOI: 10.1108/SHR-01-2019-0003

Rao, T. R., Mitra, P., Bhatt, R., & Goswami, A. (2019). The big data system, components, tools, and technologies: A survey. *Knowledge and Information Systems*, 60(3), 1165–1245. DOI: 10.1007/s10115-018-1248-0

Ras, E., Wild, F., Stahl, C., & Baudet, A. (2017, June). Bridging the skills gap of workers in Industry 4.0 by human performance augmentation tools: Challenges and roadmap. In *Proceedings of the 10th international conference on PErvasive technologies related to assistive environments* (pp. 428-432). DOI: 10.1145/3056540.3076192

Rasheed, H. (2024). Consideration of Cloud-Web-Concepts for Standardization and Interoperability: A Comprehensive Review for Sustainable Enterprise Systems, AI, and IoT Integration. *Journal of Information Technology and Informatics*, 3(2).

Reddy, K., & Muralidhar, A. (2023). Machine Learning-Based Road Safety Prediction Strategies for Internet of Vehicles (IoV) Enabled Vehicles: A Systematic Literature Review. *IEEE Access, PP*, 1–1. DOI: 10.1109/ACCESS.2023.3315852

Reddy, K. S. R., Prasad, B. V. D., & Reddy, V. V. S. M. (2021). Artificial Intelligence and Machine Learning Techniques in Disaster Management: A Review. *IEEE Transactions on Computational Social Systems*, 8(2), 389–398.

Reina, R., & Scarozza, D. (2021). Human resource management in the public administration. *Organizational development in public administration: The Italian way*, 61-101.

Rein, R., & Memmert, D. (2016). Big data and tactical analysis in elite soccer: Future challenges and opportunities for sports science. *SpringerPlus*, 5(1), 1–13. DOI: 10.1186/s40064-016-3108-2 PMID: 27610328

Rejeb, A., Rejeb, K., Simske, S. J., & Treiblmaier, H. (2021). Drones for supply chain management and logistics: A review and research agenda. *International Journal of Logistics*, 26(6), 708–731. DOI: 10.1080/13675567.2021.1981273

Rizkiani, F., Geraudy, I., & Imran, A. (2024). Solving the Vehicle Routing Problem with Multiple Trips and Simultaneous Delivery-Pickup for Drinking Water Distribution Company. *E3S Web of Conferences, 484*. DOI: 10.1051/e3sconf/202448401003

Rodríguez, M. Á., Alemany, M. M. E., Boza, A., Cuenca, L., & Ortiz, Á. (2020). Artificial intelligence in supply chain operations planning: Collaboration and digital perspectives. In Boosting Collaborative Networks 4.0: 21st IFIP WG 5.5 Working Conference on Virtual Enterprises, PRO-VE 2020, Valencia, Spain, November 23–25, 2020 *Proceedings*, 21, 365–378.

Romero-Gázquez, J. L., Cañavate-Cruzado, G., & Bueno-Delgado, M. V. (2021). IN4WOOD: A Successful European Training Action of Industry 4.0 for Academia and Business. *IEEE Transactions on Education*, 65(2), 200–209. DOI: 10.1109/TE.2021.3111696

Rout, R. R., Vemireddy, S., Raul, S. K., & Somayajulu, D. V. L. N. (2020). Fuzzy logic-based emergency vehicle routing: An IoT system development for smart city applications. *Computers & Electrical Engineering*, 88, 106839. DOI: 10.1016/j.compeleceng.2020.106839

Russell, S., & Norvig, P. (2016). *Artificial Intelligence: A Modern Approach* (3rd ed.). Pearson.

Rutkauskiene, D., & Volodzkaite, G. (2020). Design of the Platform Solutions to Increase the Employability and E-Learning Opportunities for Low Skilled Women. In *Information and Software Technologies: 26th International Conference, ICIST 2020, Kaunas, Lithuania, October 15–17, 2020Proceedings*, 26, 334–342.

Sahoo, S., & Vijayvargy, L. (2020). Green supply chain management practices and its impact on organizational performance: Evidence from Indian manufacturers.

Sahoo, S., Kumar, S., Abedin, M. Z., Lim, W. M., & Jakhar, S. K. (2023). Deep learning applications in manufacturing operations: A review of trends and ways forward. *Journal of Enterprise Information Management*, 36(1), 221–251. DOI: 10.1108/JEIM-01-2022-0025

Sahu, R. K., Gupta, M. R., & Sharma, S. C. (2021). A Framework for Disaster Management Using Artificial Intelligence and IoT. *IEEE Transactions on Network and Service Management*, 18(1), 620–632.

Sahu, S. C. K., Dutta, A. K., & Sahu, S. S. (2020). Disaster Management System Using Machine Learning and IoT. *IEEE Access : Practical Innovations, Open Solutions*, 8, 68145–68156.

Sallam, K., Mohamed, M., & Mohamed, A. W. (2023). Internet of Things (IoT) in supply chain management: Challenges, opportunities, and best practices. *Sustainable Machine Intelligence Journal*, 2, 3–1. DOI: 10.61185/SMIJ.2023.22103

Samsuddin, S., Othman, M., & Mi Yusuf, L. (2020). Utilizing Ant Colony Optimization and Intelligent Water Drop for Solving Multi Depot Vehicle Routing Problem. *IOP Conference Series. Materials Science and Engineering*, 864(1), 012095. DOI: 10.1088/1757-899X/864/1/012095

Santandreu Calonge, D., & Aman Shah, M. (2016). MOOCs, graduate skills gaps, and employability: A qualitative systematic review of the literature. *International Review of Research in Open and Distance Learning*, 17(5), 67–90.

Sarkis, J., Zhu, Q., & Lai, K. (2011). An organizational theoretic review of green supply chain management literature. *International Journal of Production Economics*, 130(1), 1–15. DOI: 10.1016/j.ijpe.2010.11.010

Sato, T., Morita, T., & Murata, T. (2010). Lagrangian relaxation method for network flow modeled crew and vehicle rescheduling. *2010 2nd International Conference on Advanced Computer Control*, 403–408. 2010 2nd International Conference on Advanced Computer Control. DOI: 10.1109/ICACC.2010.5486971

Sceulovs, D., Shatrevich, V., & Gaile-Sarkane, E. (2017). *Evaluation of E-recruitment as a Business Model through Internet of Things Approach. Journal of Business and Economics Review* (Vol. 2). JBER.

Schlegel, G. L., & Trent, R. J. (2014). *Supply chain risk management: An emerging discipline*. CRC Press.

Seidgar, H., Abedi, M., Rad, S., & Rezaeian, J. (2016). An efficient hybrid of genetic and simulated annealing algorithms for multi server vehicle routing problem with multi entry. *International Journal of Industrial and Systems Engineering*, 24(3), 333. DOI: 10.1504/IJISE.2016.079823

Senyo, P. K., Liu, K., & Effah, J. (2019). Digital business ecosystem: Literature review and a framework for future research. *International Journal of Information Management*, 47, 52–64. DOI: 10.1016/j.ijinfomgt.2019.01.002

Seth, H., Talwar, S., Bhatia, A., Saxena, A., & Dhir, A. (2020). Consumer resistance and inertia of retail investors: Development of the resistance adoption inertia continuance (RAIC) framework. *Journal of Retailing and Consumer Services*, 55, 102071. DOI: 10.1016/j.jretconser.2020.102071

Sewalkar, P., & Seitz, J. (2019). Vehicle-to-Pedestrian Communication for Vulnerable Road Users: Survey, Design Considerations, and Challenges. *Sensors (Basel)*, 19(2), 358. Advance online publication. DOI: 10.3390/s19020358 PMID: 30658392

Shafique, A. T. A., Ali, A. H., & Qadir, K. S. T. (2020). AI and Machine Learning for Disaster Management: A Review. *IEEE Access : Practical Innovations, Open Solutions*, 8, 89912–89924.

Shahbaz, U., Beheshti, A., Nobari, S., Qu, Q., Paik, H. Y., & Mahdavi, M. (2019). irecruit: Towards automating the recruitment process. In *Service Research and Innovation: 7th Australian Symposium, ASSRI 2018, Sydney, NSW, Australia, September 6, 2018, and Wollongong, NSW, Australia, December 14, 2018, Revised Selected Papers 7* (pp. 139-152). Springer International Publishing.

Shahzad, M. F., Xu, S., Lim, W. M., Yang, X., & Khan, Q. R. (2024). Artificial intelligence and social media on academic performance and mental well-being: Student perceptions of positive impact in the age of smart learning. *Heliyon*, 10(8), e29523. DOI: 10.1016/j.heliyon.2024.e29523 PMID: 38665566

Shaik, M., & Siddque, K. Q. (2023). Predictive Analytics in Supply Chain Management using SAP and AI. *Journal of Computer Sciences and Applications*, 11(1), 1–6. DOI: 10.12691/jcsa-11-1-1

Shankar, V., Kalyanam, K., Setia, P., Golmohammadi, A., Tirunillai, S., Douglass, T., Hennessey, J., Bull, J. S., & Waddoups, R. (2021). How technology is changing retail. *Journal of Retailing*, 97(1), 13–27. DOI: 10.1016/j.jretai.2020.10.006

Shao, X. (2022). Improved energy-efficient routing architecture for traffic management system using a hybrid meta-heuristic algorithm in Internet of vehicles. *Journal of High Speed Networks*, 28(4), 1–10. DOI: 10.3233/JHS-222003

Sharkey, N. (2019). The Ethical Implications of AI in Warfare. In *Artificial Intelligence and International Relations: A Security Perspective*. Routledge.

Sharma, S., & Sharma, K. (2023, June). Analyzing Employee's Attrition and Turnover at Organization Using Machine Learning Technique. In *2023 3rd International Conference on Intelligent Technologies (CONIT)* (pp. 1-7). IEEE. DOI: 10.1109/CONIT59222.2023.10205676

Sharma, E., Mahajan, R., Mahajan, R., & Mansotra, V. (2021). Automated personality prediction of social media users: A decade review. *Turkish Journal of Computer and Mathematics Education*, 12(14), 5225–5237.

Sharma, R. (2023). The transformative power of AI as future GPTs in propelling society into a new era of advancement. *IEEE Engineering Management Review*, 51(4), 215–224. DOI: 10.1109/EMR.2023.3315191

Shashi, C., Centobelli, P., Cerchione, R., & Ertz, M. (2020). Managing supply chain resilience to pursue business and environmental strategies. *Business Strategy and the Environment*, 29(3), 1215–1246. DOI: 10.1002/bse.2428

Shen, Z. M., & Sun, Y. (2023). Strengthening supply chain resilience during COVID-19: A case study of JD. com. *Journal of Operations Management*, 69(3), 359–383. DOI: 10.1002/joom.1161

Shetty, A., Puthusseri, K., Shetty, A., & Mani, R. (2018). *A Savings Algorithm Approach to the Truck Routing and Loading Problem*. 284–290. DOI: 10.1109/ICRCICN.2018.8718688

Shi, L. P., & Wang, S. (2022). Demand-side consequences of unemployment and horizontal skill mismatches across national contexts: An employer-based factorial survey experiment. *Social Science Research*, 104, 102668. DOI: 10.1016/j.ssresearch.2021.102668 PMID: 35400384

Shull, A., Twomey, P., & Yoo, C. S. (2017). Legal Mechanisms for Governing the Transition of Key Domain Name Functions to the Global Multi-stakeholder Community. Who Runs the Internet, 18-23.

Shurrab, M., Singh, S., Otrok, H., Mizouni, R., Khadkikar, V., & Zeineldin, H. (2022). An Efficient Vehicle-to-Vehicle (V2V) Energy Sharing Framework. *IEEE Internet of Things Journal, 9*(7), 5315–5328. *IEEE Internet of Things Journal*. Advance online publication. DOI: 10.1109/JIOT.2021.3109010

Silva, V., Amaral, A., & Fontes, T. (2023). Sustainable urban last-mile logistics: A systematic literature review. *Sustainability (Basel)*, 15(3), 2285. DOI: 10.3390/su15032285

Simone, C., Barile, S., & Calabrese, M. (2018). Managing territory and its complexity: A decision-making model based on the viable system approach (VsA). *Land Use Policy*, 72, 493–502. DOI: 10.1016/j.landusepol.2017.12.070

Şimşek, T., Öner, M. A., Kunday, Ö., & Olcay, G. A. (2022). A journey towards a digital platform business model: A case study in a global tech-company. *Technological Forecasting and Social Change*, 175, 1–17. DOI: 10.1016/j.techfore.2021.121372

Singh, H., & Kumar, Y. (2019). Cellular automata based model for e-healthcare data analysis. *International Journal of Information System Modeling and Design*, 10(3), 1–18. DOI: 10.4018/IJISMD.2019070101

Singh, N., Raza, M., Paranthaman, V. V., Awais, M., Khalid, M., & Javed, E. (2021). Internet of Things and cloud computing. In *Digital Health* (pp. 151–162). Academic Press. DOI: 10.1016/B978-0-12-818914-6.00013-2

Singh, P. K. (2023). Digital transformation in supply chain management: Artificial intelligence (AI) and machine learning (ML) as catalysts for value creation. *International Journal of Supply Chain Management*, 12(6), 57–63. DOI: 10.59160/ijscm.v12i6.6216

Smythe, S., Grotlüschen, A., & Buddeberg, K. (2021). The automated literacies of e-recruitment and online services. *Studies in the Education of Adults*, 53(1), 4–22. DOI: 10.1080/02660830.2020.1855870

Sobb, T., Turnbull, B., & Moustafa, N. (2020). Supply chain 4.0: A survey of cyber security challenges, solutions and future directions. *Electronics (Basel)*, 9(11), 1864. DOI: 10.3390/electronics9111864

Sodachi, M., Pirayesh, A., & Valilai, O. F. (2024). Using Markov decision process model for sustainability assessment in Industry 4.0. *IEEE Access : Practical Innovations, Open Solutions*, 12, 189417–189438. DOI: 10.1109/ACCESS.2024.3514786

Soleimani, P., Marvasti, M. R. B., & Ghorbanzadeh, P. (2020). A Hybrid Traffic Management Method Based on Combination of IOV and VANET Network in Urban Routing for Emergency Vehicles. *2020 4th International Conference on Smart City, Internet of Things and Applications (SCIOT)*, 58–65. DOI: 10.1109/SCIOT50840.2020.9250198

Solomon, M. M. (1987). Algorithms for the Vehicle Routing and Scheduling Problems with Time Window Constraints. *Operations Research*, 35(2), 254–265. DOI: 10.1287/opre.35.2.254

Song, J., Krishnamurthy, V., Kwasinski, A., & Sharma, R. (2012). Development of a Markov-chain-based energy storage model for power supply availability assessment of photovoltaic generation plants. *IEEE Transactions on Sustainable Energy*, 4(2), 491–500. DOI: 10.1109/TSTE.2012.2207135

Sparrow, R. (2020). Killing by Robot: The Ethics of Autonomous Weapons. In *The Oxford Handbook of Ethics of AI*. Oxford University Press.

Squicciarini, M., & Nachtigall, H. (2021). Demand for AI skills in jobs: Evidence from online job postings.

Srikumar, M., Finlay, R., Abuhamad, G., Ashurst, C., Campbell, R., Campbell-Ratcliffe, E., Hongo, H., Jordan, S. R., Lindley, J., Ovadya, A., & Pineau, J. (2022). Advancing ethics review practices in AI research. *Nature Machine Intelligence*, 4(12), 1061–1064. DOI: 10.1038/s42256-022-00585-2

Srivastava, S. K. (2007). Green supply-chain management: A state-of-the-art literature review. *International Journal of Management Reviews*, 9(1), 53–80. DOI: 10.1111/j.1468-2370.2007.00202.x

Stephany, F. (2022). Closing the Digital Skill Gap: The Potential of Online Platform Data For Active Labour Markets Policies. *Zenodo*. Retrieved in: https://static.ie.edu/CGC/Stephany-%20Closing%20the%20Digital%20Skill%20Gap.pdf

Stodola, P., & Kutěj, L. (2024). Multi-Depot Vehicle Routing Problem with Drones: Mathematical formulation, solution algorithm and experiments. *Expert Systems with Applications*, 241, 122483. DOI: 10.1016/j.eswa.2023.122483

Sudan, T., Taggar, R., Jena, P. K., & Sharma, D. (2023). Supply chain disruption mitigation strategies to advance future research agenda: A systematic literature review. *Journal of Cleaner Production*, 425, 138643. DOI: 10.1016/j.jclepro.2023.138643

Sultana, N., Chan, J., Abbasi, B., Sarwar, T., & Qin, A. K. (2024). Learning to Guide Local Search Optimisation for Routing Problems. *Operations Research Letters*, 55, 107136. DOI: 10.1016/j.orl.2024.107136

Sundararajan, A. (2019). *The Sharing Economy: The End of Employment and the Rise of Crowd-Based Capitalism*. MIT Press.

Sun, Q., Tyagu, T., & Noche, B. (2022). Savings matrix algorithm to address vehicle routing problem. In Ye, X., Falcone, F., & Cui, H. (Eds.), *2nd International Conference on Internet of Things and Smart City (IoTSC 2022)* (p. 163). SPIE. DOI: 10.1117/12.2637152

Suppatvech, C., Godsell, J., & Day, S. (2019). The roles of internet of things technology in enabling servitized business models: A systematic literature review. *Industrial Marketing Management*, 82, 70–86. DOI: 10.1016/j.indmarman.2019.02.016

Šuštaršič, A., Videmšek, M., Karpljuk, D., Miloloža, I., & Meško, M. (2022). Big data in sports: A bibliometric and topic study. Business Systems Research. *International Journal of the Society for Advancing Innovation and Research in Economy*, 13(1), 19–34.

Sutton, R. S., & Barto, A. G. (1998). *Reinforcement learning: An introduction*. MIT Press.

Szpytko, J., & Salgado Duarte, Y. (2020). Markov chain Monte Carlo simulation model for risk assessment of power systems for electromobility use. *Journal of KONBiN*, 50(1), 1–9. DOI: 10.2478/jok-2020-0002

Tadić, S., & Veljović, M. (2021). Home delivery: A framework for structuring. *IJTTE. International Journal for Traffic and Transport Engineering*, 11(1).

Taghipour, A. (2021). Digital transformation of supply chains during crisis: Covid19. *11th Annual International Conference on Industrial Engineering and Operations Management, IEOM 2021*. https://www.academia.edu/79144632/Digital_transformation_of_supply_chains_during_crisis_Covid19

Taghipour, A. (2023). Government impact on sustainable and responsible supply chain management. IGI Global.

Taghipour, A. (2024). Big Data: The New Gold Mine for Supermarkets. In Applications of New Technology in Operations and Supply Chain Management (pp. 98-106). IGI Global.

Taghipour, A. (2024). Supply Chains and Artificial Intelligence Chatbots. In Applications of New Technology in Operations and Supply Chain Management (pp. 260-268). IGI Global. DOI: 10.4018/979-8-3693-1578-1.ch013

Taghipour, A., & Frayret, J. M. (2011). Supply chain coordination planning: A review. Work paper, École polytechnique de Montréal.

Taghipour, A., & Frayret, J.-M. (2011). Supply chain coordination planning: A review. Work Paper, École Polytechnique de Montréal.

Taghipour, A., & Merimi, M. (2021, March). Digital transformation of supply chains during crisis: covid-19. In 11th Annual International Conference on Industrial Engineering and Operations Management, IEOM (pp. 7493-7501). DOI: 10.46254/AN11.20211298

Taghipour, A., Abed, M., & Zoghlami, N. (2015, May). Design for remanufacturing respecting reverse logistics processes: A review. In 2015 4th International Conference on Advanced Logistics and Transport (ICALT) (pp. 299-304). IEEE. DOI: 10.1109/ICAdLT.2015.7136602

Taghipour, A. (2014). Improving the plan of a manufacturing network with non-integrated business units. *International Journal of Applied Logistics*, 5(2), 1–11. DOI: 10.4018/ijal.2014040101

Taghipour, A. (Ed.). (2020). *Demand forecasting and order planning in supply chains and humanitarian logistics*. IGI Global.

Taghipour, A., & Beneteau-Piet, C. (2020). Sustainable supply chain management performance. *International Journal of Innovation, Management and Technology*, 11(6), 165–169. DOI: 10.18178/ijimt.2020.11.6.897

Taghipour, A., Foukolaei, P. Z., Ghaedi, M., & Khazaei, M. (2023a). Sustainable Multi-Objective Models for Waste-to-Energy and Waste Separation Site Selection. *Sustainability (Basel)*, 15(22), 15764. DOI: 10.3390/su152215764

Taghipour, A., & Frayret, J. M. (2013a, May). An algorithm to improve operations planning in decentralized supply chains. In *2013 International Conference on Advanced Logistics and Transport* (pp. 100-103). IEEE. DOI: 10.1109/ICAdLT.2013.6568442

Taghipour, A., & Frayret, J. M. (2013b). Coordination of operations planning in supply chains: A review. *International Journal of Business Performance and Supply Chain Modelling*, 5(3), 272–307. DOI: 10.1504/IJBPSCM.2013.055729

Taghipour, A., Lu, X., & Smahi, M. (2022). Opportunities and challenges for society 5.0 in procurement. *Blockchain Technology in Supply Chain Management for Society*, 5(0), 91–100. DOI: 10.1201/9781003177432-9

Taghipour, A., Padash, A., Etemadi, V., Khazaei, M., & Ebrahimi, S. (2024). Sustainable and Circular Hotels and the Water–Food–Energy Nexus: Integration of Agrivoltaics, Hydropower, Solar Cells, Water Reservoirs, and Green Roofs. *Sustainability (Basel)*, 16(5), 1985. DOI: 10.3390/su16051985

Taghipour, A., Ramezani, M., Khazaei, M., Roohparvar, V., & Hassannayebi, E. (2023b). Smart transportation behavior through the COVID-19 pandemic: A ride-hailing system in Iran. *Sustainability (Basel)*, 15(5), 4178. DOI: 10.3390/su15054178

Taghipour, A., Sohrabi, A., Ghaedi, M., & Khazaei, M. (2023c). A robust vaccine supply chain model in pandemics: Case of Covid-19 in Iran. *Computers & Industrial Engineering*, 183, 109465. DOI: 10.1016/j.cie.2023.109465

Taj, S., Imran, A. S., Kastrati, Z., Daudpota, S. M., Memon, R. A., & Ahmed, J. (2023). IoT-based supply chain management: A systematic literature review. *Internet of Things : Engineering Cyber Physical Human Systems*, 24, 100982. DOI: 10.1016/j.iot.2023.100982

Talari, G., Cummins, E., McNamara, C., & O'Brien, J. (2022). State of the art review of Big Data and web-based Decision Support Systems (DSS) for food safety risk assessment with respect to climate change. *Trends in Food Science & Technology*, 126, 192–204. DOI: 10.1016/j.tifs.2021.08.032

Tamym, L., Benyoucef, L., Moh, A. N. S., & El Ouadghiri, M. D. (2021). A big data based architecture for collaborative networks: Supply chains mixed-network. *Computer Communications*, 175, 102–111. DOI: 10.1016/j.comcom.2021.05.008

Tan, C. K., Gan, S. P., & Tan, D. H. K. (2021). Web-Based Disaster Management System: A Review of IoT Solutions. *IEEE Access : Practical Innovations, Open Solutions*, 9, 74060–74073.

Tan, X. (2023). Enhanced sports predictions: A comprehensive analysis of the role and performance of predictive analytics in the sports sector. *Wireless Personal Communications*, 132(3), 1613–1636. DOI: 10.1007/s11277-023-10585-z

Tavakkoli-Moghaddam, R., Rabbani, M., Saremi, A., & Safaei, N. (2005). *Solving the backhaul vehicle routing problem by genetic algorithms.*

Tay, H. L., & Loh, H. S. (2022). Digital transformations and supply chain management: A Lean Six Sigma perspective. *Journal of Asia Business Studies*, 16(2), 340–353. DOI: 10.1108/JABS-10-2020-0415

Team, A. A. C. (2025). Hidden Markov model in machine learning. Retrieved January 12, 2025, from https://www.appliedaicourse.com/blog/hidden-markov-model-in-machine-learning/

Teng, R., & Sato, K. (2024). A Fundamental Study of Reliable Vehicle-to-Cloud Communication Using Multiple Paths with Redundancy Mitigation. *Applied Sciences (Basel, Switzerland)*, 14(7), 2841. DOI: 10.3390/app14072841

The World Economic Forum. (2020). Closing the Skills Gap: Key Insights and Success Metrics.

Thirugnanam, A., & Jahir Hussain, F. B. (2023). Exploring Machine Learning Algorithms for the Prediction of Dengue: A Comprehensive Review. *Revue d'Intelligence Artificielle*, 37(5), 1281–1290. DOI: 10.18280/ria.370521

Tiwari, S. (2021). Supply chain integration and Industry 4.0: A systematic literature review. *Benchmarking*, 28(3), 990–1030. DOI: 10.1108/BIJ-08-2020-0428

Tiwari, S., Wee, H. M., & Daryanto, Y. (2018). Big data analytics in supply chain management between 2010 and 2016: Insights to industries. *Computers & Industrial Engineering*, 115, 319–330. DOI: 10.1016/j.cie.2017.11.017

Todolí-Signes, A. (2017). The 'gig economy': Employee, self-employed or the need for a special employment regulation? *Transfer: European Review of Labour and Research*, 23(2), 193–205. DOI: 10.1177/1024258917701381

Tolk, A., & Diallo, S. Y. (2019). *Modeling and Simulation Support for System of Systems Engineering Applications*. Wiley.

Torgler, B. (2023). *Big data, artificial intelligence, and quantum computing in sports. In 21st century sports: How technologies will change sports in the digital age*. Springer International Publishing.

Torres, F., Gendreau, M., & Rei, W. (2022). Crowdshipping: An open VRP variant with stochastic destinations. *Transportation Research Part C, Emerging Technologies*, 140, 103677. DOI: 10.1016/j.trc.2022.103677

Toth, P., & Vigo, D. (2001). An overview of vehicle routing problems. In *The vehicle routing problem* (pp. 1–26). Society for Industrial and Applied Mathematics.

Turban, E., & Aronson, J. E. (2001). *Decision Support Systems and Intelligent Systems* (6th ed.). Prentice Hall.

United Nations AI for Good. (2024). Cracking the code on supply chain sustainability: How AI and data are creating more responsible operations. Retrieved December 18, 2024, from https://medium.com/@UN_AIforGood/cracking-the-code-on-supply-chain-sustainability-how-ai-and-data

Upadhyay, A. K., & Khandelwal, K. (2018). Applying artificial intelligence: Implications for recruitment. *Strategic HR Review*, 17(5), 255–258. DOI: 10.1108/SHR-07-2018-0051

Urciuoli, L., & Hintsa, J. (2017). Adapting supply chain management strategies to security–an analysis of existing gaps and recommendations for improvement. *International Journal of Logistics*, 20(3), 276–295. DOI: 10.1080/13675567.2016.1219703

Valero, M. E. J. J. M., Alda, A. M. M., & Carretero, M. J. P. F. (2022). Machine Learning for Natural Disaster Prediction: A Systematic Review. *IEEE Transactions on Computational Intelligence and AI in Games*, 14(4), 623–634.

Van Alstyne, M. W., & Schrage, M. (2016). The best platforms are more than matchmakers. *Harvard Business Review*, 94(7/8), 1–6.

Veile, J. W., Schmidt, M. C., & Voigt, K. I. (2022). Toward a new era of cooperation: How industrial digital platforms transform business models in Industry 4.0. *Journal of Business Research*, 143, 387–405. DOI: 10.1016/j.jbusres.2021.11.062

Volna, E. (2016). *Genetic algorithms for the vehicle routing problem. 1738*, 120002. DOI: 10.1063/1.4951885

Volodymyr, M., & Oksana, O. (2020). World Trends in Warehousing Logistics. Intellectualization of logistics and Supply Chain Management, (2), 32-50.

Vrontis, D., Christofi, M., Pereira, V., Tarba, S., Makrides, A., & Trichina, E. (2023). Artificial intelligence, robotics, advanced technologies and human resource management: a systematic review. *Artificial Intelligence and International HRM*, 172-201.

Vural, C. A., & Aktepe, Ç. (2022). Why do some sustainable urban logistics innovations fail? The case of collection and delivery points. *Research in Transportation Business & Management*, 45, 100690. DOI: 10.1016/j.rtbm.2021.100690

Wamba, S. F., Akter, S., Edwards, A., Chopin, G., & Gnanzou, D. (2020). How 'big data' can make big impact: Findings from a systematic review and a longitudinal case study. *International Journal of Production Economics*, 165, 234–246. DOI: 10.1016/j.ijpe.2014.12.031

Wang, X., Shu-Zhi, Z., Hao, C., & Yan, L. (2016). *An improved savings method for vehicle routing problem*. 1–4. DOI: 10.1109/CCSSE.2016.7784340

Wang, P., Che, F., Fan, S., & Gu, C. (2014). Ownership Governance, Institutional Pressures and Circular Economy Accounting Information Disclosure. *Chinese Management Studies*, 8(3), 487–501. DOI: 10.1108/CMS-10-2013-0192

Wang, X. L., Wang, L., Bi, Z., Li, Y. Y., & Xu, Y. (2016). Cloud computing in human resource management (HRM) system for small and medium enterprises (SMEs). *International Journal of Advanced Manufacturing Technology*, 84(1-4), 485–496. DOI: 10.1007/s00170-016-8493-8

Wassan, N., Wassan, N., Nagy, G., & Salhi, S. (2017). The Multiple Trip Vehicle Routing Problem with Backhauls. *Computers & Operations Research*, 78(C), 454–467. DOI: 10.1016/j.cor.2015.12.017

West, D. M. (2018). *The Future of Work: Robots, AI, and Automation*. Brookings Institution Press.

Williams, P., McDonald, P., & Mayes, R. (2023). Recruitment in the gig economy: Attraction and selection on digital platforms. In *Technologically Mediated Human Resource Management* (pp. 142–168). Routledge. DOI: 10.4324/9781003388203-6

Wong, D. T., & Ngai, E. W. (2019). Critical review of supply chain innovation research (1999–2016). *Industrial Marketing Management*, 82, 158–187. DOI: 10.1016/j.indmarman.2019.01.017

Wong, S., Yeung, J. K. W., Lau, Y. Y., & Kawasaki, T. (2023). A case study of how Maersk adopts cloud-based blockchain integrated with machine learning for sustainable practices. *Sustainability (Basel)*, 15(9), 7305. DOI: 10.3390/su15097305

Wongthatsanekorn, W., Realff, M. J., & Ammons, J. C. (2010). Multi-time scale Markov decision process approach to strategic network growth of reverse supply chains. *Omega*, 38(1–2), 20–32. DOI: 10.1016/j.omega.2009.02.001

World Economic Forum (WEF). (2020). Closing the skills gap: key insights and success metrics. *White paper*. Retrieved in: https://www3.weforum.org/docs/WEF_GSC_NES_White_Paper_2020.pdf

Xie, N., Lee, X., Cai, D., Saito, Y., Asai, N., & Lau, H. C. (2024). *A Feasibility-Preserved Quantum Approximate Solver for the Capacitated Vehicle Routing Problem*. /arXiv.2308.08785DOI: 10.21203/rs.3.rs-4274870/v1

Xu, L. D., He, W., & Li, S. (2014). Internet of Things in industries: A survey. *IEEE Transactions on Industrial Informatics*, 10(4), 2233–2243. DOI: 10.1109/TII.2014.2300753

Xu, X., Lu, Y., Vogel-Heuser, B., & Wang, L. (2021). Industry 4.0 and Industry 5.0—Inception, conception and perception. *Journal of Manufacturing Systems*, 61, 530–535. DOI: 10.1016/j.jmsy.2021.10.006

Yadav, R. P., & Gupta, M. K. (2021). Utilization of Web Technologies for Disaster Risk Reduction: A Systematic Review. *IEEE Access : Practical Innovations, Open Solutions*, 9, 19880–19892.

Yafooz, W. M., Bakar, Z. B. A., Fahad, S. A., & Mithun, M. (2019). A. (2020). Business intelligence through big data analytics, data mining and machine learning. In Data Management, Analytics and Innovation *Proceedings of ICDMAI*, 2, 217–230.

Yavary, A., & Sajedi, H. (2018). *Solving Dynamic Vehicle Routing Problem with Pickup and Delivery by CLARITY Method*. 207–212. DOI: 10.1109/INES.2018.8523908

Yousefikhoshbakht, M., Didehvar, F., Rahmati, F., & Ahmed, Z. (2023). *Fixed fleet open vehicle routing problem: Mathematical model and a modified ant colony optimization.*

Youshanlouei, H. R., Gol, A. E., Ščasný, M., & Shahraki, A. M. (2024). A Qualitative Study of the Intention to Use Smart Hybrid Electric Vehicles in an Oil-Rich Country. In *Applications of New Technology in Operations and Supply Chain Management* (pp. 15–37). IGI Global. DOI: 10.4018/979-8-3693-1578-1.ch002

Yue, X., Wang, Y., Wang, Y., & Zhang, X. (2020). Blockchain technology for enhancing supply chain resilience: A literature review. *International Journal of Production Economics*, 227, 107678.

Yu, M., Nagarajan, V., & Shen, S. (2021). Improving Column Generation for Vehicle Routing Problems via Random Coloring and Parallelization. *INFORMS Journal on Computing*, 34(2), 953–973. Advance online publication. DOI: 10.1287/ijoc.2021.1105

Yu, V., Lin, S.-W., Zhou, L., & Baldacci, R. (2023). A fast simulated annealing heuristic for the multi-depot two-echelon vehicle routing problem with delivery options. *Transportation Letters*, 1–12. DOI: 10.1080/19427867.2023.2257923

Yu, V., Redi, A. A. N. P., Agustina, Y., & Wibowo, O. (2016). A Simulated Annealing Heuristic for the Hybrid Vehicle Routing Problem. *Applied Soft Computing*, 53, 119–132. Advance online publication. DOI: 10.1016/j.asoc.2016.12.027

Zarbakhshnia, N., Govindan, K., Kannan, D., & Goh, M. (2023). Outsourcing logistics operations in circular economy towards to sustainable development goals. *Business Strategy and the Environment*, 32(1), 134–162. DOI: 10.1002/bse.3122

Zenezini, G., Lagorio, A., Pinto, R., De Marco, A., & Golini, R. (2018). The collection-and-delivery points implementation process from the courier, express and parcel operator's perspective. *IFAC-PapersOnLine*, 51(11), 594–599. DOI: 10.1016/j.ifacol.2018.08.383

Zhang, G., Yang, Y., & Yang, G. (2023). Smart supply chain management in Industry 4.0: The review, research agenda and strategies in North America. *Annals of Operations Research*, 322(2), 1075–1117. DOI: 10.1007/s10479-022-04689-1 PMID: 35531562

Zhang, S., Chen, C., & Liu, H. (2020). 5G Technology and Its Applications: An Overview. *IEEE Access : Practical Innovations, Open Solutions*, 8, 34923–34936.

Zhang, Y., & Lu, Z. (2015). A survey on the integration of machine learning in expert systems. *Expert Systems with Applications*, 42(21), 7569–7585.

Zhang, Z., Zhou, R., Wang, M., & Liu, L. (2023). V2V Composite Routing Algorithm for Internet of Vehicles Based on Improved Q-Learning. *2023 China Automation Congress (CAC)*, 446–450. DOI: 10.1109/CAC59555.2023.10451920

Zhao, B., & Tu, C. (2021). Research and development of inventory management and human resource management in ERP. *Wireless Communications and Mobile Computing*, 2021(1), 3132062. DOI: 10.1155/2021/3132062

Zhao, H., & Yang, Y. (2020). The research on two phase pickup vehicle routing based on the K-means++ and genetic algorithms. *International Journal of Web Engineering and Technology*, 15(1), 32. DOI: 10.1504/IJWET.2020.107685

Zhao, J., Poon, M., Tan, V. Y. F., & Zhang, Z. (2024). A hybrid genetic search and dynamic programming-based split algorithm for the multi-trip time-dependent vehicle routing problem. *European Journal of Operational Research*, 317(3), 921–935. DOI: 10.1016/j.ejor.2024.04.011

Zhao, S. (2021). Rhetoric and reality of China's global leadership in the context of COVID-19: implications for the US-led world order and liberal globalization. In *China's Big Power Ambition under Xi Jinping* (pp. 3–18). Routledge. DOI: 10.4324/9781003198871-1

Zheng, T., Glock, C. H., & Grosse, E. H. (2022). Opportunities for using eye tracking technology in manufacturing and logistics: Systematic literature review and research agenda. *Computers & Industrial Engineering*, 171, 108444. DOI: 10.1016/j.cie.2022.108444

Zhu, H., Huberman, B. A., & Lu, Q. (2017). To Personalize or Not to Personalize: A Risk Management Perspective. *Journal of Management Information Systems*, 34(2), 578–607.

Zhu, Q., Sarkis, J., & Lai, K. (2007). Green supply chain management: Pressures, practices and performance within the Chinese automobile industry. *Journal of Cleaner Production*, 15(11), 1041–1052. DOI: 10.1016/j.jclepro.2006.05.021

Zillies, J., Westphal, S., Thakur, D., Kumar, V., Pappas, G., & Scheidt, D. (2016, October 1). *A Column Generation Approach for Optimized Routing and Coordination of a UAV Fleet*. DOI: 10.1109/SSRR.2016.7784326

Zoghlami, N., Taghipour, A., Merlo, C., & Abed, M. (2016). Management of divergent production network using decentralised multi-level capacitated lot-sizing models. *International Journal of Shipping and Transport Logistics*, 8(5), 590–604. DOI: 10.1504/IJSTL.2016.078686

Zorkany, M., Yasser, A., & Galal, A. I. A. (2020). Vehicle To Vehicle "V2V" Communication: Scope, Importance, Challenges, Research Directions and Future. *The Open Transplantation Journal*, 14(1), 86–98. DOI: 10.2174/1874447802014010086

About the Contributors

Atour Taghipour is a professor and the head of an international management master program at Normandy University in France and a visiting professor at Solbridge in South Korea. He holds a PhD in Industrial Engineering from the Polytechnic School of Montreal in Canada. He has more than ten years of experiences as a manager in automobile industries. He has published many books and research papers in international journals. His current research interests are supply chain and operations management based on quantitative and qualitative approaches.

***

Mojgan Bazhrang is a researcher in sport management, her domain of interest is sport analytics, using data analysis to enhance performance, fan engagement, and business operations. She is also interessted in researching metrics related to player performance, injury prevention, and team strategies.

Abhijit Pramod Chandratreya is currently working as Deputy Director in the Department of Post Graduate Research Centre at SCES's Indira Institute of Management, Pune for the past 16 years. He joined IIMP after working in various Industries for 17 years. He graduated in Production Engineering from Birla Vishwakarma Mahavidyalaya (Sardar Patel University), Anand, Gujarat, India. He secured a Master of Business Administration in Human Resource Development from IGNOU. He has been awarded a Ph.D. in the field of Marketing Management from Savitribai Phule Pune University, India. He has presented several papers in National and International Journals, conferences, and Symposiums and has published 15 research papers both in referred and SCOPUS / ABDC-listed Journals. He has published 10 chapters in books released by IGI Global of which 4 are indexed in SCOPUS and three chapters in books released by NOVA SCIENCE of which one is indexed in SCOPUS. He has published three patents. He is a certified trainer of

CSCP, CPTP, CLDM, MTF, and CID certified by MiddleEarth HR. His main areas of interest include Marketing, HRM, SCM, Statistics, Business Analytics, and Research Methodologies. ORCID ID https://orcid.org/0000-0001-9403-2094.

Taoufiq El Moussaoui received a master's degree from the faculty of sciences of Sidi Mohamed Ben Abdellah University, Fez, Morocco, in 2017, where he is currently pursuing the Ph.D. degree in artificial intelligence. His research interests include text mining, pattern recognition, information extraction, machine learning and artificial intelligence.

Fouad Jawab is Professor of Supply Chain at Sidi Mohamed Ben Abdellah University in Fez, Morocco. He is a researcher specializing in industrial management and logistics. He heads the MILOG research team and has directed the MIDLOG laboratory. He founded the international conference on logistics and supply chain management (LOGISTIQUA), which he is developing in collaboration with national and international research teams. In addition to his research activities, he has held a number of management positions within the school, notably as head of the STG department and logistics training courses. He is the author of several articles published in international scientific journals. He is also an expert for several renowned organizations and a reviewer for internationally renowned journals.

Mayuri K. N. is currently perusing her Doctoral degree in Department of Mathematics, School of Advanced Sciences, VIT, Vellore. She received her Master degree from Thiruvalluvar University Vellore. Her area of Interest includes Operations Research, Supply chain Management, Optimization Technique, Queueing Model, Artificial Intelligence and Machine Learning.

Ruchita Khilari is a researcher focusing on innovative solutions in disaster management and healthcare technology through advanced technological approaches.

Delna Kuriyakose is a doctoral candidate in the Department of Mathematics at the School of Advanced Sciences, VIT University, Vellore. She holds her Master's degree in Statistics from the University of Calicut Kerala. Her primary research interests lie in Stochastic processes, Mathematical Modelling of Biostatistics, Queueing models, Artificial Intelligence, and Machine Learning.

XiaoWen Lu is a dedicated PhD candidate and a proud member of Normandy University. Her academic focus lies in the field of Supply Chain Management, a domain that is becoming increasingly critical in today's interconnected and globalized world. Over the course of her doctoral studies, Lu has concentrated her efforts on exploring innovative approaches to enhance the efficiency and sustainability of supply chains. Her work addresses some of the pressing challenges faced by organizations, such as optimizing operations, reducing environmental impact, and adapting to evolving market demands. In addition to her research endeavors, Lu has contributed to the academic community by publishing several papers in reputable journals and conferences within the field of Supply Chain Management. These publications highlight her commitment to advancing knowledge and providing practical solutions to real-world problems. Lu continues to actively engage with the academic and professional communities, striving to make meaningful contributions to her area of expertise.

Gowsalya M. is currently working as an Assistant Professor in Department of Mathematics, School of Advanced Sciences, Vellore Institute of Technology (VIT). She received her Doctoral degree from College of Engineering Guindy, Anna University Chennai in the year 2017. Her area of Interest includes stochastic processes, mathematical modeling of biostatistics, Supply chain management, Artificial Intelligence and Machine Learning.

Mohammad Mirkazemi Mood, with a PhD and a Master of Science (M.Sc.) in Operations Research from the University of Tehran, combines academic excellence with practical experience. He has taught at various universities and conducted impactful workshops while publishing over 30 papers and books. His background includes roles as a project manager, market research manager, and business analyst within dynamic startup environments. Currently, as a driven entrepreneur, he has launched his own startup focused on an innovative online platform. His keen interest in complex socio-technical systems drives him to explore emerging trends and innovations that shape these critical areas.

Nilamadhab Mishra is an Associate Professor in the School of Computer Science and Engineering, AI-ML Division, at VIT Bhopal University, India. He received his Doctor of Philosophy (Ph.D.) in Computer Science & Information Engineering with a specialization in Data Science & Machine Learning from Chang Gung University, Taiwan. He has more than 22 years of national and international involvement in academic teaching and research at recognized Indian, Taiwanese, and African universities. He has over 50 publications in SCI/SCIE and SCOPUS-indexed journals, ISBN books and chapters, Indian and Australian patents, IEEE conference proceedings, and others. He has served as a reviewer, associate editor, and editorial board member for SCI/SCIE-indexed journals and conferences. Dr. Mishra worked on multiple funded research projects from the MOST, NSC, and CGU Memorial Hospital, Taiwan, during his Ph.D. and has been involved with several professional bodies. He is currently an Associate Editor for two SCOPUS-indexed journals, and his research interests span the areas of AI, Data Science, Machine Learning, and Cognitive Analytics & Applications.

Kingsly Stephen R. serves as an Assistant Professor in the Department of Computer Science and Engineering at SRM Institute of Science and Technology, Ramapuram campus. His academic journey is marked by a relentless pursuit of knowledge, as he concurrently works towards his Ph.D. in the dynamic field of Artificial Intelligence and Machine Learning, starting in 2022. With an impressive 17 years of teaching experience, he possesses a wealth of expertise and a knack for programming. His commitment to staying updated is evident through his certification in various courses from renowned platforms like Coursera and NPTEL. Furthermore, his contributions to academia extend to his publications, including articles in both National and International Journals, Conferences, and Symposiums. His research interests are predominantly centred around Computer Networks, Artificial Intelligence, and Deep Learning. In addition to his academic pursuits, he is an active member of the International Association of Engineers (IAENG), solidifying his commitment to professional growth and networking.

Kapil Sharma earned his Doctorate in Computer Science and Engineering in 2022, following a Master's degree from JUIT, Solan, in 2018. He has authored over 20+ research articles which were published in various national and international journals, conferences, and book chapters. His research focuses on Network on Chip, Optical Interconnects, Energy Efficient Routing Algorithms, Machine Learning, Wireless Sensor Networks, and Energy Efficient Interconnects. Additionally, he possesses expertise in Full Stack Development, utilizing technologies like Flutter, Laravel Framework and has vast experience in simulation software's like MATLAB, and Lumerical FDTD Simulator.

Archana T. is currently pursuing her Ph.D. in the Department of Computer Science and Engineering at SRM Institute of Science and Technology, Ramapuram Campus in Chennai. With over six years of teaching experience in esteemed institutions, she has honed her skills in imparting knowledge and shaping the minds of young students. Her primary research interests lie in the captivating fields of Artificial Intelligence and Cloud Computing, where she aspires to contribute to cutting-edge innovations and solutions. She also has an impressive track record of publishing papers in well-regarded international journals and conferences, showcasing her dedication to advancing the realm of computer science. She has completed several certified courses from NPTEL and Coursera. With an unwavering commitment to learning and research, she is eager to explore new horizons and make a meaningful impact in the world of technology.

Index

A

B

C

D

E

F

G

H

I

J

L

M

V

W

www.ingramcontent.com/pod-product-compliance
Lightning Source LLC
LaVergne TN
LVHW080353011026
849161LV00009B/2293